P9-AGV-129

3G, HSPA and FDD versus TDD Networking

3G, HSPA and FDD versus TDD Networking

Smart Antennas and Adaptive Modulation

Second Edition

L. Hanzo, *University of Southampton, UK*

J. S. Blogh, *Anritsu, UK*

S. Ni, *Panasonic Mobile Communication, UK*

IEEE PRESS

IEEE Communications Society, Sponsor

John Wiley & Sons, Ltd

Other Wiley Editorial Offices

John Wiley & Sons Inc., 111 River Street, Hoboken, NJ 07030, USA

Jossey-Bass, 989 Market Street, San Francisco, CA 94103-1741, USA

Wiley-VCH Verlag GmbH, Boschstr. 12, D-69469 Weinheim, Germany

John Wiley & Sons Australia Ltd, 42 McDougall Street, Milton, Queensland 4064, Australia

John Wiley & Sons (Asia) Pte Ltd, 2 Clementi Loop #02-01, Jin Xing Distripark, Singapore 129809

John Wiley & Sons Canada Ltd, 6045 Freemont Blvd, Mississauga, ONT, L5R 4J3, Canada

Wiley also publishes its books in a variety of electronic formats. Some content that appears in print may not be available in
electronic books.

IEEE Communications Society, Sponsor
COMMS-S Liaison to IEEE Press, Mostafa Hashem Sherif

Library of Congress Cataloging-in-Publication Data

Hanzo, Lajos, 1952-
    3G, HSPA and FDD versus TDD Networking / by L. Hanzo, J. S. Blogh and S. Ni – 2nd ed.
        p. cm.
    Includes bibliographical references and index.
    ISBN 978-0-470-75420-7 (cloth)
1. Wireless communication systems. 2. Cellular telephone systems.
I. Blogh, J. S. (Jonathan S.) II. Title.
    TK5103.2.H35 2008
    621.382'1–dc22

                                                                                    2007046621

British Library Cataloguing in Publication Data

A catalogue record for this book is available from the British Library

ISBN 978-0-470-75420-7 (HB)

Typeset by the authors using LaTeX software.
Printed and bound in Great Britain by Antony Rowe Ltd, Chippenham, England.

Contents

About the Authors

Lajos Hanzo (http://www-mobile.ecs.soton.ac.uk) FREng, FIEEE, FIET, DSc received his degree in electronics in 1976 and his doctorate in 1983. During his 31-year career in telecommunications he has held various research and academic posts in Hungary, Germany and the UK. Since 1986 he has been with the School of Electronics and Computer Science, University of Southampton, UK, where he holds the chair in telecommunications. He has co-authored 15 books on mobile radio communications totalling in excess of 10 000, published in excess of 700 research papers, acted as TPC Chair of IEEE conferences, presented keynote lectures and been awarded a number of distinctions. Currently he is directing an academic research team, working on a range of research projects in the field of wireless multimedia communications sponsored by industry, the Engineering and Physical Sciences Research Council (EPSRC) UK, the European IST Programme and the Mobile Virtual Centre of Excellence (VCE), UK. He is an enthusiastic supporter of industrial and academic liaison and he offers a range of industrial courses. He is also an IEEE Distinguished Lecturer of both the Communications Society (ComSoc) and the Vehicular Technology Society (VTS) as well as a Governor of both ComSoc and the VTS. For further information on research in progress and associated publications please refer to http://www-mobile.ecs.soton.ac.uk

Jonathan Blogh was awarded an MEng. degree with Distinction in Information Engineering from the University of Southampton, UK in 1997. In the same year he was also awarded the IEE Lord Lloyd of Kilgerran Memorial Prize for his interest in and commitment to mobile radio and RF engineering. Between 1997 and 2000 he conducted postgraduate research and in 2001 he earned a PhD in mobile communications at the University of Southampton, UK. His current areas of research include the networking aspects of FDD and TDD mode third generation mobile cellular networks. Following a spell with Radioscape, London, UK, working as a software engineer, currently he is a senior researcher with Anritsu, UK.

Song Ni received his BEng degree in Information detection and instru-
mentation from Shanghai Jiaotong University in 1999. Subsequently, he
was employed by Winbond Electronics (Shanghai) Ltd. as a Software
Engineer. His primary responsibility was telecom products R & D.
In 2001 he started a PhD on Intelligent Wireless Networking at the
University of Southampton, which was sponsored by IST SCOUT
project. During four years research, he developed a simulation platform
for the UTRA TDD network layer in the UMTS WCDMA system and
studied various technologies to enhance achievable performance of UTRA systems. Dr Song
Ni is currently a system engineer with Panasonic Mobile Communication, UK.

Other Wiley and IEEE Press Books on Related Topics[1]

- R. Steele, L. Hanzo (Ed): *Mobile Radio Communications: Second and Third Generation Cellular and WATM Systems*, John Wiley and IEEE Press, 2nd edition, 1999, ISBN 07 273-1406-8, 1064 pages.

- L. Hanzo, T.H. Liew, B.L. Yeap: *Turbo Coding, Turbo Equalisation and Space-Time Coding*, John Wiley and IEEE Press, 2002, 751 pages.

- L. Hanzo, C.H. Wong, M.S. Yee: *Adaptive Wireless Transceivers: Turbo-Coded, Turbo-Equalised and Space-Time Coded TDMA, CDMA and OFDM Systems*, John Wiley and IEEE Press, 2002, 737 pages.

- L. Hanzo, L-L. Yang, E-L. Kuan, K. Yen: *Single- and Multi-Carrier CDMA: Multi-User Detection, Space-Time Spreading, Synchronization, Networking and Standards*, John Wiley and IEEE Press, June 2003, 1060 pages.

- L. Hanzo, M. Münster, T. Keller, B-J. Choi, *OFDM and MC-CDMA for Broadband Multi-User Communications, WLANs and Broadcasting*, John-Wiley and IEEE Press, 2003, 978 pages.

- L. Hanzo, S-X. Ng, T. Keller and W.T. Webb, *Quadrature Amplitude Modulation: From Basics to Adaptive Trellis-Coded, Turbo-Equalised and Space-Time Coded OFDM, CDMA and MC-CDMA Systems*, John Wiley and IEEE Press, 2004, 1105 pages.

- L. Hanzo, T. Keller: *An OFDM and MC-CDMA Primer*, John Wiley and IEEE Press, 2006, 430 pages.

- L. Hanzo, F.C.A. Somerville, J.P. Woodard: *Voice and Audio Compression for Wireless Communications*, John Wiley and IEEE Press, 2nd edition, 2007, 858 pages.

---

[1]For detailed contents and sample chapters please refer to http://www-mobile.ecs.soton.ac.uk

- L. Hanzo, P.J. Cherriman, J. Streit: *Video Compression and Communications: H.261, H.263, H.264, MPEG4 and HSDPA-Style Adaptive Turbo-Transceivers* John Wiley and IEEE Press, 2nd edition, 2007, 680 pages.

Preface

Background and Overview

Wireless communications is experiencing an explosive growth rate. This high demand for wireless communications services requires increased system capacities. The simplest solution would be to allocate more bandwidth to these services, but the electromagnetic spectrum is a limited resource, which is becoming increasingly congested [1]. Furthermore, the frequency bands to be used for the Third-Generation (3G) wireless services have been auctioned in various European countries, such as Germany and the UK, at an extremely high price. Therefore, the efficient use of the available frequencies is paramount [1, 2].

The digital transmission techniques of the Second-Generation (2G) mobile radio networks have already improved upon the capacity and voice quality attained by the analog mobile radio systems of the first generation. However, more efficient techniques allowing multiple users to share the available frequencies are necessary. Classic techniques of supporting a multiplicity of users are frequency, time, polarization, code or spatial division multiple access [3]. In Frequency Division Multiple (FDMA) Access [4, 5] the available frequency spectrum is divided into frequency bands, each of which is used by a different user. Time Division Multiple Access (TDMA) [4,5] allocates each user a given period of time, referred to as a timeslot, over which their transmission may take place. The transmitter must be able to store the data to be transmitted and then transmit it at a proportionately increased rate during its timeslot constituting a fraction of the TDMA frame duration. Alternatively, Code Division Multiple Access (CDMA) [4, 5] allocates each user a unique code. This code is then used to spread the data over a wide bandwidth shared with all users. For detecting the transmitted data the same unique code, often referred to as the user signature, must be used.

The increasing demand for spectrally efficient mobile communications systems motivates our quest for more powerful techniques. With the aid of spatial processing at a cell site, optimum receive and transmit beams can be used for improving the system's performance in terms of the achievable capacity and the Quality of Service (QoS) measures. This approach is usually referred to as Spatial Division Multiple Access (SDMA) [3, 6], which enables multiple users in the same cell to be accommodated on the same frequency and timeslot by exploiting the spatial selectivity properties offered by adaptive antennas [7]. In contrast, if the desired signal and interferers occupy the same frequency band and timeslot, then "temporal filtering" cannot be used to separate the signal from the interference. However, the desired and interfering signals usually originate from different spatial locations and this spatial separation may be exploited in order to separate the desired signal from the interference using a "spatially selective filter" at the receiver [8–10]. As a result, given a sufficiently large distance between two users communicating in the same frequency band, there will be negligible interference between them. The higher the number of cells in a region, owing to using small cells, the more frequently the same frequency is re-used and, hence, the higher the teletraffic density per unit area that can be carried.

However, the distance between co-channel cells must be sufficiently high so that the intra-cell interference becomes lower than its maximum acceptable limit [3]. Therefore, the number of cells in a geographic area is limited by the base stations' transmission power level. A method of increasing the system's capacity is to use 120° sectorial beams at different carrier frequencies [11]. Each of the sectorial beams may serve the same number of users as supported in ordinary omni-directional cells, while the Signal-to-Interference Ratio (SIR) can be increased owing to the antenna's directionality. The ultimate solution, however, is to use independently steered high-gain beams for tracking the individual users [3] roaming in the network.

High Speed Downlink Packet Access (HSDPA)-style Adaptive Quadrature Amplitude Modulation (AQAM) [12, 13] is another technique that is capable of increasing the achievable spectral efficiency. The philosophy behind adaptive modulation is to select a specific modulation mode, from a set of modes, according to the instantaneous radio channel quality [12, 13]. Thus, if the channel quality exhibits a high instantaneous Signal-to-Interface plus Noise Ratio (SINR), then a high-order modulation mode may be employed, enabling the exploitation of the temporarily high channel capacity. In contrast, if the channel has a low instantaneous SINR, using a high-order modulation mode would result in an unacceptably high Frame Error Ratio (FER) and, hence, a more robust, but lower throughput modulation mode would be invoked. Therefore, adaptive modulation not only combats the effects of a poor quality channel, but also attempts to maximize the throughput, whilst maintaining a given target FER. Thus, there is a trade-off between the mean FER and the data throughput, which is governed by the modem mode switching thresholds. These switching thresholds define the SINRs, at which the instantaneous channel quality requires the current modulation mode to be changed, i.e. where an alternative AQAM mode must be invoked.

A more explicit representation of the wideband HSDPA-style AQAM mode switching regime is shown in Figure 1, which displays the variation of the modulation mode with respect to the near-instantaneous SINR at average channel SNRs of 10 and 20 dB. In this figure, it can be seen explicitly that the lower-order modulation modes were chosen when the pseudo-SNR was low. In contrast, when the pseudo-SNR was high, the higher-order modulation modes were selected in order to increase the transmission throughput. This figure can also be used to exemplify the application of wideband AQAM in an indoor and outdoor environment. In this respect, Figure 1(a) can be used to characterize a hostile low-SINR outdoor environment, where the average channel quality was low. This resulted in the utilization of predominantly more robust modulation modes, such as Binary Phase Shift Keying (BPSK) and 4 Quadrature Amplitude Modulation (4QAM). Conversely, a less hostile high-SINR indoor environment is exemplified by Figure 1(b), where the channel quality was consistently higher. As a result, the wideband AQAM regime can adapt by suitably invoking higher-order modula-tion modes, as evidenced by Figure 1(b). Again, this simple example demonstrated that

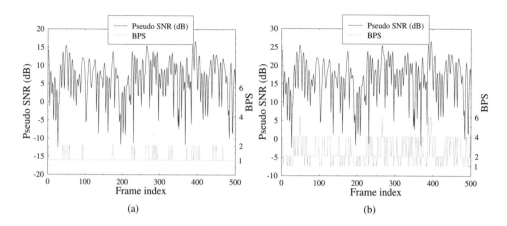

(a)   (b)

Figure 1: Modulation mode variation with respect to the pseudo-SNR evaluated at the output of the
channel equalizer of a wideband AQAM modem for transmission over the TU Rayleigh
fading channel. The Bits per symbol (BPS) throughputs of 1, 2, 4 and 6 represent BPSK,
4QAM, 16QAM and 64QAM, respectively. Channel SNR of (a) 10 dB and (b) 20 dB.

HSDPA-style wideband AQAM can be utilized in order to provide a seamless, near-
instantaneous reconfiguration for example between indoor and outdoor environments. The
most convincing argument in favor of HSDPA-style AQAM is that a fixed-mode system
would increase the required uplink (UL) or downlink (DL) transmit power for the sake of
maintaining a given user's target Bit Error Ratio (BER), hence the system is expected to
inflict a higher Multi-User Interface (MUI) upon all other system users. Therefore, all of the
other users would in turn also increase their power requirement, which may result in a system
instability. In contrast, an AQAM system would simply adjust the AQAM mode used, in order
to use the system's resources as judiciously as possible.

In this book we study the network capacity gains that may be achieved with the
advent of adaptive antenna arrays and HSDPA-style adaptive modulation techniques in both
FDMA/TDMA and CDMA-based mobile cellular networks employing Frequency Division
Duplexing (FDD) as well as Time Division Duplexing (TDD). The advantages of employing
adaptive antennas are multifold, as outlined in the following.

Reduction of Co-channel Interference

Antenna arrays employed by the base station allow the implementation of spatial filtering, as
shown in Figure 2, which may be exploited in both transmitting as well as receiving modes
in order to reduce co-channel interferences [1, 2, 14, 15] experienced in the UL and DL of
wireless systems. When transmitting with an increased antenna gain in a certain direction
of the DL, the base station's antenna is used to focus the radiated energy in order to form a
high-gain directive beam in the area where the mobile receiver is likely to be. This, in turn,
implies that there is a reduced amount of radiated energy and, hence, reduced interference
inflicted upon the mobile receivers roaming in other directions where the directive beam
has a lower gain. The co-channel interference generated by the base station in its transmit

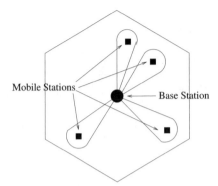

Figure 2: A cell layout showing how an antenna array can support many users on the same carrier frequency and timeslot with the advent of spatial filtering or SDMA.

mode may be further reduced by forming beams exhibiting nulls in the directions of other receivers [6, 16]. This scheme deliberately reduces the transmitted energy in the direction of co-channel receivers and, hence, requires prior knowledge of their positions.

The employment of antenna arrays at the base station for reducing the co-channel interference in its receive mode has been also reported widely [1, 2, 6, 16–18]. This technique does not require explicit knowledge of the co-channel interference signal itself, however, it has to possess information concerning the desired signal, such as the direction of its source, a reference signal, such as a channel sounding sequence, or a signal that is highly correlated with the desired signal.

Capacity Improvement and Spectral Efficiency

The spectral efficiency of a wireless network refers to the amount of traffic a given system having a certain spectral allocation could handle. An increase in the number of users of the mobile communications system without a loss of performance increases the spectral efficiency. Channel capacity refers to the maximum data rate a channel of a given bandwidth can sustain. An improved channel capacity leads to an ability to support more users of a specified data rate, implying a better spectral efficiency. The increased QoS that results from the reduced co-channel interference and reduced multipath fading [18, 19] upon using smart antennas may be exchanged for an increased number of users [2, 20].

Increase of Transmission Efficiency

An antenna array is directive in its nature, having a high gain in the direction where the beam is pointing. This property may be exploited in order to extend the range of the base station, resulting in a larger cell size or may be used to reduce the transmitted power of the mobiles. The employment of a directive antenna allows the base station to receive weaker signals than an omni-directional antenna. This implies that the mobile can transmit at a lower power and its battery recharge period becomes longer, or it would be able to use a smaller battery, resulting in a smaller size and weight, which is important for hand-held mobiles.

A corresponding reduction in the power transmitted from the base station allows the use of electronic components having lower power ratings and, therefore, lower cost.

Reduction of the Number of Handovers

When the amount of traffic in a cell exceeds the cell's capacity, cell splitting is often used in order to create new cells [2], each with its own base station and frequency assignment. The reduction in cell size leads to an increase in the number of handovers performed. By using antenna arrays for increasing the user capacity of a cell [1] the number of handovers required may actually be reduced. More explicitly, since each antenna beam tracks a mobile [2], no handover is necessary, unless different beams using the same frequency cross each other.

Avoiding Transmission Errors

When the instantaneous channel quality is low, conventional fixed-mode transceivers typi-cally inflict a burst of transmission errors. In contrast, adaptive transceivers avoid this problem by reducing the number of transmitted bits per symbol, or even by disabling transmissions temporarily. The associated throughput loss can be compensated for by transmitting a higher number of bits per symbol during the periods of relatively high channel qualities. This advantageous property manifests itself also in terms of an improved service quality, which is quantified in this book in terms of the achievable video quality.

However, realistic propagation scenarios are significantly more complex than that depicted in Figure 2. Specifically, both the desired signal and the interference sources experience multipath propagation, resulting in a high number of received uplink signals impinging upon the base station's receiver antenna array. A result of the increased number of received uplink signals is that the limited degrees of freedom of the base station's adaptive antenna array are exhausted, resulting in reduced nulling of the interference sources. A solution to this limitation is to increase the number of antenna elements in the base station's adaptive array, although this has the side effect of raising the cost and complexity of the array. In a macro-cellular system it may be possible to neglect multipath rays arriving at the base station from interfering sources, since the majority of the scatterers are located close to the mobile station [21]. In contrast, in a micro-cellular system the scatterers are located in both the region of the reduced-elevation base station and that of the mobile, and hence multipath propagation must be considered. Figure 3 shows a realistic propagation environment for both the UL and the DL, with the multipath components of the desired signal and interference signals clearly illustrated, where the UL and DL multipath components were assumed to be identical for the sake of simplicity. Naturally, this is not always the case and, hence, we investigate the potential performance gains, when the UL and DL beamforms are determined independently.

To elaborate a little further, the design of wireless networks is based on a complex interplay of the various performance metrics as well as on a range of other often contradictory trade-offs, which are summarized in the stylized illustration seen in Figure 7.4. For example, Figure 7.4 suggests that it is always possible to reduce the call dropping probability by increasing the call blocking probability, since this implies admitting less users to the system. In contrast, we may admit more users to the system for the sake of reducing the call blocking probability, which however results in an increased call dropping probability. Furthermore,

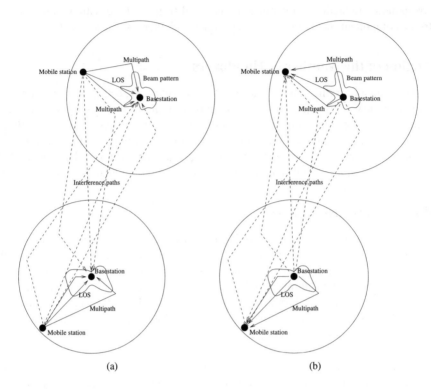

Figure 3: The multipath environments of both (a) the UL and (b) the DL, showing the multipath components of the desired signals, the line-of-sight interference and the associated base station antenna array beam patterns.

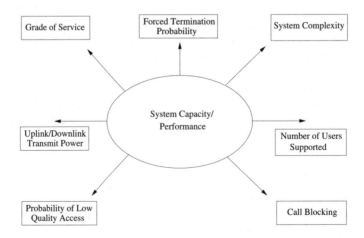

Figure 4: System capacity/performance illustration factors.

the performance of the entire system may also be improved by increasing the system's complexity upon using more intelligent, but more complex signal processing algorithms, such as the beamforming and HSDPA-style adaptive modulation aided transceiver techniques advocated throughout the book, more specifically for example in Chapters 6 and 8. Similarly, the Genetic Algorithm (GA)-based intelligent scheduling techniques of Chapter 10 may be used for reducing the co-channel interference experienced by the system and, hence, for increasing the number of users, and/or for improving the call blocking and call dropping performance. Still continuing our discourse in the spirit of Figure 4, the number of users supported may also be increased, provided that an increased probability of low-quality access value may be tolerated. A whole raft of further similar comments may be made in the context of Figure 4, which will emanate from our detailed discourse throughout the forthcoming chapters. Hence, we postpone the discussion of these detailed findings to our forthcoming chapters.

The various contributions on the network performance of the UMTS Terrestrial Radio Access (UTRA) FDD and TDD modes are summarized in Table 1.

The Outline of the Book

- **Chapter 1.** Following a brief introduction to the principles of CDMA the three most important 3G wireless standards, namely UTRA, IMT 2000 and cdma 2000 are characterized. The range of various transport and physical channels, the multiplexing of various services for transmission, the aspects of channel coding are discussed. The various options available for supporting variable rates and a range QoS are highlighted. The UL and DL modulation and spreading schemes are described and UTRA and IMT 2000 are compared in terms of the various solutions standardized. The chapter closes with a similar portrayal of the pan-American cdma 2000 system.

- **Chapter 2.** Since the standardization of the 3G systems substantial technological advances have been made in adaptive modulation and coding techniques, which may be employed to compensate for the inevitably time-variant channel quality fluctuations of wireless channels. These advances led to the definition of the HSDPA and HSUPA modes, which are detailed in this chapter. The HSDPA mode is capable of supporting a bitrates up to about 14 MBit/s with the aid of adaptive modulation. In contrast, the UL dispenses with the employment of adaptive modulation in the interest of avoiding the application of low-efficiency, power-hungry class-A amplification in the mobile terminal. It rather employs multiple spreading sequences to increase the achievable UL bitrate, which may reach about 4 MBit/s.

- **Chapter 3.** Following the portrayal of the HSDPA/High Speed Uplink Packet Access (HSUPA) standards, in this chapter the HSDPA-style adaptive modulation techniques are further detailed, which are invoked in an effort to compensate for the inevitably time-variant channel quality fluctuations of wireless channels. In this chapter we have not restricted ourselves to standardized solutions, we have rather provided an evolutionary landscape, speculating on the types of more advanced solutions that might find their way into future standards, such as the extensions of the 3GPP Long-Term Evolution (LTE) project or the IEEE 802.11 Wireless Local Area Network (WLAN) standards. We commence our discourse by briefly reviewing the state-of-the-art in

Table 1: Contributions on the network performance of UTRA FDD and TDD cellular systems.

Year	Author	Contribution
1998	Ojanpera and Prasad [22]	An overview of 3G wireless personal communications systems was presented.
	Dahlman, Gudmundson, Nilsson and Skold [23]	Wideband Code Division Multiple Access (WCDMA) was presented as a mature technology to provide the basis for the Universal Mobile Telecommunications System (UMTS)/IMT-2000 standards.
	Brand and Aghvami [24]	Multidimensional Packet Reservation Multiple Access (PRMA) was proposed as a Medium Access Control (MAC) strategy for the UL channel of the UTRA TDD/CDMA mode.
	Markoulidakis, Menolascino, Galliano and Pizarroso [25]	An efficient network planning methodology applied to the UTRA specifications was proposed.
1999	Mestre, Najar, Anton and Fonollosa [26]	A semi-blind beamforming technique was proposed for the UTRA FDD system.
	Akhtar and Zeghlache [27]	A network capacity study of the UTRA WCDMA system was presented.
	Berens, Bing, Michel, Worm and Baier [28]	The performance of low-complexity turbo-codes employed in the UTRA TDD mode was studied.
2000	Haardt and Mohr [29]	An overview of UMTS as specified by the Third Generation Partnership Project (3GPP) was presented.
	Holma, Heikkinen, Lehtinen and Toskala [30]	An interference study of the UTRA TDD system based on simulations was provided.
	Aguado, O'Farrell and Harris [31]	An investigation into the impact of mixed traffic on the UTRA system's performance was presented.
2001	Haas and McLaughlin [32]	The "TS-opposing" DCA algorithm was proposed for a TD-CDMA/TDD air–interface.
	Guenach and Vandendorpe [33]	The DL performance of the conventional Rake receiver was investigated in the context of the UTRA-WCDMA system.
	Poza, Heras, Lablanca and Lopez [34]	An analytical DL interference estimation technique was proposed for the UMTS system.
2002	Perez-Romero, Sallent, Agusti and Sanchez [35]	Congestion control mechanisms were proposed and analyzed designed for the UTRA FDD system.
	Allen, Beach and Karlsson [36]	The outage imposed by beamformer-based smart antennas was studied in a UTRA FDD macro-cell environment.
	Ruiz-Garcia, Romero-Jerez and Diaz-Estrella [37]	The effect of the MAC on QoS guarantees was investigated in order to handle multimedia traffic in the UTRA system.

Table 1: Continued

Year	Author	Contribution
2002	Ebner, Rohling, Halfmann and Lott [38]	Solutions for the synchronization of *ad hoc* networks based on the UTRA TDD system were proposed.
2003	Agnetis, Brogi, Ciaschetti Detti and Giambene [39]	A frame-by-frame exact DL scheduling algorithm considering different traffic QoS levels was proposed.
	Kao and Mar [40]	An intelligent MAC protocol based on cascade fuzzy-logic-control (CFLC) and designed for the UTRA TDD mode was presented.
	Blogh and Hanzo [41]	The adaptive antenna array and adaptive modulation-aided network performance of a UTRA FDD system was investigated.
	Rummler, Chung and Aghvami [42]	A new multicast protocol contrived for UMTS was proposed.
2004	Yang and Yum [43]	A flexible OVSF spreading code assignment designed for multirate traffic in the UTRA system was proposed.
	Sivarajah and Al-Raweshidy [44]	A comparative analysis of different Dynamic Channel Assignment (DCA) schemes conceived for supporting ongoing calls in a UTRA TDD system was presented.
	Yang and Yum [45]	A power-ramping scheme contrived for the UTRA FDD random access channel was proposed.
2005	Ni and Hanzo [46]	A genetic algorithm-aided timeslot scheduling scheme designed for UTRA TDD CDMA networks was proposed.
	Rouse, S. McLaughlin and Band [47]	A network topology was investigated that allows both peer-to-peer and non-local traffic in a TDD CDMA system.
	Zhang, Tao, Wang and Li [48]	Developments beyond 3G mobile proposed by the Chinese communications TDD Special Work Group were disseminated.

near-instantaneously adaptive modulation and introduce the associated principles. We then apply the AQAM philosophy in the context of CDMA as well as Orthogonal Frequency Division Multiplexing (OFDM) and quantify the service-related benefits of adaptive transceivers in terms of the achievable video quality. The associated application examples demonstrate the potential of the proposed adaptive techniques in terms of tangible service quality improvements.

- **Chapter 4.** The principles behind beamforming and the various techniques by which it may be implemented are presented. From this the concept of adaptive beamforming is developed, and temporal as well as spatial reference techniques are examined. Performance results are then presented for three different temporal-reference-based adaptive beamforming algorithms, namely the Sample Matrix Inversion (SMI),

Unconstrained Least Mean Squares (ULMS) and the Normalized Least Mean Squares (NLMS) algorithms.

- **Chapter 5.** A brief summary of possible methods used for modeling the performance of an adaptive antenna array is provided. This is followed by an overview of fixed and dynamic channel allocation. Multipath propagation models are then considered for use in our network simulations. Metrics are then developed for characterizing the performance of mobile cellular networks and our results are presented for simulations conducted under Line-Of-Sight (LOS) propagation conditions, both with and without adaptive antennas. Further results are then given for identical networks under multipath propagation conditions, which are then extended to power-controlled scenarios using both fixed and adaptive Quadrature Amplitude Modulation (QAM) techniques. These network capacity results are obtained for both "island" type simulation areas and for an infinite plane, using wraparound techniques.

- **Chapter 6.** In this chapter we briefly review the 3G mobile cellular network, known as the UTRA network, in order to enable readers to turn directly to the network-layer performance characterization of the system, without having to consult the previous chapters. This chapter then continues to present network capacity results obtained under various propagation conditions, in conjunction with different soft handover threshold metrics. The performance benefits of adaptive antenna arrays are then analyzed, both in a non-shadowed environment and in log-normal shadow fading conditions obeying frequencies of 0.5 and 1.0 Hz. This work is then extended by invoking HSDPA-style adaptive modulation techniques combined with beamforming, which are studied when the channel quality fluctuation is further aggravated by shadow fading.

- **Chapter 7.** We characterize the achievable system performance of a UTRA-like FDD CDMA system employing Loosely Synchronized (LS) spreading codes. The achievable network performance is quantified by simulation and is compared with that of a UTRA-like FDD/CDMA system using Orthogonal Variable Spreading Factor (OVSF) spreading codes. The trade-offs between the achievable user capacity and the cell size as well as the SINR threshold are then explored. We also examine the achievable user-load and the overall performance of a Multi-Carrier Code Division Multiple Access (MC-CDMA)-based cellular network benefiting from both adaptive antenna arrays and adaptive modulation techniques.

- **Chapter 8.** In this chapter we present FDD versus TDD network capacity results obtained under various propagation conditions. The performance benefits of adaptive beamforming and adaptive modulation techniques are analyzed. These results are then compared with those acquired when employing LS spreading codes.

- **Chapter 9.** In this chapter, we study the effects of the hard handover margin and of different power control schemes on the UTRA TDD/CDMA system's performance. Both closed-loop power control as well as open-loop power control schemes are developed based on the 3GPP standard. A frame-delay based power adjustment algorithm is proposed to overcome the channel quality variations imposed by the erratically fluctuating timeslot allocations in the different interfering radio cells.

- **Chapter 10.** In this chapter, we design a GA assisted UL/DL timeslot scheduling scheme for the sake of avoiding the severe inter-cell interference caused by using the UTRA TDD/CDMA air interface.

- **Chapter 11.** Here we give our conclusions and further work.

Contributions of the Book

- Providing an introduction to near-instantaneously adaptive modulation invoked in the context of both single- and multi-carrier modulation or OFDM, as well as CDMA.

- Quantifying the service-related benefits of HSDPA-style adaptive transceivers in the context of wireless video telephony.

- Providing an overview of the various CDMA-based 3G wireless standards.

- Study of the network performance gains using adaptive antenna arrays at the base station in an FDMA/TDMA cellular mobile network [49,50].

- Study of the network performance gains using adaptive antenna arrays in conjunction with power control at the base station in an FDMA/TDMA cellular mobile network [51, 52].

- Design of a combined power control and adaptive modulation assisted channel allocation algorithm, and characterization of its performance in an FDMA/TDMA cellular mobile network [52,53].

- Comparing the performance of various UTRA/HSDPA-style soft-handover techniques.

- Quantifying the UTRA network capacity under various channel conditions.

- Evaluating the network performance of UTRA with the aid of adaptive antenna arrays.

- Demonstrating the benefits of adaptive modulation in the context of both FDMA/TDMA and CDMA cellular mobile networks.

Our hope is that the book offers you a range of interesting topics in the era of the imminent introduction of 3G wireless networks. We have attempted to provide an informative technological roadmap, allowing the reader to quantify the achievable network capacity gains with the advent of introducing more powerful enabling technologies in the physical layer. Analyzing the associated system design trade-offs in terms of network complexity and network capacity is the basic aim of this book. We aimed for underlining the range of contradictory system design trade-offs in an unbiased fashion, with the motivation of providing you with sufficient information for solving your own particular wireless networking problems. Most of all, however, we hope that you will find this book an enjoyable and relatively effortless reading, providing you with intellectual stimulation.

Lajos Hanzo, Jonathan Blogh and Song Ni

Acknowledgements

We are indebted to our many colleagues who have enhanced our understanding of the subject, in particular to Prof. Emeritus Raymond Steele. These colleagues and valued friends, too numerous to be mentioned, have influenced our views concerning various aspects of wireless multimedia communications. We thank them for the enlightenment gained from our collaborations on various projects, papers, and books. We are grateful to Jan Brecht, Marco Breiling, Marco del Buono, Sheng Chen, Peter Cherriman, Stanley Chia, Byoung Jo Choi, Joseph Cheung, Peter Fortune, Sheyam Lal Dhomeja, Lim Dongmin, Dirk Didascalou, Stephan Ernst, Eddie Green, David Greenwood, Hee Thong How, Thomas Keller, Ee Lin Kuan, W. H. Lam, Matthias Münster, C. C. Lee, M. A. Nofal, Xiao Lin, Chee Siong Lee, Tong-Hooi Liew, Jeff Reeve, Vincent Roger-Marchart, Redwan Salami, David Stewart, Clare Sommerville, Jeff Torrance, Spyros Vlahoyiannatos, William Webb, Stefan Weiss, John Williams, Jason Woodard, Choong Hin Wong, Henry Wong, James Wong, Lie-Liang Yang, Bee-Leong Yeap, Mong-Suan Yee, Kai Yen, Andy Yuen, and many others with whom we enjoyed an association.

We also acknowledge our valuable associations with the Virtual Centre of Excellence in Mobile Communications, in particular with its chief executive, Dr. Walter Tuttlebee, and other members of its Executive Committee, namely Dr. Keith Baughan, Prof. Hamid Aghvami, Prof. Mark Beach, Prof. John Dunlop, Prof. Barry Evans, Prof. Steve MacLaughlin, Prof. Joseph McGeehan and many other valued colleagues. Our sincere thanks are also due to John Hand and Nafeesa Simjee EPSRC, UK for supporting our research. We would also like to thank Dr. Joao Da Silva, Dr Jorge Pereira, Bartholome Arroyo, Bernard Barani, Demosthenes Ikonomou, and other valued colleagues from the Commission of the European Communities, Brussels, Belgium, as well as Andy Aftelak, Mike Philips, Andy Wilton, Luis Lopes, and Paul Crichton from Motorola ECID, Swindon, UK, for sponsoring some of our recent research. Further thanks are due to Tim Wilkinson at HP in Bristol for funding some of our research efforts.

Similarly, our sincere thanks are due to Katharine Unwin, Mark Hammond, Sarah Hinton and their colleagues at Wiley in Chichester, UK, as well as Denise Harvey, who assisted us during the production of the book. Finally, our sincere gratitude is due to the numerous authors listed in the Author Index—as well as to those, whose work was not cited due to space limitations—for their contributions to the state of the art, without whom this book would not have materialized.

Lajos Hanzo, Jonathan Blogh and Song Ni

Chapter 1

Third-generation CDMA Systems

K. Yen and L. Hanzo

1.1 Introduction

Although the number of cellular subscribers continues to grow worldwide [54], the predominantly speech-, data- and e-mail-oriented services are expected to be enriched by a whole host of new services in the near future. Thus the performance of the recently standardized Code Division Multiple Access (CDMA) third-generation (3G) mobile systems is expected to become comparable to, if not better than, that of their wired counterparts.

These ambitious objectives are beyond the capabilities of the present second-generation (2G) mobile systems such as the Global System for Mobile Communications known as GSM [55], the Interim Standard-95 (IS-95) Pan-American system, or the Personal Digital Cellular (PDC) system [56] in Japan. Thus, in recent years, a range of new system concepts and objectives were defined, and these will be incorporated in the 3G mobile systems. Both the European Telecommunications Standards Institute (ETSI) and the International Telecommunication Union (ITU) are defining a framework for these systems under the auspices of the Universal Mobile Telecommunications System (UMTS) [54,56–60] and the International Mobile Telecommunications scheme in the year 2000 (IMT-2000)[1] [57,58,61].

Their objectives and the system concepts will be discussed in more detail in later sections. CDMA is the predominant multiple access technique proposed for the 3G wireless communications systems worldwide. CDMA was already employed in some 2G systems, such as the IS-95 system and it has proved to be a success. Partly motivated by this success, both the Pan-European UMTS and the IMT-2000 initiatives have opted for a CDMA-based system, although the European system also incorporates an element of TDMA. In this chapter, we provide a rudimentary introduction to a range of CDMA concepts. Then the European,

---

[1]Formerly known as Future Public Land Mobile Telecommunication Systems.

3G, HSPA and FDD versus TDD Networking   Second Edition
L. Hanzo, J. S. Blogh and S. Ni   © 2008 John Wiley & Sons, Ltd

American and Japanese CDMA-based 3G mobile system concepts are considered, followed by a research-oriented outlook on potential future systems.

The chapter is organized as follows. Section 1.2 offers a rudimentary introduction to CDMA in order to make this chapter self-contained, whereas Section 1.3 focuses on the basic objectives and system concepts of the 3G mobile systems, highlighting the European, American and Japanese CDMA-based third-generation system concepts. Finally, our conclusions are presented in Section 1.4.

1.2   Basic CDMA System

CDMA is a spread spectrum communications technique that supports simultaneous digital transmission of several users' signals in a multiple access environment. Although the development of CDMA was motivated by user capacity considerations, the system capacity provided by CDMA is similar to that of its more traditional counterparts, frequency division multiple access (FDMA), and time division multiple access (TDMA) [62]. However, CDMA has the unique property of supporting a multiplicity of users in the same radio channel with a graceful degradation in performance due to multi-user interference. Hence, any reduction in interference can lead to an increase in capacity [63]. Furthermore, the frequency reuse factor in a CDMA cellular environment can be as high as unity, and being a so-called wideband system, it can coexist with other narrowband microwave systems, which may corrupt the CDMA signal's spectrum in a narrow frequency band without inflicting significant interference [64]. This eases the problem of frequency management as well as allowing a smooth evolution from narrowband systems to wideband systems. But perhaps the most glaring advantage of CDMA is its ability to combat or in fact to benefit from multipath fading, as it will become explicit during our further discourse.

In the forthcoming sections, we introduce our nomenclature, which will be used throughout the subsequent sections. Further in-depth information on CDMA can be found in a range of excellent research papers [62, 64, 65] and textbooks [66–69].

1.2.1   Spread Spectrum Fundamentals

In spread spectrum transmission, the original information signal, which occupies a bandwidth of B Hz, is transmitted after spectral spreading to a bandwidth N times higher, where N is known as the processing gain. In practical terms the processing gain is typically in the range of $10 - 30$ dB [64]. This frequency-domain spreading concept is illustrated in Figure 1.1. The power of the transmitted spread spectrum signal is spread over N times the original bandwidth, while its spectral density is correspondingly reduced by the same amount. Hence, the processing gain is given by:

$$N = \frac{B_s}{B},\tag{1.1}$$

where B_s is the bandwidth of the spread spectrum signal while B is the bandwidth of the original information signal. As we shall see during our further discourse, this unique technique of spreading the information spectrum is the key to improving its detection in a mobile radio environment, and it also allows narrowband signals exhibiting a significantly higher spectral density to share the same frequency band [64].

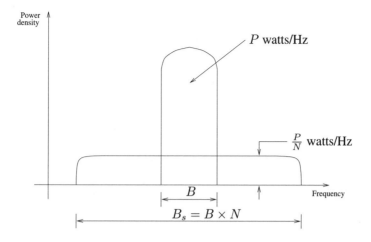

Figure 1.1: Power spectral density of signal before and after spreading.

There are basically two main types of spread spectrum (SS) systems [62]:

- Direct Sequence (DS) SS systems and

- Frequency Hopping (FH) SS systems.

1.2.1.1 Frequency Hopping

In FH spreading, which was invoked in the 2G GSM system the narrowband signal is transmitted using different carrier frequencies at different times. Thus, the data signal is effectively transmitted over a wide spectrum. There are two classes of frequency hopping patterns. In fast frequency hopping, the carrier frequency changes several times per transmitted symbol, while in slow frequency hopping, the carrier frequency changes typically after a number of symbols or a transmission burst. In the GSM system each transmission burst of 114 channel-coded speech bits was transmitted on a different frequency and since the TDMA frame duration was 4.615 ms, the associated hopping frequency was its reciprocal, that is, 217 hops/s. The exact sequence of frequency hopping will be made known only to the intended receiver so that the frequency hopped pattern may be dehopped in order to demodulate the signal [64]. Direct sequence (DS) spreading is more commonly used in CDMA. Hence, our forthcoming discussions will be in the context of direct sequence spreading.

1.2.1.2 Direct Sequence

In DS spreading, the information signal is multiplied by a high-frequency signature sequence, also known as a spreading code or spreading sequence. This user signature sequence facilitates the detection of different users' signals in order to achieve a multiple access capability in CDMA. Although in CDMA this user "separation" is achieved using orthogonal spreading codes, in FDMA and TDMA orthogonal frequency slots or timeslots are provided, respectively.

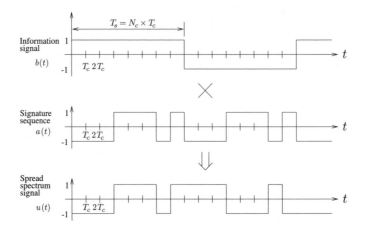

Figure 1.2: Time-domain waveforms involved in generating a direct sequence spread signal.

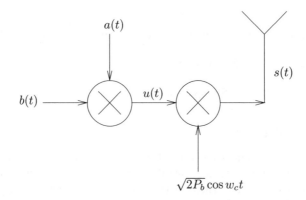

Figure 1.3: BPSK modulated DS-SS transmitter.

We can see from Figure 1.2 that each information symbol of duration T_s is broken into N_c equi-spaced subintervals of duration T_c, each of which is multiplied with a different chip of the spreading sequence. Hence, $N_c = \frac{T_s}{T_c}$. The resulting output is a high-frequency sequence.

For binary signaling $T_s = T_b$, where T_b is the data bit duration. Hence, N_c is equal to the processing gain N. However, for M-ary signaling, where $M > 2$, $T_s \neq T_b$ and hence $N_c \neq N$. An understanding of the distinction between N_c and N is important, since the values of N_c and N have a direct effect on the bandwidth efficiency and performance of the CDMA system.

The block diagram of a typical binary phase shift keying (BPSK) modulated DS-SS transmitter is shown in Figure 1.3. We will now express the associated signals mathematically.

The binary data signal may be written as:

$$b(t) = \sum_{j=-\infty}^{\infty} b_j \Gamma_{T_b}(t - jT_b), \tag{1.2}$$

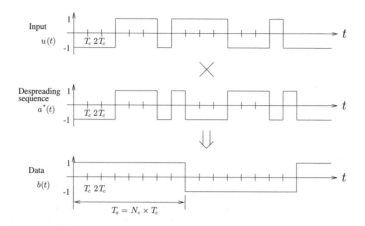

Figure 1.4: Time-domain waveforms involved in decoding a direct sequence signal.

where T_b is the bit duration, $b_j \in \{+1, -1\}$ denotes the jth data bit, and $\Gamma_{T_b}(t)$ is the pulse shape of the data bit. In practical applications, $\Gamma_\tau(t)$ has a bandlimited waveform, such as a raised cosine Nyquist pulse. However, for analysis and simulation simplicity, we will assume that $\Gamma_\tau(t)$ is a rectangular pulse throughout this chapter, which is defined as:

$$\Gamma_\tau(t) = \begin{cases} 1, & 0 \le t < \tau, \\ 0, & \text{otherwise.} \end{cases} \tag{1.3}$$

Similarly, the spreading sequence may be written as

$$a(t) = \sum_{h=-\infty}^{\infty} a_h \Gamma_{T_c}(t - hT_c), \tag{1.4}$$

where $a_h \in \{+1, -1\}$ denotes the hth chip and $\Gamma_{T_c}(t)$ is the chip-pulse with a chip duration of T_c. The energy of the spreading sequence over a bit duration of T_b is normalized according to:

$$\int_0^{T_b} |a(t)|^2 dt = T_b. \tag{1.5}$$

As seen in Figure 1.3, the data signal and spreading sequence are multiplied, and the resultant spread signal is modulated on a carrier in order to produce the wideband signal $s(t)$ at the output:

$$s(t) = \sqrt{2P_b} b(t) a(t) \cos w_c t, \tag{1.6}$$

where P_b is the average transmitted power. At the intended receiver, the signal is multiplied by the conjugate of the transmitter's spreading sequence, which is known as the despreading sequence, in order to retrieve the information. Ideally, in a single-user, nonfading, noiseless environment, the original information can be decoded without errors. This is seen in Figure 1.4.

In reality, however, the conditions are never so perfect. The received signal will be corrupted by noise, interfered by both multipath fading—resulting in intersymbol interference

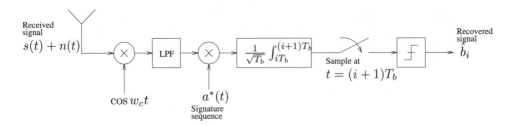

Figure 1.5: BPSK DS-SS receiver for AWGN channel.

(ISI)—and by other users, generating multi-user interference. Furthermore, this signal is delayed by the time-dispersive medium. It is possible to reduce the interference due to multipath fading and other users by innovative signal processing methods, which will be discussed in more detail in later sections.

Figure 1.5 shows the block diagram of the receiver for a noise-corrupted channel using a correlator for detecting the transmitted signal, yielding:

$$\hat{b}_i = \text{sgn}\left\{ \frac{1}{\sqrt{T_b}} \int_{iT_b}^{(i+1)T_b} a^*(t)[s(t) + n(t)] \cos w_c t \, dt \right\}$$

$$= \text{sgn}\left\{ \sqrt{\frac{\xi_b}{2}} b_i + \frac{1}{\sqrt{T_b}} \int_{iT_b}^{(i+1)T_b} a^*(t) n(t) \cos w_c t \, dt \right\}, \qquad (1.7)$$

where $\xi_b = T_b \times P_b$ is the bit energy and $\text{sgn}(x)$ is the signum function of x, which returns a value of 1, if $x > 0$ and returns a value of -1, if $x < 0$. In a single-user Additive White Gaussian Noise (AWGN) channel, the receiver shown in Figure 1.5 is optimum. In fact, the performance of the DS-SS system discussed so far is the same as that of a conventional BPSK modem in an AWGN channel, whereby the probability of bit error $Pr_b(\epsilon)$ is given by:

$$Pr_b(\epsilon) = Q\left(\sqrt{\frac{2\xi_b}{N_0}} \right), \qquad (1.8)$$

where

$$Q(x) = \frac{1}{\sqrt{2\pi}} \int_x^\infty e^{-y^2/2} dy \qquad (1.9)$$

is the Gaussian Q-function. The advantages of spread spectrum communications and CDMA will only be appreciated in a multipath multiple access environment. The multipath aspects and how the so-called Rake receiver [5, 70] can be used to overcome the multipath effects will be highlighted in the next section.

1.2.2   The Effect of Multipath Channels

In this section, we present an overview of the effects of the multipath wireless channels encountered in a digital mobile communication system, which was treated in depth for example in [11]. Interested readers may also refer to the recent articles written by Sklar in [71, 72] for a brief overview on this subject.

Since the mobile station is usually close to the ground, the transmitted signal is reflected, refracted, and scattered from objects in its vicinity, such as buildings, trees, and mountains [62]. Therefore, the received signal is comprised of a succession of possibly overlapping, delayed replicas of the transmitted signal. Each replica is unique in its arrival time, power, and phase [73]. As the receiver or the reflecting objects are not stationary, such reflections will be imposed fading on the received signal, where the fading causes the signal strength to vary in an unpredictable manner. This phenomenon is referred to as multipath propagation [11].

There are typically two types of fading in the mobile radio channel [71]:

- long-term fading

- short-term fading.

As argued in [11] long-term fading is caused by the terrain configuration between the base station and the mobile station, such as hills and clumps of buildings, which result in an average signal power attenuation as a function of distance. For our purposes the channel can be described in terms of its average pathloss, typically obeying an inverse fourth power law [62] and a log-normally distributed variation around the mean. Thus, long-term shadow fading was also referred to as log-normal fading in [11, 71] .

On the other hand, short-term fading refers to the dramatic changes in signal amplitude and phase as a result of small changes in the spatial separation between the receiver and transmitter, as we noted in [11, 71].

Furthermore, the motion between the transmitter and receiver results in propagation path changes, such that the channel appears to be time-variant. The time-variant frequency-selective channel was modeled as a tapped delay line in [11], where the complex low-pass impulse response can be modeled as:

$$\tilde{h}(t) = \sum_{l=1}^{L} |\alpha_l(t)| e^{j\phi_l(t)} \delta(t - \tau_l), \qquad (1.10)$$

where $|\alpha_l(t)|$, $\phi_l(t)$ and τ_l are the amplitude, phase, and delay of the lth path, respectively, and L is the total number of multipath components. It was argued in [11] that the rate of signal level fluctuation is determined by the Doppler frequency, f_D, which in turn is dependent on the carrier frequency, f_c, and the speed of the mobile station v according to (see also page 16 of [74]):

$$f_D = v\frac{f_c}{c}, \qquad (1.11)$$

where c is the speed of light.

Typically, the short-term fading phenomenon is modeled statistically by a Rayleigh, Rician, or Nakagami-m distribution [75]. The Rayleigh and Rician distributions were characterized for example in [11]. There have been some contrasting views in the literature as to which of these distributions best describes the fast-fading channel statistically. Although empirical results have shown that the fading statistics are best described by a Nakagami distribution [76], in most cases a Rayleigh-distributed fading is used for analysis and simulation because of simplicity, and it serves as a useful illustrative example in demonstrating the effects of fading on transmission. Moreover, the Rayleigh distribution is a special case

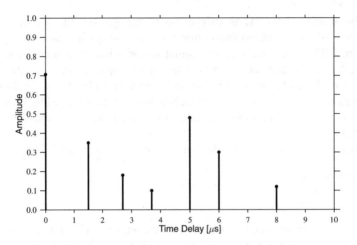

Figure 1.6: COST 207 BU impulse response.

of the Nakagami distribution, when, m, known as the fading parameter, is equal to unity (see page 48 in [5]). The Rician distribution is more applicable to satellite communication, due to the presence of a dominant signal component known as the specular component [71], than to large-cell terrestrial communication, where often there is no Line-of-Sight (LOS) path between the terrestrial base station and the mobile station. However, in small microcells often the opposite is true. In our investigations in this chapter, Rayleigh-distributed frequency selective fading is assumed.

The delay is proportional to the length of the corresponding signal path between the transmitter and receiver. The delay spread due to the path-length differences between the multipath components causes Intersymbol Interference (ISI) in data transmission, which becomes particularly dominant for high data rates.

A typical radio channel impulse response is shown in Figure 1.6. This channel impulse response is known as the COST 207 bad urban (BU) impulse response [77]. It can be clearly seen that the response consists of two main groups of delayed propagation paths: a main profile and a smaller echo profile following the main profile at a delay of 5 μs. The main profile is caused by reflections of the signal from structures in the vicinity of the receiver with shorter delay times. On the other hand, the echo profile could be caused by several reflections from a larger but more distant object, such as a hill [78]. In either case, we can see that both profiles approximately follow a negative exponentially decaying function with respect to the time-delay.

Figure 1.7 shows the impairments of the spread spectrum signal travelling over a multipath channel with L independent paths, yielding the equivalent baseband received signal of:

$$r(t) = \sum_{l=1}^{L} \alpha_l(t)\tilde{s}(t - \tau_l) + n(t), \qquad (1.12)$$

where $\alpha_l(t)$ is the time-variant complex channel gain, which is given by $|\alpha_l(t)|e^{j\phi_l(t)}$ in Equation 1.10 with a Rayleigh-distributed amplitude, uniformly distributed phase over the

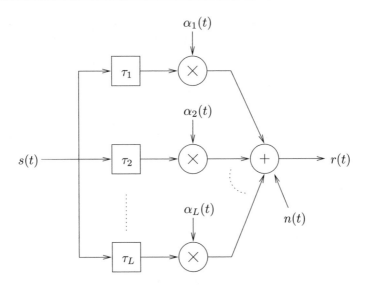

Figure 1.7: Multipath propagation model of the transmitted signal.

interval $[-\pi \ldots \pi]$ and $\tilde{s}(t - \tau_l)$ is the equivalent baseband transmitted spread spectrum signal from Equation 1.6 delayed by τ_l. The above equation shows that the lth path is attenuated by the channel coefficient $\alpha_l(t)$ and delayed by τ_l. Without intelligent diversity techniques [5], these paths are added together at the receiver and any phase or delay difference between the paths may result in a severely multipath interfered signal, corrupted by dispersion-induced intersymbol interference (ISI).

Figure 1.8 shows the effect of a nonfading channel and a fading channel on the bit error probability of BPSK-modulated CDMA. Without diversity, the bit error rate (BER) in a fading channel decreases approximately according to $Pr_b(\epsilon) \approx \frac{1}{4\bar{\gamma}_c}$, where $\bar{\gamma}_c$ is the average Signal-to-Noise Ratio (SNR), and hence plotted on a logarithmic scale according to $\log Pr_b(\epsilon) = -\log 4\bar{\gamma}_c$ we have a near-linear curve [5]. This is different from a nonfading, or AWGN, channel, whereby the BER decreases exponentially with increasing the SNR. Thus, in a fading channel, a high transmitted power is required to obtain a low probability of error. As we shall see in the next section, diversity techniques can be used to overcome this impediment.

1.2.3 Rake Receiver

As mentioned previously, spread spectrum techniques can take advantage of the multipath nature of the mobile channel in order to improve reception. This is possible due to the signal's wideband nature, which has a significantly higher bandwidth than the multipath channel's coherence bandwidth [79]. In this case, the channel was termed a frequency selective fading channel, since different transmitted frequencies faded differently if their separation was higher than the previously mentioned coherence bandwidth. Suppose that the spread spectrum has a bandwidth of B_s and the channel's coherence bandwidth is B_c, such that $B_s \gg B_c$. Thus, the number of resolvable independent paths—that is, the paths that fade

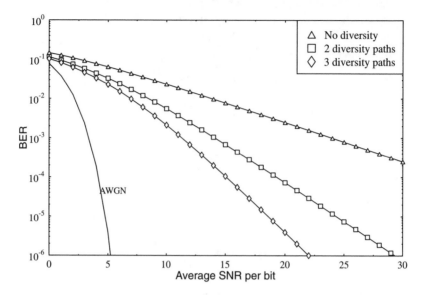

Figure 1.8: Performance of BPSK modulated CDMA over various Rayleigh-fading channels. The
curves were obtained using perfect channel estimation, and there was no self-interference
between diversity paths.

near-independently—L_R is equal to

$$L_R = \left\lfloor \frac{B_s}{B_c} \right\rfloor + 1, \tag{1.13}$$

where $\lfloor x \rfloor$ is the largest integer that is less than or equal to x. The number of resolvable
paths L_R varies according to the environment, and it is typically higher in urban than in
suburban areas, since in urban areas the coherence bandwidth is typically lower due to the
typically higher delay-spread of the channel [62]. More explicitly, this is a consequence
of the more dispersive impulse response, since the coherence bandwidth is proportional to
the reciprocal of the impulses responses delay spread, as it was argued in [79]. Similarly
to frequency diversity or space diversity, these L_R resolvable paths can be employed in
multipath diversity schemes, which exploit the fact that statistically speaking, the different
paths cannot be in deep fades simultaneously; hence, there is always at least one propagation
path, which provides an unattenuated channel. These multipath components are diversity
paths.

Multipath diversity can only be exploited in conjunction with wideband signals. From
Equation 1.13, for a narrowband signal, where no deliberate signal spreading takes place,
the signal bandwidth B_s is significantly lower than B_c. In this case, the channel was termed
frequency nonselective in [79]. Hence, no resolvable diversity paths can be observed, unlike
in a wideband situation, and this renders TDMA and FDMA potentially less robust in a
narrowband mobile radio channel than CDMA.

Multipath diversity is achieved, for example, by a receiver referred to as the Rake
receiver invented by Price and Green [70]. This is the optimum receiver for wideband fading

multipath signals. It inherited its name from the analogy of a garden rake, whereby the fingers constitute the resolvable paths. The point where the handle and fingers meet is where diversity combining takes place. There are four basic methods of diversity combining, namely [80]:

- Selection Combining (SC).

- Maximal Ratio Combining (MRC).

- Equal Gain Combining (EGC).

- Combining of the n best signals (SCn).

The performance analysis of selection combining in CDMA can be found in [81, 82], while a general comparison of the various diversity combining techniques can be found in [80] for Rayleigh-fading channels. Maximal ratio combining gives the best performance, while selection combining is the simplest to implement. The number of resolvable paths that are combined at the receiver, represents the order of diversity of the receiver, which is denoted here as L_P. We note, however, that in practical receivers not all resolvable multipath components are combined due to complexity reasons, that is, $L_P \leq L_R$.

There are two basic demodulation techniques, namely coherent and noncoherent demodulation [5]. We will highlight the basics of coherent demodulation in this section in the context of CDMA. However, before demodulation can take place, synchronization between the transmitter and the intended receiver has to be achieved.

Synchronization in DS-CDMA is performed by a process known as code acquisition and tracking. Acquisition is usually carried out by invoking correlation techniques between the receiver's own copy of the signature sequence and the received signature sequence and searching for the displacement between them—associated with a specific chip epoch—that results in the high correlation [64, 83, 84]. Once acquisition has been accomplished, usually a code tracking loop [85] is employed to achieve fine alignment of the two sequences and to maintain their alignment. The details of code acquisition and tracking are beyond the scope of this chapter. Interested readers may refer to [86–89] and the references therein for an in-depth treatise on this subject. Hence, in this chapter, we will assume that the transmitter and the intended receiver are perfectly synchronized.

For optimum performance of the Rake receiver using coherent demodulation, the path attenuation and phase must be accurately estimated. This estimation is performed by another process known as channel estimation, which will be elaborated on in Section 1.2.6. In typical low-complexity applications, known pilot symbols can be inserted in the transmitted sequence in order to estimate the channel's attenuation and phase rotation. However, for now, let us assume perfect channel estimation in order to assess the performance of the Rake diversity combiner.

Figure 1.9 shows the block diagram of the BPSK Rake receiver. The received signal is first multiplied by the estimated channel coefficients $\alpha_1(t), \ldots, \alpha_{L_P}(t)$ in each Rake branch tuned to each resolvable path. For optimum performance of the Rake receiver using maximal ratio combining, these channel coefficient estimates should be the conjugates of the actual coefficients of the appropriate paths in order to invert the channel effects.[2] Note that for equal gain combining only the phase is estimated, and the various path contributions are

---

[2]$\alpha_l e^{j\phi_l} \times \alpha_l e^{-j\phi_l} = \alpha_l^2.$

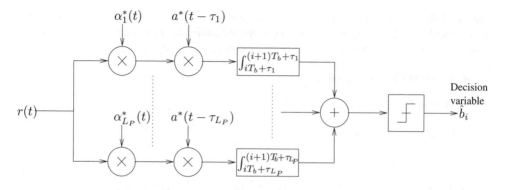

Figure 1.9: Rake receiver.

multiplied by a unity gain before summation. The resulting signals in each Rake branch are then multiplied by the conjugate signature sequences as we have seen in Figure 1.3, delayed accordingly by the code acquisition process. After despreading by the conjugate signature sequences $a^*(t - \tau_1), \ldots, a^*(t - \tau_{L_P})$, the outputs of the correlators in Figure 1.9 are combined in order to obtain the decoded symbol of:[3]

$$
\begin{aligned}
\hat{b}_i &= \mathrm{sgn} \left\{ \sum_{l=1}^{L_P} \left[\frac{1}{\sqrt{T_b}} \int_{iT_b+\tau_l}^{(i+1)T_b+\tau_l} \alpha_l^*(t) r(t) a^*(t - \tau_l)\, dt \right] \right\} \\
&= \mathrm{sgn} \left\{ \sum_{l=1}^{L_P} \left[\sqrt{\frac{P_b}{T_b}} \int_{iT_b+\tau_l}^{(i+1)T_b+\tau_l} |\alpha_l(t)|^2 b(t - \tau_l) a(t - \tau_l) a^*(t - \tau_l)\, dt \right. \right. \\
&\qquad\qquad \left. \left. + \frac{1}{\sqrt{T_b}} \int_{iT_b+\tau_l}^{(i+1)T_b+\tau_l} \alpha_l^*(t) n(t) a^*(t - \tau_l)\, dt \right] \right\} \\
&= \mathrm{sgn} \left\{ \sum_{l=1}^{L_P} \left[|\alpha_l(t)|^2 \sqrt{\xi_b} b_i \right. \right. \\
&\qquad\qquad \left. \left. + \frac{1}{\sqrt{T_b}} \int_{iT_b+\tau_l}^{(i+1)T_b+\tau_l} \alpha_l^*(t) n(t) a^*(t - \tau_l)\, dt \right] \right\}.
\end{aligned}
\tag{1.14}
$$

Normally, the first term of Equation 1.14, which contains the useful information, is much larger than the despread, noise-related second term. This is because the first term is proportional to the sum of the absolute values of the channel coefficients, whereas the second term in Equation 1.14 is proportional to the vectorial sum of the complex-valued channel coefficients. Hence, the real part of the first term is typically larger than that of the second term. Thus, the Rake receiver can enhance the detection of the data signal in a multipath environment.

Referring back to the BER curves of Figure 1.8, we can see that the performance of the system is improved when multipath diversity is used. Better performance is observed by

---

[3]Here we assumed that there is no multipath interference. This interference can be considered as part of multi-user interference, which will be discussed in the next section.

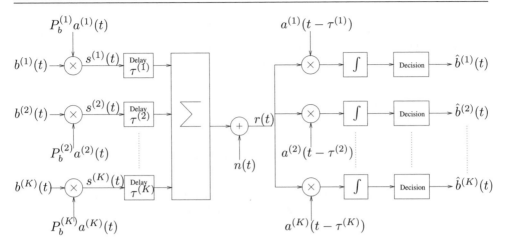

Figure 1.10: CDMA system model.

increasing the number of diversity paths L_P. However, this also increases the complexity of the receiver, since the number of correlators has to be increased, which is shown in Figure 1.9.

1.2.4 Multiple Access

So far, only single-user transmission was considered. The system is simple and straight-forward to implement. Let us now consider how multiple user transmission can affect the performance of the system.

Multiple access is achieved in DS-CDMA by allowing different users to share a common frequency band. Each transmitter and its intended receiver are assigned a distinct user signature sequence. Only the receivers having the explicit knowledge of this distinct sequence are capable of detecting the transmitted signal. Consider a CDMA scenario with K number of active users, transmitting simultaneously. The baseband equivalent system model is shown in Figure 1.10. For simplicity, it is assumed that there is no multipath propagation and perfect power control is maintained.

The mathematical representation of the kth user's data signal is similar to that shown in Equation 1.2, except for an additional superscript, denoting multi-user transmission. Hence, it is written as:

$$b^{(k)}(t) = \sum_{j=-\infty}^{\infty} b_j^{(k)} \Gamma_{T_b}(t - jT_b), \qquad (1.15)$$

where $b_j^{(k)} \in \{+1, -1\}$. There is a distinct user signature sequence $a^{(k)}(t)$ associated with the kth user, which is similar to that of Equation 1.4, with the exception of a superscript, differentiating between users:

$$a^{(k)}(t) = \sum_{h=-\infty}^{\infty} a_h^{(k)} \Gamma_{T_c}(t - hT_c). \qquad (1.16)$$

The kth user's data signal $b^{(k)}(t)$ and signature sequence $a^{(k)}(t)$ are multiplied in order to produce an equivalent baseband wideband signal, namely,

$$s^{(k)}(t) = \sqrt{P_b^{(k)}} b^{(k)}(t) a^{(k)}(t), \tag{1.17}$$

where $P_b^{(k)}$ is the average transmit power of the kth user's signal. The composite multi-user baseband received signal is:

$$r(t) = \sum_{k=1}^{K} \sqrt{P_b^{(k)}} b^{(k)}(t - \tau^{(k)}) a^{(k)}(t - \tau^{(k)}) + n(t), \tag{1.18}$$

where $\tau^{(k)}$ is the propagation delay plus the relative transmission delay of the kth user with respect to other users, and $n(t)$ is the AWGN with a double-sided power spectral density of $\frac{N_0}{2}$ W/Hz.

1.2.4.1   DL Interference

In the DL (base station to mobile), the base station is capable of synchronizing the transmission of all the users' signals, such that the symbol durations are aligned with each other. Hence the composite signal is received at each mobile station with $\tau^{(k)} = 0$ for $k = 1, 2, \ldots, K$. This scenario is also known as symbol-synchronous transmission. Using the conventional so-called single-user detector, each symbol of the jth user is retrieved from the received signal $r(t)$ by correlating it with the jth user's spreading code in order to give:

$$\hat{b}_i^{(j)} = \text{sgn} \left\{ \frac{1}{\sqrt{T_b}} \int_{iT_b}^{(i+1)T_b} r(t) a^{(j)^*}(t) dt \right\}. \tag{1.19}$$

Substituting Equation 1.18 into Equation 1.19 yields:

$$\hat{b}_i^{(j)} = \text{sgn} \left\{ \frac{1}{\sqrt{T_b}} \int_{iT_b}^{(i+1)T_b} \left[\sum_{k=1}^{K} \sqrt{P_b^{(k)}} b^{(k)}(t) a^{(k)}(t) + n(t) \right] a^{(j)^*}(t) \, dt \right\}$$

$$= \text{sgn} \left\{ \frac{1}{\sqrt{T_b}} \int_{iT_b}^{(i+1)T_b} \sqrt{P_b^{(j)}} b^{(j)}(t) a^{(j)}(t) a^{(j)^*}(t) \, dt \right.$$

$$+ \frac{1}{\sqrt{T_b}} \int_{iT_b}^{(i+1)T_b} \sum_{\substack{k=1 \\ k \neq j}}^{K} \sqrt{P_b^{(k)}} b^{(k)}(t) a^{(k)}(t) a^{(j)^*}(t) \, dt$$

$$+ \frac{1}{\sqrt{T_b}} \int_{iT_b}^{(i+1)T_b} n(t) a^{(j)^*}(t) \, dt \Bigg\}$$

$$= \text{sgn} \Bigg\{ \underbrace{\sqrt{\xi_b^{(j)}} b_i^{(j)}}_{\text{wanted signal}} + \underbrace{\sum_{\substack{k=1 \\ k \neq j}}^{K} \sqrt{\xi_b^{(k)}} b_i^{(k)} R_{jk}}_{\text{multiple access interference}} + \underbrace{n^{(j)}}_{\text{white noise}} \Bigg\}, \tag{1.20}$$

where R_{jk} is the cross-correlation of the spreading codes of the kth and jth user for $iT_b \leq t \leq (i+1)T_b$, which is given by:

$$R_{jk} = \frac{1}{T_b} \int_0^{T_b} a^{(j)}(t) a^{(k)}(t)\, dt. \tag{1.21}$$

There will be no interference from the other users if the spreading codes are perfectly orthogonal to each other. That is, $R_{jk} = 0$ for all $k \neq j$. However, designing orthogonal codes for a large number of users is extremely complex. The so-called Walsh-Hadamard codes [90] used in the IS-95 system excel in terms of achieving orthogonality.

1.2.4.2 Uplink Interference

In contrast to the previously considered DL scenario, in practical systems perfect orthogonality cannot be achieved in the UL (mobile to base station), since there is no coordination in the transmission of the users' signals. In CDMA, all users transmit in the same frequency band in an uncoordinated fashion. Hence, $\tau^{(k)} \neq 0$, and the corresponding scenario is referred to as an asynchronous transmission scenario. In this case, the time-delay $\tau^{(k)}$, $k = 1, \ldots, K$, has to be included in the calculation. Without loss of generality, it can be assumed that $\tau^{(1)} = 0$ and that $0 < \tau^{(2)} < \tau^{(3)} < \cdots < \tau^{(K)} < T_b$. In contrast to the synchronous DL scenario of Equation 1.19, the demodulation of the ith symbol of the jth user is performed by correlating the received signal $r(t)$ with $a^{(j)^*}(t)$ delayed by $\hat{\tau}^{(j)}$, yielding:

$$\hat{b}_i^{(j)} = \mathrm{sgn}\left\{ \frac{1}{\sqrt{T_b}} \int_{iT_b + \hat{\tau}^{(j)}}^{(i+1)T_b + \hat{\tau}^{(j)}} r(t) a^{(j)^*}(t - \hat{\tau}^{(j)}) dt \right\}, \tag{1.22}$$

where $\hat{\tau}^{(j)}$ is the estimated time-delay at the receiver.

Substituting Equation 1.18 into Equation 1.22 and assuming perfect code acquisition and tracking yield:[4]

$$\hat{b}_i^{(j)} = \mathrm{sgn}\left\{ \frac{1}{\sqrt{T_b}} \int_{iT_b + \tau^{(j)}}^{(i+1)T_b + \tau^{(j)}} \left[\sum_{k=1}^{K} \sqrt{P_b^{(k)}} b^{(k)}(t - \tau^{(k)}) a^{(k)}(t - \tau^{(k)}) + n(t) \right] \right.$$

$$\left. \cdot\, a^{(j)^*}(t - \tau^{(j)}) dt \right\}$$

$$= \mathrm{sgn}\left\{ \frac{1}{\sqrt{T_b}} \left[\int_{iT_b + \tau^{(j)}}^{(i+1)T_b + \tau^{(j)}} \sqrt{P_b^{(j)}} b^{(j)}(t - \tau^{(j)}) a^{(j)}(t - \tau^{(j)}) \right.\right.$$

$$\times\, a^{(j)^*}(t - \tau^{(j)}) dt$$

$$\left.\left. + \sum_{k=1}^{j-1} \int_{(i+1)T_b + \tau^{(j)}}^{(i+1)T_b + \tau^{(k)}} \sqrt{P_b^{(k)}} b^{(k)}(t - \tau^{(k)}) a^{(k)}(t - \tau^{(k)}) \right.\right.$$

---

[4]For perfect code acquisition and tracking, $\hat{\tau}^{(j)} = \tau^{(j)}$.

$$\times\ a^{(j)^*}(t-\tau^{(j)})dt$$

$$+\sum_{k=1}^{j-1}\int_{iT_b+\tau^{(k)}}^{(i+1)T_b+\tau^{(j)}}\sqrt{P_b^{(k)}}b^{(k)}(t+T_b-\tau^{(k)})a^{(k)}(t+T_b-\tau^{(k)})$$

$$\times\ a^{(j)^*}(t-\tau^{(j)})dt$$

$$+\sum_{k=j+1}^{K}\int_{iT_b+\tau^{(j)}}^{iT_b+\tau^{(k)}}\sqrt{P_b^{(k)}}b^{(k)}(t-T_b-\tau^{(k)})a^{(k)}(t-T_b-\tau^{(k)})$$

$$\times\ a^{(j)^*}(t-\tau^{(j)})dt$$

$$+\sum_{k=j+1}^{K}\int_{iT_b+\tau^{(k)}}^{(i+1)T_b+\tau^{(j)}}\sqrt{P_b^{(k)}}b^{(k)}(t-\tau^{(k)})a^{(k)}(t-\tau^{(k)})$$

$$\times\ a^{(j)^*}(t-\tau^{(j)})dt+\int_{iT_b+\tau^{(j)}}^{(i+1)T_b+\tau^{(j)}}n(t)a^{(j)^*}(t-\tau^{(j)})dt\Bigg]\Bigg\} \quad\quad (1.23)$$

$$\hat{b}_i^{(j)}=\mathrm{sgn}\left\{\underbrace{\sqrt{\xi_b^{(j)}}b_i^{(j)}}_{\text{wanted signal}}+\underbrace{\sum_{k=1}^{j-1}\sqrt{\xi_b^{(k)}}b_i^{(k)}R_{jk}(0)}_{\text{multiple access interference}}\right.$$

$$+\underbrace{\sum_{k=1}^{j-1}\sqrt{\xi_b^{(k)}}b_{i+1}^{(k)}\hat{R}_{jk}(+1)+\sum_{k=j+1}^{K}\sqrt{\xi_b^{(k)}}b_{i-1}^{(k)}R_{jk}(-1)}_{\text{multiple access interference}}$$

$$\left.+\underbrace{\sum_{k=j+1}^{K}\sqrt{\xi_b^{(k)}}b_i^{(k)}\hat{R}_{jk}(0)}_{\text{multiple access interference}}+\underbrace{n^{(j)}}_{\text{white noise}}\right\}, \quad\quad (1.24)$$

where $R_{jk}(i)$ and $\hat{R}_{jk}(i)$, $i\in\{+1,0,-1\}$ represent the cross-correlation of the spreading codes due to asynchronous transmissions, which are given by [91]:

$$R_{jk}(i)=\frac{1}{T_b}\int_{\tau^{(j)}}^{\tau^{(k)}}a^{(j)}(t-\tau^{(j)})a^{(k)}(t+iT_b-\tau^{(k)})dt \quad\quad (1.25)$$

and

$$\hat{R}_{jk}(i)=\frac{1}{T_b}\int_{\tau^{(k)}}^{T_b+\tau^{(j)}}a^{(j)}(t-\tau^{(j)})a^{(k)}(t+iT_b-\tau^{(k)})dt \qu\quad (1.26)$$

and is limited to $+1,0,-1$, since the maximum path delay is assumed to be limited to one symbol duration, as mentioned in Section 1.2.2.

Equations 1.24 and 1.20 represent the estimated demodulated data symbol of the jth user at the base station and mobile station, respectively. Both contain the desired symbol of the jth user. However, this is corrupted by noise and interference from the other users. This interference is known as multiple access interference (MAI). It contains the undesired interfering signals from the other $(K - 1)$ users. The MAI arises due to the nonzero cross-correlation of the spreading codes. Ideally, the spreading codes should satisfy the orthogonality property such that

$$R_{jk}(\tau) = \frac{1}{T_b} \int_0^{T_b} a^{(k)}(t)a^{(j)}(t - \tau)dt = \left\{ \begin{array}{ll} 1 & \text{for } k = j, \tau = 0 \\ 0 & \text{for all } k \text{ and all } \tau. \end{array} \right. \tag{1.27}$$

However, it is impossible to design codes that are orthogonal for all possible time offsets imposed by the asynchronous UL transmissions. Thus there will always be MAI in the UL. These observations are augmented by comparing the terms of Equations 1.20 and 1.24.

On the other hand, multipath interference is always present in both the forward and reverse link. Multipath interference is due to the different arrival times of the same signal via the different paths at the receiver. This is analogous to the signals transmitted from other users; hence, multipath interference is usually analyzed in the same way as MAI.

As the number of users increases, the MAI increases too. Thus, the capacity of CDMA is known to be interference limited. CDMA is capable of accommodating additional users at the expense of a gradual degradation in performance in a fixed bandwidth, whereas TDMA or FDMA would require additional bandwidth to accommodate additional users. Intensive research has been carried out to find ways of mitigating the effects of MAI. Some of the methods include voice activity control, spreading code design, power control schemes, and sectored/adaptive antennas [92]. These methods reduce the MAI to a certain extent.

The most promising UL method so far has been in the area of *multi-user detection*, which was first proposed by Verdú [93]. Multi-user detection [94–96], which will be discussed in more depth in the next chapter, invokes the knowledge of all users' signature sequences and all users' channel impulse response estimates in order to improve the detection of each individual user. The employment of this algorithm is more feasible for the UL, because all mobiles transmit to the base station and the base station has to detect all the users' signals anyway. The topic of multi-user detection is however beyond the scope of this chapter, since it will be discussed in a little more detail in the next chapter, namely in Chapter 3. For a more indepth treatment the interested readers are referred to Verdu's excellent book [97], which provides a comprehensive discussion on the topic. A general review of the various multi-user detection schemes and further references can also be found, for example, in Moshavi's contribution [92].

Another shortcoming of CDMA systems is their susceptibility to the near–far problem to be highlighted below. If all users transmit at equal power, then signals from users near the base station are received at a higher power than those from users at a higher distance due to their different pathlosses. The effects of fading highlighted in Section 1.2.2 also contribute to the power variation. Hence, according to Equation 1.24, if the jth user is transmitting from the cell border and all other users are transmitting near the base station, then the desired jth user's signal will be masked by the other users' stronger signals, which results in a high bit error rate. In order to mitigate this so-called near–far problem, power control is used to ensure that all signals from the users are received at near-equal power, regardless of their distance from the base station.

There are typically two basic types of power control [65]:

- open-loop power control

- closed-loop power control.

Open-loop power control is usually used to overcome the variation in power caused by pathloss. On the other hand, closed-loop power control is used to overcome shadow fading caused by multipath. The details of the various power control techniques will not be elaborated on in this chapter. Readers may refer to [98] for more information.

1.2.4.3   Gaussian Approximation

In order to simplify any analysis involving multi-user transmission in CDMA, the MAI is usually assumed to be Gaussian distributed by virtue of the central limit theorem [99–101]. This assumption is fairly accurate even for $K < 10$ users, when the BER is 10^{-3} or higher. We will use the standard Gaussian approximation theory presented by Pursley [99] to represent the MAI. When the desired user sequence is chip- and phase-synchronous with all the interfering sequences, where the phase-synchronous relationship is defined as in the absence of noise, the worst-case probability of error $Pr_b(\epsilon)$ performance was given by Pursley [99] as:

$$Pr_b(\epsilon) = Q\left[\sqrt{\frac{N_c}{(K-1)}}\right],\tag{1.28}$$

where $Q(\cdot)$ is the Gaussian Q-function of Equation 1.9, since the synchronous transitions do not generate pure random Gaussian-like impairments. This formula would be characteristic of the synchronous DL scenario of Section 1.2.4.1. However, in practical UL situations as augmented in Section 1.2.4.2, there is always some delay among the users, and each received signal will be phase-shifted independently. In this case, according to Pursley, the probability of error in the absence of noise will be [99]:

$$Pr_b(\epsilon) = Q\left[\sqrt{\frac{3N_c}{(K-1)}}\right].\tag{1.29}$$

Equation 1.29 represents the best performance corresponding to Gaussian-like impairments. In between these two extremes are situations whereby, in the first case, the desired sequence and the interfering sequence are chip synchronous but not phase synchronous. The probability of error in the absence of noise is given by [99]:

$$Pr_b(\epsilon) = Q\left[\sqrt{\frac{2N_c}{(K-1)}}\right].\tag{1.30}$$

In the second case, the desired sequence and interfering sequence are phase synchronous but not chip synchronous. Hence, the probability of error in the absence of noise is given by [99]:

$$Pr_b(\epsilon) = Q\left[\sqrt{\frac{3N_c}{2(K-1)}}\right].\tag{1.31}$$

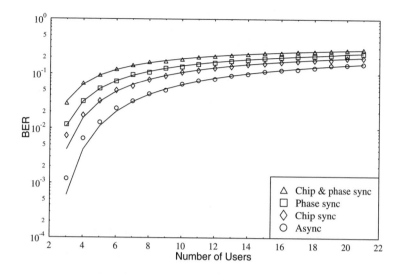

Figure 1.11: Probability of error against number of users using Equations 1.28, 1.29, 1.30, and 1.31. Markers: Simulation; solid line: Numerical computation. The processing gain is 7.

Analyzing the above equations, it can be seen that by increasing the number of chips N_c per symbol, the performance of the system will be improved. However, there is a limitation to the rate of the spreading sequence based on Digital Signal Processing (DSP) technology. Figure 1.11 compares the simulated results with the numerical results given by Equations 1.28 to 1.31 for a binary system with a processing gain of 7. The figure shows that the assumption of Gaussian distributed MAI is valid, especially for a high number of users. It also demonstrates that *CDMA attains its best possible performance in an asynchronous multi-user transmission system*. This is an advantage over TDMA and FDMA because TDMA and FDMA require some coordination among the transmitting users, which increases the complexity of the system.

1.2.5 Spreading Codes

As seen previously, the choice of spreading codes plays an important role in DS-CDMA. The main criteria for selecting a particular set of user signature sequences in CDMA applications are that the number of possible different sequences in the set for any sequence length must be high in order to accommodate a high number of users in a cell. The spreading sequences must also exhibit low cross-correlations for the sake of reducing the multi-user interference during demodulation. A high autocorrelation main-peak to secondary-peak ratio—as indicated by Equation 1.27—is also essential, in order to minimize the probability of so-called false alarms during code acquisition. This also reduces the self-interference among the diversity paths. Below we provide a brief overview of a few different spreading sequences.

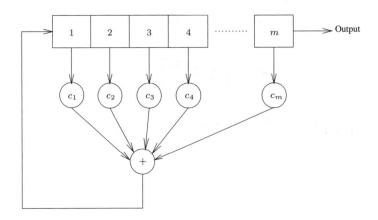

Figure 1.12: m-stage shift register with linear feedback.

Table 1.1: Properties of m- and Gold-sequences. ©McGraw-Hill, 1995 [5].

m	Number of m-sequences	Peak cross-correlation	Number of Gold sequences	Peak cross-correlation
3	2	5	$2^m + 1 = 9$	5
4	2	9	$2^m + 1 = 17$	9
5	6	11	$2^m + 1 = 33$	9
6	6	23	$2^m + 1 = 65$	17
7	18	41	$2^m + 1 = 129$	17
8	16	95	$2^m + 1 = 257$	33

1.2.5.1 m-sequences

Perhaps the most popular set of codes known are the m-sequences [5]. An m-sequence with a periodicity of $n = 2^m - 1$ can be readily generated by an m-stage shift register with linear feedback, as shown in Figure 1.12.

The tap coefficients $c_1, c_2, \ldots, c_m$ can be either 1 (short circuit) or 0 (open circuit). Information on the shift register feedback polynomials, describing the connections between the register stages and the modulo-2 adders can be found, for example, in [5]. Note that in spread spectrum applications, the output binary sequences of 0,1 are mapped into a bipolar sequence of $-1, 1$, respectively. Table 1.1 shows the total number of m-sequences and the associated chip-synchronous peak cross-correlation for $m = 3, 4, 5, 6, 7$, and 8.

In this context, the peak cross-correlation quantifies the maximum number of identical chips in a pair of different spreading codes. It is desirable to have as low a number of code pairs as possible, which exhibit this peak cross-correlation. Furthermore, the peak cross-correlation has to be substantially lower than the codes' autocorrelation, which is given by the length of the code. In general, the cross-correlations of m-sequences are too high to be useful in CDMA. Another set of spreading codes, which exhibit fairly low chip-synchronous cross-correlations are the Gold sequences [5], which will be elaborated on in the next section.

1.2.5.2 Gold Sequences

Gold sequences [5] with a period of $n = 2^m - 1$ are derived from a pair of m-sequences having the same period. Out of the total number of possible m-sequences having a periodicity or length of n, there exists a pair of m-sequences, whose chip-synchronous cross-correlation equals to either $-1, -t(m)$ or $[t(m) - 2]$, where

$$t(m) = \begin{cases} 2^{(m+1)/2} + 1 & \text{odd } m \\ 2^{(m+2)/2} + 1 & \text{even } m. \end{cases} \tag{1.32}$$

This unique pair of m-sequences is commonly known as the pair of preferred codes. A set of $n = 2^m - 1$ sequences can be constructed by cyclically shifting a preferred code one chip at a time and then taking the modulo-2 summation with the other code for every chip shift. The resulting set of $n = 2^m - 1$ sequences together with the two preferred codes constitute a set of Gold sequences. Table 1.1 compares the total number of Gold sequences for $m = 3$, 4, 5, 6, 7, and 8, and their corresponding peak cross-correlation with the same parameters of m-sequences.

Table 1.1, shows that the Gold sequences exhibit equal or lower peak cross-correlation between different sequences of the set, in comparison to m-sequences for all m. There are also more Gold sequences than m-sequences for all values of m. Thus, Gold sequences are always preferred to m-sequences in CDMA applications, despite having a poorer asynchronous autocorrelation peak, which is a disadvantage in terms of both code acquisition and detection by correlators. Pseudo Noise (PN) sequences, such as m-sequences and Gold sequences, have periods of $N = 2^l - 1$ where l is the sequence length, which is a rather awkward number to match to the system clock requirements. Extended m-sequences having periods of 2^l solved this problem, an issue augmented below.

1.2.5.3 Extended *m*-sequences [102]

Extended m-sequences are derived from an m-sequence, generated by a linear feedback shift register, by adding an element into each period of the m-sequence. We will follow the notation, whereby the binary sequences of 0 and 1 are mapped to the corresponding bipolar sequences of -1 and $+1$, respectively. In order to arrive at zero-balanced extended m-sequences, which have a zero DC-component, the element to be inserted must be chosen so that the number of -1 s and $+1$ s within a period is the same. There are $2^m - 1$ positions in a period, where the additional element can be inserted. In [102], the element is inserted into the longest run of -1 s in a period. In an m-sequence of period $2^m - 1$, the longest run of -1 s is $n - 1 = 2^m - 2$. It was shown in [102] that the off-peak autocorrelation of extended sequences was similar to that of Gold sequences. However, the cross-correlation of different extended m-sequences at even-indexed chip-positions—that is, time-domain displacements—is similar to that of the m-sequences, which is much higher than that of the Gold sequences in Table 1.1. Thus, the extended m-sequences are not suitable in a multi-user environment, where the cross-correlation between the codes of different users is required to be as low as possible. Since this has a high impact on the user-capacity of cellular mobile systems, the additional hardware needed to synchronize the $N = 2^l - 1$ chip-duration m-sequences or Gold sequences with the system clock has to be tolerated and hence, extended m-sequences are not recommended in CDMA. Section 1.3.2.6 highlights the various spreading codes proposed for employment in the forthcoming 3G systems. In the next section we provide a rudimentary introduction to channel estimation for CDMA systems.

1.2.6   Channel Estimation

As mentioned earlier, accurate estimation of the channel parameters is vital in optimizing the coherent demodulation. This channel parameter estimation process is an integral part of coherent demodulation, particularly in a multipath mobile radio environment. This is because the mobile radio channel changes randomly as a function of time, and thus the channel estimates have to be continuously estimated. This section describes various techniques used to estimate the channel path gains and phases, which will be referred to as channel coefficients. There are basically three practical channel coefficient estimation methods, each with their advantages and disadvantages, namely:

- Pilot-channel assisted, [103–105]

- Pilot-symbol assisted and [106]

- Pilot-symbol assisted decision-directed channel estimation [107],

which we briefly characterized in the following subsections.

1.2.6.1   DL Pilot-assisted Channel Estimation

Channel estimation using a pilot channel/tone was proposed, for example, in [103–105], where a channel is dedicated solely for the purpose of estimating the multipath channel attenuations and delays. In order to prevent the pilot channel from interfering with the data channel, the pilot channel must either be allocated to a dedicated portion of the spectrum or share the spectrum with the data channel, but a spectral notch has to be created for accommodating the pilot. The former technique is known as the pilot tone-above-band (TAB) regime, while the latter is referred to as the transparent tone-in-band (TTIB) technique [103], both of which have been used in conventional single-carrier modems [12].

However, CDMA is more amenable to employing the TAB or TTIB techniques and their various derivatives, since the pilot signal can be transmitted in the same frequency band as the data signal by invoking orthogonal or quasi-orthogonal spreading codes. Hence, the pilot signal is treated as part of the MAI, and no notch filtering or additional pilot frequency band is required. In some 2G mobile systems, such as the IS-95 system this method is used on the DL but not on the UL. This is because it would be inefficient to have every mobile station transmitting their own pilot channel.

In 3G mobile systems, however, it was proposed [108] that a separate dedicated user control channel be transmitted simultaneously with the information channel, which could also be used as an alternative to the pilot channel – an issue to be elaborated on at a later stage. Suffice to say here that the main advantage of pilot-channel based channel estimation is that since the pilot channel is always present, the channel coefficients can be continuously estimated for every data symbol's demodulation. Hence, it is particularly useful for channels that are highly time-variant.

The block diagram of the channel estimator is shown in Figure 1.13, where $r(t)$ is

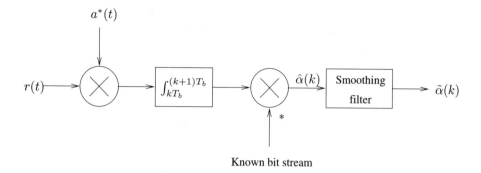

Figure 1.13: Structure of the channel estimator using known transmitted pilot symbols or bits.

the received signal and $a(t)$ is the spreading code. Assume that the known bit-stream is a continuous sequence of binary 1 s, then

$$\hat{\alpha}(k) = \frac{1}{T_b} \int_{kT_b}^{(k+1)T_b} r(t)a^*(t)dt$$

$$= \frac{1}{T_b} \int_{kT_b}^{(k+1)T_b} [\alpha(t)a(t) + n(t)]a^*(t)dt$$

$$= \alpha(k) + \frac{1}{T_b} \int_{kT_b}^{(k+1)T_b} n(t)a^*(t)dt, \qquad (1.33)$$

where $\alpha(k)$ is the complex channel coefficient in the bit interval $kT_b \leq t < (k+1)T_b$. The variable $\hat{\alpha}(k)$ is termed the noisy channel estimate derived from the received signal contaminated by the noise element in the second term of Equation 1.33, while $\tilde{\alpha}(k)$ are estimates obtained from the output of the smoothing filter in Figure 1.13, which assists in averaging out the random effects of channel noise. Assuming that $n(t)$ is the AWGN having a zero mean (any MAI can be fairly accurately modeled also as AWGN [109]), averaging a large number of these noisy estimates will suppress the noise's influence. Several proposals have been published in the literature regarding the smoothing algorithm used in channel estimation, such as moving average [110, 111], least squares line fitting [112], low-pass filtering [106, 107, 112], and adaptive linear smoothing [113]. A more in-depth discourse on the TTIB technique was also given in Section 10.3.1 of [12] in the context of QAM. A compromise in terms of complexity and accuracy has to be made in selecting a particular algorithm. So far, only the DL channel estimation has been elaborated on. The associated UL issues are discussed next.

1.2.6.2 UL Pilot-symbol Assisted Channel Estimation

Pilot-symbol assisted channel estimation was first proposed by Moher and Lodge [106], and the first detailed analysis of this technique was carried out by Cavers [113]. Since then, several papers have been published, which analyzed its effect on system performance [111,112,114]. This technique is the time-domain equivalent of the frequency-domain pilot channel-assisted TTIB method mentioned in Section 1.2.6.1, which was detailed in Section 10.3.2 of [12].

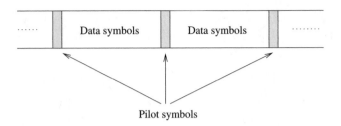

Figure 1.14: Data stream with embedded pilot symbols.

The advantage of this technique is that it dispensed with the use of a notch filter in the context of QAM modems, and so it did not result in an expanded bandwidth. However, for this technique, several parameters such as the number of pilot symbols or their periodicity has to be carefully chosen in order to trade-off the accuracy of estimation against the required pilot overhead. More explicitly, the pilot-spacing has to be sufficiently low to satisfy the Nyquist sampling theorem for the fading Doppler frequency encountered. This technique can be used for efficient coherent demodulation on the UL, and Section 1.3.2.3 highlights how UL channel estimation is carried out in the context of 3G systems.

The pilot symbols are multiplexed with the data stream periodically, as shown in Figure 1.14. This multiplexed stream is then transmitted to the base station from every communicating mobile station. The base station will extract the channel estimates from the known demodulated pilot symbols, and using, for example, ideal low-pass or simple linear interpolation [112], it will generate a channel magnitude and phase estimate for each UL symbol. These channel estimates will then be used to "de-fade", "de-rotate", and demodulate the data symbols.

If the channel has a slow fading characteristic, such that it is more or less constant between consecutive pilot symbols, this method can be fairly accurate and of low complexity. However, the bandwidth efficiency is slightly compromised, since again, a sufficiently high number of pilots has to be incorporated in order to satisfy the Nyquist sampling criterion corresponding to the normalized Doppler frequency of the fading channel. For more information on this subject we refer to Section 10.3.2 of [12]. The above pilot-symbol assisted (PSA) concept is further developed in the next section.

1.2.6.3   Pilot-symbol Assisted Decision-directed Channel Estimation

Pilot-symbol assisted decision-directed channel estimation was first proposed by Irvine and McLane [107], and it was shown that it improves the accuracy of the estimation as compared with the original pilot symbol-assisted method of Section 1.2.6.2. It extends the concept of the pilot-symbol assisted channel estimation technique by using the detected data symbols in order to obtain the subsequent channel parameters, since in the absence of channel errors these demodulated data symbols can be considered to be known pilot symbols.

A decision-directed pilot-symbol assisted (PSA) scheme is illustrated in Figure 1.15, where $s(k)$ is the kth received symbol and $b(k)$ is the kth detected symbol. The signal is still transmitted in a transmission burst or frame format, similarly to that shown in Figure 1.14. At the beginning of the frame, the pilot symbols will be used to estimate the channel parameters

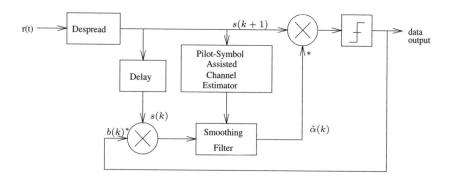

Figure 1.15: Receiver structure of PSA decision-directed channel estimation.

in order to demodulate the data symbol immediately following the pilot symbol. This is performed by the pilot symbol-assisted channel estimator block of Figure 1.15. The detected data symbol $b(k)$ is then fed back and multiplied with its original but delayed received version $s(k)$, as seen in Figure 1.15. If this symbol is detected correctly, then it is analogous to a known pilot symbol and the channel coefficient corresponding to this received symbol can be estimated in the same way. This estimated channel coefficient is then passed through the smoothing filter of Figure 1.15 in order to obtain a smoothed estimate $\tilde{\alpha}(k)$ to be used in its conjugate form for de-fading and de-rotating the next symbol, as portrayed in Figure 1.15.

If the decision is wrong, obviously the estimated channel coefficient would be inaccurate. The effect of erroneous decisions is mitigated by the smoothing filter, which will suppress the effects of an occasional glitch due to the incorrect channel estimates. In the event that the smoothing filter is unable to average out the channel coefficient errors and its output is a complex channel coefficient, which is far from the actual value, then this error may propagate through the data stream, since the correct decoding of each data symbol is dependent on the accuracy of the previous channel coefficient estimates. In order to prevent this from happening, the smoothing process is reset when the next block of pilot symbols arrives. The averaging process will recommence with the pilot symbol-assisted channel estimates.

The schematic diagram shown in Figure 1.15 is only one of the few possibilities of implementing a decision-directed PSA channel estimation arrangement. This structure is also known as a decision-feedback PSA channel estimator because the estimated channel coefficient is used for compensating the channel's effects for the next symbol. In another version of this algorithm, shown in Figure 1.16, a tentative decision, $\hat{b}(k)$, is carried out concerning the current symbol, $s(k)$, using the pilot symbol-assisted estimate, $\tilde{\alpha}^*(k)$. Using this tentative decision concerning the received symbol $s(k)$, its corresponding channel coefficient estimate, $\hat{\alpha}(k)$, is derived from the product of $\hat{b}^*(k)$ and $s(k)$ in Figure 1.15 and averaged or smoothed with the aid of the previous estimates. The output of the smoothing filter is then multiplied with the received signal $s(k)$ again, in order to compensate the channel attenuation and phase rotation and hence to obtain the final decision, $b(k)$. Such an estimator is known as a feedforward estimator. This implementation is slightly more complicated but has the advantage of using the current estimate on the current symbol rather than tolerating a latency in the channel estimation process.

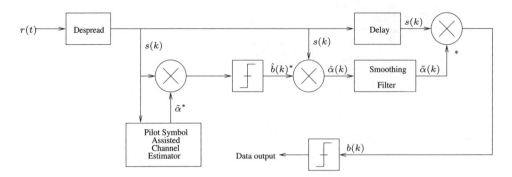

Figure 1.16: Receiver structure using decision-feedforward PSA channel estimation.

1.2.7   Summary

In this section we have briefly studied the fundamentals of a CDMA system. We have seen that several processes are vital in optimizing the performance, such as spreading, channel estimation, code synchronization, and power control. In the subsequent sections, we will make certain assumptions that will ease our analysis and simulation. These assumptions are:

- Perfect code acquisition and tracking. Hence, the transmitter and the intended receiver will always be synchronized for every path.

- Perfect channel estimation. This assumption will be used unless our focus is on the effects of imperfect estimation.

- Gaussian approximation of multi-user and multipath interference. This assumption will be used only in analysis and numerical computation, and will be validated by simulations performed in actual multi-user and multipath transmission scenarios. This also implies that random sequences will be considered instead of the deterministic sequences introduced in Section 1.2.5.

- On the UL, the number of paths encountered by each user's signal is equal.

- Perfect power control is used. This implies that all users' signals will be received at the base station with equal power.

Following the above rudimentary considerations on PSA channel estimation, let us now review the third-generation (3G) mobile system proposals in the next section.

1.3   Third-generation Systems

1.3.1   Introduction

The evolution of third-generation (3G) wireless systems began in the late 1980s when the International Telecommunication Union's Radiocommunication Sector (ITU-R) Task Group 8/1 defined the requirements for the 3G mobile radio systems. This initiative was then known as Future Public Land Mobile Telecommunication System (FPLMTS) [54,61]. The frequency spectrum for FPLMTS was identified on a worldwide basis during the World Administrative Radio Conference (WARC) in 1992 [61], as the bands 1885–2025 MHz and 2110–2200 MHz.

The tongue-twisting acronym of FPLMTS was also aptly changed to IMT-2000, which refers to the International Mobile Telecommunications system in the year 2000. Besides possessing the ability to support services from rates of a few kbps to as high as 2 Mbps in a spectrally efficient way, IMT-2000 aimed to provide a seamless global radio coverage for global roaming. This implied the ambitious goal of aiming to connect virtually any two mobile terminals worldwide. The IMT-2000 system was designed to be sufficiently flexible in order to operate in any propagation environment, such as indoor, outdoor to indoor, and vehicular scenarios. It is also aiming to be sufficiently flexible to handle circuit as well as packet mode services and to handle services of variable data rates. In addition, these requirements must be fulfilled with a QoS comparable to that of the current wired network at an affordable cost.

Several regional standard organizations—led by the European Telecommunications Standards Institute (ETSI) in Europe, the Association of Radio Industries and Businesses (ARIB) in Japan, and the Telecommunications Industry Association (TIA) in the United States—have been dedicating their efforts to specifying the standards for IMT-2000. A total of 15 Radio Transmission Technology (RTT) IMT-2000 proposals were submitted to ITU-R in June 1998, five of which are satellite-based solutions, while the rest are terrestrial solutions. Table 1.2 shows a list of the terrestrial-based proposals submitted by the various organizations and their chosen radio access technology.

As shown in Table 1.2 most standardization bodies have based their terrestrial oriented solutions on *Wideband-CDMA* (W-CDMA), due to its advantageous properties, which satisfy most of the requirements specified for 3G mobile radio systems. W-CDMA is aiming to provide improved coverage in most propagation environments in addition to an increased user capacity. Furthermore, it has the ability to combat—or to benefit from—multipath fading through Rake multipath diversity combining [66–68]. W-CDMA also simplifies frequency planning due to its unity frequency reuse. A rudimentary discourse on the RTT proposals submitted by ETSI, ARIB, and TIA can be found in [11].

Recently, several of the regional standard organizations have agreed to cooperate and jointly prepare the Technical Specifications (TS) for the 3G mobile systems in order to assist as well as to accelerate the ITU process for standardization of IMT-2000. This led to the formation of two Partnership Projects (PPs), which are known as 3GPP1 [115] and 3GPP2 [116]. 3GPP1 was officially launched in December 1998 with the aim of establishing the TS for IMT-2000 based on the evolved Global System for Mobile Telecommunications (GSM) [55] core networks and the UMTS[5] Terrestrial Radio Access (UTRA) RTT proposal. There are six organizational partners in 3GPP1: ETSI, ARIB, the China Wireless Telecommunication Standard (CWTS) group, the Standards Committee T1 Telecommunications (T1, USA), the Telecommunications Technology Association (TTA, Korea), and the Telecommunication Technology Committee (TTC, Japan). The first set of specifications for UTRA was released in December 1999, which contained detailed information on not just the physical layer aspects for UTRA, but also on the protocols and services provided by the higher layers. Here we will

---

[5]UMTS, an abbreviation for Universal Mobile Telecommunications System, is a term introduced by ETSI for the 3G wireless mobile communication system in Europe.

Table 1.2: Proposals for the radio transmission technology of terrestrial IMT-2000 (obtained from ITU's web site: http://www.itu.int/imt).

Proposal	Description	Multiple access	Source
DECT	Digital Enhanced Cordless Telecommunications	Multicarrier TDMA (TDD)	ETSI Project (EP) DECT
UWC-136	Universal Wireless Communications	TDMA (FDD and TDD)	USA TIA TR45.3
WIMS W-CDMA	Wireless Multimedia and Messaging Services Wideband CDMA	Wideband CDMA (FDD)	USA TIA TR46.1
TD-CDMA	Time Division Synchronous CDMA	Hybrid with TDMA/CDMA/ SDMA (TDD)	Chinese Academy of Telecommunication Technology (CATT)
W-CDMA	Wideband CDMA	Wideband DS-CDMA (FDD and TDD)	Japan ARIB
CDMA II	Asynchronous DS-CDMA	DS-CDMA (FDD)	South Korean TTA
UTRA	UMTS Terrestrial Radio Access	Wideband DS-CDMA (FDD and TDD)	ETSI SMG2
NA: W-CDMA	North America Wideband CDMA	Wideband DS-CDMA (FDD and TDD)	USA T1P1-ATIS
cdma2000	Wideband CDMA (IS-95)	DS-CDMA (FDD and TDD)	USA TIA TR45.5
CDMA I	Multiband synchronous DS-CDMA	Multiband DS-CDMA	South Korean TTA

concentrate on the UTRA physical layer specifications, and a basic familiarity with CDMA principles is assumed.

In contrast to 3GPP1, the objective of 3GPP2 is to produce the TS for IMT-2000 based on the evolved ANSI-41 core networks, the cdma2000 RTT. 3GPP2 is spearheaded by TIA, and its members include ARIB, CWTS, TTA, and TTC. Despite evolving from completely diversified core networks, members from the two PPs have agreed to cooperate closely in order to produce a globally applicable TS for the 3G mobile systems.

This chapter serves as an overview of the UTRA specifications, which is based on the evolved GSM core network. However, information given here is by no means the final specifications for UTRA or indeed for IMT-2000. It is very likely that the parameters and technologies presented in this chapter will evolve further. Readers may also want to refer to a recent book by Ojanperä and Prasad [117], which addresses W-CDMA 3G mobile radio systems in more depth.

1.3.2 UMTS Terrestrial Radio Access (UTRA) [59, 115, 117–124]

Research activities for UMTS [54, 56, 58, 60, 118, 119, 125] within ETSI have been spear-headed by the European Union's (EU) sponsored programmes, such as the Research in Advanced Communication Equipment (RACE) [108, 126] and the Advanced Communications Technologies and Services (ACTS) [118, 125, 126] initiative. The RACE programme, which is comprised of two phases, commenced in 1988 and ended in 1995. The objective of this programme was to investigate and develop testbeds for the air interface technology candidates. The ACTS programme succeeded the RACE programme in 1995. Within the ACTS Future Radio Wideband Multiple Access System (FRAMES) project, two multiple access modes have been chosen for intensive study, as the candidates for UMTS terrestrial radio access (UTRA). They are based on Time Division Multiple Access (TDMA) with and without spreading, and on W-CDMA [57, 59, 127].

As early as January 1997, ARIB decided to adopt W-CDMA as the terrestrial radio access technology for its IMT-2000 proposal and proceeded to focus its activities on the detailed specifications of this technology [58]. Driven by a strong support behind W-CDMA worldwide and this early decision from ARIB, ETSI reached a consensus agreement in January 1998 to adopt W-CDMA as the terrestrial radio access technology for UMTS. In this section, we highlight the key features of the physical layer aspects of UTRA that have been developed since then. Most of the material in this section is based on an amalgam of [59, 115, 117–124].

1.3.2.1 Characteristics of UTRA

The proposed spectrum allocation for UTRA is shown in Figure 1.17. As can be seen, UTRA is unable to utilize the full frequency spectrum allocated for the 3G mobile radio systems during the WARC'92, since those frequency bands have also been partially allocated to the Digital Enhanced Cordless Telecommunications (DECT) systems. Also, the allocated frequency spectrum was originally based on the assumption that speech and low data rate transmission would be the dominant services offered by IMT-2000. However, this assumption has become invalid, as the trend has shifted toward services that require high-speed data transmission, such as Internet access and multimedia services. A study conducted by the UMTS Forum [128] forecasted that the current frequency bands allocated for IMT-2000 are only sufficient for the initial deployment until the year 2005. According to the current demand estimates, it was foreseen that an additional frequency spectrum of 187 MHz is required for IMT-2000 in high-traffic demand areas by the year 2010. This extension band will be identified during the World Radio Conference (WRC)-2000. Among the many candidate extension bands, the band 2520–2670 MHz has been deemed by many people to be the most likely to be chosen. Unlike other bands, which have already been allocated for use in other applications, this band was allocated to mobile services in all regions. Furthermore, the 150 MHz bandwidth available is sufficiently wide to satisfy most of the forecasted spectrum requirements.

The radio access supports both *Frequency Division Duplex* (FDD) and *Time Division Duplex (TDD)* operations. The operating principles of these two schemes are augmented here in the context of Figure 1.18.

Specifically, the UL and DL signals are transmitted using different carrier frequencies f_1 and f_2, respectively, separated by a frequency guard band in FDD mode. On the other

| DECT | W-CDMA (TDD) | W-CDMA Uplink (FDD) | MS | W-CDMA (TDD) | | W-CDMA Downlink (FDD) | MS |

1885   1900   1920   1980 2010   2025   2110   2170 2200

Frequency (MHz)

MS : Mobile satellite application
DECT : Digital Enhanced Cordless Telecommunications
FDD : Frequency Division Duplex
TDD : Time Division Duple
DECT frequency band : 1880 – 1900 MHz

Figure 1.17: The proposed spectrum allocation in UTRA.

hand, the UL and DL messages in the TDD mode are transmitted using the same carrier frequency f_c, but in different timeslots, separated by a guard period. As seen from the spectrum allocation in Figure 1.17, the paired bands of 1920–1980 MHz and 2110–2170 MHz are allocated for FDD operation in the UL and DL, respectively, whereas the TDD mode is operated in the remaining unpaired bands [118]. The parameters designed for FDD and TDD operations are mutually compatible so as to ease the implementation of a dual-mode terminal capable of accessing the services offered by both FDD and TDD operators.

We note furthermore that recent research advocates the TDD mode quite strongly in the context of burst-by-burst adaptive CDMA modems [96], in order to adjust the modem parameters, such as the spreading factor or the number of bits per symbol on a burst-by-burst basis. This allows the system to more efficiently exploit the time-variant wireless channel capacity, hence maintaining a higher bits/s/Hz bandwidth efficiency. Furthermore, there have been proposals in the literature for allowing TDD operation in certain segments of the FDD spectrum as well, since FDD is incapable of surrendering the UL or DL frequency band of the duplex link, when the traffic demand is basically simplex. In fact, segmenting the spectrum in FDD/TDD bands inevitably results in some inefficiency in bandwidth utilization terms, especially in case of asymmetric or simplex traffic, when only one of the FDD bands is required. Hence, the more flexible TDD link could potentially double the link's capacity by allocating all timeslots in one direction. The idea of eliminating the dedicated TDD band was investigated [129], where TDD was invoked within the FDD band by simply allowing TDD transmissions in either the UL or DL frequency band, depending on which one was less interfered. This flexibility is unique to CDMA, since as long as the amount of interference is not excessive, FDD and TDD can share the same bandwidth. This would be particularly feasible in the indoor scenario of [129], where the surrounding outdoor cell could be using FDD, while the indoor cell would reuse the same frequency band in TDD mode. The buildings' walls and partitions could mitigate the interference between the FDD/TDD schemes.

Table 1.3 shows the basic parameters of the UTRA. Some of these parameters are discussed during our further discourse, but significantly more information can be gleaned concerning the UTRA system by carefully studying the table.

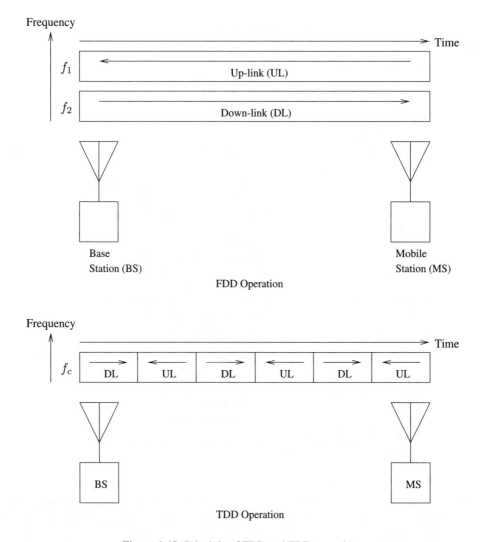

Figure 1.18: Principle of FDD and TDD operation.

The UTRA system is operated at a basic chip rate of 3.84 Mcps,[6] giving a nominal bandwidth of 5 MHz, when using root-raised cosine Nyquist pulse-shaping filters with a rolloff factor of 0.22. UTRA fulfilled the requirements of 3G mobile radio systems by offering a range of user bit rates up to 2 Mbps. Various services having different bit rates and QoS can be readily supported using Orthogonal Variable Spreading Factor (OVSF) codes [130], which will be highlighted in Section 1.3.2.6.1, and service multiplexing, which will be discussed in Section 1.3.2.4. A key feature of the UTRA system, which was absent in the second-generation (2G) IS-95 system [90] was the use of a dedicated pilot sequence embedded in the users' data stream. These can be invoked in order to support the operation of adaptive

---

[6]In the UTRA RTT proposal submitted by ETSI to ITU, the chip rate was actually set at 4.096 Mcps.

Table 1.3: UTRA basic parameters.

Radio access technology	FDD:DS-CDMA
	TDD : TDMA/CDMA
Operating environments	Indoor/Outdoor to indoor/Vehicular
Chip rate (Mcps)	3.84
Channel bandwidth (MHz)	5
Nyquist rolloff factor	0.22
Duplex modes	FDD and TDD
Channel bit rates (kbps)	FDD (UL) : 15/30/60/120/240/480/960
	FDD (DL) : 15/30/60/120/240/480/960/1920
	TDD (UL)[†] : variable, from 366 to 6624
	TDD (DL)[†] : 366/414/5856/6624
Frame length	10 ms
Spreading factor	FDD (UL) : variable, 4 to 256
	FDD (DL) : variable, 4 to 512
	TDD (UL) : variable, 1 to 16
	TDD (DL) : 1, 16
Detection scheme	Coherent with time-multiplexed pilot symbols
	Coherent with common pilot channel
Intercell operation	FDD : Asynchronous
	TDD : Synchronous
Power control	Inner-loop
	Open loop (TDD UL)
Transmit power dynamic range	80 dB (UL), 30 dB (DL)
Handover	Soft handover
	Inter-frequency handover

[†] Channel bit rate per timeslot.

antennas at the base station (BS), which was not facilitated by the common pilot channel of the IS-95 system. However, a common pilot channel was still retained in UTRA in order to provide the demodulator's phase reference for certain common physical channels, when embedding pilot symbols for each user is not feasible.

Regardless of whether a common pilot channel is used or dedicated pilots are embedded in the data, they facilitate the employment of *coherent detection*. Coherent detection is known to provide better performance than noncoherent detection [5]. Furthermore, the inclusion of short spreading codes enables the implementation of various performance enhancement techniques, such as interference cancellers and joint-detection algorithms, which results in excessive complexity in conjunction with long spreading codes. In order to support flexible system deployment in indoor and outdoor environments, *inter-cell-asynchronous operation* is used in the FDD mode. This implies that no external timing source, such as a reference signal or the Global Positioning System (GPS) is required. However, in the TDD mode inter-cell synchronization is required in order to be able to seamlessly access the timeslots offered by adjacent BSs during handovers. This is achieved by maintaining synchronization between the BSs.

Table 1.4: UTRA transport channels.

Dedicated Transport Channel	Common Transport Channel
Dedicated CHannel (DCH) (UL/DL)	Broadcast CHannel (BCH) (DL)
	Forward Access CHannel (FACH) (DL)
	Paging CHannel (PCH) (DL)
	Random Access CHannel (RACH) (UL)
	Common Packet CHannel (CPCH) (UL)
	DL Shared CHannel (DSCH) (DL)

1.3.2.2 Transport Channels

Transport channels are offered by the physical layer to the higher Open Systems Inter-connection (OSI) layers, and they can be classified into two main groups, as shown in Table 1.4 [59, 118]. The Dedicated transport CHannel (DCH) is related to a specific Mobile Station (MS)-BS link, and it is used to carry user and control information between the network and an MS. Hence, the DCHs are bidirectional channels. There are six transport channels within the common transport channel group, as shown in Table 1.4. The Broadcast CHannel (BCH) is used to carry system- and cell-specific information on the DL to all MSs in the entire cell. This channel conveys information, such as the initial UL transmit power of the MS during a random access transmission and the cell-specific scrambling code, as we shall see in Section 1.3.2.7. The Forward Access CHannel (FACH) of Table 1.4 is a DL common channel used for carrying control information and short user data packets to MSs, if the system knows the serving BS of the MS. On the other hand, the Paging CHannel (PCH) of Table 1.4 is used to carry control information to an MS if the serving BS of the MS is unknown, in order to page the MS, when there is a call for the MS. The Random Access CHannel (RACH) of Table 1.4 is UL channel used by the MS to carry control information and short user data packets to the BS, in order to support the MS's access to the system, when it wishes to set up a call. The Common Packet CHannel (CPCH) is UL channel used for transmitting bursty data traffic in a contention-based random access manner. Lastly, as its name implies, the DL Shared CHannel (DSCH) is a DL channel that is shared by several users.

The philosophy of these channels is fairly plausible, and it is informative as well as enlightening to explore the differences between the somewhat less flexible control regime of the 2G GSM [55] system and the more advanced 3G proposals, which we leave for the motivated reader due to lack of space. Unfortunately it is not feasible to design the control regime of a sophisticated mobile radio system by "direct synthesis" and so some of the solutions reviewed throughout this section in the context of the 3G proposals may appear somewhat heuristic and quite ingenious. These solutions constitute an amalgam of the wireless research community's experience in the design of the existing 2G systems and of the lessons learned from their operation. Further contributing factors in the design of the 3G systems were based on solving the signaling problems specific to the favored physical layer traffic channel solutions, namely, CDMA. In order to mention only one of them, the TDMA-based GSM system [55] was quite robust against power control inaccuracies, while the Pan-American IS-95 CDMA system [90] required an accurate power control. As we will see in

Figure 1.19: UTRA physical channel structure.

Section 1.3.2.8, the power control problem was solved quite elegantly in the 3G proposals. We will also see that statistical multiplexing schemes—such as ALOHA, the original root of the recently more familiar Packet Reservation Multiple Access (PRMA) procedure—found their way into public mobile radio systems. A variety of further interesting solutions have also found applications in these 3G proposals, which are the results of the past decade of wireless system research. Let us now review the range of physical channels in the next section.

1.3.2.3   Physical Channels

The transport channels are transmitted using the physical channels. The physical channels are typically organized in terms of radio frames and timeslots, as shown in Figure 1.19. The philosophy of this hierarchical frame structure is also reminiscent to a certain degree of the GSM TDMA frame hierarchy of [55]. However, while in GSM each TDMA user had an exclusive slot allocation, in W-CDMA the number of simultaneous users supported is dependent on the users' required bit rate and their associated spreading factors. The MSs can transmit continuously in all slots or discontinuously, for example, when invoking a voice activity detector (VAD). Some of these issues will be addressed in Section 1.3.2.4.

As seen in Figure 1.19, there are 15 timeslots within each radio frame. The duration of each timeslot is $\frac{2}{3}$ ms, which gives a duration of 10 ms for the radio frame. As we shall see later in this section, the configuration of the information in the timeslots of the physical channels differs from one another in the UL and DL, as well as in the FDD and TDD modes. The 10 ms frame duration also conveniently coincides, for example, with the frame length of the ITU's G729 speech codec for speech communications, while it is a "submultiple" of the GSM system's various full- and half-rate speech codecs' frame durations [55]. We also note that a convenient mapping of the video stream of the H.263 videophone codec can be arranged on the 10 ms-duration radio frames for supporting interactive video services, while on the move. Furthermore, the spreading factor (SF) can be varied on a 10 ms burst-by-burst (BbB) basis, in order to adapt the transmission mode in harmony with channel quality fluctuations, while maintaining a given target bit error rate. Although it is not part of the standard proposal, we found that it was more beneficial to adapt the number of bits per symbol on a BbB basis than varying the SF [96].

In the FDD mode, a DL physical channel is defined by its spreading code and frequency. Furthermore, in the UL, the modem's orthogonal in-phase (I) and quadrature-phase (Q)

Dedicated Physical Channels	Transport Channels
Dedicated Physical Data CHannel (DPDCH) (UL/DL) [†] ——————— DCH	
Dedicated Physical Control CHannel (DPCCH) (UL/DL)	

Common Physical Channels	Transport Channels
Physical Random Access CHannel (PRACH) (UL) ——————— RACH	
Physical Common Packet CHannel (PCPCH) (UL) ——————— CPCH	
Common PIlot CHannel (CPICH) (DL)	
Primary Common Control Physical CHannel (P-CCPCH) (DL) ——— BCH	
Secondary Common Control Physical CHannel (S-CCPCH) (DL) ——— FACH	
Synchronisation CHannel (SCH) (DL) PCH	
Physical Downlink Shared CHannel (PDSCH) (DL) ——————— DSCH	
Acquisition Indication CHannel (AICH) (DL)	
Page Indication CHannel (PICH) (DL)	

[†]On the DL, the DPDCH and DPCCH are time-multiplexed in each time slot to form
a single Dedicated Physical CHannel (DPCH).

Table 1.5: Mapping the transport channels of Table 1.4 to the UTRA physical channels.

branches are used to deliver the data and control information simultaneously in parallel (as will be augmented in Figure 1.37). Thus, knowledge of the relative carrier phase, namely whether the I or Q branch is involved, constitutes part of the physical channel's identifier. On the other hand, in the TDD mode, a physical channel is defined by its spreading code, frequency, and timeslot.

Similarly to the transport channels of Table 1.4, the physical channels in UTRA can also be classified as dedicated and common channels. Table 1.5 shows the type of physical channels and the corresponding mapping of transport channels on the physical channels in UTRA.

1.3.2.3.1 Dedicated Physical Channels.   The dedicated physical channels of UTRA shown in Table 1.5 consist of the Dedicated Physical Data CHannel (DPDCH) and Dedicated Physical Control CHannel (DPCCH), both of which are bidirectional. The timeslot structures of the UL and DL dedicated physical channels are shown in Figures 1.20 and 1.21, respectively. Notice that on the DL, as illustrated by Figure 1.21, the DPDCH and DPCCH are interspersed by time-multiplexing to form a single Dedicated Physical CHannel (DPCH), as will be augmented in the context of Figure 1.38. On the other hand, the DPDCH and DPCCH on the UL are transmitted in parallel on the I and Q branches of the modem, as will become more explicit in the context of Figure 1.37 [59]. The reason for the parallel transmission on the UL is to avoid Electromagnetic Compatibility (EMC) problems due to Discontinuous Transmission (DTX) of the DPDCH of Table 1.5 [58]. DTX occurs when temporarily there are no data to transmit, but the link is still maintained by the DPCCH. If the UL DPCCH is time-multiplexed with the DPDCH, as in the DL of Figure 1.21, this can create short, sharp

DPCCH : Dedicated Physical Control CHannel
DPDCH : Dedicated Physical Data CHannel
TFCI : Transport-Format Combination Indicator
FBI : FeedBack Information
TPC : Transmit Power Control

Figure 1.20: UTRA UL FDD dedicated physical channels timeslot configuration, which is mapped to the timeslots of Figure 1.19. The UL DPDCH and DPCCH messages are transmitted in parallel on the I and Q branches of the modem of Figure 1.37. By contrast, the DPDCH and DPCCH bursts are time-multiplexed on the DL as shown in Figure 1.21.

DPDCH : Dedicated Physical Data CHannel
DPCCH : Dedicated Physical Control CHannel
TFCI : Transport-Format Combination Indicator
TPC : Transmit Power Control

Figure 1.21: UTRA DL FDD dedicated physical channels timeslot configuration, which is mapped to the timeslots of Figure 1.19. The DPDCH and DPCCH messages are time-multiplexed on the DL, as it will be augmented in Figure 1.38. By contrast, the UL DPDCH and DPCCH bursts are transmitted in parallel on the I and Q branches of the modem as shown in Figure 1.20.

energy spikes. Since the MS may be located near sensitive electrical equipment, these spikes may affect this equipment.

The DPDCH is used to transmit the DCH information between the BS and MS, while the DPCCH is used to transmit the Layer 1 information, which includes the pilot bits, Transmit Power Control (TPC) commands, and an optional Transport-Format Combination Indicator (TFCI), as seen in Figures 1.20 and 1.21. In addition, on the UL the Feedback Information (FBI) is also mapped to the DPDCH in Figure 1.20. The pilot bits are used to facilitate coherent detection on both the UL and DL as well as to enable the implementation of performance enhancement techniques, such as adaptive antennas and interference cancellation. Since the pilot sequences are known, they can also be used as frame synchronization words in order to maintain transmission frame synchronization between the BS and MS. The TPC commands support an agile and efficient power control scheme, which is essential in DS-CDMA using the techniques to be highlighted in Section 1.3.2.8. The TFCI carries information concerning the instantaneous parameters of each transport channel multiplexed on the physical channel in the associated radio frame. The FBI is used to provide the capability to support certain transmit diversity techniques. The FBI field is further divided into two smaller fields as shown in Figure 1.20, which are referred to as the S field and D field. The S field is used to support the *Site Selection DiversiTy* (SSDT), which can reduce the amount of interference caused by multiple transmissions during a soft handover operation, while assisting in fast cell selection. On the other hand, the D field is used to provide attenuation and phase information in order to facilitate *closed-loop transmit diversity*, a technique highlighted in Section 1.3.4.1.3. Given that the TPC and TFCI segments render the transmission packets "self-descriptive", the system becomes very flexible, supporting burst-by-burst adaptivity, which substantially improves the system's performance [96], although this side-information is vulnerable to transmission errors.

The parameter k in Figures 1.20 and 1.21 determines the number of bits in each timeslot, which in turn corresponds to the bit rate of the physical channel. Therefore, the channel bit rates available for the UL DPDCH are 15/30/60/120/240/480/960 kbps, due to the associated "payload" of 10×2^k bits per $\frac{2}{3}$ ms burst in Figure 1.20, where $k = 0 \ldots 6$. Note that the UL DPCCH has a constant channel bit rate of 15 kbps. Similarly, the channel bit rates available for the DL DPCH are 15/30/60/120/240/480/960 and 1920 kbps. However, since the user data is time-multiplexed with the Layer 1 control information, the actual user data rates on the DL will be slightly lower than those mentioned above. Even higher channel bit rates can be achieved using a technique known as multicode transmission [131], which will be highlighted in more detail in the context of Figure 1.35 in Section 1.3.2.5. Let us now consider the common physical channels summarized in Table 1.5.

1.3.2.3.2 Common Physical Channels

1.3.2.3.2.1 Common Physical Channels of the FDD Mode.
The Physical Random Access CHannel (PRACH) of Table 1.5 is used to carry the RACH message on the UL. A random access transmission is activated whenever the MS has data to transmit and wishes to establish a connection with the local BS. Although the procedure of this transmission will be elaborated on in Section 1.3.2.7, here we will briefly highlight the structure of a random access transmission burst. Typically, a random access burst consists of one or several so-called preambles and a message. Each preamble contains a signature that is constructed of 256 repetitions of a 16-chip Hadamard code, which yields a $256 \times 16 = 4096$-chip-long signature. Similarly to the UL dedicated physical channels of Figure 1.20, the message part of the random access transmission consists of data information and control information that are transmitted in parallel on the I/Q channels of the modulator, as shown in Figure 1.22. The channel bit rates available for the data part of the message are 15/30/60/120 kbps. By contrast, the control information, which contains an 8-bit pilot and a 2-bit TFCI, is transmitted at a fixed rate of 15 kbps. Obviously in this case, no FBI and TPC commands are required, since transmission is initiated by the MS.

TFCI : Transport-Format Combination Indicator

Figure 1.22: The timeslot configuration of the message part during a random access transmission in UTRA, which are mapped to the frame structure of Figure 1.19. The data and control information are multiplexed on the I/Q channels of the modulator and the frame is transmitted at the beginning of an access slot, as it will be augmented in Section 1.3.2.7.1.

P-CCPCH: Primary Common Control Physical CHannel
SCH: Synchronization CHannel

Figure 1.23: UTRA DL FDD Primary Common Control Physical CHannel (P-CCPCH) timeslot configuration, which is mapped to the timeslots of Figure 1.19.

The Physical Common Packet CHannel (PCPCH) of Table 1.5 is used to carry the CPCH message on the UL, based on a Digital Sense Multiple Access-Collision Detection (DSMA-CD) random access technique. A CPCH random access burst consists of one or several Access Preambles (A-P), one Collision Detection Preamble (CD-P), a DPCCH Power Control Preamble (PC-P), and a message. The length of both the A-P and CD-P spans a total of 4096 chips, while the duration of the PC-P can be equivalent to either 0 or 8 timeslots. Each timeslot of the PC-P contains the pilot, the FBI, and the TPC bits. The message part of the CPCH burst consists of a data part and a control part, which is identical to the UL dedicated physical channel shown in Figure 1.20 in terms of its structure and available channel bit rates. A 15 kbps DL DPCH is always associated with an UL PCPCH. Hence, both the FBI and TPC information are included in the message of a CPCH burst in order to facilitate a DL transmit diversity and power control, unlike a RACH burst. The procedure of a CPCH transmission will be further elaborated in Section 1.3.2.7.

The DL Primary Common Control Physical CHannel (P-CCPCH) of Table 1.5 is used by the BS in order to broadcast the BCH information at a fixed rate of 30 kbps to all MSs in the cell. The P-CCPCH is transmitted only after the first 256 chips of each slot, as shown in Figure 1.23. During the first 256 chips of each slot, the Synchronization CHannel (SCH) message is transmitted instead, as will be discussed in Section 1.3.2.9. The P-CCPCH is used as a timing reference directly for all the DL physical channels and indirectly for all the UL physical channels. Hence, as long as the MS is synchronized to the DL P-CCPCH

S-CCPCH: Secondary Common Control Physical CHannel
TFCI: Transport-Format Combination Indicator

Figure 1.24: UTRA DL FDD Secondary Common Control Physical CHannel (S-CCPCH) timeslot configuration, which is mapped to the timeslots of Figure 1.19.

of a specific cell, it is capable of detecting any DL messages transmitted from that BS by listening at the predefined times. For example, the DL DPCH will commence transmission at an offset, which is a multiple of 256 chips from the start of the P-CCPCH radio frame seen in Figure 1.23. Upon synchronization with the P-CCPCH, the MS will know precisely when to begin receiving the DL DPCH. The UL DPDCH/DPCCH is transmitted 1024 chips after the reception of the corresponding DL DPCH.

The Secondary Common Control Physical CHannel (S-CCPCH) of Table 1.5 carries the FACH and PCH information of Table 1.4 on the DL, and they are transmitted only when data is available for transmission. The S-CCPCH will be transmitted at an offset, which is a multiple of 256 chips from the start of the P-CCPCH message seen in Figure 1.23. This will allow the MS to know exactly when to detect the S-CCPCH, as long as the MS is synchronized to the P-CCPCH. The timeslot configuration of the S-CCPCH is shown in Figure 1.24. Notice that the S-CCPCH message can be transmitted at a variable bit rate, namely, at 30/60/120/240/480/960/1920 kbps.

At this stage it is worth mentioning that the available control channel rates are significantly higher in the 3G systems than in their 2G counterparts. For example, the maximum BCH signaling rate in GSM [55] is more than an order of magnitude lower than the above-mentioned 30 kbps UTRA BCH rate. In general, this increased control channel rate will support a significantly more flexible system control than the 2G systems.

The Physical DL Shared CHannel (PDSCH) of Table 1.5 is used to carry the DSCH message at rates of 30/60/120/240/480/960/1920 kbps. The PDSCH is shared among several users based on code multiplexing. The Layer 1 control information is transmitted using the associated DL DPCH.

The Acquisition Indicator CHannel (AICH) of Table 1.5 and the Page Indicator CHannel (PICH) are used to carry Acquisition Indicator (AI) and Page Indicator (PI) messages, respectively. More specifically, the AI is a response to a PRACH or PCPCH transmission, and it corresponds to the signature used by the associated PRACH preamble, a PCPCH A-P or a PCPCH CD-P, which were defined above. The AICH consists of a repeated sequence of 15 consecutive Access Slots (AS). Each AS consists of a 32-symbol AI part and an eight-symbol unused part, as shown in Figure 1.25. The AS#0 will commence at the start of every other 10 ms P-CCPCH radio frame seen in Figure 1.19, since its duration is 20 ms.

A PI message is used to signal to the MS on the associated S-CCPCH that there are data addressed to it, in order to facilitate a power-efficient sleep-mode operation. A PICH, illustrated in Figure 1.26, is a 10 ms frame consisting of 300 bits, out of which 288 bits are used to carry PIs, while the remaining 12 bits are unused. Each PICH frame can carry a total of N PIs, where $N = 18, 36, 72$, and 144. The PICH is also transmitted at an offset

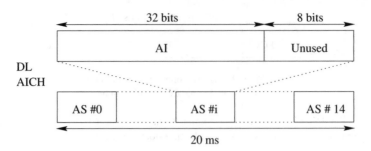

AICH: Acquisition Indicator CHannel
AI: Acquisition Indicator
AS: Access Slot

Figure 1.25: UTRA DL Acquisition Indicator CHannel (AICH) Access Slot (AS) configuration, which is mapped to the corresponding AS of the AICH. Due to its duration of 20 ms, it is mapped to every other 10 ms frame in Figure 1.19.

PICH: Page Indicator CHannel
PI: Page Indicator

Figure 1.26: UTRA DL Page Indicator CHannel (PICH) configuration. Each PICH frame can carry a total of N PIs, where $N = 18, 36, 72$, and 144.

with respect to the start of the P-CCPCH, which is a multiple of 256 chips. The associated S-CCPCH will be transmitted 7680 chips later.

Finally, the Common PIlot CHannel (CPICH) of Table 1.5 is a 30 kbps DL physical channel that carries a predefined bit sequence. It provides a phase reference for the SCH, P-CCPCH, AICH, and PICH, since these channels do not carry pilot bits, as shown in Figures 1.23, 1.25, and 1.26, respectively. The PICH is transmitted synchronously with the P-CCPCH.

1.3.2.3.2.2 Common Physical Channels of the TDD Mode. In contrast to the previous FDD structures of Figures 1.20–1.26, in TDD operation the burst structure of Figure 1.27 is used for all the physical channels, where each timeslot's transmitted information can be arbitrarily allocated to the DL or UL, as shown in the three possible TDD allocations in Figure 1.28. Hence, this flexible allocation of the UL and DL burst in the TDD mode enables the use of an adaptive modem [96, 132] whereby the modem parameters, such as the spreading factor or the number of bits per symbol can be adjusted on a burst-by-burst basis to optimize the link quality. A symmetric UL/DL allocation refers to a scenario

2/3 ms

| Data | Midamble | Data | GP |

Burst Type 1 : Data = 976 chips, Midamble = 512 chips
Burst Type 2 : Data = 1104 chips, Midamble = 256 chips
GP : Guard Period = 96 chips

Figure 1.27: Burst configuration mapped on the TDD burst structure of Figure 1.28 in the UTRA TDD mode. Two different types of TDD bursts are defined in UTRA, namely, Burst Type 1 and Burst Type 2.

10 ms

2/3 ms

(a) Symmmetric UL/DL allocation with multiple switching points

(b) Asymmetric UL/DL allocation with multiple switching points

(c) Asymmetric UL/DL allocation with a single switching point

↓ : Downlink   ↑ : Uplink

Figure 1.28: UL/DL allocation examples for the 15 slots in UTRA TDD operation using the timeslot configurations of Figure 1.27.

in which an approximately equal number[7] of DL and UL bursts are allocated within a TDD frame, while in asymmetric UL/DL allocation, there is an unequal number of UL and DL bursts, such as, for example, in "near-simplex" file download from the Internet or video-on-demand.

In UTRA, two different TDD burst structures, known as *Burst Type 1* and *Burst Type 2*, are defined, as shown in Figure 1.27. The Type 1 burst has a longer midamble (512 chips) than the Type 2 burst (256 chips). However, both types of bursts have an identical *Guard Period* (GP) of 96 chips. The midamble sequences that are allocated to the different TDD bursts in

---

[7]Since there are 15 timeslots per frame, there will always be one more additional DL or UL burst per frame in a symmetric allocation.

Burst structure with TFCI information only

Burst structure with TFCI and TPC information

GP : Guard Period
TFCI : Transport Format Combination Indicator
TPC : Transmit Power Control

Figure 1.29: Burst configuration mapped on the TDD burst configuration of Figure 1.28 in the UTRA TDD mode incorporating TFCI and/or TPC information.

each timeslot belong to a so-called *midamble code set*. The codes in each midamble code set are derived from a unique *Basic Midamble Code*. Adjacent cells are allocated different midamble code sets, that is, different basic midamble code. This can be exploited to assist in cell identification.

Unlike in the FDD mode, there is only one type of Dedicated Physical CHannel (DPCH) in the TDD mode. Hence, the Layer 1 control information—such as the TPC command and the TFCI information—will be transmitted in the data field of Figure 1.27, if required. The TDD burst structures that incorporate the TFCI information as well as the TFCI+TPC information are shown in Figure 1.29. This should be contrasted with their corresponding FDD allocations in Figures 1.20 and 1.21. The TFCI field is divided into two parts, which reside immediately before and after the midamble (or after the TPC command, if power control is invoked) in the data field. The TPC command is always transmitted immediately after the midamble, as portrayed in Figure 1.29. As a result of these control information segments, the amount of user data is reduced in each timeslot. Note that the TPC command is only transmitted on the UL and only once per 10 ms frame for each MS.

In contrast to the FDD mode, the SCH in the TDD mode is not time-multiplexed onto the P-CCPCH of Table 1.5. Instead, the SCH messages are transmitted on one or two timeslots per frame.[8] The P-CCPCH will be code-multiplexed with the first SCH timeslot in each frame.

Having highlighted the basic features of the various UTRA channels, let us now consider how the various services are error protected, interleaved, and multiplexed on to the physical channels. This issue is discussed with reference to Figures 1.30 and 1.31 in the context of UTRA.

---

[8]If two timeslots are allocated to the SCH per frame, they will be spaced seven slots apart.

1.3.2.4 Service Multiplexing and Channel Coding in UTRA

Service multiplexing is employed when multiple services of identical or different bit rates requiring different QoS belonging to the same user's connection are transmitted. An example would be the simultaneous transmission of a voice and video service for a multimedia application. Each service is represented by its corresponding transport channels, as described in Section 1.3.2.2. The coding and multiplexing of the transport channels are performed in sets of transport blocks that arrived from the higher layers at fixed intervals of 10, 20, 40 or 80 ms. These intervals are known as the *Transmission Time Interval* (TTI). Note that the number of bits on each transport channel can vary between different TTIs, as well as between different transport channels. A possible method of transmitting multiple services is by using code-multiplexing with the aid of orthogonal codes. Every service could have its own DPDCH and DPCCH, each assigned to a different orthogonal code. This method is not very efficient, however, since a number of orthogonal codes would be reserved by a single user, while on the UL it would also inflict self-interference when the multiple DPDCH and DPCCH codes' orthogonality is impaired by the fading channel. Alternatively, these services can be time-multiplexed into one or several DPDCHs, as shown in Figures 1.30 and 1.31 for the UL and DL, respectively. The function of the individual processing steps is detailed below.

1.3.2.4.1 CRC Attachment.

A Cyclic Redundancy Checksum (CRC) is first calculated for each incoming transport block within a TTI. The CRC consists of either 24, 16, 12, 8, or 0 parity bits, which is decided by the higher layers. The CRC is then attached to the end of the corresponding transport block in order to facilitate reliable error detection at the receiver. This facility is very important, for example, for generating the video packet acknowledgement flag in wireless video telephony using standard video codecs, such as H.263 [133].

1.3.2.4.2 Transport Block Concatenation.

Following the CRC attachment, the incoming transport blocks within a TTI are serially concatenated in order to form a code block. If the number of bits exceeds the maximum code block length, denoted as Z, then the code block is segmented into shorter ones and filler bits (zeros) are added to the last code block, if necessary, in order to generate code blocks of the same length. The maximum code block length Z is dependent on the type of channel-coding invoked. For convolutional coding $Z = 504$, while for turbo coding $Z = 5114$, since turbo codes require a long coded block length [134]. If no channel-coding is invoked, then the code block can be of unlimited length.

1.3.2.4.3 Channel-coding.

Each of the code blocks is then delivered to the channel-coding unit. Several Forward Error Correction (FEC) techniques are proposed for channel-coding. The FEC technique used is dependent on the QoS requirement of that specific transport channel. Table 1.6 shows the various types of channel-coding techniques invoked for different transport channels. Typically, *convolutional coding* is used for services with a bit error rate requirement on the order of 10^{-3}, for example, for voice services. For services requiring a lower BER, namely, on the order of 10^{-6}, *turbo coding* is applied. Turbo coding is known to guarantee a high performance [135] over AWGN channels at the cost of increased interleaving-induced latency or delay. The implementational complexity of the turbo codec (TC) does not necessarily have to be higher than that of the convolutional codes (CC), since

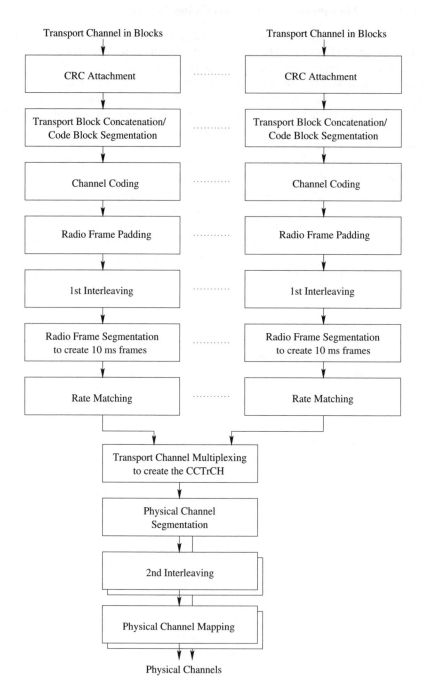

Figure 1.30: Transport channel-coding/multiplexing flowchart for the UL in UTRA.

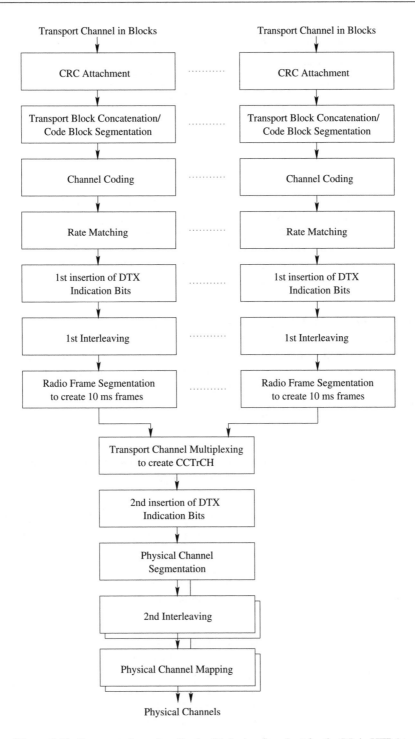

Figure 1.31: Transport channel-coding/multiplexing flowchart for the DL in UTRA.

Table 1.6: UTRA Channel-coding parameters for the channels of Table 1.4.

Transport channels	Channel-coding schemes	Coding rate
BCH, PCH, RACH	Convolutional code	1/2
CPCH,DCH,DSCH,FACH	Convolutional code	1/3, 1/2
	Turbo code	1/3
	No coding	

a constraint-length $K = 7$ or $K = 9$ CC is often invoked, while the constraint-length of the turbo codes employed may be as low as $K = 3$. In somewhat simplistic but plausible terms, one could argue that a $K = 3$ TC using two decoding steps per iteration and employing four iterations has a similar complexity to a $K = 6$ CC, since they are associated with the same number of trellis states. The encoded code blocks within a TTI are then serially concatenated after the channel-coding unit, as seen in Figures 1.30 and 1.31.

1.3.2.4.4 Radio Frame Padding. Radio frame padding is only performed on the UL whereby the input bit sequence (the concatenated encoded code blocks from the channel-coding unit) is padded in order to ensure that the output can be segmented equally into (TTI/10 ms) number of 10 ms radio frames. Note that radio frame padding is not required on the DL, since DTX is invoked, as seen in Figure 1.31. This process was termed Radio Frame Equalization in the standard. However, in order to avoid confusion with channel equalization, we used the terminology "padding".

1.3.2.4.5 First Interleaving. The depth of this first interleaver seen in Figures 1.30 and 1.31 may range from one radio frame (10 ms) to as high as 80 ms, depending on the TTI.

1.3.2.4.6 Radio Frame Segmentation. The input bit sequence after the first interleaving is then segmented into consecutive radio frames of 10 ms duration, as highlighted in Section 1.3.2.3. The number of radio frames required is equivalent to (TTI/10). Because of the Radio Frame Padding step performed prior to the segmentation on the UL in Figure 1.30 and also because of the Rate Matching step on the DL in Figure 1.31, the input bit sequence can be conveniently divided into the required number of radio frames.

1.3.2.4.7 Rate Matching. The rate matching process of Figures 1.30 and 1.31 implies that bits on a transport channel are either repeated or punctured in order to ensure that the total bit rate after multiplexing all the associated transport channels will be identical to the channel bit rate of the corresponding physical channel, as highlighted in Section 1.3.2.3. Thus, rate matching must be coordinated among the different coded transport channels, so that the bit rate of each channel is adjusted to a level that fulfills its minimum QoS requirements [118]. On the DL, the bit rate is also adjusted so that the total instantaneous transport channel bit rate approximately matches the defined bit rate of the physical channel, as listed in Table 1.3.

1.3.2.4.8 Discontinuous Transmission Indication. On the DL, the transmission is interrupted if the bit rate is less than the allocated channel bit rate. This is known as discontinuous transmission (DTX). DTX indication bits are inserted into the bit sequence in order to indicate when the transmission should be turned off. The first insertion of the DTX indication bits shown in Figure 1.31 is performed only if the position of the transport channel in the radio frame is fixed. In this case, a fixed number of bits is reserved for each transport channel in the radio frame. For the second insertion step shown in Figure 1.31, the DTX indication bits are inserted at the end of the radio frame.

1.3.2.4.9 Transport Channel Multiplexing. One radio frame from each transport channel that can be mapped to the same type of physical channel is delivered to the transport channel multiplexing unit of Figures 1.30 and 1.31, where they are serially multiplexed to form a *Coded Composite Transport CHannel* (CCTrCH). At this point, it should be noted that the bit rate of the multiplexed radio frames may be different for the various transport channels. In order to successfully de-multiplex each transport channel at the receiver, the TFCI—which contained information about the bit rate of each multiplexed transport channel—can be transmitted together with the CCTrCH information (which will be mapped to a physical channel), as highlighted in Section 1.3.2.3. Alternatively, *blind transport format detection* can be performed at the receiver without the explicit knowledge of the TFCI, where the receiver acquires the transport format combination through some other means, such as, for example, the received power ratio of the DPDCH to the DPCCH.

1.3.2.4.10 Physical Channel Segmentation. If more than one physical channel is required in order to accommodate the bits of a CCTrCH, then the bit sequence is segmented equally into different physical channels, as seen in Figures 1.30 and 1.31. A typical example of this scenario would be where the bit rate of the CCTrCH exceeds the maximum allocated bit rate for the particular physical channel. Thus, multiple physical channels are required for its transmission. Furthermore, restrictions are imposed on the number of transport channels that can be multiplexed onto a CCTrCH. Hence, several physical channels are required to carry any additional CCTrCHs.

1.3.2.4.11 Second Interleaving. The depth of the second interleaving stage shown in Figures 1.30 and 1.31 is equivalent to one radio frame. Hence, this process does not increase the system's delay.

1.3.2.4.12 Physical Channel Mapping. Finally, the bits are mapped to their respective physical channels summarized in Table 1.5, as portrayed in Figures 1.30 and 1.31.

Having highlighted the various channel-coding and multiplexing techniques as well as the structures of the physical channels illustrated by Figures 1.20–1.27, let us now consider how the services of different bit rates are mapped on the UL and DL dedicated physical data channels (DPDCH) of Figures 1.20 and 1.21, respectively. In order to augment the process, we will present three examples. Specifically, we consider the mapping of two multirate services on a UL DPDCH and an example of the mapping of a 4.1 kbps data service on a DL DPDCH in the FDD mode. We will then use the same parameters as employed in the first

Table 1.7: Parameters for the multimedia communication example of Section 1.3.2.4.13.

	Service 1, DCH#1	Service 2, DCH#2
Transport Block Size	640 bits	164 bits
Transport Block Set Size	4 * 640 bits	1 * 164 bits
TTI	40 ms	40 ms
Bit Rate	64 kbps	4.1 kbps
CRC	16 bits	16 bits
Coding	Turbo	Convolutional
	Rate: 1/3	Rate: 1/3

example and show how the multirate services can be mapped to the corresponding UL DPCH in TDD mode.

1.3.2.4.13 Mapping Several Multirate Services to the UL Physical Channels in FDD Mode [115].

In this example, we assume that a 4.1 kbps speech service and a 64 kbps video service are to be transmitted simultaneously on the UL. The parameters used for this example are shown in Table 1.7. As illustrated in Figure 1.32, a 16-bit CRC checksum is first attached to each transport block of DCH#1, that is, #1a,...,#1d, as well as the transport block of DCH#2 for the purpose of error detection. As a result, the number of bits in the transport block of Service 1 and Service 2 is increased to $640 + 16 = 656$ bits and $164 + 16 = 180$ bits, respectively. The four transport blocks of Service 1 are then concatenated, as illustrated in Figure 1.32. Notice that no code block segmentation is invoked, since the total number of bits in the concatenated transport block is less than $Z = 5114$ for turbo coding, as highlighted in Section 1.3.2.4.2. Since the video service typically requires a low BER—unless specific measures are invoked for mitigating the video effects of transmission errors [132]—turbo coding is invoked, using a coding rate of $\frac{1}{3}$. Hence, after turbo coding and the attachment of tailing bits, the resulting 40 ms segment would contain $(656 \times 4) \times 3 + 12 = 7884$ bits, as shown in Figure 1.32. By contrast, the speech service can tolerate a higher BER. Hence, convolutional coding is invoked. First, a block of $4 + 4 = 8$ tail bits is concatenated to the transport block in order to flush the assumed constraint-length $K = 5$ shift registers of the convolutional encoder. Thus, a total of $180 + 8 = 188$ bits are conveyed to the convolutional encoder of DCH#1, as shown in Figure 1.32. Again, no code block segmentation is invoked, since the total number of bits in the transport channel is less than $Z = 504$ for convolutional coding, as highlighted in Section 1.3.2.4.2. A coding rate of $\frac{1}{3}$ is used for the convolutional encoding of DCH#1, as exemplified in Table 1.7. The output of the convolutional encoder of DCH#1 will have a total of $188 \times 3 = 564$ bits per 40 ms segment. Since the TTI of these transport channels is 40 ms, four radio frames are required to transmit the associated data. At this stage, notice that there are a total of 7884 bits and 564 bits for DCH#1 and DCH#2, respectively. Since these numbers are divisible by four, they can be divided equally into four radio frames. Thus, no padding is required as illustrated in the Radio Frame Padding step of Figure 1.32. Interleaving is then performed across the 40 ms segment for each transport channel before being segmented into four 10 ms radio frames.

Table 1.8: Parameters for the example of Section 1.3.2.4.14.

	Service 1, DCH#1
Transport Block Size	164 bits
TTI	40 ms
Bit Rate	4.1 kbps
CRC	16 bits
Coding	Convolutional
	Rate: 1/3

At this point, we note that these two transport channels can be mapped to the same DPDCH, since they belong to the same MS. Hence, the 10 ms radio frames, marked "A" in Figure 1.32 will be multiplexed, in order to form a CCTrCH. Similarly, the frames marked "B", "C" (not shown in Figure 1.7 due to lack of space), and "D" will be multiplexed, in order to form another three CCTrCHs. The rate of these CCTrCHs must be matched to the allocated channel bit rate of the physical channel. Without rate matching, the bit rate of these CCTrCHs is $(1971 + 141)/10\,\text{ms} = 211.2\,\text{kbps}$, which does not fit any of the available channel bit rates of the UL DPDCH, as listed in Table 1.3. Hence, the Rate Matching step of Figures 1.30, 1.31, and 1.32 must be invoked in order to adapt the multiplexed bit rate to one of the available UL DPDCH bit rates of Table 1.3. Let us assume that the allocated channel bit rate is 240 kbps. Thus, a number of bits must be punctured or repeated for each service, in order to increase the total number of bits per 10 ms segment after multiplexing from 2171 to 2400. This would require coordination among the different services, as it was highlighted in Section 1.3.2.4.7. After multiplexing the transport channels, a second interleaving is performed across the 10 ms radio frame before finally mapping the bits to the UL DPDCH.

1.3.2.4.14   Mapping of a 4.1 Kbps Data Service to the DL DPDCH in FDD Mode.   The parameters for this example are shown in Table 1.8. In this context, we assume that a single DCH consisting of one transport block within a TTI duration of 40 ms is to be transmitted on the DL. As illustrated in Figure 1.33, a 16-bit CRC sum segment is appended to the transport block. A $4 + 4 = 8$-bit tailing block is then attached to the end of the segment in order to form a 188-bit code block. Similarly to the previous example, the length of the code block is less than $Z = 504$, since CC is used. Hence, no segmentation is invoked. The 188-bit data block is convolutional coded at a rate of $\frac{1}{3}$, which results in a $3 \times 188 = 564$-bit segment. According to Figure 1.31, rate matching is invoked for the encoded block. Since the TTI duration is 40 ms, four radio frames are required to transmit the data. Without rate matching, the bit rate per radio frame is $564/40\,\text{ms} = 14.1\,\text{kbps}$, which does not fit any of the available bit rates listed in Table 1.4 for the DL. Note that for the case of the DL dedicated physical channels, the channel bit rate will include the additional bits required for the pilot and TPC, as shown explicitly in Figure 1.21. Since there is only one transport channel in this case, no TFCI bits are required. We assume that an 8-bit pilot and a 2-bit TPC per slot are assigned to this transmission, which yields a total rate of 15 kbps for the DPCCH. Hence all the bits in the encoded block will be repeated in order to increase its bit rate of 15 kbps to 30 kbps

Figure 1.32: Mapping of several multimedia services to the UL dedicated physical data channel of Figure 1.20 in FDD mode. The corresponding schematic diagram is seen in Figure 1.30.

for the DL DPCH. In this case the number of padding bits appended becomes $N = 36$. After the second interleaving stage of Figure 1.31, the segmented radio frames are mapped to the corresponding DPDCH, which are then multiplexed with the DPCCH, as shown in Figure 1.33.

1.3.2.4.15 Mapping Several Multirate Services to the UL Physical Channels in TDD Mode [115].

In this example, we will demonstrate how the multirate multimedia services, considered previously in the example of Section 1.3.2.4.13 in an FDD context, are mapped to the corresponding dedicated physical channels (DPCH) in the TDD mode.

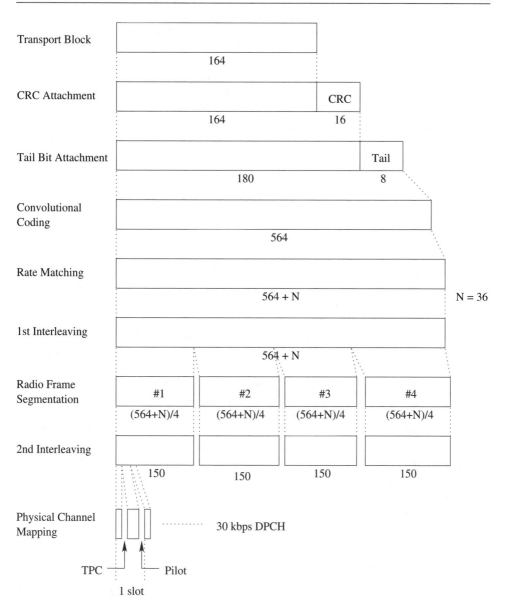

Figure 1.33: Mapping of a 4.1 kbps data service to the DL dedicated physical channel of Figure 1.21 in FDD mode. The corresponding schematic diagram is seen in Figure 1.31.

The channel-coding/multiplexing process is identical in the FDD and TDD mode, and so both are based on Figures 1.30 and 1.31. The only difference is in the mapping of the transport channels to the corresponding physical channels seen at the bottom of Figures 1.30 and 1.31, since the FDD and TDD modes have a different frame structure, as shown previously in Figures 1.20–1.26 and Figure 1.27, respectively. In this example, we are only interested in the process of service mapping to the physical channel, which follows the second interleaving stage of Figure 1.34. Here we assumed that for the TDD UL scenario of Table 1.7 the total number of bits per segment after DCH multiplexing is 2186 as a result of rate matching. In the FDD example of Section 1.3.2.4.13, this was 2400. Each segment is divided into two bursts, which can be transmitted either by orthogonal code multiplexing onto a single timeslot, or using two timeslots within a 10 ms radio frame. Note that only one burst in each segment is required to carry the TFCI and the TPC information.

Following these brief discussions on service multiplexing, channel coding, and interleaving, let us now concentrate on the aspects of variable-rate and multicode transmission in UTRA in the next section.

1.3.2.5   Variable-rate and Multicode Transmission in UTRA

Three different techniques have been proposed in the literature for supporting variable-rate transmission, namely, multicode-, modulation-division multiplexing- (MDM), and multiple processing gain (MPG)-based techniques [136]. UTRA employs a number of different processing gains, or variable spreading factors, in order to transmit at different channel bit rates, as highlighted previously in Section 1.3.2.3. The spreading factor (SF) has a direct effect on the performance and capacity of a DS-CDMA system. Since the chip rate is constant, the SF—which is defined as the ratio of the spread bandwidth to the original information bandwidth—becomes lower, as the bit rate increases. Hence, there is a limit to the value of the SF used, which is SF = 4 in FDD mode in the proposed UTRA standards. Multicode transmission [131,136,137] is used if the total bit rate to be transmitted exceeds the maximum bit rate supported by a single DPDCH, which was stipulated as 960 kbps for the UL and 1920 kbps for the DL. When this happens, the bit rate is split among a number of spreading codes and the information is transmitted using two or more codes. However, only one DPCCH is transmitted during this time. Thus, on the UL one DPCCH and several DPDCH are code-multiplexed and transmitted in parallel, as it will be augmented in the context of Figure 1.37. On the DL, the DPDCH and DPCCH are time-multiplexed on the first physical channel associated with the first spreading code as seen in Figure 1.35. If more physical channels are required, the DPCCH part in the slot will be left blank again, as shown in Figure 1.35. The transmit power of the DPDCH is also reduced.

1.3.2.6   Spreading and Modulation

It is well known that the performance of DS-CDMA is interference limited [99]. The majority of the interference originates from the transmitted signals of other users within the same cell, as well as from neighboring cells. This interference is commonly known as *Multiple Access Interference* (MAI). Another source of interference, albeit less dramatic, is a result of the wideband nature of CDMA, yielding several delayed replicas of the transmitted signal, which reach the receiver at different time instants, thereby inflicting what is known as *interpath interference*. However, the advantages gained from wideband transmissions, such as multipath diversity and the noise-like properties of the interference, outweigh the drawbacks.

The choice of the spreading codes [138,139] used in DS-CDMA will have serious implications for the amount of interference generated. Suffice to say that the traditional measures used in comparing different codes are their *cross-correlations* (CCL) and *autocorrelation*

Figure 1.34: Mapping of several multirate multimedia services to the UL dedicated physical data channel of Figure 1.20 in TDD mode. The corresponding schematic diagram is seen in Figure 1.30.

Figure 1.35: DL FDD slot format for multicode transmission in UTRA, based on Figure 1.21, but dispensing with transmitting DPCCH over all multicode physical channels.

(ACL). If the CCL of the spreading codes of different users is nonzero, this will increase their interference, as perceived by the receiver. Thus a low CCL reduces the MAI. The so-called out-of-phase ACL of the codes, on the other hand, plays an important role during the initial synchronization between the BS and MS, which has to be sufficiently low to minimize the probability of synchronizing to the wrong ACL peak.

In order to reduce the MAI and thereby improve the system's performance and capacity, the UTRA physical channels are spread using two different codes, namely, the *channelization code* and a typically longer so-called *scrambling code*. In general, the channelization codes are used to maintain orthogonality between the different physical channels originating from the same source. On the other hand, the scrambling codes are used to distinguish between different cells, as well as between different MSs. All the scrambling codes in UTRA are in complex format. Complex-valued scrambling balances the power on the I and Q branches. This can be shown by letting c_s^I and c_s^Q be the I and Q branch scrambling codes, respectively. Let $d(t)$ be the complex-valued data of the transmitter, which can be written as:

$$d(t) = d_I + jd_Q, \tag{1.34}$$

where d_I and d_Q represent the data on the I and Q branches, respectively. Let us assume for the sake of argument that the power level in the I and Q branches is unbalanced due to, for instance, their different bit rates or different QoS requirements. If only real-valued scrambling is used, then the output becomes:

$$s(t) = c_s^I \left(d_I + jd_Q \right), \tag{1.35}$$

Table 1.9: UL/DL spreading and modulation parameters in UTRA.

	Channelization Codes	Scrambling Codes
Type of codes	OVSF (Section 1.3.2.6.1)	UL : Gold codes, S(2) codes (Section 1.3.2.6.2) DL : Gold codes (Section 1.3.2.6.3)
Code length	Variable	UL : 10 ms of $(2^{25} - 1)$-chip Gold code DL : 10 ms of $(2^{18} - 1)$-chip Gold code
Type of spreading	BPSK (UL/DL)	QPSK (UL/DL)

which is also associated with an unbalanced power level on the I and Q branches. By contrast, if complex-valued scrambling is used, then the output would become:

$$s(t) = (d_I + jd_Q) \cdot \left(c_s^I + jc_s^Q\right) \tag{1.36}$$

$$= c_s^I \cdot d_I - c_s^Q \cdot d_Q + j\left(c_s^Q \cdot d_I + c_s^I \cdot d_Q\right). \tag{1.37}$$

As can be seen, the power on the I and Q branches after complex scrambling is the same, regardless of the power level of the unscrambled data on the I and Q branches. Hence, complex scrambling potentially improves the power amplifier's efficiency by reducing the peak-to-average power fluctuation. This also relaxes the linearity requirements of the UL power amplifier used.

Table 1.9 shows the parameters and techniques used for spreading and modulation in UTRA, which will be discussed in depth in the following sections.

1.3.2.6.1 Orthogonal Variable Spreading Factor Codes.

The channelization codes used in the UTRA systems are derived from a set of orthogonal codes known as *Orthogonal Variable Spreading Factor* (OVSF) codes [130]. OVSF codes are generated from a tree-structured set of orthogonal codes, such as the Walsh-Hadamard codes, using the procedure shown in Figure 1.36. Specifically, each channelization code is denoted by $c_{N,n}$, where $n = 1, 2, \ldots, N$ and $N = 2^x, x = 2, 3, \ldots 8$. Each code $c_{N,n}$ is derived from the previous code $c_{(N/2),n}$ as follows [130]:

$$\begin{bmatrix} c_{N,1} \\ c_{N,2} \\ c_{N,3} \\ \vdots \\ c_{N,N} \end{bmatrix} = \begin{bmatrix} c_{(N/2),1} | c_{(N/2),1} \\ c_{(N/2),1} | \bar{c}_{(N/2),1} \\ c_{(N/2),2} | c_{(N/2),2} \\ \vdots \\ c_{(N/2),(N/2)} | \bar{c}_{(N/2),(N/2)} \end{bmatrix}, \tag{1.38}$$

where $[|]$ at the right-hand side of Equation 1.38 denotes an augmented matrix and $\bar{c}_{(N/2),n}$ is the binary complement of $c_{(N/2),n}$. For example, according to Equation 1.38 and Figure 1.36, $c_{N,1} = c_{8,1}$ is created by simply concatenating $c_{(N/2),1}$ and $c_{(N/2),1}$, which doubles the number of chips per bit. By contrast, $c_{N,2} = c_{8,2}$ is generated by attaching $\bar{c}_{(N/2),1}$ to $c_{(N/2),1}$. From Equation 1.38, we see that, for example, $c_{N,1}$ and $c_{N,2}$ at the left-hand side of Equation 1.38 are not orthogonal to $c_{(N/2),1}$, since the first half of both was derived

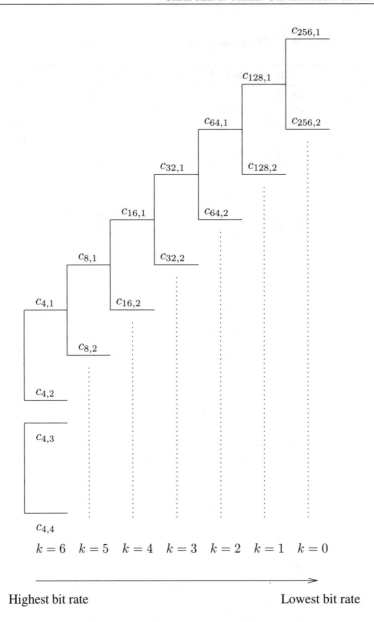

$$k = 6 \quad k = 5 \quad k = 4 \quad k = 3 \quad k = 2 \quad k = 1 \quad k = 0$$

Highest bit rate   Lowest bit rate

Figure 1.36: Orthogonal variable-spreading factor code tree in UTRA according to Equation 1.38. The parameter k in the figure is directly related to that found in Figures 1.20–1.24.

from $c_{(N/2),1}$ in Figure 1.36, but they are orthogonal to $c_{(N/2),n}, n = 2, 3, \ldots, (N/2)$. The code $c_{(N/2),1}$ in Figure 1.36 is known as the mother code of the codes $c_{N,1}$ and $c_{N,2}$, since these two codes are derived from $c_{(N/2),1}$. The codes on the "highest"-order branches ($k = 6$) of the tree at the left of Figure 1.36 have a spreading factor of 4, and they are used for transmission at the highest possible bit rate for a single channel, which is 960 kbps.

On the other hand, the codes on the "lowest"-order branches ($k = 0$) of the tree at the right of Figure 1.36 result in a spreading factor of 256, and these are used for transmission at the lowest bit rate, which is 15 kbps. It is worth noting here that an intelligent BbB adaptive scheme may vary its SF on a 10 ms frame basis in an attempt to adjust the SF on a near-instantaneous channel-quality motivated basis [96, 132]. Orthogonality between parallel transmitted channels of the same bit rate is preserved by assigning each channel a different orthogonal code accordingly. For channels with different bit rates transmitting in parallel, orthogonal codes are assigned, ensuring that no code is the mother-code of the other. Thus, OVSF channelization codes provide total isolation between different users' physical channels on the DL that are transmitted synchronously and hence eliminate MAI among them. OVSF channelization codes also provide orthogonality between the different physical channels seen in Figure 1.35 during multicode transmission.

Since there is only a limited set of OVSF codes, which is likely to be insufficient to support a large user-population, while also allowing identification of the BSs by the MSs on the DL, *each cell will reuse the same set of OVSF codes*. Statistical multiplexing schemes such as packet reservation multiple access (PRMA) can be used to allocate and de-allocate the OVSF codes on a near-instantaneous basis, for example, depending on the users' voice activity in the case of DTX-based communications [140]. However, orthogonal codes, such as the orthogonal OVSF codes, in general exhibit poor out-of-phase ACL and CCL properties [141]. Therefore, the correlations of the OVSFs of adjacent asynchronous BSs will become unacceptably high, degrading the correlation receiver's performance at the MS. On the other hand, certain long codes such as Gold codes exhibit low CCL, which is advantageous in CDMA applications [66]. Hence in UTRA, cell-specific long codes are used in order to reduce the inter-cell interference on the DL. On the UL, MAI is reduced by assigning different scrambling codes to different users.

1.3.2.6.2 Uplink Scrambling Codes. The UL scrambling codes in UTRA can be classified into long scrambling codes and short scrambling codes. A total of 2^{24} UL scrambling codes can be generated for both the long and short codes.

Long scrambling codes are constructed from two m-sequences using the polynomials of $1+X^3+X^{25}$ and $1+X+X^2+X^3+X^{25}$, following the procedure highlighted by Proakis [5] in order to produce a set of *Gold codes* for the I branch. The Q-branch Gold code is a shifted version of the I-branch Gold code, where a shift of 16,777,232 chips was recommended. Gold codes are rendered different from each other by assigning a unique initial state to one of the shift registers of the m-sequence. The initial state of the other shift register is a sequence of logical 1. Although the Gold codes generated have a length of $2^{25} - 1$ chips, only 38,400 chips (10 ms at 3.84 Mcps) are required in order to scramble a radio frame.

Short scrambling codes are defined from a family of periodically extended S(2) codes. This 256-chip S(2) code was introduced to ease the implementation of multi-user detection at the BS [58]. The multi-user detector has to invert the so-called system matrix [95], the dimension of which is proportional to the sum of the channel impulse response duration and the spreading code duration. Thus, using a relatively short scrambling code is an important practical consideration in reducing the size of the system-matrix to be inverted.

1.3.2.6.3 Downlink Scrambling Codes. Unlike the case for the UL, only Gold codes are used on the DL. The DL Gold codes on the I branch are constructed from two m-sequences using the polynomials of $1 + X^7 + X^{18}$ and $1 + X^5 + X^7 + X^{10} + X^{18}$. These Gold codes are shifted by 131,072 chips in order to produce a set of Gold codes for the Q branch.

Although a total of $2^{18} - 1 = 262,143$ Gold codes can be generated, only 8192 of them will be used as the DL scrambling code. These codes are divided into 512 groups, each of which contains a *primary scrambling code* and 15 *secondary scrambling codes*. Altogether there are 512 primary scrambling codes and $8192 - 512 = 7680$ secondary scrambling codes. Each cell is allocated one primary scrambling code, which is used on the CPICH and P-CCPCH channels of Table 1.5. This primary scrambling code will be used to identify the BS for the MS. All the other physical channels belonging to this cell can use either the primary scrambling code or any of the 15 secondary scrambling codes that belong to the same group, as the primary scrambling code. In order to facilitate fast cell or BS identification, the set of 512 primary scrambling codes is further divided into 64 subsets, each consisting of eight primary scrambling codes, as will be shown in Section 1.3.2.9.

1.3.2.6.4   Uplink Spreading and Modulation.   A model of the UL transmitter for a single DPDCH is shown in Figure 1.37 [59]. We have seen in Figure 1.20 that the DPDCH and DPCCH are transmitted in parallel on the I and Q branches of the UL, respectively. Hence, to avoid I/Q channel interference in case of I/Q imbalance of the quadrature carriers, different orthogonal spreading codes are assigned to the DPDCH and DPCCH on the I and Q branch, respectively. These two channelization codes for DPDCH and DPCCH, denoted by $c_{D,1}$ and c_C in Figure 1.37, respectively, are allocated in a predefined order. From Figure 1.20, we know that the SF of the DPCCH is 256. Hence, $c_C = c_{256,1}$ in the context of Figure 1.36. This indicates that the high SF of the DPCCH protects the vulnerable control channel message against channel impairments. On the other hand, we have $c_{D,1} = c_{SF,2}$, depending on the SF of the DPDCH. In the event of multicode transmission portrayed by the dashed lines in Figure 1.37, different additional orthogonal channelization codes, namely, $c_{D,2}$ and $c_{D,3}$, are assigned to each DPDCH for the sake of maintaining orthogonality, and they can be transmitted on either the I or Q branch. In this case, the BS and MS have to agree on the number of channelization codes to be used. After spreading, the BPSK modulated I and Q branch signals are summed in order to produce a complex Quadrature Phase Shift Keying (QPSK) signal. The signal is then scrambled by the complex scrambling code, c_{scramb}. The pulse-shaping filters, $p(t)$, are root-raised cosine Nyquist filters using a roll-off factor of 0.22.

The transmitter of the UL PRACH and PCPCH message part is also identical to that shown in Figure 1.37. As we have mentioned in Section 1.3.2.3.2 in the context of Figure 1.22, the PRACH and the CPCH message consist of a data part and a control part. In this case, the data part will be transmitted on the I branch, and the control part on the Q branch. The choice of the channelization codes for the data and control part depends on the signature of the preambles transmitted beforehand. As highlighted in Section 1.3.2.3.2, the preamble signature is a 256-chip sequence generated by the repetition of a 16-chip Hadamard code. This 16-chip code actually corresponds to one of the OVSF codes, namely, to $c_{16,n}$, where $n = 1, \ldots, 16$. The codes in the subtree of Figure 1.36 below this specific 16-chip code n will be used as the channelization codes for the data part and control part.

1.3.2.6.5   Downlink Spreading and Modulation.   The schematic diagram of the DL transmitter is shown in Figure 1.38. All the DL physical channel bursts (except for the SCH) are first QPSK modulated in order to form the I and Q branches, before spreading to the chip

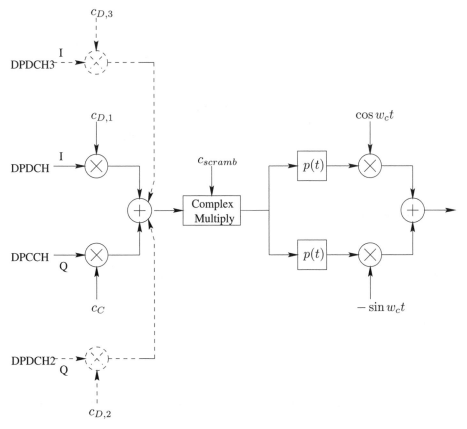

Figure 1.37: UL transmitter in UTRA using the frame structure of Figure 1.20. Multicode transmissions are indicated by the dashed lines.

rate. In contrast to the UL of Figure 1.37, the same OVSF channelization code c_{ch} is used on the I and Q branches. Different physical channels are assigned different channelization codes in order to maintain their orthogonality. For instance, the channelization codes for the CPICH and P-CCPCH of Table 1.5 are fixed to the codes $c_{256,1}$ and $c_{256,2}$ of Figure 1.36, respectively. The channelization codes for all the other physical channels are assigned by the network.

The resulting signal in Figure 1.38 is then scrambled by a cell-specific scrambling code c_{scramb}. Similarly to the DL, the pulse-shaping filters are root-raised cosine Nyquist filters using a rolloff factor of 0.22.

In TDD mode, the transmitter structure for both the UL and DL are similar to that of a FDD DL transmitter of Figure 1.38. Since each timeslot can be used for transmitting several TDD bursts from the same source or from different sources, the OVSF codes are invoked in order to maintain orthogonality between the burst of different TDD/CDMA users/messages. An advantage of the TDD/CDMA mode is that the user population is separated in both the time and the code domain. In other words, only a small number of CDMA users/services will be supported within a TDD timeslot, which dramatically reduces the complexity of the multi-

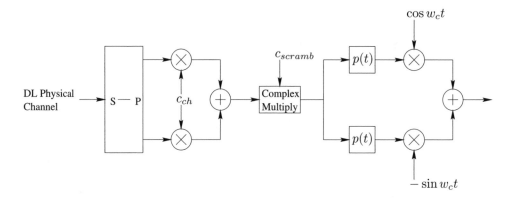

Figure 1.38: DL transmitter in UTRA using the frame structure of Figure 1.21.

Figure 1.39: ALOHA-based physical UL random access slots in UTRA.

user detector that can be used in both the UL and DL for mitigating the MAI or multi-code interference.

1.3.2.7 Random Access

1.3.2.7.1 Mobile-initiated Physical Random Access Procedures. If data transmission is initiated by an MS, it is required to send a random access request to the BS. Since such requests can occur at any time, collisions may result when two or more MSs attempt to access the network simultaneously. Hence, in order to reduce the probability of a collision, the random access procedure in UTRA is based on the slotted ALOHA technique [118].

Random access requests are transmitted to the BS via the PRACH of Table 1.5. Each random access transmission request may consist of one or several preambles and a message part, whose timeslot configuration was shown in Figure 1.22. According to the regime of Figure 1.39, the preambles and the message part can only be transmitted at the beginning of one of those 15 so-called *access slots*, which span two radio frames (i.e., 20 ms). Thus, each access slot has a length equivalent to 5120 chips or $\frac{4}{3}$ ms.

Before any random access request can be transmitted, the MS has to obtain certain information via the DL BCH transmitted on the P-CCPCH of Table 1.5 according to the format of Figure 1.23. This DL BCH/PCCPCH information includes the identifier of the cell-specific scrambling code for the preamble and message part of Figure 1.22, the available preamble signatures, the available access slots of Figure 1.39, which can be contended for in ALOHA mode, the initial preamble transmit power, the preamble power ramping factor, and the maximum number of preamble retransmissions necessitated by their decoding failure due to collisions at the BS. All this information may become available once synchronization is achieved, as will be discussed in Section 1.3.2.9. After acquiring all the necessary information, the MS will randomly select a preamble signature from the available signatures and transmit a preamble at the specific power level specified by the BS on a randomly selected access slot chosen from the set of available access slots seen in Figure 1.39. Note that the preamble is formed by multiplying the selected signature with the preamble scrambling code.

After the preamble is transmitted, the MS will listen for the acknowledgement of reception transmitted from the BS on the AICH of Table 1.5. Note that the AICH is also transmitted at the beginning of an access slot and the phase reference for coherent detection is obtained from the DL CPICH of Table 1.5. The acknowledgement is represented by an AI in the AICH of Table 1.5 that corresponds to the selected preamble signature. If a negative acknowledgement is received, the random access transmission will recommence in a later access slot. If a positive acknowledgement is received, the MS will proceed to transmit the message part at the beginning of a predefined access slot. However, if the MS fails to receive any acknowledgement after a predefined time-out, it will retransmit the preamble in another randomly selected access slot of Figure 1.39 with a newly selected signature, provided that the maximum number of preamble retransmissions was not exceeded. The transmit power of the preamble is also increased, as specified by the above-mentioned preamble power ramping factor. This procedure is repeated until either an acknowledgement is received from the BS or the maximum number of preamble retransmissions is reached.

1.3.2.7.2 Common Packet Channel Access Procedures.

The transmission of the CPCH of Table 1.5 is somewhat similar to that of the RACH transmission regime highlighted in Figure 1.39. Before commencing any CPCH transmission, the MS must acquire vital information from the BCH message transmitted on the P-CCPCH. This information includes the scrambling codes, the available signatures and the access slots for both the A-P and CD-P messages introduced in Section 1.3.2.3.2.1, the scrambling code of the message part, the DL AICH and the associated DL DPCCH channelization code, the initial transmit power of the preambles, the preamble power ramping factor, and the maximum allowable number of retransmissions.

The procedure of the A-P transmission is identical to that of the random access transmission highlighted in Section 1.3.2.7.1. We will accordingly omit the details here.

Once a positive acknowledgement is received from the BS on the DL AICH, the MS will transmit the CD-P on a randomly selected access slot of Figure 1.39 using a randomly selected signature. Upon receiving a positive acknowledgment from the BS on the AICH, the MS will begin transmitting the PC-P followed immediately by the message part shown in Figure 1.20 at a predefined access slot of Figure 1.39.

1.3.2.8 Power Control

Accurate power control is essential in CDMA in order to mitigate the so-called *near–far problem* [142, 143]. Furthermore, power control has a dramatic effect on the coverage and capacity of the system: we will therefore consider the UTRA power control issues in detail.

1.3.2.8.1   Closed-loop Power Control in UTRA.   Closed-loop power control is employed on both the UL and DL of the FDD mode through the TPC commands that are conveyed in the UL and DL according to the format of Figures 1.20 and 1.21, respectively. Since the power control procedure is the same on both links, we will only elaborate further on the UL procedure.

UL closed-loop power control is invoked in order to adjust the MS's transmit power such that the received Signal-to-Interference Ratio (SIR) at the BS is maintained at a given target SIR. The value of the target SIR depends on the required quality of the connection. The BS measures the received power of the desired UL transmitted signal for both the DPDCH and the DPCCH messages shown in Figure 1.20 after Rake combining, and it also estimates the total received interference power in order to obtain the estimated received SIR. This SIR estimation process is performed every $\frac{2}{3}$ ms, or a timeslot duration, in which the SIR estimate is compared to the target SIR. According to the values of the estimated and required SIRs, the BS will generate a TPC command, which is conveyed to the MS using the burst of Figure 1.21. If the estimated SIR is higher than the target SIR, the TPC command will instruct the MS to lower the transmit power of the DPDCH and DPCCH of Figure 1.20 by a step size of Δ_{TPC} dB. Otherwise, the TPC command will instruct the MS to increase the transmit power by the same step size. The step size Δ_{TPC} is typically 1 dB or 2 dB. Transmitting at an unnecessarily high power reduces the battery life, while degrading other users' reception quality, who—as a consequence—may request a power increment, ultimately resulting in an unstable overall system operation.

In some cases, BS-diversity combining may take place, whereby two or more BSs transmit the same information to the MS in order to enhance its reception quality. These BSs are known as the active BS set of the MS. The received SIR at each BS will be different and so the MS may receive different TPC commands from its active set of BSs. In this case, the MS will adjust its transmit power according to a simple algorithm, increasing the transmit power only if the TPC commands from all the BSs indicate an "increase power" instruction. Similarly, the MS will decrease its transmit power if all the BSs issue a "decrease power" TPC command. Otherwise, the transmit power remains the same. In this way, the multi-user interference will be kept to a minimum without significant deterioration of the performance, since at least one BS has a good reception. Again, the UL and DL procedures are identical, obeying the TPC transmission formats of Figures 1.20 and 1.21, respectively.

1.3.2.8.2   Open-loop Power Control in TDD Mode.   As mentioned previously in Section 1.3.2.3, in contrast to the closed-loop power control regime of the FDD mode, no TPC commands are transmitted on the DL in TDD mode. Instead, open-loop power control is used to adjust the transmit power of the MS. Prior to any data burst transmission, the MS would have acquired information about the interference level measured at the BS and also about the BS's P-CCPCH transmitted signal level, which are conveyed to the MS via the BCH according to the format of Figure 1.27. At the same time, the MS would also measure the power of the received P-CCPCH. Hence, with knowledge of the transmitted and received power of the P-CCPCH, the DL pathloss can be found. Since the interference level and the estimated pathloss are now known, the required transmitted power of the TDD burst can be readily calculated based on the required SIR level. Let us now consider how the MS identifies the different cells or BSs with which it is communicating.

1.3.2.9 Cell Identification

1.3.2.9.1 Cell Identification in the FDD Mode. System- and cell-specific information is conveyed via the BCH transmitted by the P-CCPCH of Table 1.5 in the context of Figure 1.23 in UTRA. This information has to be obtained before the MS can access the network. The P-CCPCH information broadcast from each cell is spread by the system-specific OVSF channelization code $c_{256,2}$ of Figure 1.36. However, each P-CCPCH message is scrambled by a cell-specific primary scrambling code as highlighted in Section 1.3.2.6.3 in order to minimize the inter-cell interference as well as to assist in identifying the corresponding cell. Hence, the first step for the MS is to recognize this primary scrambling code and to synchronize with the corresponding BS.

As specified in Section 1.3.2.6.3, there are a total of 512 DL primary scrambling codes available in the network. Theoretically, it is possible to achieve scrambling code identification by cross-correlating the P-CCPCH broadcast signal with all the possible 512 primary scrambling codes. However, this would be an extremely tedious and slow process, unduly delaying the MS's access to the network. In order to achieve a fast cell identification by the MS, UTRA adopted a three-step approach [144], which invoked the SCH broadcast from all the BSs in the network. The SCH message is transmitted during the first 256 chips of the P-CCPCH, as illustrated in Figure 1.23. The concept behind this three-step approach is to divide the set of 512 possible primary scrambling codes into 64 subsets, each containing a smaller set of primary scrambling codes, namely, eight codes. Once knowledge of which subset the primary scrambling code of the selected BS belongs to is acquired, the MS can proceed to search for the correct primary scrambling code from a smaller subset of the possible codes.

The frame structure of the DL SCH message seen in Figure 1.23 is shown in more detail in Figure 1.40. It consists of two subchannels, the *Primary SCH* and *Secondary SCH*, transmitted in parallel using code multiplexing. As seen in Figure 1.40, in the Primary SCH a so-called *Primary Synchronization Code* (PSC), based on a generalized hierarchical Golay sequence [145] of length 256 chips, is transmitted periodically at the beginning of each slot, which is denoted by c_p in Figure 1.40. The same PSC is used by all the BSs in the network. This allows the MS to establish slot-synchronization and to proceed to the frame-synchronization phase with the aid of the secondary SCH. On the secondary SCH, a sequence of 15 *Secondary Synchronization Codes* (SSCs), each of length 256 chips, is transmitted with a period of one 10 ms radio frame duration, that is, 10 ms, as seen in Figure 1.40. An example of this 15-SSC sequence would be:

$$c_1^1 \ c_1^2 \ c_2^3 \ c_8^4 \ c_9^5 \ c_{10}^6 \ c_{15}^7 \ c_8^8 \ c_{10}^9 \ c_{16}^{10} \ c_2^{11} \ c_7^{12} \ c_{15}^{13} \ c_7^{14} \ c_{16}^{15}, \tag{1.39}$$

where each of these 15 SSCs is selected from a set of 16 legitimate SSCs. The specific sequence of 15 SSCs denoted by $c_i^1, \ldots, c_i^{15}$—where $i = 1, \ldots, 16$ in Figure 1.40—is used as a code in order to identify and signal to the MS which of the 64 subsets the primary scrambling code used by the particular BS concerned belongs to. The parameter a in Figure 1.40 is a binary flag used to indicate the presence ($a = +1$) or absence ($a = -1$) of a Space Time Block Coding Transmit Diversity (STTD) encoding scheme [146] in the P-CCPCH, as will be discussed in Section 1.3.4.1.1. Specifically, when each of the 16 legitimate 256-chip SSCs can be picked for any of the 15 positions in Figure 1.40 and assuming no other

c_p : Primary Synchronization Code
$c_i^j, i = 1, \ldots, 16; j = 1, \ldots, 15$: Secondary Synchronization Code
P-CCPCH : Primary Common Control Physical CHannel
BCH : Broadcast CHannel
SCH : Synchronisation CHannel

Figure 1.40: Frame structure of the UTRA DL synchronization channel (SCH), which is mapped to the first 256 chips of the P-CCPCH of Figure 1.23. The primary and secondary SCH are transmitted in parallel using code multiplexing. The parameter a is a gain factor used to indicate the presence ($a = +1$) or absence ($a = -1$) of STTD encoding in the P-CCPCH.

further constraints, one could construct

$$
\begin{aligned}
c_{i,j}^{\text{repeated}} &= \binom{i + j - 1}{j} \\
&= \frac{(i + j - 1)!}{j!(i - 1)!} \\
&= \frac{30!}{15! \cdot 15!} \\
&= 155, 117, 520
\end{aligned}
\tag{1.40}
$$

different such sequences, where $i = 16$ and $j = 15$. However, the 15 different 256-chip SSCs of Figure 1.40 must be constructed so that their cyclic shifts are also unique, since these sequences have to be uniquely recognized before synchronization. In other words, none of the cyclic shifts of the 64 required $15 \times 256 = 3840$-chip sequences can be identical to any of the other sequences' cyclic shifts. Provided that these conditions are satisfied, the 15 specific 256-chip secondary SCH sequences can be recognized within one 10 ms-radio frame-duration of 15 slots. Thus, both slot and frame synchronization can be established within the particular 10 ms frame received. Using this technique, initial cell identification and synchronization can be carried out in the following three basic steps.

Step 1: The MS uses the 256-chip PSC of Figure 1.40 to perform cross-correlation with all the received Primary SCHs of the BSs in its vicinity. The BS with the highest correlator output is then chosen, which constitutes the best cell site associated with the lowest pathloss. Several periodic correlator output peaks have to be identified in order to achieve a high BS detection reliability, despite the presence of high-level interference. *Slot synchronization is also achieved* in this step by recognizing the 15 consecutive c_p sequences, providing 15 periodic correlation peaks.

Step 2: Once the best cell site is identified, the primary scrambling code subset of that cell site is found by cross-correlating the Secondary SCH with the 16 possible SSCs in each of the 15 timeslots of Figure 1.40. This can be easily implemented using 16 correlators, since the timing of the SSCs is known from Step 1. Hence, there are a total of $15 \times 16 = 240$ correlator outputs. From these outputs, a total of $64 \times 15 = 960$ decision variables corresponding to the 64 possible secondary SCH sequences and 15 cyclic shifts of each $15 \times 256 = 3840$-chip sequence are obtained. The highest decision variable determines the primary scrambling code subset. Consequently, *frame synchronization is also achieved.*

Step 3: With the primary scrambling code subset identified and frame synchronization achieved, the primary scrambling code itself is acquired in UTRA by cross-correlating the received CPICH signal—which is transmitted synchronously with the P-CCPCH—on a symbol-by-symbol basis with the eight possible primary scrambling codes belonging to the identified primary scrambling code subset. Note that the CPICH is used in this case, because it is scrambled by the same primary scrambling code as the P-CCPCH and also uses a predefined pilot sequence and so it can be detected more reliably. By contrast, the P-CCPCH carries the unknown BCH information. Once the exact primary scrambling code is identified, the BCH information of Table 1.5, which is conveyed by the P-CCPCH of Figure 1.23, can be detected.

1.3.2.9.2 Cell Identification in the TDD Mode.

The procedure of cell identification in the TDD mode is somewhat different from that in FDD mode. In the TDD mode, a combination of three 256-chip SSCs out of 16 unique SSCs are used to identify one of 32 SSC code groups allocated to that cell. Each code group contains four different scrambling codes and four corresponding long (for Type 1 burst) and short (for Type 2 burst) basic midamble codes, which were introduced in the context of Figure 1.27. Each code group is also associated with a specific time offset, t_{offset}. The three SSCs, c_i^1, c_i^2, and c_i^3, are transmitted in parallel with the PSC, c_p, at a time offset t_{offset} measured from the start of a timeslot, as shown in Figure 1.41. Similarly to the FDD mode, the PSC is based on a so-called generalized hierarchical Golay sequence [145], which is common to all the cells in the system. Initial cell identification and synchronization in the TDD mode can also be carried out in three basic steps.

Step 1: The MS uses the 256-chip PSC of Figure 1.41 to perform cross-correlation with all the received PSC of the BSs in its vicinity. The BS associated with the highest correlator output is then chosen, which constitutes the best cell site exhibiting the lowest pathloss. Slot synchronization is also achieved in this step. If only one timeslot per frame is used to transmit the SCH as outlined in the context of Figure 1.27, then frame synchronization is also achieved.

Step 2: Once the PSC of the best cell site is identified, the three SSCs transmitted in parallel with the PSC in Figure 1.41 can be identified by cross-correlating the received signal with the 16 possible prestored SSCs. The specific combination of the three SSCs will identify the code group used by the corresponding cell. The specific frame timing of that cell also becomes known from the time offset t_{offset} associated with that code group.

If two timeslots per frame are used to transmit the SCH as outlined in the context of Figure 1.27, then the second PSC must be detected at an offset of seven or eight timeslots with respect to the first one in order to achieve frame synchronization.

Step 3: As mentioned in Section 1.3.2.3, each basic midamble code defined in the context of Figure 1.27 is associated with a midamble code set. The P-CCPCH of Table 1.5 is always associated with the first midamble of that set. Hence, with the code group identified and frame synchronization achieved, the cell-specific scrambling code and the associated basic midamble code are acquired in the TDD mode of UTRA by cross-correlating the four possible

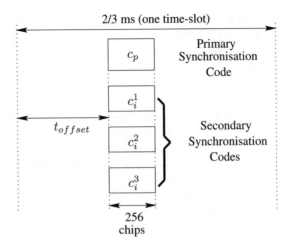

Figure 1.41: Timeslot structure of the UTRA TDD DL synchronization channel (SCH), which obeys the format of Figure 1.19. The primary and three secondary synchronization codes are transmitted in parallel at a time offset t_{offset} from the start of a timeslot.

midamble codes with the P-CCPCH. Once the exact basic midamble code is identified, the associated scrambling code will be known, and the BCH information of Table 1.5, which is conveyed by the P-CCPCH of Figure 1.23, can be detected. Having highlighted the FDD and TDD UTRA cell-selection and synchronization solutions, let us now consider some of the associated handover issues.

1.3.2.10 Handover

In this section, we consider the handover issues in the context of the FDD mode, since the associated procedures become simpler in the TDD mode, where the operations can be carried out during the unused timeslots. Theoretically, DS-CDMA has a frequency reuse factor of one [147]. This implies that neighboring cells can use the same carrier frequency without interfering with each other, unlike in TDMA or FDMA. Hence, seamless uninterrupted handover can be achieved when mobile users move between cells, since no switching of carrier frequency and synthesizer retuning is required. However, in *hierarchical cell structures* (HCS)[9] catering, for example, for high-speed mobiles with the aid of a macrocell oversailing a number of microcells, using a different carrier frequency is necessary in order to reduce the inter-cell interference. In this case, inter-frequency handover is required. Furthermore, because the various operational GSM systems used different carrier frequencies, handover from UTRA systems to GSM systems will have to be supported during the transitory migration phase, while these systems will coexist. Thus, handovers in terrestrial UMTSs can be classified into inter-frequency and intra-frequency handovers.

1.3.2.10.1 Intra-frequency Handover or Soft Handover. Soft handover [148, 149] involves no frequency switching because the new and old cell use the same carrier frequency.

---

[9]Microcells overlaid by a macrocell.

One radio frame (10 ms)

N > M

Figure 1.42: UL frame structure in compressed mode operation during UTRA handovers.

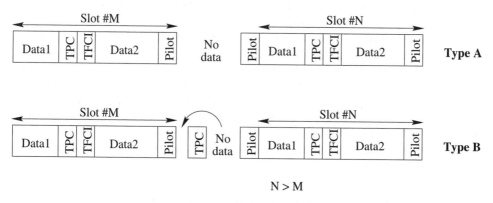

N > M

Figure 1.43: DL frame structure in compressed mode operation during UTRA handovers using the transmission formats of Figure 1.19.

The MS will continuously monitor the received signal levels from the neighboring cells and compares them against a set of thresholds. This information is fed back to the network. Based on this information, if a weak or strong cell is detected, the network will instruct the MS to drop or add the cell from/to its active BS set. In order to ensure a seamless handover, a new link will be established before relinquishing the old link, using the *make before break* approach.

1.3.2.10.2 Inter-frequency Handover or Hard Handover. In order to achieve handovers between different carrier frequencies without affecting the data flow, a technique known as *compressed mode* can be used [150]. With this technique, the UL data, which normally occupies the entire 10 ms frame of Figure 1.19 is time-compressed, so that it only occupies a portion of the frame, that is, slot#1-slot#M and slot#N-slot#15, while no data is transmitted during the remaining portion, that is, slot#(M+1)-slot#(N-1). The latter interval is known as the idle period, as shown in Figure 1.42. There are two types of frame structures for the DL compressed mode, as shown in Figure 1.43. In the Type A structure, shown at the top of Figure 1.43, no data is transmitted after the pilot field of slot#M until the start of the pilot field of slot#(N-1) in order to maximize the transmission gap length. By contrast, in the Type B structure shown at the bottom of Figure 1.43, a TPC command is transmitted in slot#(M+1) during the idle period in order to optimize the power control.

The idle period has a variable duration, but the maximum period allowable within a 15-slot, 10 ms radio frame is seven slots. The idle period can occur either at the centre of a

10 ms frame or at the end and the beginning of two consecutive 10 ms frames, such that the idle period spans over two frames. However, in order to maintain the seamless operation of all MSs occupying the uncompressed 15-slot, 10 ms frame, the duration of all timeslots has to be shortened by "compressing" their data. The compression of data can be achieved by channel-code puncturing, a procedure that obliterates some of the coded parity bits, thereby slightly reducing the code's error correcting power, or by adjusting the spreading factor. In order to maintain the quality of the link, the instantaneous power is also increased during the compressed mode operation. After receiving the data, the MS can use this idle period in the 10 ms frame, to switch to other carrier frequencies of other cells and to perform the necessary link-quality measurements for handover.

Alternatively, a twin-receiver can be used in order to perform inter-frequency handovers. One receiver can be tuned to the desired carrier frequency for reception, while the other receiver can be used to perform handover link-quality measurements at other carrier frequencies. This method, however, results in a higher hardware complexity at the MS.

The 10 ms frame length of UTRA was chosen so that it is compatible with the multiframe length of 120 ms in GSM. Hence, the MS is capable of receiving the Frequency Correction Channel (FCCH) and Synchronization Channel (SCH) messages in the GSM [55] frame using compressed mode transmission and to perform the necessary handover link-quality measurements [117].

1.3.2.11 Intercell Time Synchronization in the UTRA TDD Mode

Time synchronization between BSs is required when operating in the TDD mode in order to support seamless handovers. A simple method of maintaining inter-cell synchronization is by periodically broadcasting a reference signal from a source to all the BSs. The propagation delay can be easily calculated, and hence compensated, from the fixed distance between the source and the receiving BSs. There are three possible ways of transmitting this reference signal, namely, via the terrestrial radio link, via the physical wired network, or via the Global Positioning System (GPS).

Global time synchronization in 3G mobile radio systems is achieved by dividing the synchronous coverage region into three areas, namely, the so-called subarea, main area and coverage area, as shown in Figure 1.44. Intercell synchronization within a sub-area is provided by a subarea reference BS. Since the subarea of Figure 1.44 is smaller than the main area, transmitting the reference signal via the terrestrial radio link or the physical wired network is more feasible. All the subarea reference BSs in a main area are in turn synchronized by a main-area reference BS. Similarly, the reference signal can be transmitted via the terrestrial radio link or the physical wired network. Finally, all the main-area reference BSs are synchronized using the GPS. The main advantage of dividing the coverage regions into smaller areas is that each lower hierarchical area can still operate on its own, even if the synchronization link with the higher hierarchical areas is lost.

1.3.3 The cdma2000 Terrestrial Radio Access [151–153]

The current 2G mobile radio systems standardized by TIA in the United States are IS-95-A and IS-95-B [151]. The radio access technology of both systems is based on narrowband DS-CDMA with a chip rate of 1.2288 Mcps, which gives a bandwidth of 1.25 MHz. IS-95-A

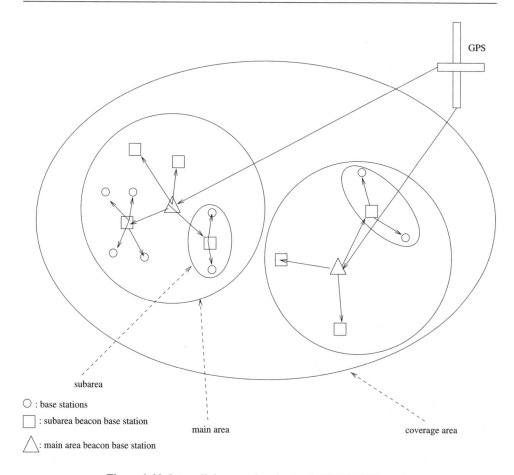

Figure 1.44: Intercell time synchronization in UTRA TDD mode.

was commercially launched in 1995, supporting circuit and packet mode transmissions at a maximum bit rate of only 14.4 kbps [151]. An enhancement to the IS-95-A standards, known as IS-95-B, was developed and introduced in 1998 in order to provide higher data rates, on the order of 115.2 kbps [58]. This was feasible without changing the physical layer of IS-95-A. However, this still falls short of the 3G mobile radio system requirements. Hence, the technical committee TR45.5 within TIA has proposed cdma2000, a 3G mobile radio system that is capable of meeting all the requirements laid down by ITU. One of the problems faced by TIA is that the frequency bands allocated for the 3G mobile radio system, identified during WARC'92 to be 1885–2025 MHz and 2110–2200 MHz, have already been allocated for Personal Communications Services (PCS) in the United States from 1.8 GHz to 2.2 GHz. In particular, the CDMA PCS based on the IS-95 standards has been allocated the frequency bands of 1850–1910 MHz and 1930–1990 GHz. Hence, the 3G mobile radio systems have to fit into the allocated bandwidth without imposing significant interference on the existing applications. Thus, the framework for cdma2000 was designed so that it can be overlaid on IS-95 and it is backwards compatible with IS-95. Most of this section is based on [151–153].

Table 1.10: The cdma2000 basic parameters.

Radio Access Technology	DS-CDMA, Multicarrier CDMA
Operating environments	Indoor/Outdoor to Indoor/Vehicular
Chip rate (Mcps)	1.2288/3.6864/7.3728/11.0592/14.7456
Channel bandwidth (MHz)	1.25/3.75/7.5/11.25/15
Duplex modes	FDD and TDD
Frame length	5 and 20 ms
Spreading factor	Variable, 4 to 256
Detection scheme	Coherent with common pilot channel
Intercell operation	FDD : Synchronous
	TDD : Synchronous
Power control	Open and closed loop
Handover	Soft-handover
	Inter-frequency handover

1.3.3.1 Characteristics of cdma2000

The basic parameters of cdma2000 are shown in Table 1.10. The cdma2000 system has a basic chip rate of 3.6864 Mcps, which is accommodated in a bandwidth of 3.75 MHz. This chip rate is in fact three times the chip rate used in the IS-95 standards, which is 1.2288 Mcps. Accordingly, the bandwidth was also trebled. Hence, the existing IS-95 networks can also be used to support the operation of cdma2000. Higher chip rates on the order of $N \times 1.2288$ Mcps, $N = 6, 9, 12$ are also supported. These are used to enable higher bit rate transmission. The value of N is an important parameter in determining the channel-coding rate and the channel bit rate. In order to transmit the high chip-rate signals ($N > 1$), two modulation techniques are employed. In the *direct-spread modulation mode*, the symbols are spread according to the chip rate and transmitted using a single carrier, giving a bandwidth of $N \times 1.25$ MHz. This method is used on both the UL and DL. In *multicarrier (MC) modulation*, the symbols to be transmitted are de-multiplexed into separate signals, each of which is then spread at a chip rate of 1.2288 Mcps. N different carrier frequencies are used to transmit these spread signals, each of which has a bandwidth of 1.25 MHz. This method is used for the DL only, because in this case, transmit diversity can be achieved by transmitting the different carrier frequencies over spatially separated antennas.

By using multiple carriers, cdma2000 is capable of overlaying its signals on the existing IS-95 1.25 MHz channels and its own channels, while maintaining orthogonality. An example of an overlay scenario is shown in Figure 1.45. Higher chip rates are transmitted at a lower power than lower chip rates, thereby keeping the interferences to a minimum.

Similarly to UTRA and IMT-2000, cdma2000 also supports TDD operation in unpaired frequency bands. In order to ease the implementation of a dual-mode FDD/TDD terminal, most of the techniques used for FDD operation can also be applied in TDD operation. The difference between these two modes is in the frame structure, whereby an additional guard time has to be included for TDD operation.

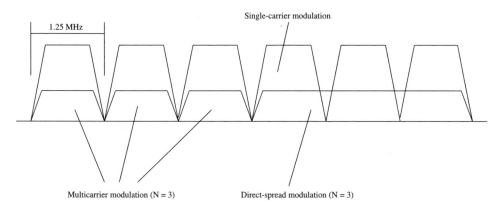

Figure 1.45: Example of an overlay deployment in cdma2000. The multicarrier mode is only used in the DL.

In contrast to UTRA and IMT-2000, where the pilot symbols of Figure 1.21 are time-multiplexed with the dedicated data channel on the DL, cdma2000 employs a common code multiplexed continuous pilot channel on the DL, as in the IS-95 system. The advantage of a common DL pilot channel is that no additional overhead is incurred for each user. However, if adaptive antennas are used, then additional pilot channels have to be transmitted from each antenna.

Another difference with respect to UTRA and IMT-2000 is that the base stations are operated in synchronous mode in cdma2000. As a result, the same PN code but with different phase offsets can be used to distinguish the base stations. Using one common PN sequence can expedite cell acquisition as compared to a set of PN sequences, as we have seen in Section 1.3.2.9 for IMT-2000/UTRA. Let us now consider the cdma2000 physical channels.

1.3.3.2 Physical Channels in cdma2000

The physical channels (PHCH) in cdma2000 can be classified into two groups, namely Dedicated Physical Channels (DPHCH) and Common Physical Channels (CPHCH). DPHCHs carry information between the base station and a single mobile station, while CPHCHs carry information between the base station and several mobile stations. Table 1.11 shows the collection of physical channels in each group. These channels will be elaborated on during our further discourse. Typically, all physical channels are transmitted using a frame length of 20 ms. However, the control information on the so-called Fundamental Channel (FCH) and Dedicated Control Channel (DCCH) can also be transmitted in 5 ms frames.

Each base station transmits its own DL Pilot Channel (PICH), which is shared by all the mobile stations within the coverage area of the base station. Mobile stations can use this common DL PICH in order to perform channel estimation for coherent detection, soft handover, and fast acquisition of strong multipath rays for Rake combining. The PICH is transmitted orthogonally along with all the other DL physical channels from the base station by using a unique orthogonal code (Walsh code 0) as in the IS-95 system The optional Common Auxiliary Pilot Channels (CAPICH) and Dedicated Auxiliary Pilot Channels (DAPICH) are used to support the implementation of antenna arrays. CAPICHs provide spot

Table 1.11: The cdma2000 physical channels.

Dedicated Physical Channels (DPHCH)	Common Physical Channels (CPHCH)
Fundamental Channel (FCH) (UL/DL)	Pilot Channel (PICH) (DL)
Supplemental Channel (SCH) (UL/DL)	Common Auxiliary Pilot Channel (CAPICH) (DL)
Dedicated Control Channel (DCCH) (UL/DL)	Forward Paging Channel (PCH) (DL)
Dedicated Auxiliary Pilot Channel (DAPICH) (DL)	Sync Channel (SYNC) (DL)
Pilot Channel (PICH) (UL)	Access Channel (ACH) (UL)
	Common Control Channel (CCCH) (UL/DL)

Figure 1.46: UL pilot channel structure in cdma2000 for a 1.25 ms duration PCG, where $N = 1, 3, 6, 9, 12$ is the rate-control parameter.

coverage shared among a group of mobile stations, while a DAPICH is directed toward a particular mobile station. Every mobile station also transmits an orthogonal code-multiplexed UL pilot channel (PICH), which enables the base station to perform coherent detection in the UL as well as to detect strong multipaths and to invoke power control measurements. This differs from IS-95, which supports only noncoherent detection in the UL due to the absence of a coherent UL reference. In addition to the pilot symbols, the UL PICH also contains time-multiplexed power control bits assisting in DL power control. A power control bit is multiplexed onto the 20 ms frame every 1.25 ms, giving a total of 16 power control bits per 20 ms frame or 800 power updates per second, implying a very agile, fast response power control regime. Each 1.25 ms duration is referred to as a Power Control Group, as shown in Figure 1.46.

The use of two dedicated data physical channels, namely, the so-called Fundamental (FCH) and Supplemental (SCH) channels, optimizes the system during multiple simultaneous service transmissions. Each channel carries a different type of service and is coded and interleaved independently. However, in any connection, there can be only one FCH, but several SCHs can be supported. For a FCH transmitted in a 20 ms frame, two sets of uncoded data rates, denoted as Rate Set 1 (RS1) and Rate Set 2 (RS2), are supported. The data rates in RS1 and RS2 are 9.6/4.8/2.7/1.5 kbps and 14.4/7.2/3.6/1.8 kbps, respectively. Regardless of the uncoded data rates, the coded data rate is 19.2 kbps and 38.4 kbps for RS1 and RS2, respectively, when the rate-control parameter is $N = 1$. The 5 ms frame only supports one data rate, which is 9.6 kbps. The SCH is capable of transmitting higher data rates than the

Figure 1.47: The cdma2000 TDD frame structure.

FCH. The SCH supports variable data rates ranging from 1.5 kbps for $N = 1$ to as high as 2073.6 kbps, when $N=12$. Blind rate detection [154] is used for SCHs not exceeding 14.4 kbps, while rate information is explicitly provided for higher data rates. The dedicated control physical channel has a fixed uncoded data rate of 9.6 kbps on both 5 ms and 20 ms frames. This control channel rate is more than an order of magnitude higher than that of the IS-95 system hence it supports a substantially enhanced system control.

The Sync Channel (SYCH)—note the different acronym in comparison to the SCH abbreviation in UTRA/IMT-2000—is used to aid the initial synchronization of a mobile station to the base station and to provide the mobile station with system-related information, including the Pseudo Noise (PN) sequence offset, which is used to identify the base stations and the long code mask, which will be defined explicitly in Section 1.3.3.4. The SYCH has an uncoded data rate of 1.2 kbps and a coded data rate of 4.8 kbps.

Paging functions and packet data transmission are handled by the DL Paging Channel (PCH) and the DL Common Control Channel (CCCH). The uncoded data rate of the PCH can be either 4.8 kbps or 9.6 kbps. The CCCH is an improved version of the PCH, which can support additional higher data rates, such as 19.2 and 38.4 kbps. In this case, a 5 ms or 10 ms frame length will be used. The PCH is included in cdma2000 in order to provide IS-95-B functionality.

In TDD mode, the 20 ms and 5 ms frames are divided into 16 and 4 timeslots, respectively. This gives a duration of 1.25 ms per timeslot, as shown in Figure 1.47. A guard time of 52.08 μs and 67.44 μs is used for the DL in multicarrier modulation and for direct-spread modulation, respectively. In the UL, the guard time is 52.08 μs. Having described the cdma2000 physical channels of Table 1.11, let us now consider the service multiplexing and channel-coding aspects.

Table 1.12: The cdma2000 channel-coding parameters.

	Convolutional	Turbo
Rate	1/2 or 1/3 or 1/4	1/2 or 1/3 or 1/4
Constraint length	9	4

1.3.3.3   Service Multiplexing and Channel Coding

Services of different data rates and different QoS requirements are carried by different physical channels, namely, by the FCH and SCH of Table 1.11. This differs from UTRA and IMT-2000, whereby different services were time-multiplexed onto one or more physical channels, as highlighted in Section 1.3.2.4. These channels in cdma2000 are code-multiplexed using Walsh codes. Two types of coding schemes are used in cdma2000, as shown in Table 1.12. Basically, all channels use convolutional codes for forward error correction. However, for SCHs at rates higher than 14.4 kbps, turbo coding [135] is preferable. The rate of the input data stream is matched to the given channel rate by either adjusting the coding rate or using symbol repetition with and without symbol puncturing, or alternatively, by sequence repetition. Tables 1.13 and 1.14 show the coding rate and the associated rate matching procedures for the various DL and UL physical channels, respectively, when $N = 1$. Following the above brief notes on the cdma2000 channel coding and service multiplexing issues, let us now turn to the spreading and modulation processes.

1.3.3.4   Spreading and Modulation

There are generally three layers of spreading in cdma2000, as shown in Table 1.15. Each user's UL signal is identified by different offsets of a long code, a procedure that is similar to that of the IS-95 system portrayed in [155]. As seen in Table 1.15, this long code is an m-sequence with a period of $2^{42} - 1$ chips. The construction of m-sequences was highlighted by proakis [5]. Different user offsets are obtained using a long code mask. Orthogonality between the different physical channels of the same user belonging to the same connection in the UL is maintained by spreading using Walsh codes.

In contrast to the IS-95 DL of Figure 1.42 of [155], whereby Walsh code spreading is performed prior to QPSK modulation, the data in cdma2000 is first QPSK modulated before spreading the resultant I and Q branches with the same Walsh code. In this way, the number of Walsh codes available is increased twofold due to the orthogonality of the I and Q carriers. The length of the UL/DL channelization Walsh codes of Table 1.15 varies according to the data rates. All the base stations in the system are distinguished by different offsets of the same complex DL m-sequence, as indicated by Table 1.15. This DL m-sequence code is the same as that used in IS-95, which has a period of $2^{15} = 32768$, and it is derived from m-sequences. The feedback polynomials of the shift registers for the I and Q sequences are $X^{15} + X^{13} + X^9 + X^8 + X^7 + X^5 + 1$ and $X^{15} + X^{12} + X^{11} + X^{10} + X^6 + X^5 + X^4 + X^3 + 1$, respectively. The offset of these codes must satisfy a minimum value, which is equal to $N \times 64 \times$ Pilot_Inc, where Pilot_Inc is a code reuse parameter, which depends on

Table 1.13: The cdma2000 DL physical channel (see Table 1.11) coding parameters for $N = 1$, where repetition $\times 2$ implies transmitting a total of two copies.

Physical Channel	Data Rate (kbps)	Code Rate	Repetition	Puncturing	Channel Rate (ksps)
SYCH	1.2	1/2	×2	0	4.8
PCH	4.8	1/2	×2	0	19.2
	9.6	1/2	×1	0	19.2
CCCH	9.6	1/2	×1	0	19.2
	19.2	1/2	×1	0	38.4
	38.4	1/2	×1	0	76.8
FCH	1.5	1/2	×8	1 of 5	19.2
	2.7	1/2	×4	1 of 9	19.2
	4.8	1/2	×2	0	19.2
	9.6	1/2	×1	0	19.2
	1.8	1/3	×8	1 of 9	38.4
	3.6	1/3	×4	1 of 9	38.4
	7.2	1/3	×2	1 of 9	38.4
	14.4	1/3	×1	1 of 9	38.4
SCH	9.6	1/2	×1	0	19.2
	19.2	1/2	×1	0	38.4
	38.4	1/2	×1	0	76.8
	76.8	1/2	×1	0	153.6
	153.6	1/2	×1	0	307.2
	307.2	1/2	×1	0	614.4
	14.4	1/3	×1	1 of 9	38.4
	28.8	1/3	×1	1 of 9	76.8
	57.6	1/3	×1	1 of 9	153.6
	115.2	1/3	×1	1 of 9	307.2
	230.4	1/3	×1	1 of 9	614.4
DCCH	9.6	1/2	×1	0	19.2

the topology of the system, analogously to the frequency reuse factor in FDMA. Let us now focus on DL spreading issues more closely.

1.3.3.4.1 Downlink Spreading and Modulation. Figure 1.48 shows the structure of a DL transmitter for a physical channel. In contrast to the IS-95 DL transmitter shown in [155], the data in the cdma2000 DL transmitter shown in Figure 1.48 are first QPSK modulated before spreading using Walsh codes. As a result, the number of Walsh codes available is increased twofold due to the orthogonality of the I and Q carriers, as mentioned previously.

Table 1.14: The cdma2000 UL physical channel (see Table 1.11) coding parameters for $N = 1$, where repetition $\times 2$ implies transmitting a total of two copies.

Physical Channel	Data Rate (kbps)	Code Rate	Repetition 1	Puncturing	Repetition 2	Channel Rate (ksps)
CCCH	19.2	1/4	×1	0	×4	307.2
	38.4	1/4	×1	0	×2	307.2
FCH	1.5	1/4	×8	1 of 5	×8	307.2
	2.7	1/4	×4	1 of 9	×8	307.2
	4.8	1/4	×2	0	×8	307.2
	9.6	1/4	×1	0	×8	307.2
	1.8	1/4	×16	1 of 3	×4	307.2
	3.6	1/4	×8	1 of 3	×4	307.2
	7.2	1/4	×4	1 of 3	×4	307.2
	14.4	1/4	×2	1 of 3	×4	307.2
SCH	9.6	1/4	×1	0	×16	614.4
	19.2	1/4	×1	0	×8	614.4
	38.4	1/4	×1	0	×4	614.4
	76.8	1/4	×1	0	×2	614.4
	153.6	1/4	×1	0	×1	614.4
	307.2	1/2	×1	0	×1	614.4
ACH	4.8	1/4	×1	0	×8	307.2
	9.6	1/4	×1	0	×4	307.2
DCCH	9.6	1/4	×1	0	×4	307.2

The user data is first scrambled by the long scrambling code by assigning a different offset to different users for the purpose of improving user privacy, which is then mapped to the I and Q channels. This long, scrambling code is identical to the UL user-specific scrambling code given in Table 1.15. The DL pilot channels of Table 1.11 (PICH, CAPICH, DAPICH) and the SYNC channel are not scrambled with a long code since there is no need for user-specificity. The UL power control symbols are inserted into the FCH at a rate of 80 Hz, as shown in Figure 1.48. The I and Q channels are then spread using a Walsh code and complex multiplied with the cell-specific complex PN sequence of Table 1.15, as portrayed in Figure 1.48. Each base station's DL channel is assigned a different Walsh code in order to eliminate any intra-cell interference since all Walsh codes transmitted by the serving base station are received synchronously. The length of the DL channelization Walsh code of Table 1.15 is determined by the type of physical channel and its data rate. Typically for $N = 1$, DL FCHs with data rates belonging to RS1, that is, those transmitting at 9.6/4.8/2.7/1.5 kbps, use a 128-chip Walsh code, and those in RS2, transmitting at 14.4/7.2/3.6/1.8 kbps, use a 64-chip Walsh code. Walsh codes for DL SCHs can range from 4-chip to 128-chip Walsh codes. The DL

Table 1.15: Spreading parameters in cdma2000.

	Channelization Codes (UL/DL)	User-specific Scrambling Codes (UL)	Cell-specific Scrambling Codes (DL)
Type of codes	Walsh codes	Different offsets of a real m-sequence	Different offsets of a complex m-sequence
Code length	Variable	$2^{42} - 1$ chips	2^{15} chips
Type of Spreading	BPSK	BPSK	QPSK
Data Modulation	DL : QPSK UL : BPSK		

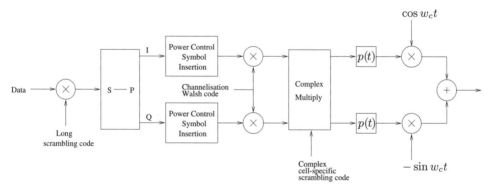

Figure 1.48: The cdma2000 DL transmitter. The long scrambling code is used for the purpose of improving user privacy. Hence, only the paging channels and the traffic channels are scrambled with the long code. The common pilot channel and the SYNC channel are not scrambled by this long code (the terminology of Table 1.15 is used).

PICH is an unmodulated sequence (all 0 s) spread by Walsh code 0. Finally, the complex spread data in Figure 1.48 are baseband filtered using the Nyquist filter impulse responses $p(t)$ in Figure 1.48 and modulated on a carrier frequency.

For the case of multicarrier modulation, the data is split into N branches immediately after the long code scrambling of Figure 1.48 which was omitted in the figure for the sake of simplicity. Each of the N branches is then treated as a separate transmitter and modulated using different carrier frequencies.

1.3.3.4.2 Uplink Spreading and Modulation. The UL cdma2000 transmitter is shown in Figure 1.49. The UL PICH and DCCH of Table 1.11 are mapped to the I data channel, while the UL FCH and SCH of Table 1.11 are mapped to the Q channel in Figure 1.49. Each of these UL physical channels belonging to the same user is assigned different Walsh channelization codes in order to maintain orthogonality, with higher rate channels using shorter Walsh codes. The I and Q data channels are then spread by complex multiplication

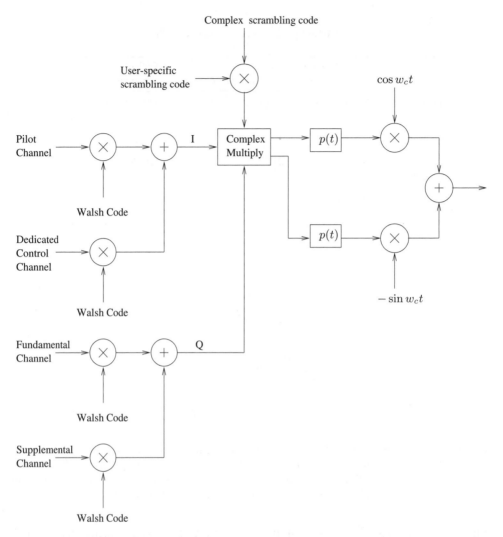

Figure 1.49: The cdma2000 UL transmitter. The complex scrambling code is identical to the DL cell-
specific complex scrambling code of Table 1.15 used by all the base stations in the system
(the terminology of Table 1.15 is used).

with the user-specifically offset real m-sequence based scrambling code of Table 1.15 and a
complex scrambling code, which is the same for all the mobile stations in the system, as seen
at the top of Figure 1.49. However, this latter complex scrambling code is not explicitly shown
in Table 1.15, since it is identical to the DL cell-specific scrambling code. This complex
scrambling code is only used for the purpose of quadrature spreading. Thus, in order to reduce
the complexity of the base station receiver, this complex scrambling code is identical to the
cell-specific scrambling code of Table 1.15 used on the DL by all the base stations.

Figure 1.50: An access attempt by a mobile station in cdma2000 using the access probe of Figure 1.51.

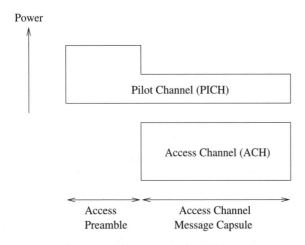

Figure 1.51: A cdma2000 access probe transmitted using the regime of Figure 1.50.

1.3.3.5 Random Access

The mobile station initiates an access request to the network by repeatedly transmitting a so-called access probe until a request acknowledgement is received. This entire process of sending a request is known as an access attempt. Within a single access attempt, the request may be sent to several base stations. An access attempt addressed to a specific base station is known as a subattempt. Within a subattempt, several access probes with increasing power can be sent. Figure 1.50 shows an example of an access attempt. The access probe transmission follows the slotted ALOHA algorithm, which is a relative of PRMA. An access probe can be divided into two parts, as shown in Figure 1.51. The access preamble carries a nondata-bearing pilot channel at an increased power level. The so-called access channel message capsule carries the data-bearing Access Channel (ACH) or UL Common Control Channel (CCCH) messages of Table 1.11 and the associated nondata-bearing pilot channel. The structure of the pilot channel is similar to that of the UL pilot channel (PICH) of Figure 1.46 except that in this case there are no time-multiplexed power control bits. The preamble length

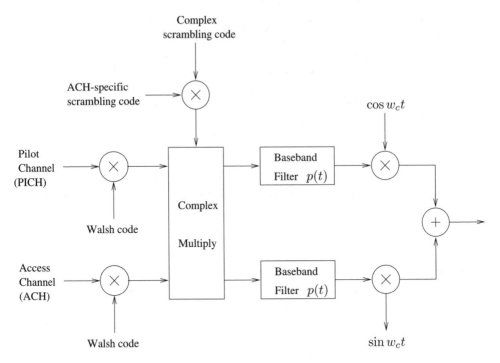

Figure 1.52: The cdma2000 access channel modulation and spreading. The complex scrambling code is identical to the DL cell-specific complex scrambling code of Table 1.15 used by all the base stations in the system (the terminology of Table 1.15 is used).

in Figure 1.51 is an integer multiple of the 1.25 ms slot intervals. The specific access preamble length is indicated by the base station, which depends on how fast the base station can search the PN code space in order to recognize an access attempt. The ACH is transmitted at a fixed rate of either 9.6 or 4.8 kbps, as seen in Table 1.14. This rate is constant for the duration of the access probe of Figure 1.50. The ACH or CCCH and their associated pilot channel are spread by the spreading codes of Table 1.15, as shown in Figure 1.52. Different ACHs or CCCHs and their associated pilot channels are spread by different long codes.

The access probes of Figures 1.50 and 1.51 are transmitted in predefined slots, where the slot length is indicated by the base station. Each slot is sufficiently long in order to accommodate the preamble and the longest message of Figure 1.51. The transmission must begin at the start of each 1.25 ms slot. If an acknowledgement of the most recently transmitted probe is not received by the mobile station after a time-out period, another probe is transmitted in another randomly chosen slot, obeying the regime of Figure 1.50.

Within a subattempt of Figure 1.50, a sequence of access probes is transmitted until an acknowledgement is received from the base station. Each successive access probe is transmitted at a higher power compared to the previous access probe, as shown in Figure 1.53. The initial power (IP) of the first probe is determined by the open-loop power control plus a nominal offset power that corrects for the open-loop power control imbalance between UL and DL. Subsequent probes are transmitted at a power level higher than the previous probe.

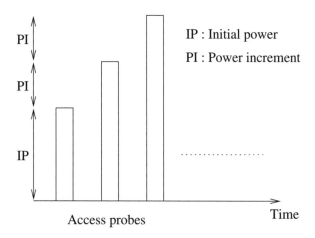

Figure 1.53: Access probes within a subattempt of Figure 1.50.

This increased level is indicated by the Power Increment (PI). Let us now highlight some of the cdma2000 handover issues.

1.3.3.6 Handover

Intra-frequency or soft-handover is initiated by the mobile station. While communicating, the mobile station may receive the same signal from several base stations. These base stations constitute the Active Set of the mobile station. The mobile station will continuously monitor the power level of the received pilot channels (PICH) transmitted from neighboring base stations, including those from the mobile station's active set. The power levels of these base stations are then compared to a set of thresholds according to an algorithm, which will be highlighted later in this chapter. The set of thresholds consists of the static thresholds, which are maintained at a fixed level, and the dynamic thresholds, which are dynamically adjusted based on the total received power. Subsequently, the mobile station will inform the network when any of the monitored power levels exceed the thresholds.

Whenever the mobile station detects a PICH, whose power level exceeds a given static threshold, denoted as T_1, this PICH will be moved to a candidate set and will be searched and compared more frequently against a dynamically adjusted threshold denoted as T_2. This value of T_2 is a function of the received power levels of the PICHs of the base stations in the active set. This process will determine whether the candidate base station is worth adding to the active set. If the overall power level in the active set is weak, then adding a base station of higher power will improve the reception. By contrast, if the overall power level in the active set is relatively high, then adding another high-powered base station may not only be unnecessary, but may actually utilize more network resources.

For the base stations that are already in the active set, the power level of their corresponding PICH is compared to a dynamically adjusted threshold, denoted as T_3, which is also a function of the total power of the PICH in the active set, similar to T_2. This is to ensure that each base station in the active set is contributing sufficiently to the overall power level. If any of the PICH's power level dropped below T_3 after a specified period of time allowed

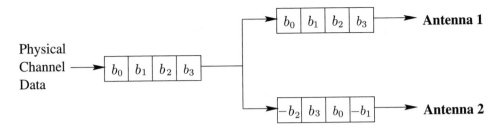

Figure 1.54: Transmission of a physical channel using Space Time block coding Transmit Diversity (STTD).

in order to eliminate any uncertainties due to fading which may have caused fluctuations in the power level, the base station will again be moved to the candidate set where it will be compared with a static threshold T_4. At the same time, the mobile station will report to the network the identity of the low-powered base station in order to allow the corresponding base station to increase its transmit power. If the power level decreases further below a static threshold, denoted as T_4, then the mobile station will again report this to the network and the base station will subsequently be dropped from the candidate set.

Inter-frequency or hard-handovers can be supported between cells having different carrier frequencies. Here we conclude our discussions on the cdma2000 features and provide some rudimentary notes on a number of advanced techniques, which can be invoked in order to improve the performance of the 3G W-CDMA systems.

1.3.4   Performance-enhancement Features

The treatment of adaptive antennas, multi-user detection, interference cancellation, or the portrayal of transmit diversity techniques is beyond the scope of this chapter. Here we simply provide a few pointers to the associated literature.

1.3.4.1   Downlink Transmit Diversity Techniques

1.3.4.1.1   Space Time Block Coding-based Transmit Diversity. Further diversity gain can be provided for the mobile stations by upgrading the base station with the aid of Space Time block coding assisted Transmit Diversity (STTD) [146], which can be applied to all the DL physical channels with the exception of the SCH. Typically the data of physical channels are encoded and transmitted using two antennas, as shown in Figure 1.54.

1.3.4.1.2   Time-switched Transmit Diversity. Time-Switched Transmit Diversity (TSTD) [156] is only applicable to the SCH, and its operation becomes explicit in Figure 1.55.

1.3.4.1.3   Closed-loop Transmit Diversity. Closed-loop transmit diversity is only applicable to the DPCH and PDSCH messages of Table 1.5 on the DL, which is illustrated in Figure 1.56. The weights w_1 and w_2 are related to the DL channel's estimated phase and attenuation information, which are determined and transmitted by the MS to the BS using the FBI D field, as portrayed in Figure 1.20. The weights for each antenna are

c_p : Primary Synchronization Code
$c_i^j, i = 1, \ldots, 16; j = 1, \ldots, 15$: Secondary Synchronization Code
BCH : Broadcast CHannel
P-CCPCH : Primary Common Control Physical CHannel

Figure 1.55: Frame structure of the UTRA DL synchronization channel (SCH), transmitted by a TSTD scheme. The primary and secondary SCH are transmitted alternatively from Antennas 1 and 2. The parameter a is a binary flag used to indicate the presence ($a = +1$) or absence ($a = -1$) of STTD encoding in the P-CCPCH.

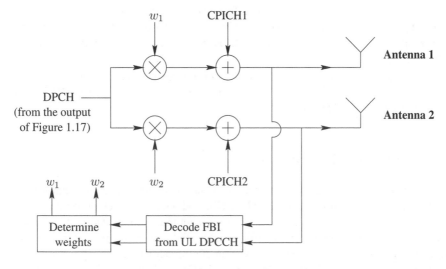

DPCH : Dedicated Physical CHannel
FBI : FeedBack Information
CPICH : Common PIlot CHannel

Figure 1.56: Transmission of the DL DPCH using a closed-loop transmit diversity technique.

independently measured by the MS using the corresponding pilot channels CPICH1 and CPICH2.

1.3.4.2 Adaptive Antennas

The transmission of time-multiplexed user-specific pilot symbols on both the UL and DL as seen for UTRA in Figures 1.20–1.24 facilitates the employment of adaptive antennas. Adaptive antennas are known to enhance the capacity and coverage of the system [157, 158].

1.3.4.3 Multi-user Detection/Interference Cancellation

Following Verdú's seminal paper [93], extensive research has shown that Multi-user Detection (MUD) [92, 95, 159–164] and Interference Cancellation techniques [91, 165–175] can substantially improve the performance of the CDMA link in comparison to conventional Rake receivers. However, using long scrambling codes increases the complexity of the MUD [58]. As a result, UTRA introduced an optional short scrambling code, namely, the S(2) code of Table 1.9, as mentioned in Section 1.3.2.6.4, in order to reduce the complexity of MUD [118]. Another powerful technique is invoking burst-by-burst adaptive CDMA [96, 132] in conjunction with MUD.

However, interference cancellation and MUD schemes require accurate channel estimation, in order to reproduce and deduct or cancel the interference. Several stages of cancellation are required in order to achieve a good performance, which in turn increases the canceller's complexity. It was shown that recursive channel estimation in a multistage interference canceller improved the accuracy of the channel estimation and hence gave improved BER performance [111].

Because of the complexity of the multi-user or interference canceller detectors, they were originally proposed for the UL. However, recently reduced-complexity DL MUD techniques have also been proposed [176].

1.3.5 Summary of 3G Systems

We have presented an overview of the terrestrial radio transmission technology of 3G mobile radio systems proposed by ETSI, ARIB, and TIA. All three proposed systems are based on Wideband-CDMA. Despite the call for a common global standard, there are some differences in the proposed technologies, notably, the chip rates and inter-cell operation. These differences are partly due to the existing 2G infrastructure already in use all over the world, and are specifically due to the heritage of the GSM and the IS-95 systems. Huge capital has been invested in these current 2G mobile radio systems. Therefore, the respective regional standard bodies have endeavored to ensure that the 3G systems are compatible with the 2G systems. Because of the diversified nature of these 2G mobile radio systems, it is not an easy task to reach a common 3G standard that can maintain perfect backwards compatibility. Non-coherent M-ary orthogonal CDMA is described in the next chapter.

1.4 Summary and Conclusions

Following the rudimentary introduction of Sections 1.1–1.2.6, Section 1.3 reviewed the 3G WB-CDMA standard proposals. The 3G systems are more amenable to the transmission of interactive video signals than their more rigid 2G counterparts. This is due partly to the higher supported bit rate and partly to the higher variety of available transmission integrities and bit rates. During our further discourse we will rely on this chapter and quantify the network performance of various joint-detection-based CDMA systems.

Chapter 2

High Speed Downlink and Uplink Packet Access

T-H. Liew and L. Hanzo

2.1 Introduction

The work on the standardization of third-generation (3G) mobile communication systems commenced in the early 1990s. In 1997, the choice of Wideband Code Division Multiple Access (WCDMA) was ratified by the different regions of the globe as the core technology of 3G systems. However, numerous regional versions of WCDMA emerged and in order to avert the risk of arriving at incompatible solutions, a collaboration agreement was established in December 1998 in order to form the 3rd Generation Partnership Project (3GPP). The 3GPP brings together the following regional telecommunications standardization bodies:

- Association of Radio Industries and Businesses (ARIB) from Japan;

- China Communications Standards Association (CCSA) from China;

- European Telecommunications Standard Institute (ETSI) from Europe;

- Alliances for Telecommunications Industry Solutions (ATIS) from the USA;

- Telecommunications Technology Association (TTA) from South Korea;

- Telecommunication Technology Committee (TTC) from Japan.

At the end of 1999, the 3GPP research activities reached a major milestone with the publication of WCDMA Release 99. The maximum theoretical DL speed is 936 Kbps [177] for a single physical channel and 2.3 Mbps, if three physical channels are multiplexed. In contrast, the maximum UL speed is 480 Kbps [177] for a single physical channel and 2.3 Mbps, provided that six physical channels are multiplexed.

3G, HSPA and FDD versus TDD Networking   Second Edition
L. Hanzo, J. S. Blogh and S. Ni   © 2008 John Wiley & Sons, Ltd

HS-DPCCH : High Speed Dedicated Physical Control Channel
HS-SCCH   : High Speed Shared Control Channel
HS-PDSCH : High Speed Physical Downlink Shared Channel

Figure 2.1: HSDPA physical channels.

Unfortunately, the bitrate offered by the WCDMA Release 99 specifications does not satisfy the requirements of seamless wireless Internet services and file download, for example. Therefore, the quest for higher bitrates has initiated the formation of a new 3GPP working group in 2000. A year later, this resulted in the standardization of the High Speed Downlink Packet Access (HSDPA) mode of the 3G systems in Release 5 of the 3GPP. This evolution to the HSDPA mode has increased the maximum DL speed to 13.976 Mbps. As HSDPA only improves the throughput of the DL, there is a demand for improving the UL throughput as well. Hence, another 3GPP working group was started in 2002. It took about one and half years to release the standardization of High Speed Uplink Packet Access (HSUPA) in Release 6 of 3GPP, which increased the maximum UL speed to 5.742 Mbps.

2.2   High Speed Downlink Packet Access

Three powerful techniques have been employed in HSDPA in order to increase the achievable throughput, namely:

- adaptive modulation, multiple spreading codes and variable rate channel coding;

- Hybrid Automatic Repeat Request (HARQ) techniques;

- fast packet scheduling.

Let us highlight how these techniques may be used to improve the attainable DL throughput. When a HSDPA-enabled User Equipment (UE), i.e. the Mobile Station (MS) shown in Figures 2.1 and 2.2 is switched on, it registers itself with the network. When the UE

Figure 2.2: HSDPA time diagram.

starts an application, such as file downloading for example, which requires a high DL bandwidth, the protocol initiates HSDPA transmission invoking the protocol handshake seen in Figure 2.2. Then the UE estimates the DL Signal-to-Interference Ratio (SIR) encountered by the reference channel's Common Pilot Channel (CPICH) transmitted from Node B. Node B is the term used in the 3G systems for the Base Station (BS). The resultant DL SIR measurements are mapped to the so-called Channel Quality Indicator (CQI) value, which has a range spanning from 0 to 30. The DL CQI value is then transmitted in the UL control channel referred to as the High Speed Dedicated Physical Control Channel (HS-DPCCH) to Node B, as shown in Figures 2.1 and 2.2, in order to inform the BS about the MS's channel which determines the achievable bitrate of the BS's transmission. More explicitly, based on the CQI value received from the UE, Node B determines the transport block size to be transmitted to the UE in the DL data channel termed as the High Speed Physical Downlink Shared Channel (HS-PDSCH).[1] Table 2.1 is reproduced from [178] and it shows the CQI mapping table used by Node B for UE category 10. The UE category is further detailed at

---

[1]The terminology of "shared" is used, because in the DL the BS is capable of serving numerous UEs by sharing a single channel amongst them, which in the UL a dedicated channel has to be used in order to avoid the potentially excessive complexity UL scheduling and synchronization of all of the UEs.

Table 2.1: CQI mapping table for UE category 10 (see [178]).

CQI value	Transport block size	Number of multicodes	Modulation scheme	Coding rate	Throughput (Mbps)
0	NA		Out of range		
1	137	1	QPSK	0.14	0.069
2	173	1	QPSK	0.18	0.087
⋮	⋮	⋮	⋮	⋮	⋮
7	650	2	QPSK	0.24	0.325
⋮	⋮	⋮	⋮	⋮	⋮
10	1262	3	QPSK	0.44	0.631
⋮	⋮	⋮	⋮	⋮	⋮
13	2279	4	QPSK	0.59	1.140
14	2583	4	QPSK	0.67	1.292
15	3319	5	QPSK	0.69	1.660
16	3565	5	16QAM	0.37	1.783
⋮	⋮	⋮	⋮	⋮	⋮
22	7168	5	16QAM	0.75	3.584
23	9719	7	16QAM	0.72	4.860
24	11 418	8	16QAM	0.74	5.709
25	14 411	10	16QAM	0.76	7.206
26	17 237	12	16QAM	0.75	8.619
27	21 754	15	16QAM	0.76	10.877
⋮	⋮	⋮	⋮	⋮	⋮
30	25 558	15	16QAM	0.89	12.779

a later stage in this section. It is shown in Table 2.1 that logically a larger transport block size is used when the CQI value is higher, i.e. a higher channel SIR is encountered. On the other hand, a smaller transport block size is employed when the CQI signals a low SIR in order to activate more robust modulation and coding modes and hence avoid inflicting a huge number of errors. In order to accommodate diverse transport block sizes spanning from 137 to 27 952 bits [179], according to the wide range of SIRs encountered, the number of HS-PDSCH spreading codes transmitted in parallel varies from 1 to 15. Furthermore, Adaptive Modulation and Coding (AMC) is used, where both QPSK and 16QAM schemes are employed in conjunction with various channel coding rates.

Note that Table 2.1 shows the CQI mapping table for UE category 10. Depending on the UE's specific capabilities, they are grouped in 12 different HSDPA-enabled UE categories. These categories are listed in the 3GPP standard [180] and are reproduced in Table 2.2. It is shown in Table 2.2 that the UE category 10 is potentially capable of supporting a transport block size as high as 27 952 bits, when employing 15 parallel spreading codes and 16QAM, which is equivalent to the HSDPA mode's maximum DL speed of 13.976 Mbps. Table 2.1 shows that the UE is operating close to its limit when the CQI value is 30. The second to last

Table 2.2: HSDPA enabled UE categories (see [180]).

UE category	Maximum transport block size	Maximum number of multicodes	Highest order modulation scheme	Total number of soft channel bits	Throughput (Mbps)
1	7298	5	16QAM	19 200	3.649
2	7298	5	16QAM	28 800	3.649
3	7298	5	16QAM	28 800	3.649
4	7298	5	16QAM	38 400	3.649
5	7298	5	16QAM	57 600	3.649
6	7298	5	16QAM	67 200	3.649
7	14 411	10	16QAM	115 200	7.206
8	14 411	10	16QAM	134 400	7.206
9	20 251	15	16QAM	172 800	10.126
10	27 952	15	16QAM	172 800	13.976
11	3630	5	QPSK	14 400	1.815
12	3630	5	QPSK	28 800	1.815

column of Table 2.2 shows the total memory space in the UE and its purpose is explained during our further discourse in Section 2.2.1.1.

The error-free reception of the control information transport block conveying the specific modulation scheme activated by the BS for its DL transmission, the number of spreading codes and other transmission-rate related parameters are essential for successfully decoding the HS-PDSCH by the HSDPA-enabled UEs. The corresponding side-information is also masked with the identity of the destination UE, which is a unique UE identity used for exclusively informing the UE of the specific parameters to be used for decoding the transport block. This side-information is then signaled to the UE with the aid of the High Speed Shared Control Channel (HS-SCCH), as shown in Figures 2.1 and 2.2. The HSDPA-enabled UE is configured to simultaneously monitor up to four HS-SCCHs, but only one of them shall carry the required control information masked with its UE identity. Using the received DL control information decoded from the HS-SCCH, the UE has to appropriately configure itself in order to decode the HS-PDSCH's message. As shown in Figure 2.2, Cyclic Redundancy Checking (CRC) is performed at the end of each decoding process in order to ascertain that the data in HS-PDSCH was received error free. Depending on the result of the CRC, the UE shall inform Node B as to whether the packet has been received successfully or not. It does so by sending a positive Acknowledgement (ACK) or a Negative Acknowledgement (NACK) UL message along with the associated CQI value in the HS-PDSCH shown in Figure 2.2.

Based on the UL Hybrid Automatic Repeat Request (HARQ) result received on the HS-PDSCH, the BS shall activate the HARQ procedure, as follows. It transmits a new DL packet, if the UL ACK message was received and it performs a DL packet retransmission when the UL NACK message is received. When a DL packet retransmission was requested, two different HARQ techniques may be activated. The first entails the identical DL retransmission of the corrupted message, where the same systematic and parity bits are retransmitted in their entirety. At the UE, the received soft decision bits of the previous and current replicas of the message shall be added together. This technique is referred to as Chase [181] or soft-combining. Naturally, this technique improves the reliability of the soft decision based bit values input to the turbo decoder and hence increases the probability of successful decoding.

The second HARQ technique invokes a more intelligent, non-identical retransmission technique, where incremental or additional redundancy is transmitted. This may take the form of initially transmitting only a low number of parity bits of a high-rate punctured code. Then, in the event of receiving corrupted bits and hence failing the CRC test, the previously unsent bits of a punctured mother code will be transmitted. Depending on the availability of unused DL payload transmission capability, the same original systematic information bits may be transmitted as well. At the UE, the original systematic information bits are soft-combined with those received during the previous transmission and the extra parity bits are used along with the parity bits of the previous transmission. Given the extra redundancy, the turbo decoder would have a better chance of successfully decoding the error-infested packet. This technique requires extra memory space at the UE to store the extra parity bits received during the retransmissions. As shown in the last column of Table 2.2, different UE categories have different amounts of memory space for incremental redundancy decoding and Node B must take the UE's memory into consideration when performing DL packet retransmissions using this technique.

In the DL Dedicated Physical Channel (DPCH) of release 99 of the 3GPP standard the channel capacity resource allocation remains fixed, once it has been set up. The bandwidth assigned is occupied by a UE, even if it is not transmitting at the maximum allocated bit rate. Changing the maximum allocated bit rate involves tearing down the existing DPCH and setting up a new DPCH with the new bit rate and this takes time. In contrast to the Release 99 DL DPCH, the HSDPA system has the beneficial feature of dynamic resource allocation, which enables fast packet scheduling. In a typical scenario, Node B might be serving several HSDPA enabled UEs. Since the UEs are independent of each other, their reported CQI value profiles are different and their maximum tolerable bit rates may also be different, depending on the services they support. By exploiting this information, Node B might allocate the entire DL bandwidth required for transmitting 15 HS-PDSCH spreading codes to a particular UE, when its reported DL CQI value is high, provided that the corresponding DL transmit buffer is not empty, i.e. there are a lot of data packets to be delivered to the UE. When the DL CQI value reported by the UE to the BS is low or the DL data packet buffer has been "flushed", Discontinuous Transmission (DTX) is activated for the UE. The benefit of this DTX mode is that the carrier is disabled during the instances of having an empty DL buffer, which mitigates the interference imposed. When no data is found in the transmit buffer, the associated bandwidth is reallocated to other UEs. The prompt scheduling of packets for DL transmission is facilitated by having a short Transmission Time Interval (TTI) of only three slot durations, corresponding to 2 ms which results in five subframes in Figure 2.3. The 3GPP radio frame of Figure 2.3 has 15 timeslots, which span over a frame duration of 10 ms.

2.2.1 Physical Layer

As discussed in the previous section and portrayed in Figure 2.1, the HSDPA uses the HS-PDSCH and HS-SCCH physical channels in the DL and HS-DPCCH in the UL. Again, Figure 2.3 shows the timing diagram of the HSDPA physical channels with respect to the Primary Common Control Physical Channel (P-CCPCH), which is the DL broadcast channel

P-CCPCH   : Primary Common Control Physical Channel
HS-SCCH   : High Speed Shared Control Channel
HS-PDSCH : High Speed Physical Downlink Shared Channel
HS-DPCCH : High Speed Dedicated Physical Control Channel

Figure 2.3: Timing diagram of the HSDPA physical channels.

of Node B. It carries information such as the System Frame Number (SFN) for the UE to synchronize with Node B. The propagation delay between Node B and the UE is assumed to be insignificant, which implies having small cell sizes since they may be expected to have sufficiently better propagation conditions for supporting a high bit rate. As an additional measure, the classic timing advance control technique known from the Global System of Mobile communications known as GSM may be invoked, which allows the BS to measure the turn-around delay of its transmitted signal, when it prompts the MS for a turn-around delay measurement response. The BS may then advance its transmissions by the estimated DL propagation delay, so that its transmitted signal arrives at the MS within a significantly shorter window, thereby reducing the guard-time-related wastage of active information transmission time.

It can be seen from Figure 2.3 that the role of the five subframes is that of creating 2 ms duration transmission frames for the sake of supporting low-delay delivery of delay-sensitive information. The data from Layer 2 is split into blocks of various sizes, such as those shown in Table 2.1. Each block of data is then passed through the coding chain shown in Figure 2.4. The coded block is then mapped onto one of the HARQ processes, for example HARQ process 0 of Figure 2.3. The classic Stop And Wait (SAW) method is employed, where no further transmission takes place during the HARQ process 0 until the corresponding ACK or NACK control channel message is received from the intended UE. However, the Node B is still allowed to transmit different block of coded data on other HARQ processes. The total number of HARQ processes was chosen to be eight. This allows sufficient processing time in between the consecutive HARQ action of the same process number for both the UE and Node B.

Figure 2.3 also shows that the required control information bearing HS-SCCH's message is transmitted two timeslots before the data-bearing HS-PDSCH. This gives sufficient time for the UE to decode the necessary control information and to configure itself appropriately in preparation for decoding the HS-PDSCH message, namely the data. The success or failure of the HS-PDSCH decoding operation results in the UL transmission of the ACK or NACK message using the HS-DPCCH 5 ms after the UE has received the HS-PDSCH subframe, as seen in Figure 2.3. On the other hand, Node B has only 1.3 ms to determine, whether its next DL transmission will have to be a new transmission packet or a retransmitted packet.

2.2.1.1 High Speed Physical Downlink Shared Channel (HS-PDSCH)

The HS-PDSCH has a subframe structure, where the transmit time interval used for fast packet scheduling is constituted by three slots corresponding to 2 ms, as shown in Figure 2.3. In release 99 of the 3GPP standard the Dedicated Physical Data Channel (DPDCH) may opt for using various Spreading Factors (SF), such as $SF = 4, 8, 16, \ldots, 256$ and 512, depending on the slot format of the transmission. Naturally, increasing the SF improves the spreading gain, but proportionally reduces the maximum data rate. In contrast, the SF of HS-PDSCH is fixed to $SF = 16$ to avoid signaling and hence simplifying the receiver algorithm. Adaptive modulation and multicodes utilization in HS-PDSCH have already enabled various transmission rates. In adaptive modulation, there are two possible modulation schemes, namely QPSK and 16QAM [182]. For multicodes utilization, only the channelization codes 1 to 15 are used because the channelization code 0 is reserved for the transmission of other messages, such as the CPICH message used by the UE for SIR measurements and channel estimation. The 15 other channelization codes can be allocated dynamically to any UE, resulting in the transmission of different number of codes, as shown in Table 2.1 for supporting different bit rates.

The transport channel referred to as the High Speed Downlink Shared Channel (HS-DSCH) carries the HSDPA data packets and it is mapped to the physical channel HS-PDSCH with the aid of a single or multiple spreading codes. Its coding chain, which was extracted from [183] is shown in Figure 2.4. When the UE activates its continuous, i.e. non-DTX-style 2 ms TTI transmission mode mapped to the subframes of Figure 2.3, there is always a HS-DSCH transport block having a variable size ranging from 137 to 27 952 bits [179], as shown in the second column of Table 2.1. During the first stage of the channel coding process, the CRC bits are attached to the transport block. The CRC-protected transport block is then scrambled. Since a rate-$1/3$ turbo code is employed, the scrambled transport block is segmented into a number of transmission blocks of the same size, where the maximum size is 5114 bits. The turbo interleaver's memory does not extend beyond this block size.

There are two rate-matching stages in the physical layer's HARQ functionality. The task of the first rate matching stage is to match the number of turbo encoded bits to be transmitted in the DL to the memory of the UE, which is determined by its capability or class, as defined in Table 2.2. If the UE's available memory is higher than the number of turbo encoded bits, the first rate-matching stage should be transparent. The second rate-matching stage matches the number of output bits generated by the first stage to the available physical channel capacity. The DL physical channel capacity depends on both the number of HS-PDSCH spreading codes and on the modulation scheme activated, which in turn depend on the bit rate requirement and the SIR experienced. In addition to rate matching, the second stage of the HARQ functionality seen in Figure 2.3 additionally allows the selection of both systematic and parity bits during both the first transmission attempt as well as during the subsequent retransmissions. The DL transmission of the systematic and parity bits is controlled with the aid of the so-called Redundancy Version (RV) parameters. This enables the UE to perform Chase combining or incremental redundancy decoding, as briefly described earlier.

Figure 2.4: Coding chain for HS-DSCH [183].

Following the action of the HARQ functionality, the rate-matched bits are segmented into the required number of bit-sequences, which are then mapped to the HS-PDSCH spreading codes, as shown in Figure 2.4. The appropriately segmented bits are channel-interleaved over $960 \times n$ bits for QPSK and $1920 \times n$ bits for 16QAM, where n is the number of multicodes. Then the channel interleaved bits are re-arranged for 16QAM modulation. There are four possible re-arrangements which involve bits swapping and logical values inversion. In the case of QPSK, this functionality is transparent. Finally, as portrayed in Figure 2.4, they are then mapped to the HS-PDSCH physical channel for transmission.

2.2.1.2   High Speed Shared Control Channel (HS-SCCH)

As seen in Figure 2.3, the control information associated with the HS-PDSCH message is carried by the HS-SCCH. Similarly to the HS-PDSCH message, it is transmitted in a subframe, as shown in Figure 2.3. The spreading factor is fixed to SF $= 128$. Figure 2.5 shows the coding chain for the HS-SCCH [183]. Observe from Figure 2.5 that the HS-SCCH consists of two parts, which are encoded by two separate coding chains. This enables the UE to decode the first part without receiving the second part. Therefore the first part, which is constituted by the first slot of the three-slot, 2 ms subframe of Figure 2.3, contains essential information required for configuring the UE, which are as follows:

- first HS-PDSCH channelization code number and the number of parallel codes, x_{ccs} (7 bits);

- modulation scheme, x_{ms} (1 bit).

The remaining two slots of the subframe constitute the second part, which contains the following information:

- transport block size, x_{tbs} (6 bits);

- HARQ process number, x_{hap} (3 bits);

- redundancy version parameters and 16QAM constellation re-arrangement, x_{rv} (3 bits);

- HARQ new data packet transmission indicator, x_{nd} (1 bit).

Both of the aforementioned parts are encoded with the aid of two separate coding chains, as shown in Figure 2.5. The first part constituted by the 8-bit x_{ccs} and x_{ms} messages mapped to the first slot, is rate-1/3 convolutional encoded and rate-matched in order to fit into a slot of 40 bits. The rate-matched bits are then masked with the UE identity of the intended UE. This is because the UE can be configured to monitor up to four HS-SCCHs and masking with the target UE's identity x_{ue} assists the UE to detect the first part. In Section 2.4.1 a range of different UE identity detection algorithms will be explored. In order to assist the UE in correctly identifying the HS-SCCH that is intended for itself, the CRC bits are attached to the second part, before they are masked with the UE's identity. Similarly to the first part, the CRC-protected and masked second part of Figure 2.5 is then also rate-1/3 convolutional encoded. The resultant encoded bits are then rate-matched in order to fit them into two slots with a total of 80 bits. Finally, both parts are mapped to the physical channel, ready for transmission, as seen in Figure 2.5.

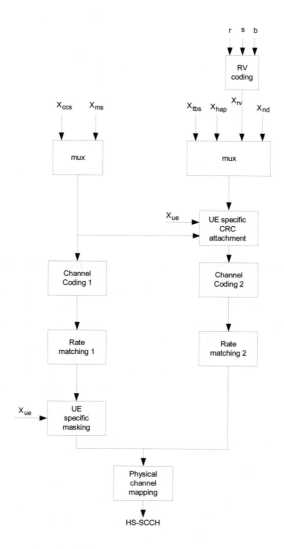

Figure 2.5: Coding chain for HS-SCCH [183].

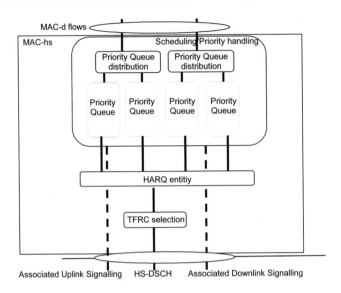

Figure 2.6: MAC-hs details at Node B [179].

2.2.1.3 High Speed Dedicated Physical Control Channel (HS-DPCCH)

The HARQ result of decoding success or failure indicated by the ACK or NACK messages, as well as the channel-quality related information CQI are transmitted with the aid of the HS-DPCCH in a three-slot subframe, as shown in Figure 2.3. The spreading factor is fixed to $SF = 256$. Similarly to the HS-SCCH, the HS-DPCCH message is also constituted by two parts. The single-bit HARQ result of ACK/NACK is repetitively encoded and mapped to the first slot, which constitutes the first part of the three-slot HS-DPCCH message. The CQI value is protected with the aid of a short block code and the coded bits are mapped to the last two slots of the HS-DPCCH subframe of Figure 2.3.

2.2.2 Medium Access Control (MAC) Layer

In order to support the HSDPA mode, the so-called high-speed functionality MAC-hs was added to the existing MAC layer specifications of both the UE and BS side. This functionality is responsible for handling the data transmitted on the transport channel HS-DSCH, which is then mapped to the physical channel HS-PDSCH. The MAC-hs functionality also manages the physical resources allocated to HS-DSCH. Figure 2.6 was adopted from the 3GPP standard [179] and it portrays some of the MAC-hs functionality details at Node B.

To elaborate a little further in the context of Figure 2.6, there are four MAC-hs functional entities. The *flow control* is responsible for the data flow between the MAC-hs and other entities within the MAC layer. Naturally, it is beneficial to limit the associated MAC layer signaling latency as well as to reduce the amount of discarded and retransmitted data, which may be imposed as a result of HS-DSCH congestion. The *scheduling/priority handling* seen in Figure 2.6 checks the HARQ decision of ACK/NACK received in the HS-DPCCH message and determines whether a new transmission or a retransmission action should be authorized.

E-DPCCH : E-DCH Dedicated Physical Control Channel
E-DPDCH : E-DCH Dedicated Physical Data Channel
E-HICH   : E-DCH Hybrid ARQ Indicator Channel
E-RGCH   : E-DCH Relative Grant Channel
E-AGCH   : E-DCH Absolute Grant Channel

Figure 2.7: HSUPA physical channels.

Furthermore, the scheduling/priority handling ensures that the new data packets are transmitted in accordance with their priority class. The role of the *HARQ functionality* seen in Figure 2.6 has already been made plausible and, finally, the *Transport Format and Resource Combination (TFRC)* functionality selects the appropriate transport block size as well as the number of spreading codes used for conveying the data to be transmitted on the HS-DSCH. The corresponding inverse operations of the MAC-hs are carried out at the UE side.

2.3   High Speed Uplink Packet Access

When the standardization of the HSUPA mode commenced, the plausible choice of techniques to be used encompassed those, which became well-established during the HSDPA standardization. However, not all HSDPA techniques described in Section 2.2 were adopted. The specific techniques used in HSUPA are as follows:

- multiple spreading codes and variable rate channel coding;

- HARQ;

- fast packet scheduling.

Figure 2.8: HSUPA time diagram.

Adaptive modulation is not employed in the HSUPA mode, because the employment of high-order modulation schemes, such as 16QAM, increases the crest factor [184] of the UL signal, which would require a high-linearity class-A or linearized class-AB amplifier [182]. These high-linearity amplifiers typically have a low power efficiency and hence would adversely affect the power consumption of the UE. In addition, they typically impose an increased RF amplifier complexity, hence adding to the cost of the UE. Finally, high-order modulation schemes typically require an increased energy per bit and this drains the battery more rapidly.

Table 2.3: HSUPA enabled UE categories (see [180]).

UE category	Maximum number of multicodes	Minimum spreading factor	TTI support (ms)	Maximum transport block size (Throughput in Mbps) 10 ms	2 ms
1	1	SF4	10	7110 (0.711)	N/A
2	2	SF4	2, 10	14484 (1.448)	2798 (1.399)
3	2	SF4	10	14484 (1.448)	N/A
4	2	SF2	2, 10	20000 (2.000)	5772 (2.886)
5	2	SF2	10	20000 (2.000)	N/A
6	4	SF2	2, 10	20000 (2.000)	11484 (5.742)

Let us now detail further how the standardized HSUPA techniques may be used to improve the achievable UL throughput by considering a real-life scenario. When a HSUPA enabled UE is switched on, as shown in Figures 2.7 and 2.8, it registers itself with the network. When it is granted access to the network, it is informed of the maximum total power that it can transmit in the UL. When the UE starts an application, such as file uploading, which requires a high UL bandwidth, the protocol initiates HSUPA transmission. As shown in Figure 2.8, when the UE is configured for HSUPA transmission, the UE is also granted a certain maximum transmit power budget by the serving BS for HSUPA transmission. This granted power may be lower than or equal to the total transmit power limit of the UE. Given the amount of energy required to transmit a data bit, the employment of a larger transport block size would inherently require a higher total transmit power. Therefore, at the beginning of each HSUPA transmission, the UE determines the maximum transport block size that it can transmit in the UL E-DCH Dedicated Physical Data Channel (E-DPDCH) without exceeding its given "grant" and the maximum total transmit power, as shown in Figure 2.8.

The transport block size ranges from 18 bits to 20 000 bits [179]. In order to accommodate this wide-ranging transport block size variation, the number of E-DPDCH spreading codes employed varies from 1 to 4. In addition, various channel coding rates are employed. Unlike HSDPA HS-PDSCH that employs adaptive modulation and fixed spreading factor of 16, E-DPDCH employs the fixed modulation scheme of BPSK, but the spreading factor varies from $SF = 2$ to 256. Table 2.3 shows that there are six different HSUPA enabled UE categories, as listed in [180]. Note that for HSUPA there are two possible TTIs having a duration of either 2 or 10 ms, whereas HSDPA always employs 2 ms TTIs. Recall that the HSDPA transmission frame structure was shown in Figure 2.3. Each TTI has its corresponding maximum transport block size.

The error-free reception of the control information, such as the transport block size, is vital for the successful decoding of the E-DPDCH message by Node B. The related control information is carried by the E-DCH Dedicated Physical Control Channel (E-DPCCH), as shown in Figures 2.7 and 2.8. Based on the CRC result of the decoded E-DPDCH, Node B will inform the UE whether the packet has been received successfully. The HARQ ACK or NACK message is then carried in the DL E-DCH Hybrid ARQ Indicator Channel (E-HICH). Similarly to HSDPA, the UE carries out the required HARQ procedures and activates either the transmission of a new packet or the appropriately configured retransmission of the previous one based on the E-HICH results received from Node B. At Node B, Chase combining or incremental redundancy decoding can then be performed in the interest of reducing the associated BER. As seen in Figure 2.7, the neighboring or non-serving BSs might be able to correctly decode the E-DPDCH message and hence may generate the ACK message in the E-HICH as well. The UE will then perform a new transmission, even though the serving BS transmits NACK. This is due to the fact that an ACK message is received from the non-serving Node B.

Recall that in HSDPA, fast packet scheduling is achieved by having a shared HS-PDSCH, which conveys the data of all UEs. This may be readily achieved, because DL data transmission takes place from a single Node B to many UEs. In contrast, when a single UL physical channel is shared by many UEs, there is a lot of signaling involved in synchronizing the data transmission of many UEs to a single Node B. Therefore, each UE has to have its own dedicated physical channel E-DPDCH. However, this poses a problem for Node B, if all UEs are transmitting their E-DPDCH at the peak rate, because the interference level experienced at the receiver of Node B will be excessive for the data to be decoded correctly. The situation becomes even worse when the power control algorithm starts to increase the transmit power of a particular UE in the interest of maintaining its target BER. This transmit power-surge may inflict increased interference upon the other UEs and hence the power control algorithm has to increase the transmit power of the other UEs. This may result in an avalanche effect, where all UEs may eventually transmit at their full power and yet no data can be successfully decoded at the BS.

As shown in Figure 2.8, in order to avoid the above-mentioned avalanche-like power-boosting problem at Node B and achieve prompt scheduling of packets for their transmission, Node B should limit the maximum transmit power of the UEs by issuing an "absolute grant" or power assignment to a single UE or a group of UEs. This absolute grant power level is the same as that issued when the UE is switched on and it also limits the maximum transport block size of E-DPDCH messages, which ultimately limits the transmit power of the UE. The absolute grant message is carried by the E-DCH Absolute Grant Channel (E-AGCH), as shown in Figures 2.7 and 2.8. Given the current grant, the UE shall generate a feedback to Node B reporting for example its buffer occupancy status and the estimated time required to clear the buffer, etc. Based on the feedback provided by the UE, Node B might issue another absolute grant to the UE in order to increase or decrease its current grant. Alternatively, Node B might issue a "relative grant" command to the UE in order to increase or decrease its current grant. The relative grant command is carried on E-DCH Relative Grant Channel (E-RGCH), as shown in Figures 2.7 and 2.8.

Occasionally, the UE's transmit power assigned to the E-DPDCH message might impose excessive interference on the active links in the neighboring cell. Therefore, as shown in Figure 2.7 the neighboring or non-serving Node B might issue a power-down command in E-RGCH to the UE in order to decrease its grant. Note that the non-serving Node B is not allowed to increase the UE's current grant.

2.3.1 Physical Layer

As described in the previous section and in Figure 2.7, the HSUPA system employs the E-AGCH, E-RGCH and E-HICH in the DL. In contrast, in the UL it employs the E-DPDPCH and E-DPCCH. Figure 2.9 shows the timing diagram of the 2 ms subframe-duration HSUPA

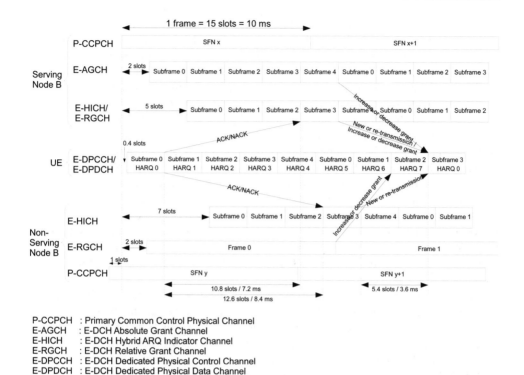

P-CCPCH   : Primary Common Control Physical Channel
E-AGCH   : E-DCH Absolute Grant Channel
E-HICH   : E-DCH Hybrid ARQ Indicator Channel
E-RGCH   : E-DCH Relative Grant Channel
E-DPCCH   : E-DCH Dedicated Physical Control Channel
E-DPDCH   : E-DCH Dedicated Physical Data Channel

Figure 2.9: Timing diagram of 2 ms TTI HSUPA physical channels.

physical channels with respect to the broadcast channel P-CCPCH of the serving Node B. The timing diagram of the P-CCPCH and HSUPA DL physical channels of the non-serving Node B is also shown in Figure 2.9. Note that the timing difference between the broadcast channel P-CCPCH of the serving and non-serving BSs is chosen to be one slot-duration. This limits the maximum delay imposed on the non-serving Node B, within which it has to provide the E-HICH result. For the UE, there is a minimum delay of 0.4 slots [185] between the serving BS's P-CCPCH message and the commencement of the E-DPCCH and E-DPDCH transmission. The offset of the E-AGCH from the serving Node B is fixed to 2 slots [185] with respect to P-CCPCH. In contrast, the delay of E-HICH and E-RGCH with respect to P-CCPCH varies depending on the UE's E-DPCCH delay. In this example, the delay is calculated to be 5 slots [185], as shown in Figure 2.9. No DL E-AGCH message is sent from the non-serving Node B, but there is an E-RGCH DL message, which has a fixed delay of 2 slots with respect to the serving Node B's P-CCPCH. Using the same calculation for the E-HICH DL transmission from the serving Node B, the delay of E-HICH with respect to the serving Node B's P-CCPCH is found to be 7 slots, as shown in Figure 2.9. Again, the propagation delay from the serving and non-serving Node Bs to the UE is assumed to be insignificant.

Figure 2.9 indicates that the control information bearing E-DPCCH message is transmitted at the same time as the data packets of the E-DPDCH. This regime is different to that of the HSDPA solution, where the HS-SCCH messages are transmitted two slots in advance of the HS-PDSCH messages in order to allow sufficient time for the configuration of the adaptive modulation modes, for example. In contrast, no adaptive modulation is employed in the HSUPA mode and, hence, the simultaneous reception of the E-DPCCH and E-DPDCH messages does not result in any implementational problems. Moreover, there is no need for the Node B to monitor more than one E-DPCCH. This philosophy is different from that of HSDPA, where the UE has to monitor up to four HS-SCCHs. As shown in Figure 2.9, there are a maximum of eight HARQ processes for the HSUPA E-DPDCH messages in conjunction with a 2 ms TTI, which is similar to the corresponding transmission regime of HSDPA. However, the number of HARQ processes is reduced to four for the 10 ms duration TTI. This is because it has a significantly longer TTI and, hence, there is sufficient processing time in between the same consecutive HARQ process number for both the UE and Node B. Again, the SAW method is employed in HSUPA, where the UE sends a packet on a specific HARQ process and it waits for the ACK or NACK message, before there is any further transmission on the same HARQ process.

Observe in Figure 2.9 that the HARQ process 0 is highlighted, in order to show the HARQ SAW procedure. After the E-DPDCH message is received, the serving Node B has about 7.2 ms before sending the ACK or NACK feedback in the E-HICH message. Since there is a single slot difference between the transmission of the P-CCPCH message of serving and non-serving Node B, the non-serving Node B has a longer time, about 8.4 ms, before sending its E-HICH message to the UE. The E-HICH message received from both the serving and non-serving Node B must be decoded by the UE. The slightly higher 8.4 ms delay of the E-HICH message received from the non-serving Node B inherently reduces the UE's HARQ processing time before the UE transmits a new packet or retransmits the E-DPDCH message in the UL. In the example shown in Figure 2.9, the UE has only about 3.6 ms before the next transmission takes place on the same HARQ process.

In addition to the DL E-HICH message received from the serving and non-serving Node Bs, there are E-AGCH and E-RGCH messages to be decoded by the UE. As shown in Figure 2.8, these messages affect the grant and, hence, the transmit power. As for the example shown in Figure 2.9, the E-AGCH message of subframe 4 and the E-RGCH message of subframe 3 will limit the transmit power of E-DPDCH transmission of the HARQ process 0 in subframe 3 and all of its subsequent HARQ processes. Note that the power limitation is only applicable when there is a new E-DPDCH transmission in the HARQ process. There is no E-AGCH transmission from the non-serving Node B, but there is E-RGCH transmission in the 10 ms TTI. Owing to the timing of the non-serving Node B and the E-RGCH transmission during a TTI of 10 ms, the E-RGCH command is applied earlier to HARQ process 7 in subframe 2 and all of its subsequent HARQ processes.

2.3.1.1 E-DCH Dedicated Physical Data Channel (E-DPDCH)

The UL E-DPDCH is quite similar to the Release 99 UL DPDCH. They both use BPSK modulation. They are dedicated to a specific UE in order to avoid that rather complex scheduling of messages which would be required by a shared UL channel owing to different distances of the UEs from the Node B. They avoid the employment of high-order modulation

Figure 2.10: Coding chain for E-DCH [183].

schemes for the sake of employing high-efficiency class-C power amplifiers by the UE. They use multiple spreading codes and variable spreading factors for supporting high data rates. The spreading factors of the DPDCH messages are SF $= 4, 8, \ldots, 128$ and 256 indicating that the ratio of the maximum and minimum rate is as high as 64. The E-DPDCH has the same set of spreading factors, with the addition of the extra spreading factor of SF $= 2$. In order to support high data rates, the E-DPDCH messages use the spreading factor combinations arranged in the order of increasing bit rates, such as the rates associated with SF $= 256, 128, \ldots, 2 \times$ SF4, $2 \times$ SF2 and $2 \times$ SF4 $+ 2 \times$ SF2.

The transport channel referred to as the Enhanced Dedicated Channel (E-DCH) carries the HSUPA data packets and it is mapped to the physical channel E-DPDCH using a single code or multiple codes associated with different spreading factors. Its coding chain was extracted from [183] and it is shown in Figure 2.10. Compared with the HS-DSCH coding chain of Figure 2.4, it can be seen that the E-DPDCH coding chain is a simplified version of the HS-DSCH regime. For each continuous mode, i.e. non-DTX 2 or 10 ms TTI regime, there is always an E-DCH transport block having a packet size spanning from 18 to 20 000 bits

[179]. Similarly to the HS-DSCH coding chain, CRC bits are used for detecting any potential errors in the transport block at the first stage of the coding process of Figure 2.10. The CRC-protected transport block is segmented into identical-length data blocks having a maximum size of 5114 bits and the resultant blocks are then encoded using a rate-$1/3$ turbo code. Recall from Section 2.2.1.1 that the first stage of the HSDPA HS-DSCH's HARQ functionality is to match the number of turbo encoded bits about to be transmitted to the available memory of the UE. In contrast, in HSUPA no restrictions are imposed on the BS receiver's memory space and, hence, the first-stage rate-matching process of HSDPA portrayed in Figure 2.4 is not needed in the HSUPA E-DCH's HARQ functionality. Therefore, only a single rate-matching stage is required in the E-DCH physical layer's HARQ functionality of Figure 2.10, which matches the number of turbo encoded bits to the available physical channel capacity. The physical channel's capacity depends on the number of E-DPDCH spreading codes used as well as on the specific choice of the spreading factor.

In addition to rate matching, the Redundancy Version parameters RV previously introduced in the context of the HSDPA mode are also used in the HSUPA mode to control the selection of systematic and parity bits during both the first transmission of a packet as well as in subsequent retransmission attempts. This enables the BS to perform Chase combining or incremental redundancy decoding. Following the action of the HARQ functionality, the rate-matched bits are segmented into the required number of sequences for further mapping to the corresponding number of E-DPDCH spreading codes as shown in Figure 2.10. The segmented bits of Figure 2.10 are interleaved and mapped to the E-DPDCH, ready for UL transmission.

2.3.1.2 E-DCH Dedicated Physical Control Channel (E-DPCCH)

The control information associated with the E-DPDCH is carried in the UL E-DPCCH. Its spreading factor is fixed to SF $= 256$ which ensures a high spreading gain and, hence, a high integrity. The information contained in E-DPCCH is as follows:

- Retransmission Sequence Number (RSN), X_{rsn} (2 bits);

- E-DCH Transport Format Combination Indicator (E-TFCI), which represents the transport block size used, X_{tfci} (7 bits);

- "Happy" bit which indicates whether the UE is satisfied with the current transmission rate, X_h (1 bit).

As shown in Figure 2.11, the resultant $2+7+1 = 10$ information bits are first multiplexed and then they are Reed–Muller channel encoded. After channel encoding, they are mapped to the physical channel E-DPCCH and then they are ready for transmission.

2.3.1.3 EDCH HARQ Indicator Channel (E-HICH)

The HARQ result ACK or NACK (1 bit) is carried in the DL E-HICH and its spreading factor is fixed to SF $= 128$. In order to reduce spreading code tree usage, 40-bit long orthogonal signature sequence [185] is used to multiplex 40 different 1-bit HARQ result bits in 1 slot on a single spreading factor 128-code channel. Therefore, one single spreading factor 128-code channel is capable of supporting up to 40 users' E-HICH. Different signature sequences are

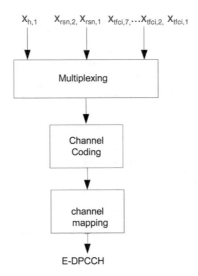

$X_{h,1}$ $\quad$ $X_{rsn,2}, X_{rsn,1}$ $\quad$ $X_{tfci,7}, \cdots X_{tfci,2}, X_{tfci,1}$

Multiplexing

Channel Coding

channel mapping

E-DPCCH

Figure 2.11: Coding chain for E-DPCCH [183].

used for each of the 3 slots by following a hopping pattern [185], but the signature sequence pattern repeats itself after 3 slots.

2.3.1.4 E-DCH Absolute Grant Channel (E-AGCH)

The spreading factor of the DL E-AGCH is fixed to SF = 256 and it carries the following information:

- absolute grant value index (5 bits);

- absolute grant scope, which indicates whether the absolute grant value is applied to a single or to all HARQ processes (1 bit).

The absolute grant value index ranges from 1 to 31 and represents the 31 different E-DPDCH transmit power levels. The absolute grant value index of 0 is to activate or de-activate the HARQ process. Both the absolute grant value index and the corresponding scope are multiplexed and are protected with the aid of CRC bits, which facilitate the detection of transmission errors. The CRC-protected bits are then masked with the UE identity and rate-1/3 convolutional encoded. The encoded bits are then rate matched and mapped to the physical channel E-AGCH, again, ready for transmission.

2.3.1.5 E-DCH Relative Grant Channel (E-RGCH)

The relative grant value (1 bit) is carried in the DL E-RGCH message and its spreading factor is fixed to SF = 128. Similarly to the E-HICH, the relative grant bit is then mapped to a 40-bit signature sequence and it follows the same hopping patterns as well.

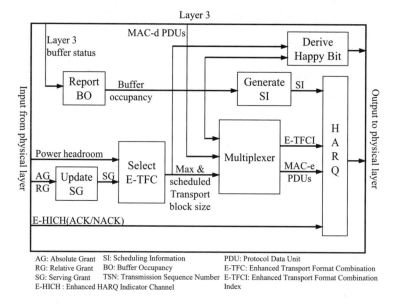

Figure 2.12: MAC-es/e block diagram at the UE side. The E-TFCI and MAC-e PDUs are generated according to Figure 2.13 and the output of the HARQ block becomes the input of the coding chain seen in Figure 2.10.

2.3.2 MAC Layer

In order to support the HSUPA mode, the so-called MAC-es/e functionality was added to the existing MAC layer of both the UE and the BS. The MAC-es/e functionality is more complicated than the MAC-hs functionality of HSPDA, because the packet scheduling control functionality of the E-DPDCH messages is split between Node B and the UE. Moreover, the scheduling control has to coordinate the transmissions of numerous UEs to a single Node B. Figure 2.12 shows the block diagram of the MAC-es/e functionality at the UE. The left-hand side of the diagram portrays the control information provided by the physical layer for the MAC-es/e functionality. The HARQ result ACK or NACK is passed to the HARQ block. If the NACK message is received, the HARQ block increments the Retransmission Sequence Number (RSN) and invokes a retransmission. In contrast, if ACK is received or the maximum affordable number of retransmissions has been exhausted, the HARQ block initiates the transmission of a new packet.

As shown in Figure 2.12, the Update Serving Grant (SG) block updates the serving grant based on the Absolute Grant (AG) and Relative Grant (SG) received from the physical layer. The serving grant represents the maximum transport block size that is allocated for a scheduled packet's transmission. The updated serving grant is then passed to the Select E-DCH Transport Format Combination (E-TFC) block of Figure 2.12.

Before any HSUPA transmission commences, Node B provides the UE with a set of reference E-TFC indices and their corresponding transmission power. These parameters serve as a reference for the Select E-TFC block of Figure 2.12 to calculate the required transmission power for each of the various transport block sizes, so that Node B becomes capable of successfully decoding them. The Select E-TFC block compares the updated serving grant to the pre-calculated required transmission power for all available transport block sizes. The maximum transport block size is specifically chosen to ensure that its required transmission power does not exceed the serving grant. The selected transport block size is referred to as the scheduled transport block size. In addition to determining the scheduled transport block size, the Select E-TFC block also selects the maximum transport block size, which limits the length of both scheduled and non-scheduled transmissions. The difference between scheduled and non-scheduled transmissions will be explained later. The maximum transport block size is determined by the power headroom which is calculated within the physical layer and forwarded to the MAC layer, as shown in Figure 2.12. The power headroom is the difference between the UE's maximum allowed transmit power and the sum of the power allocated to all of its active physical channels, except for the E-DPDCH.

The Report Buffer Occupancy (BO) block of Figure 2.12 keeps track of the buffer occupancy of the ISO Layer 3, in order to avoid encountering a MAC buffer overflow. Based on the information provided by the Report BO and the Select E-TFC blocks, the Multiplexer block seen in Figure 2.12 extracts the so-called MAC-d Protocol Data Units (PDUs)[2] from Layer 3 and constructs the so-called MAC-e PDUs. The length of the MAC-e PDUs constructed is rounded up to the nearest transport block size, which is quantified and signaled in terms of the corresponding Enhanced Transport Format Combination Index (E-TFCI) parameter forwarded to the physical layer, as seen in Figure 2.12. The resultant MAC-e PDUs and the E-TFCI are then loaded into the buffer of the HARQ block.

The buffer occupancy message of the Report BO block is also passed to the Generate Scheduling Information (SI) and Derive Happy Bit block of Figure 2.12. The associated scheduling information contains a detailed report of the total MAC-es/e buffer occupancy, the power headroom available for the UE and the ID of the highest priority UL channel associated with a non-zero buffer occupancy. In contrast to the Generate SI block, the Derive Happy bit block of Figure 2.12 generates the Happy bit, indicating whether the UE is satisfied with the current UL transmission rate and whether the current total MAC-es/e buffer status can be reset to its default value within a certain time limit. Note in Figure 2.12 that the Happy bit is not loaded into the HARQ message and it is regenerated for every transmission attempt, regardless of whether a new transmission or a retransmission has to be authorized. On the other hand, the SI can be retransmitted.

Let us now explore further the inner working of the Multiplex block of the MAC-es/e functionality by using the example shown in Figure 2.13. In this example, there are five logical channels, which have the logical channel IDs 1, 2, 3, 4 and 5, respectively.[3] As can be seen in Figure 2.13, these logical channels are grouped into two MAC-d flow types: scheduled and non-scheduled. The scheduled logical channels are used for delay-insensitive non-real-time applications, such as file uploading etc. The associated bandwidth is limited by the serving grant, which is controlled by the received absolute and relative grant at every TTI.

---

[2]There are many entities within the MAC layer. When an entity receives data from a higher layer or from another entity within the MAC layer, it adds its header information to the data block. The resultant packet is then referred to as a PDU. In this case, MAC-d receives data from Layer 3 and constructs the MAC-d PDUs. Then, these MAC-d PDUs are used to construct the MAC-e PDUs.

[3]These five logical channels have to be appropriately mapped to the physical channels and are provided for the sake of conveying different type of information in support of flexible lip-synchronized video and audio, for example, while also facilitating high-rate wireless Internet services, such as file upload, web browsing, etc.

Figure 2.13: MAC-e PDU construction at the Multiplexer block of the MAC-es/e. The transport block seen at the bottom of the figure is then loaded into the HARQ block of Figure 2.12.

On the other hand, the non-scheduled channels are invoked for delay-sensitive traffic, such as the control signals of higher layers The associated data rate is determined by the non-scheduled grant, which is configured during the initialization of the HSUPA transmission session and remains constant for the entire duration of the HSUPA connection. In the example shown in Figure 2.13, logical channel ID 1 and 2 are non-scheduled logical channels. The remaining logical channel ID 3, 4 and 5 are scheduled logical channels. Since logical channel ID 1 and 2 carry delay-sensitive signals, they should have a higher priority than the scheduled logical channels. Therefore, in the example shown in Figure 2.13 logical channel ID 1 has the highest priority and logical channel ID 4 has the lowest priority. Since logical channel ID 1 has the highest priority, its buffer content would have the highest priority to become cleared. Therefore, its buffer is empty most of the time. The example seen in Figure 2.13 shows the buffer content of a particular TTI. The logical channel IDs 1, 2, 3, 4 and 5 happen to have the buffer occupancies of 0, 100, 100, 200 and 300 bits, respectively, which have to be mapped to the current transmission packet.

As defined in [179], not all logical channels can be mapped to all HARQ processes. By mapping higher priority channels to a small set of HARQ processes, this prevents the higher priority channels from reserving all of the available bandwidth by ensuring that the lower priority channels can transmit in pre-determined HARQ processes. These mapping restrictions are only applicable to 2 ms TTIs. In 10 ms TTIs, all logical channels are mapped to all HARQ processes. In the example shown in Figure 2.13, 2 ms TTIs are employed and, hence, Node B is configured the UE to ensure that only the specific logical channels having the IDs 1, 3 and 4 are mapped to the HARQ process 0, while all logical channels can be mapped to the HARQ process 1. Let us assume that the current transmission is the HARQ process 1 in Figure 2.13 and that all logical channels are allowed to transmit. Logical channel ID 1 would have the highest priority, if it had some data for transmission, but in this example it has zero buffer occupancy, i.e. it is empty. Therefore, logical channel ID 2 is the highest priority channel having a non-zero buffer occupancy. Lets us assume that the MAC-d flow multiplexing rule of the highest priority channel ID 2 defines that only logical channels ID 3 and 4 are allowed to be multiplexed with it.

This is because logical channel ID 2 has the highest priority and its unique MAC-d flow profile should be used to maintain a certain QoS. Since logical ID 5 requires a higher QoS, it must not be multiplexed with logical channel ID 2. Therefore, logical channel ID 5 is not allowed to transmit in this TTI. On the other hand, logical channels ID 3 and 4 require a similar or lower QoS than logical channel ID 2, hence they are allowed to transmit in this TTI. In short, the mapping and multiplexing rules then limit the current transmission to convey the messages of logical channels ID 2, 3 and 4 only, as seen in Figure 2.13.

It is shown in Figure 2.13 that the "non-scheduled grant" only allows 90 non-scheduled bits to be transmitted. Since only logical channel ID 2 is non-scheduled, the entire 90-bit payload is assigned to logical channel ID 2. In addition, the current "serving grant" seen in Figure 2.13 permits the transmission of 120 scheduled bits and these have to be shared between logical channel ID 3 and 4. Since logical channel ID 3 has the higher priority of the two, the available bandwidth is used to clear all 100 bits in its buffer and the remaining 20 bits are allocated for logical channel ID 4.

At this point, the sum of the non-scheduled 90 bits and the scheduled 120 bits is a total of 210 bits. However, in the example of Figure 2.13 this has exceeded the power headroom available for E-DPDCH transmission. Therefore, another tentative bit mapping process has to be performed for mapping the services to the available capacity as best as possible. Again, priority is given to logical channel ID 2 and this time the UE tentatively assigns 80 bits out of the 90 bits of non-scheduled grant. This is because the minimum number of bits required for the MAC-es PDU and for header construction amounts to 80 bits in Figure 2.13. Inserting another MAC-d PDU into the current MAC-es PDU would require more than 90 bits. Since the power headroom[4] this time only allows 130 bits to be transmitted, logical channel ID 3 is only allowed to transmit 50 bits, even though the serving grant permits it to transmit 100 bits.

If SI is transmitted, SI is inserted at the end of the MAC-e PDU. The MAC-e PDU length is then quantized to that specific E-TFCI, which has a transport block size equal to or larger than the combined length of the MAC-es PDUs and headers plus the SI. If there is extra space, according to Figure 2.13 padding will be added at the end of the block. The transport block is then loaded into the buffer of the HARQ block of Figure 2.12, which then becomes the input block of the E-DCH coding chain in Figure 2.10.

---

[4]The power headroom can be changed by the action of the power control, since the channels' powers are increased or decreased and this will reduce or increase the power headroom accordingly.

2.4   Implementation Issues

In this section we briefly discuss some of the implementation issues involved in designing a HSDPA or/and HSUPA enabled UE.

2.4.1   HS-SCCH Detection Algorithm

It was argued in Section 2.2 that Node B may employ up to 15 HS-PDSCH channelization codes in a particular TTI. These 15 HS-PDSCH channelization codes are shared by numerous UEs within the cell. Not all HS-PDSCH transmissions of a specific TTI are intended for a particular UE. Therefore, there will be no HS-SCCH transmissions and this lack of transmission is used to inform the intended UE that there is a HS-PDSCH transmission destined for it. If every UE within the cell were to be allocated a separate HS-SCCH, this would result in an inefficient exploitation of the spreading code space. In order to be able to appropriately proportion the reserved spreading code space, only a handful of HS-SCCHs are provided, which are shared by all UEs within the cell. Since the allocation of HS-SCCHs is not fixed to a specific UE, the UEs have to monitor up to four HS-SCCHs during each TTI. It was highlighted in Figure 2.5 of Section 2.2.1.2 that a HS-SCCH is constituted by two parts. The first part is masked with the intended UE's identity so that only the intended UE can successfully detect it. However, all of the UEs including the intended UE have to attempt its detection, for example with the aid of the following detection algorithms [186]:

- Viterbi's Path Metric Difference (VPMD) algorithm;

- Yamamoto–Itoh (YI) algorithm [187].

Naturally, the detection algorithms have a limited integrity or reliability and, hence, sometimes more than one HS-SCCH is detected during a TTI. Therefore, a so-called tie-breaking algorithm has to be applied in order to select the most likely HS-SCCH that was intended for a specific UE. In [186], a range of different tie-breaking algorithms have been proposed:

- Minimum Path Metric Difference (MPMD);

- Average Path Metric Difference (APMD);

- Frequency of Path Metric Difference (FPMD);

- Last Path Metric Difference (LPMD).

Both the above-mentioned detection and tie-breaking algorithms are briefly described in the following subsections.

2.4.1.1   Viterbi's Path Metric Difference Algorithm

Figure 2.14 shows a typical example of Viterbi decoding with the all-zero path as the survivor path, shown as dashed lines. Along the winning path, there are three competing paths at three consecutive trellis states, which are drawn as continuous lines. These competing paths are discarded, because their path metrics are smaller than that of the all-zero survivor path. The path metric difference between the surviving path and the discarded path is Δ_k, where $k = 1, 2, 3$.

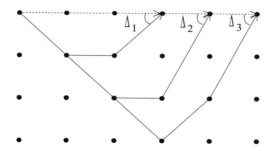

Figure 2.14: Viterbi decoding, where the all-zero path is the survivor path, shown as dashed lines.

In the VPMD algorithm, the path metric difference Δ_3 observed at the last trellis stage is compared with a threshold Δ_{VPMD}. If it is higher than the threshold, the decoding is deemed to be successful, i.e. the decoded HS-SCCH is deemed to be intended for the UE. Otherwise, it is declared as a failure and the decoded HS-SCCH is classified as not intended for the UE. The same decoding process is then applied to all monitored HS-SCCHs. The rationale of this regime is that when the difference of the metric associated with the merging paths is low, our decision is likely to be unreliable.

2.4.1.2 Yamamoto–Itoh Algorithm

The philosophy of the YI algorithm is similar to that of the VPMD algorithm outlined in the previous section, although it is more sophisticated and hence it typically yields a more reliable decoding result. Again, let us refer to Figure 2.14 for the explanation of the YI algorithm. In the VPMD algorithm, only the path metric difference recorded at the last trellis stage is compared with a threshold. In the YI algorithm, all path metric differences, Δ_k, where $k = 1, 2, 3$, are compared with a threshold Δ_{YI}. The number of path metric differences N, where we have $\Delta_k > \Delta_{YI}$ are calculated and are compared to a given parameter N_{YI}. If we have $N > N_{YI}$, the decoding is deemed to be successful, i.e. the decoded HS-SCCH message is deemed to be intended for the UE. In [187], the value of $N_{YI} + 1$ is fixed and it is equal to the number of paths that merge with the survivor path. In the example shown in Figure 2.14, we have $N_{YI} + 1 = 3$.

In certain situations there may be more than one decoded HS-SCCHs being declared as successful, regardless of whether the VPMD or the YI algorithm is employed. As mentioned above, in such scenarios a tie-breaking algorithm is used to determine the most likely HS-SCCH, some of which are highlighted in the next section.

2.4.1.3 Minimum Path Metric Difference Algorithm

In the MPMD algorithm, the minimum path metric difference between the survivor path and the competing paths is retained. Let us briefly refer to Figure 2.14 for highlighting the rationale of this algorithm. The minimum path metric difference is as follows:

$$\Delta_{min} = \min(\Delta_k), \tag{2.1}$$

where we have $k = 1, 2, 3$ in this example. When more than one HS-SCCHs is declared to have been successfully detected, the minimum path metric differences Δ_{min} of all of the detected HS-SCCHs are compared. In general, the larger the minimum path metric difference, the more reliable the decoding process. Therefore, the detected HS-SCCH associated with the largest Δ_{min} value will be selected as the serving HS-SCCH, which carries the HS-PDSCH control information for the UE concerned.

2.4.1.4 Average Path Metric Difference Algorithm

The APMD algorithm calculates the average path metric difference between the surviving path and all of the merging paths. In the example shown in Figure 2.14, the average path metric difference is calculated as follows:

$$\Delta_{average} = \frac{\sum_{k=1}^{T=3} \Delta_k}{T = 3} \tag{2.2}$$

where $T = 3$ is the number of competing paths in this example. When there is more than one HS-SCCH, which are deemed to be successfully detected, the average path metric differences $\Delta_{average}$ of all of the detected HS-SCCHs are compared. The one associated with the largest $\Delta_{average}$ is selected as the winner.

2.4.1.5 Frequency of Path Metric Difference Algorithm

The FPMD algorithm is similar to the YI algorithm. Each path metric difference Δ_k between the surviving path and the competing path is compared with a threshold Δ_{FPMD}. The number of path metric differences, where we have $\Delta_k > \Delta_{FPMD}$, is calculated and this value is referred to as the frequency of path metric difference N_{FPMD}. When there is more one HS-SCCH being declared to be successfully detected, the frequency N_{FPMD} of all of the detected HS-SCCHs is compared. The one associated with the largest N_{FPMD} value will be selected as the winner. Again, this algorithm is similar to the YI algorithm, since the threshold Δ_{FPMD} is the same as the threshold Δ_{YI} in the YI algorithm, while the frequency N_{FPMD} is the same as the parameter N in the YI algorithm.

2.4.1.6 Last Path Metric Difference Algorithm

The LPMD algorithm has a lower complexity compared with the previous three algorithms. In this algorithm, the last path metric difference Δ_{last} between the surviving path and the last competing path is stored. In our example shown in Figure 2.14, the last path metric difference is $\Delta_{last} = \Delta_3$. When more than one HS-SCCH is declared as successfully decoded, the detected HS-SCCH associated with the largest path metric difference recorded at the last trellis state will be selected as the winner.

2.4.1.7 Detection Algorithm Performances

In [186], the simulation results characterizing the VPMD and YI algorithms were presented. It was shown that the YI algorithm outperforms the VPMD algorithm by about 0.5 dB, when there is only one HS-SCCH to be monitored. The performance of the detection algorithm was also investigated, when the power of the four monitored HS-SCCHs was different. It was

Figure 2.15: QPSK and 16QAM constellation points.

found that if the intended HS-SCCH's power is lower than that of the others, the detection performance degrades. However, if the intended HS-SCCH power becomes equal to or higher than that of the others, the detection performance may slightly improve.

As shown in Figure 2.5, UE-specific CRC attachments are used in the second part of the HS-SCCH message. Therefore, if the detection and tie-breaking algorithms result in a falsely detected HS-SCCH, the UE can still check the CRC bits in order to rule out falsely detected HS-SCCHs.

2.4.2   16QAM

In HSDPA, adaptive modulation is employed, where the higher-order modulation scheme used is 16QAM. The achievable bit rate is doubled compared with those of the QPSK modes, when using the same channel codec. However, the price to pay for this increased throughput is a substantially increased complexity. Some of the issues associated with 16QAM are discussed in the following subsections.

2.4.2.1   Amplitude and Phase Estimation

As described in Section 2.2.1.1, turbo coding is used in the HS-DSCH channel coding chain of Figure 2.4. At the UE, the turbo decoder requires log-likelihood ratio calculations at its input. Therefore, this involves demapping of the received symbols to bits. Figure 2.15 shows the constellation points of both QPSK and 16QAM. Two bits per symbol are transmitted in QPSK, one bit is mapped to the in-phase component of the constellation and the corresponding log-likelihood ratio of the bit is estimated purely based on the in-phase component. Similarly, the second bit of QPSK is mapped to the quadrature phase component and the corresponding log-likelihood ratio is calculated.

On the other hand, four bits/symbol are transmitted in 16QAM. The calculation of the log-likelihood ratio of the bits is no longer as straightforward as in QPSK, where the bits' log-likelihood ratio only depends on either the in-phase or on the quadrature-phase component of the constellation. More explicitly, the log-likelihood ratio of the bits in 16QAM depends on both the in-phase and quadrature-phase of the constellation, i.e. on both the amplitude and phase. Therefore, accurate amplitude and phase estimation is required for HSDPA-enabled UEs, in order to separate the constellation points and provide an accurate log-likelihood ratio estimate of the bits.

2.4.2.2 Equalizer

It is specified in the standard [188] that the minimum requirement for Release 99 DPCH's reception is that the DPCH received power level has to be -19 dB with respect to the total received power level. On the other hand, the HSDPA HS-PDSCH using 16QAM requires that the received power level has to be -3 dB with respect to the total received power. In other words, the received signal level requirement of HSDPA HS-PDSCH using 16QAM is 16 dB higher than that of the Release 99 DPCH. If a traditional Rake receiver differing with a power-thirsty channel equalizer is used in a multipath environment, reliable reception is only possible if the second received signal path is at least 20 dB [189] below the first propagation path imposing no significant dispersion and no other source of interference is present. However, it is very unlikely that we encounter such an almost entirely non-dispersive multipath propagation channel in outdoor scenarios. Therefore, typically more advanced receivers are required in order to mitigate the Inter-Symbol Interference (ISI) caused by the multipath propagation channel. In [189, 190], Minimum Mean Square Error (MMSE) and Normalized Mean Squares (NLMS) equalizers were proposed for employment in the HSDPA enabled UE. The simulation results of [190] demonstrate that the MMSE equalizer outperforms the Rake receiver by about 10 dB for the case of no transmit diversity and by about 3 dB, when using closed loop transmit diversity.

There are three main components in the MMSE equalizer, namely channel estimation, equalizer-weight generation and adaptive filtering, mimicking the inverse of the channel's effects. The equalizer weight generation component imposes a high implementational complexity, because it involves the inversion of a sizeable matrix and it requires about 50 000 [190] complex-valued multiplication using the Cholesky decomposition. These operations typically occupy about three slots in an FPGA implementation, when using a clock rate of 16 times the chip rate. Therefore, the equalizer weights used for adaptive filtering of the received signal have to be frequently updated. When the channel is imposing fast fading, this may adversely affect the performance of the equalizer. Although it is acceptable to delay the non-interactive data while carrying out channel equalization, this imposes additional delay on the already tight processing schedule of generating the HARQ result in the UL, as shown in Figure 2.3.

The LMS equalizer is more simple than the above-mentioned equalizers and, hence, it is more popular in low-budget implementations. Although its weights can be evaluated using matrix inversion, this complex operation may be avoided by using an iterative approach [190]. However, the convergence rate of the resultant equalizer is slow and it is therefore only effective in slow-fading channels.

2.4.3 HARQ Result Processing Time

As seen in Figure 2.3, only 5 ms is available to decode the received HSDPA DL HS-PDSCH message, before generating a HARQ result in the UL. This imposes a very tight constraint on designing a HSDPA-enabled UE, since a lot of intensive processing has to be completed within this period. The received radio-frequency HS-PDSCH's symbols are sampled and down-converted to the baseband, where the initial processing is carried out at the chip rate, followed by bit-rate processing. The first stage of processing, which is implemented at the chip rate is that of buffering the data, while the channel's impulse response is being estimated and the corresponding equalizer weights are calculated. Then, the data is filtered with the aid of the adaptive equalizer using the regularly updated equalizer weights. The filtered data is descrambled and despread using the appropriate scrambling and spreading codes. After this, the received non-binary symbols are demapped to bits and the log-likelihood ratio of the bits is calculated. Then the remaining processing is carried out by processing the signal at the bit rate. The processing steps carried out at the bit rate involve the reverse operations of the coding chain shown in Figure 2.4. At the end of the bit rate processing, the HARQ result is generated based on the associated CRC check. In the whole processing chain spanning from capturing the radio-frequency signal to generating the HARQ result, the turbo decoding and equalization operations are the most time-critical, since they are the most complex.

On the other hand, Figure 2.9 shows that on receiving the HSUPA HARQ result from the non-serving cell, there is only 3.6 ms before a new transmission or retransmission of the E-DPDCH message should ensue in the UL. Similarly to HSDPA-enabled UEs, the response timing is also a major issue in a HSUPA-enabled UE. After the E-HICH, E-AGCH and E-RGCH's received symbols are captured, they are down-converted to the baseband for chip-rate and bit-rate processing at the physical layer. Since no channel equalizer is required and only convolutional decoding is used, the physical layer processing is less time-critical. The processed E-HICH, E-AGCH and E-RGCH results are then passed to the MAC-es/e functionality, which resides in Layer 2. Typically, the physical layer is implemented in hardware specifically designed for signal processing. In contrast, the Layer 2 and higher-layer functions mostly involve the processing of digital control information, therefore these operations may be implemented using general-purpose CPUs. Since physical layer and Layer 2 and higher layers are implemented in separate hardware, there will be delay in transporting data like MAC-e PDUs. Both the physical layer and Layer 2 communicate with the upper OSI layers through a specific transport layer, when conveying the E-HICH, E-AGCH and E-RGCH messages, for example. At the MAC-es/e functionality shown in Figure 2.13, the MAC-e PDU is constructed, which is then transported back to the physical layer. At the physical layer, the MAC-e PDU block is channel coded according to the coding chain shown in Figure 2.10, before it is transmitted in the UL E-DPDCH message. In the whole process, a lot of the available time budget will be consumed during the transport of the MAC-e PDU from Layer 2 to the physical layer, especially when the transport block size is high. Furthermore, both the turbo encoding and the HARQ functionality of Figure 2.10 are implementationally complex as well.

2.4.4 Crest Factor

In HSDPA, the HS-DPCCH message transmitted in the UL increases the crest factor of the Release 99 UL DPCH only slightly. In contrast, the crest factor is increased quite significantly, when the HSUPA E-DPCCH and E-EDPDCH messages are mapped to four spreading codes in the HSUPA UL. Given this increased crest factor, more expensive linear or linearized class-A or class-AB linear power amplifiers having an increased dynamic range have to be employed and this further increases the cost of the UE.

Chapter 3

HSDPA-style Burst-by-Burst Adaptive Wireless Transceivers

L. Hanzo, P.J. Cherriman, C.H. Wong, E.L. Kuan, T. Keller[1]

3.1 Motivation

In recent years the concept of intelligent multi-mode, multimedia transceivers (IMMT) has emerged in the context of wireless systems [94, 191–193] and the range of various existing solutions that have found favor in existing standard systems was summarized in the excellent overview by Nanda *et al.* [194]. *The aim of these adaptive transceivers is to provide mobile users with the best possible compromise amongst a number of contradicting design factors, such as the power consumption of the hand-held portable station (PS), robustness against transmission errors, spectral efficiency, teletraffic capacity, audio/video quality and so forth [193].* In this introductory chapter we have to limit our discourse to a small subset of the associated wireless transceiver design issues, referring the reader for a deeper exposure to the literature cited [192]. A further advantage of the IMMTs of the near future is that due to their flexibility they are likely to be able to reconfigure themselves in various operational modes in order to ensure backwards compatibility with existing, so-called second generation standard wireless systems, such as the Japanese Digital Cellular [195], the Pan-American IS-54 [196] and IS-95 [197] systems, as well as the Global System of Mobile Communications (GSM) [11] standards.

The fundamental advantage of burst-by-burst adaptive IMMTs is that—regardless of the propagation environment encountered—when the mobile roams across different environments subject to pathloss, shadow—and fast-fading, co-channel-, intersymbol- and multi-user

---

[1]This chapter is based on L. Hanzo, C.H. Wong, P.J. Cherriman: Channel-adaptive wideband wireless video telephony, ©IEEE Signal Processing Magazine, July 2000; Vol. 17, No. 4, pp 10–30 and on L. Hanzo, P.J. Cherriman, Ee Lin Kuan: Interactive cellular and cordless video telephony: State-of-the-art, system design principles and expected performance, ©IEEE Proceedings of the IEEE, Sept. 2000, pp 1388–1413.

3G, HSPA and FDD versus TDD Networking   Second Edition
L. Hanzo, J. S. Blogh and S. Ni   © 2008 John Wiley & Sons, Ltd

interference, while experiencing power control errors, the system will always be able to configure itself in the highest possible throughput mode, whilst maintaining the required transmission integrity. Furthermore, whilst powering up under degrading channel conditions may disadvantage other users in the system, invoking a more robust—although lower throughput—transmission mode will not. The employment of the above burst-by-burst adaptive modems in the context of Code Division Multiple Access (CDMA) is fairly natural and it is motivated by the fact that all three third-generation mobile radio system proposals employ CDMA [11, 151, 198].

3.2   Narrowband Burst-by-Burst Adaptive Modulation

In Burst-by-Burst Adaptive Quadrature Amplitude Modulation (BbB-AQAM) a high-order, high-throughput modulation mode is invoked, when the instantaneous channel quality is favorable [13]. By contrast, a more robust lower order BbB-AQAM mode is employed, when the channel exhibits inferior quality, for improving the average BER performance. In order to support the operation of the BbB-AQAM modem, a high-integrity, low-delay feedback path has to be invoked between the transmitter and receiver for signaling the estimated channel quality perceived by the receiver to the remote transmitter. This strongly protected message can be for example superimposed on the reverse-direction messages of a duplex interactive channel. The transmitter then adjusts its AQAM mode according to the instructions of the receiver in order to be able to meet its BER target.

A salient feature of the proposed BbB-AQAM technique is that regardless of the channel conditions, the transceiver achieves always the best possible multi-media source-signal representation quality—such as video, speech or audio quality—by automatically adjusting the achievable bit rate and the associated multimedia source-signal representation quality in order to match the channel quality experienced. The AQAM modes are adjusted on a near-instantaneous basis under given propagation conditions in order to cater for the effects of pathloss, fast-fading, slow-fading, dispersion, co-channel interference (CCI), multi-user interference, etc. Furthermore, when the mobile is roaming in a hostile outdoor—or even hilly terrain—propagation environment, typically low-order, low-rate modem modes are invoked, while in benign indoor environments predominantly the high-rate, high source-signal representation quality modes are employed.

BbB-AQAM has been originally suggested by Webb and Steele [199], stimulating further research in the wireless community for example by Sampei *et al.* [200], showing promising advantages, when compared to fixed modulation in terms of spectral efficiency, BER performance and robustness against channel delay spread. Various systems employing AQAM were also characterized in [13]. The numerical upper bound performance of narrowband BbB-AQAM over slow Rayleigh flat-fading channels was evaluated by Torrance and Hanzo [201], while over wide-band channels by Wong and Hanzo [202]. Following these developments, the optimization of the BbB-AQAM switching thresholds was carried employing Powell-optimization using a cost-function, which was based on the combination of the target BER and target Bit Per Symbol (BPS) performance [203]. Adaptive modulation was also studied in conjunction with channel coding and power control techniques by Matsuoka *et al.* [204] as well as Goldsmith and Chua [205].

In the early phase of research more emphasis was dedicated to the system aspects of adaptive modulation in a narrow-band environment. A reliable method of transmitting the modulation control parameters was proposed by Otsuki *et al.* [206], where the parameters were embedded in the transmission frame's mid-amble using Walsh codes. Subsequently, at the receiver the Walsh sequences were decoded using maximum likelihood detection. Another technique of estimating the required modulation mode used was proposed by Torrance and Hanzo [207], where the modulation control symbols were represented by unequal error protection 5-PSK symbols. The adaptive modulation philosophy was then extended to wideband multi-path environments by Kamio *et al.* [208] by utilizing a bi-directional Decision Feedback Equalizer (DFE) in a micro- and macro-cellular environment. This equalization technique employed both forward and backward oriented channel estimation based on the pre-amble and post-amble symbols in the transmitted frame. Equalizer tap gain interpolation across the transmitted frame was also utilized, in order to reduce the complexity in conjunction with space diversity [208]. The authors concluded that the cell radius could be enlarged in a macro-cellular system and a higher area-spectral efficiency could be attained for micro-cellular environments by utilizing adaptive modulation. The latency effect, which occurred when the input data rate was higher than the instantaneous transmission throughput was studied and solutions were formulated using frequency hopping [209] and statistical multiplexing, where the number of slots allocated to a user was adaptively controlled.

In [210] symbol rate adaptive modulation was applied, where the symbol rate or the number of modulation levels was adapted by using $\frac{1}{8}$-rate 16QAM, $\frac{1}{4}$-rate 16QAM, $\frac{1}{2}$-rate 16QAM as well as full-rate 16QAM and the criterion used to adapt the modem modes was based on the instantaneous received signal-to-noise ratio and channel delay spread. The slowly varying channel quality of the UL and DL was rendered similar by utilizing short frame duration Time Division Duplex (TDD) and the maximum normalized delay spread simulated was 0.1. A variable channel coding rate was then introduced by Matsuoka *et al.* in conjunction with adaptive modulation in [204], where the transmitted burst incorporated an outer Reed Solomon code and an inner convolutional code in order to achieve high-quality data transmission. The coding rate was varied according to the prevalent channel quality using the same method, as in adaptive modulation in order to achieve a certain target BER performance. A so-called channel margin was introduced in this contribution, which adjusted the switching thresholds in order to incorporate the effects of channel quality estimation errors. As mentioned above, the performance of channel coding in conjunction with adaptive modulation in a narrow-band environment was also characterized by Goldsmith and Chua [205]. In this contribution, trellis and lattice codes were used without channel interleaving, invoking a feedback path between the transmitter and receiver for modem mode control purposes. The effects of the delay in the feedback path on the adaptive modem's performance were studied and this scheme exhibited a higher spectral efficiency, when compared to the non-adaptive trellis coded performance.

Subsequent contributions by Suzuki *et al.* [211] incorporated space-diversity and power-adaptation in conjunction with adaptive modulation, for example in order to combat the effects of the multi-path channel environment at a 10 Mbits/s transmission rate. The maximum tolerable delay-spread was deemed to be one symbol duration for a target mean BER performance of 0.1%. This was achieved in a Time Division Multiple Access (TDMA) scenario, where the channel estimates were predicted based on the extrapolation of previous channel quality estimates. Variable transmitted power was then applied in combination with adaptive modulation in [205], where the transmission rate and power adaptation was optimized in order to achieve an increased spectral efficiency. In this treatise, a slowly varying channel was assumed and the instantaneous received power required in order to achieve a certain upper bound performance was assumed to be known prior to transmission. Power control in conjunction with a pre-distortion type non-linear power amplifier compensator was studied in the context of adaptive modulation in [212]. This method was used to mitigate the non-linearity effects associated with the power amplifier, when QAM modulators were used.

Results were also recorded concerning the performance of adaptive modulation in conjunction with different multiple access schemes in a narrow-band channel environment. In a TDMA system, dynamic channel assignment was employed by Ikeda et al., where in addition to assigning a different modulation mode to a different channel quality, priority was always given to those users in reserving timeslots, which benefitted from the best channel quality [213]. The performance was compared to fixed channel assignment systems, where substantial gains were achieved in terms of system capacity. Furthermore, a lower call termination probability was recorded. However, the probability of intra-cell hand-off increased as a result of the associated dynamic channel assignment (DCA) scheme, which constantly searched for a high-quality, high-throughput timeslot for the existing active users. The application of adaptive modulation in packet transmission was introduced by Ue, Sampei and Morinaga [214], where the results showed improved data throughput. Recently, the performance of adaptive modulation was characterized in conjunction with an automatic repeat request (ARQ) system in [215], where the transmitted bits were encoded using a cyclic redundant code (CRC) and a convolutional punctured code in order to increase the data throughput.

A recent treatise was published by Sampei, Morinaga and Hamaguchi [216] on laboratory test results concerning the utilization of adaptive modulation in a TDD scenario, where the modem mode switching criterion was based on the signal-to-noise ratio and on the normalized delay-spread. In these experimental results, the channel quality estimation errors degraded the performance and consequently a channel estimation error margin was devised, in order to mitigate this degradation. Explicitly, the channel estimation error margin was defined as the measure of how much extra protection margin must be added to the switching threshold levels, in order to minimize the effects of the channel estimation errors. The delay-spread also degraded the performance due to the associated irreducible BER, which was not compensated by the receiver. However, the performance of the adaptive scheme in a delay-spread impaired channel environment was better than that of a fixed modulation scheme. Lastly, the experiment also concluded that the AQAM scheme can be operated for a Doppler frequency of $f_d = 10$ Hz with a normalized delay spread of 0.1 or for $f_d = 14$ Hz with a normalized delay spread of 0.02, which produced a mean BER of 0.1% at a transmission rate of 1 Mbits/s.

Lastly, the latency and interference aspects of AQAM modems were investigated in [209, 217]. Specifically, the latency associated with storing the information to be transmitted during severely degraded channel conditions was mitigated by frequency hopping or statistical multiplexing. As expected, the latency is increased, when either the mobile speed or the channel SNR are reduced, since both of these result in prolonged low instantaneous SNR intervals. It was demonstrated that as a result of the proposed measures, typically more than 4 dB SNR reduction was achieved by the proposed adaptive modems in comparison to the conventional fixed-mode benchmark modems employed. However, the achievable gains depend strongly on the prevalent co-channel interference levels and hence interference cancellation was invoked in [217] on the basis of adjusting the demodulation decision boundaries after estimating the interfering channel's magnitude and phase.

Having reviewed the developments in the field of narrowband AQAM, let us now consider wideband AQAM modems in the next section.

3.3 Wideband Burst-by-Burst Adaptive Modulation

In the above narrow-band channel environment, the quality of the channel was determined by the short-term SNR of the received burst, which was then used as a criterion in order to choose the appropriate modulation mode for the transmitter, based on a list of switching threshold levels, l_n [199–201]. However, in a wideband environment, this criterion is not an accurate measure for judging the quality of the channel, where the existence of multipath components produces not only power attenuation of the transmission burst, but also intersymbol interference. Consequently, appropriate channel quality criteria have to be defined, in order to estimate the wideband channel quality for invoking the most appropriate modulation mode.

3.3.1 Channel Quality Metrics

The most reliable channel quality estimate is the BER, since it reflects the channel quality, irrespective of the source or the nature of the quality degradation. The BER can be estimated with a certain granularity or accuracy, provided that the system entails a channel decoder or—synonymously—Forward Error Correction (FEC) decoder employing algebraic decoding [11, 218]. If the system contains a so-called soft-in-soft-out (SISO) channel decoder, such as a turbo decoder [134], the BER can be estimated with the aid of the Logarithmic Likelihood Ratio (LLR), evaluated either at the input or the output of the channel decoder. Hence a particularly attractive way of invoking LLRs is employing powerful turbo codecs, which provide a reliable indication of the confidence associated with a particular bit decision. The LLR is defined as the logarithm of the ratio of the probabilities associated with a specific bit being binary zero or one. Again, this measure can be evaluated at both the input and the output of the turbo channel codecs and both of them can be used for channel quality estimation.

In the event that no channel encoder/decoder (codec) is used in the system, the channel quality expressed in terms of the BER can be estimated with the aid of the mean-squared error (MSE) at the output of the channel equalizer or the closely related metric, the Pseudo-Signal-to-Noise-Ratio (Pseudo-SNR) [202]. The MSE or pseudo-SNR at the output of the channel equalizer have the important advantage that they are capable of quantifying the severity of the Inter-Symbol-Interference (ISI) and/or CCI experienced, in other words quantifying the Signal-to-Interference-plus-Noise-Ratio (SINR).

In our proposed systems the wideband channel-induced degradation is combated not only by the employment of adaptive modulation but also by equalization. In following this line of thought, we can formulate a two-step methodology in mitigating the effects of the dispersive wideband channel. In the first step, the equalization process will eliminate most of the intersymbol interference based on a Channel Impulse Response (CIR) estimate derived using the channel sounding midamble and consequently, the signal-to-noise and residual interference ratio at the output of the equalizer is calculated.

We found that the residual channel-induced ISI at the output of the DFE is near-Gaussian distributed and that if there are no decision feedback errors, the pseudo-SNR at the output of the DFE, γ_{dfe} can be calculated as [94, 202, 219]:

$$
\gamma_{dfe} = \frac{\text{Wanted Signal Power}}{\text{Residual ISI Power + Effective Noise Power}}
$$
$$
= \frac{E\left[\left|S_k\sum_{m=0}^{N_f-1}C_mh_m\right|^2\right]}{\sum_{q=-(N_f-1)}^{-1}E\left[\left|f_qS_{k-q}\right|^2\right] + N_o\sum_{m=0}^{N_f-1}\left|C_m\right|^2},
$$

$$(3.1)$$

where C_m and h_m denotes the DFE's feed-forward coefficients and the channel impulse response, respectively. The transmitted signal and the noise spectral density is represented by S_k and N_o. Lastly, the number of DFE feed-forward coefficients is denoted by N_f. By utilizing the pseudo-SNR at the output of the equalizer, we are ensuring that the system performance is optimized by employing equalization and AQAM [13] in a wideband environment according to the following switching regime:

$$
\text{Modulation Mode} = \begin{cases} \text{No TX} & \text{if } \gamma_{DFE} < f_0 \\ \text{BPSK} & \text{if } f_0 < \gamma_{DFE} < f_1 \\ \text{4QAM} & \text{if } f_1 < \gamma_{DFE} < f_2 \\ \text{16QAM} & \text{if } f_2 < \gamma_{DFE} < f_3 \\ \text{64QAM} & \text{if } \gamma_{DFE} > f_3, \end{cases}
$$

$$(3.2)$$

where $f_n, n = 0 \ldots 3$ are the pseudo-SNR thresholds levels, which are set according to the system's integrity requirements and the modem modes may assume $0 \ldots 6$ bits/symbol transmissions corresponding to no transmissions (No TX), Binary Phase Shift Keying (BPSK), as well as 4- 16- and 64QAM [13]. We note, however that in the context of the interactive BbB-AQAM videophone schemes introduced during our later discourse for quantifying the service-related benefits of such adaptive transceivers we refrained from employing the No Tx mode. This allowed us to avoid the associated latency of the buffering required for storing the information, until the channel quality improved sufficiently for allowing transmission of the buffered bits.

In [220, 221] a range of novel Radial Basis Function (RBF) assisted BbB-AQAM channel equalizers have been proposed, which exhibit a close relationship with the so-called Bayesian schemes. Decision feedback was introduced in the design of the RBF equalizer in order to reduce its computational complexity. The RBF DFE was found to give similar performance to the conventional DFE over Gaussian channels using various BbB-AQAM schemes, while requiring a lower feedforward and feedback order. Over Rayleigh-fading channels similar findings were valid for binary modulation, while for higher order modems the RBF-based DFE required increased feedforward and feedback orders in order to outperform the conventional MSE DFE scheme. Then turbo BCH codes were invoked [220] for improving the associated BER and BPS performance of the scheme, which was shown to give a significant improvement in terms of the mean BPS performance compared to that of the

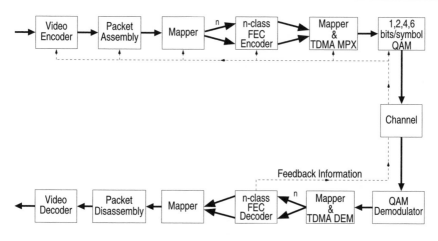

Figure 3.1: Reconfigurable transceiver schematic diagram.

uncoded RBF equalizer assisted adaptive modem. Finally, a novel turbo equalization scheme was presented in [221], which employed an RBF DFE instead of the conventional trellis-based equalizer, which was advocated in most turbo equalizer implementations. The so-called Jacobian logarithmic complexity reduction technique was proposed, which was shown to achieve an identical BER performance to the conventional trellis-based turbo equalizer, while incurring a factor 4.4 lower "per-iteration" complexity in the context of 4QAM.

In summary, in contrast to the narrowband, statically reconfigured multimode systems of [192], in this section wideband, near-instantaneously reconfigured or burst-by-burst adaptive modulation was invoked, in order to quantify the achievable service-related benefits, as perceived by users of such systems. More specifically, the achievable video performance benefits of wireless BbB-AQAM video transceivers will be quantified in this section, when using the H.263 video encoder [192]. Similar BbB-AQAM speech and audio transceivers were portrayed in [222].

It is an important element of the system that when the binary BCH [11, 218] or turbo codes [134, 218] protecting the video stream are overwhelmed by the plethora of transmission errors, the systems refrains from decoding the video packet in order to prevent error propagation through the reconstructed frame buffer [192]. Instead, these corrupted packets are dropped and the reconstructed frame buffer will not be updated, until the next packet replenishing the specific video frame area arrives. The associated video performance degradation is fairly minor for packet dropping or frame error rates (FER) below about 5%. These packet dropping events are signaled to the remote decoder by superimposing a strongly protected one-bit packet acknowledgement flag on the reverse-direction packet, as outlined in [192]. In the proposed scheme we also invoked the adaptive rate control and packetization algorithm of [192], supporting constant Baud-rate operation.

Having reviewed the basic features of adaptive modulation, in the forthcoming section we will characterize the achievable service-related benefits of BbB-AQAM video transceivers, as perceived by the users of such systems.

Table 3.1: Modulation and channel parameters.

Parameter	Value
Carrier Frequency	1.9 GHz
Vehicular Speed	30 mph
Doppler frequency	85 Hz
Norm. Doppler fr.	3.27×10^{-5}
Channel type	COST 207 Typ. Urban (Figure 3.2)
No. of channel paths	4
Data modulation	Adaptive QAM
	(BPSK, 4-QAM, 16-QAM, 64-QAM)
	Decision Feedback Equalizer
Receiver type	No. of Forward Filter Taps = 35
	No. of Backward Filter Taps = 7

3.4   Wideband BbB-AQAM Video Transceivers

Again, in this section we set out to demonstrate the service-quality related benefits of a wideband BbB-AQAM in the context of a wireless videophone system employing the programmable H.263 video codec in conjunction with an adaptive packetizer. The system's schematic diagram is shown in Figure 3.1, which will be referred to in more depth during our further discourse.

In these investigations 176x144 pixel QCIF-resolution, 30 frames/s video sequences were transmitted, which were encoded by the H.263 video codec [192, 223] at bit rates resulting in high perceptual video quality. Table 3.1 shows the modulation- and channel parameters employed. The COST207 [77] four-path typical urban (TU) channel model was used, which is characterized by its CIR in Figure 3.2. We used the Pan-European FRAMES proposal [224] as the basis for our wideband transmission system, invoking the frame structure shown in Figure 3.3. Employing the FRAMES Mode A1 (FMA1) so-called non-spread data burst mode required a system bandwidth of 3.9 MHz, when assuming a modulation excess bandwidth of 50% [13]. A range of other system parameters are shown in Table 3.2. Again, it is important to note that the proposed AQAM transceiver of Figure 3.1 requires a duplex system, since the AQAM mode required by the receiver during the next received video packet has to be signaled to the transmitter. In this system we employed TDD and the feedback path is indicated by the dashed line in the schematic diagram of Figure 3.1.

Again, the proposed video transceiver of Figure 3.1 is based on the H.263 video codec [223]. The video coded bitstream was protected by near-half-rate binary BCH coding [11] or by half-rate turbo coding [134] in all of the burst-by-burst adaptive wideband AQAM modes [13]. The AQAM modem can be configured either under network control on a more static basis, or under transceiver control on a near-instantaneous basis, in order to operate as a 1, 2, 4 and 6 bits/symbol scheme, while maintaining a constant signaling rate. This allowed us to support an increased throughput expressed in terms of the average number of bits per symbol (BPS), when the instantaneous channel quality was high, leading ultimately to an increased video quality in a constant bandwidth.

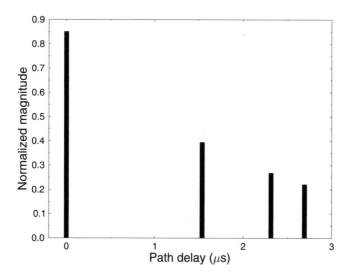

Figure 3.2: Normalized channel impulse response for the COST 207 [77] four-path Typical Urban (TU) channel.

Figure 3.3: Transmission burst structure of the FMA1 non-spread data burst mode of the FRAMES proposal [224].

The transmitted bit rate for all four modes of operation is shown in Table 3.3. The unprotected bit rate before approximately half-rate BCH coding is also shown in the table. The actual useful bit rate available for video is slightly less than the unprotected bit rate due to the required strongly protected packet acknowledgement information and packetization information. The effective video bit rate is also shown in the table.

In order to be able to invoke the inherently error-sensitive variable-length coded H.263 video codec in a high-BER wireless scenario, a flexible adaptive packetization algorithm was necessary, which was highlighted in [192]. The technique proposed exhibits high flexibility, allowing us to drop corrupted video packets, rather than allowing erroneous bits to contaminate the reconstructed frame buffer of the H.263 codec. This measure prevents the propagation of errors to future video frames through the reconstructed frame buffer of the H.263 codec. More explicitly, corrupted video packets cannot be used by either the local or

Table 3.2: Generic system features of the reconfigurable multi-mode video transceiver, using the non-spread data burst mode of the FRAMES proposal [224] shown in Figure 3.3.

Features	Value
Multiple access	TDMA
Duplexing	TDD
No. of Slots/Frame	16
TDMA frame length	4.615 ms
TDMA slot length	$288\mu s$
Data Symbols/TDMA slot	684
User Data Symbol Rate (KBd)	148.2
System Data Symbol Rate (MBd)	2.37
Symbols/TDMA slot	750
User Symbol Rate (KBd)	162.5
System Symbol Rate (MBd)	2.6
System Bandwidth (MHz)	3.9
Eff. User Bandwidth (kHz)	244

Table 3.3: Operational-mode specific transceiver parameters.

Features	Multi-rate System			
Mode	BPSK	4QAM	16QAM	64QAM
Bits/Symbol	1	2	4	6
FEC		Near Half-rate BCH		
Transmission bit rate (kbit/s)	148.2	296.4	592.8	889.3
Unprotected bit rate (kbit/s)	75.8	151.7	303.4	456.1
Effective Video-rate (kbit/s)	67.0	141.7	292.1	446.4
Video fr. rate (Hz)		30		

the remote H.236 decoder, since that would result in unacceptable video degradation over a prolonged period of time due to the error propagation inflicted by the associated motion vectors and run-length coding. Upon dropping the erroneous video packets, both the local and remote H.263 reconstruction frame buffers are updated by a blank packet, which corresponds to assuming that the video block concerned was identical to the previous one.

A key feature of our proposed adaptive packetization regime is therefore the provision of a strongly error protected binary transmission packet acknowledgement flag [192], which instructs the remote decoder not to update the local and remote video reconstruction buffers in the event of a corrupted packet. This flag can be for example conveniently repetition-coded, in order to invoke Majority Logic Decision (MLD) at the decoder. Explicitly, the binary flag is repeated an odd number of times and at the receiver the MLD scheme counts the number of binary ones and zeros and opts for the logical value, constituting the majority of the received bits. These packet acknowledgement flags are then superimposed on the forthcoming reverse-

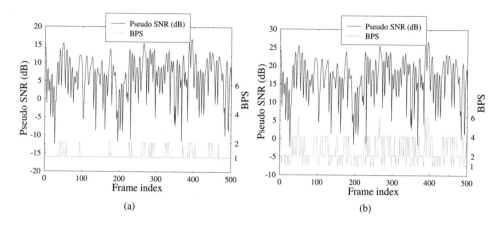

Figure 3.4: Modulation mode variation with respect to the pseudo-SNR at Channel SNRs of (a) 10 dB and (b) 20 dB; defined by Equation 3.1 over the TU Rayleigh fading channel. The BPS throughputs of 1, 2, 4 and 6 represent BPSK, 4QAM, 16QAM and 64QAM, respectively.

direction packet in our advocated Time Division Duplex (TDD) regime [192] of Table 3.2, as seen in the schematic diagram of Figure 3.1.

The proposed BbB-AQAM modem maximizes the system capacity available by using the most appropriate modulation mode for the current instantaneous channel conditions. As stated before, we found that the pseudo-SNR at the output of the channel equalizer was an adequate channel quality measure in our burst-by-burst adaptive wide-band modem. A more explicit representation of the wideband AQAM regime is shown in Figure 3.4, which displays the variation of the modulation mode with respect to the pseudo-SNR at channel SNRs of 10 and 20 dB. In these figures, it can be seen explicitly that the lower-order modulation modes were chosen, when the pseudo-SNR was low. In contrast, when the pseudo-SNR was high, the higher-order modulation modes were selected in order to increase the transmission throughput. These figures can also be used to exemplify the application of wideband AQAM in an indoor and outdoor environment. In this respect, Figure 3.4(a) can be used to characterize a hostile outdoor environment, where the perceived channel quality was low. This resulted in the utilization of predominantly more robust modulation modes, such as BPSK and 4QAM. Conversely, a less hostile indoor environment is exemplified by Figure 3.4(b), where the perceived channel quality was high. As a result, the wideband AQAM regime can adapt suitably by invoking higher-order modulation modes, as evidenced by Figure 3.4(b). Again, this simple example demonstrated that wideband AQAM can be utilized, in order to provide a seamless, near-instantaneous reconfiguration between for example indoor and outdoor environments.

3.5 BbB-AQAM Performance

The mean BER and BPS performances were numerically calculated [202] for two different target BER systems, namely for the High-BER and Low-BER schemes, respectively. The results are shown in Figure 3.5 over the COST207 TU Rayleigh fading channel of Figure 3.2.

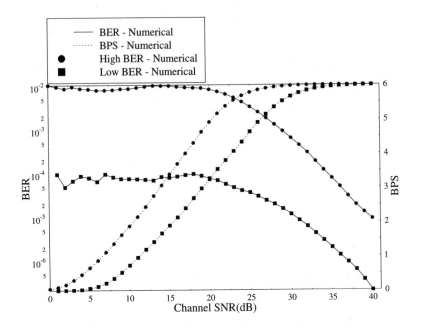

Figure 3.5: Numerical mean BER and BPS performance of the wideband equalized AQAM scheme for the High-BER and Low-BER regime over the COST207 TU Rayleigh fading channel.

The targeted mean BERs of the High-BER and Low-BER regime of 1% and 0.01% was achieved for all average channel SNRs investigated, since this scheme also invoked a no-transmission mode, when the channel quality was extremely hostile. In this mode only dummy data was transmitted, in order to facilitate monitoring the channel's quality.

At average channel SNRs below 20 dB the lower-order modulation modes were dominant, producing a robust system in order to achieve the targeted BER. Similarly, at high average channel SNRs the higher-order modulation mode of 64QAM dominated the transmission regime, yielding a lower mean BER than the target, since no higher-order modulation mode could be legitimately invoked. This is evidenced by the modulation mode probability results shown in Figure 3.6 for the COST207 TU Rayleigh fading channel of Figure 3.2. The targeted mean BPS values for the High-BER and Low-BER regime of 4.5 and 3 were achieved at approximately 19 dB channel SNR for the COST207 TU Rayleigh fading channels. However, at average channel SNRs below 3 dB the above-mentioned no-transmission or transmission blocking mode was dominant in the Low-BER system and thus the mean BER performance was not recorded for that range of average channel SNRs.

The transmission throughput achieved for the High-BER and Low-BER transmission regimes is shown in Figure 3.7. The transmission throughput for the High-BER transmission regime was higher than that of the Low-BER transmission regime for the same transmitted signal energy due to the more relaxed BER requirement of the High-BER transmission regime, as evidenced by Figure 3.7. The achieved transmission throughput of the wideband AQAM scheme was higher than that of the BPSK, 4QAM and 16QAM schemes for the same average channel SNR. However, at higher average channel SNRs the throughput

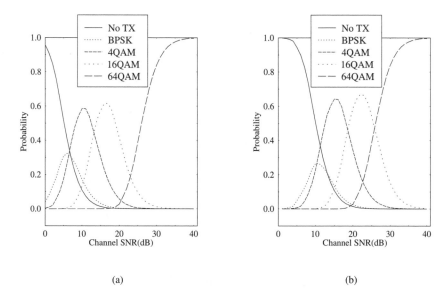

Figure 3.6: Numerical probabilities of each modulation mode utilized for the wideband AQAM and DFE scheme over the TU Rayleigh Fading channel for the (a) High-BER Transmission regime and (b) Low-BER Transmission regime.

performance of both schemes converged, since 64QAM became the dominant modulation mode for the wideband AQAM scheme. SNR gains of 1–3 dB and 7–9 dB were recorded for the High-BER and Low-BER transmission schemes, respectively. These gains were considerably lower than those associated with narrow-band AQAM, where 5–7 dB and 10–18 dB of gains were reported for the High-BER and Low-BER transmission scheme, respectively [209, 217]. This was expected, since in the narrow-band environment the fluctuation of the instantaneous SNR was more severe, resulting in increased utilization of the modulation switching mechanism. Consequently, the instantaneous transmission throughput increased, whenever the fluctuations yielded a high received instantaneous SNR. Conversely, in a wideband channel environment the channel quality fluctuations perceived by the DFE were less severe due to the associated multi-path diversity, which was exploited by the equalizer.

Having characterized the wideband BbB-AQAM modem's performance, let us now consider the entire video transceiver of Figure 3.1 and Tables 3.1–3.3 in the next section.

3.6 Wideband BbB-AQAM Video Performance

As a benchmarker, the statically reconfigured modems of [192] were invoked in Figure 3.8, in order to indicate how a system would perform, which cannot act on the basis of the near-instantaneously varying channel quality. As it can be inferred from Figure 3.8, such a statically reconfigured transceiver switches its mode of operation from a lower-order modem

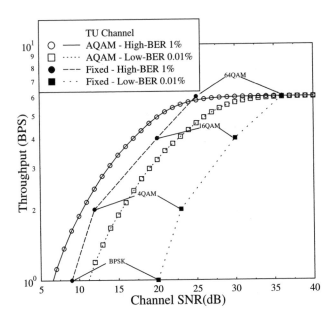

Figure 3.7: Transmission throughput of the wideband AQAM and DFE scheme and fixed modulation modes over the TU Rayleigh Fading channel for both the High-BER and Low-BER transmission regimes.

Figure 3.8: Transmission FER (or packet loss ratio) versus Channel SNR comparison of the four fixed modulation modes (BPSK, 4QAM, 16QAM, 64QAM) with 5% FER switching and adaptive burst-by-burst modem (AQAM). AQAM is shown with a realistic one TDMA frame delay between channel estimation and mode switching, and a zero delay version is included as an upper bound. The channel parameters were defined in Table 3.1 and near-half-rate BCH coding was employed [225] Cherriman, Wong, Hanzo, 2000 ©IEEE.

mode, such as for example BPSK to a higher-order mode, such as 4QAM, when the channel quality has improved sufficiently for the 4QAM mode's FER to become lower than 5% after reconfiguring the transceiver in this more long-term 4QAM mode.

In order to assess the effects of imperfect channel estimation on BbB-AQAM we considered two scenarios. In the first scheme the adaptive modem always chose the perfectly estimated AQAM modulation mode, in order to provide a maximum upper bound performance. In the second scenario the modulation mode was based upon the perfectly estimated AQAM modulation mode for the previous burst, which corresponded to a delay of one TDMA frame duration of 4.615 ms. This second scenario represents a practical burst-by-burst adaptive modem, where the one-frame channel quality estimation latency is due to superimposing the receiver's required AQAM mode on a reverse-direction packet, for informing the transmitter concerning the best mode to be used for maintaining the target performance.

Figure 3.8 demonstrates on a logarithmic scale that the "one-frame channel estimation delay" AQAM modem manages to maintain a similar FER performance to the fixed rate BPSK modem at low SNRs, although we will see during our further discourse that AQAM provides increasingly higher bit rates, reaching six times higher values than BPSK for high channel SNRs, where the employment of 64QAM is predominant. In this high-SNR region the FER curve asymptotically approaches the 64QAM FER curve for both the realistic and the ideal AQAM scheme, although this is not visible in the figure for the ideal scheme, since this occurs at SNRs outside the range of Figure 3.8. Again, the reason for this performance discrepancy is the occasionally misjudged channel quality estimates of the realistic AQAM scheme. Additionally, Figure 3.8 indicates that the realistic AQAM modem exhibits a near-constant 3% FER at medium SNRs. The issue of adjusting the switching thresholds in order to achieve the target FER will be addressed in detail at a later stage in this section and the thresholds invoked will be detailed with reference to Table 3.4. Suffice to say at this stage that the average number of bits per symbol—and potentially also the associated video quality—can be increased upon using more "aggressive" switching thresholds. However, this results in an increased FER, which tends to decrease the video quality, as it will be discussed later in this section. Having shown the effect of the BbB-AQAM modem on the transmission FER, let us now demonstrate the effects of the AQAM switching thresholds on the system's performance in terms of the associated FER performance.

3.6.1 AQAM Switching Thresholds

The set of switching thresholds used in all the previous graphs was the "standard" set shown in Table 3.4, which was determined on the basis of the required channel SINR for maintaining the specific target video FER. In order to investigate the effect of different sets of switching thresholds, we defined two new sets of thresholds, a more "conservative" set, and a more "aggressive" set, employing less robust, but more bandwidth-efficient modem modes at lower SNRs. The more conservative switching thresholds reduced the transmission FER at the expense of a lower effective video bit rate. By contrast, the more aggressive set of thresholds increased the effective video bit rate at the expense of a higher transmission FER. The transmission FER performance of the realistic burst-by-burst adaptive modem, which has a one TDMA frame delay between channel quality estimation and mode switching is shown in Figure 3.9 for the three sets of switching thresholds of Table 3.4. It can be seen that the more

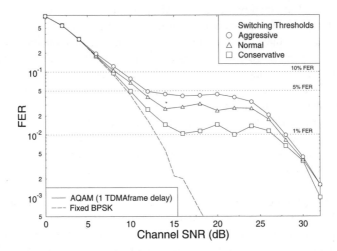

Figure 3.9: Transmission FER (or packet loss ratio) versus Channel SNR comparison of the fixed BPSK modulation mode and the adaptive burst-by-burst modem (AQAM) for the three sets of switching thresholds described in Table 3.4. AQAM is shown with a realistic one TDMA frame delay between channel estimation and mode switching. The channel parameters were defined in Table 3.1 [225] Cherriman, Wong, Hanzo, 2000 ©IEEE.

Figure 3.10: Video bit rate versus channel SNR comparison for the adaptive burst-by-burst modem (AQAM) with a realistic one TDMA frame delay between channel estimation and mode switching for the three sets of switching thresholds as described in Table 3.4. The channel parameters were defined in Table 3.1 [225] Cherriman, Wong, Hanzo, 2000 ©IEEE.

Table 3.4: SINR estimate at output of the equalizer required for each modulation mode in burst-by-burst adaptive modem, i.e. switching thresholds.

	BPSK	4QAM	16QAM	64QAM
Standard	<10 dB	≥ 10 dB	≥ 18 dB	≥ 24 dB
Conservative	<13 dB	≥ 13 dB	≥ 20 dB	≥ 26 dB
Aggressive	<9 dB	≥ 9 dB	≥ 17 dB	≥ 23 dB

"conservative" switching thresholds reduce the transmission FER from about 3% to about 1% for medium channel SNRs, while the more "aggressive" thresholds increase the transmission FER from about 3% to 4–5%. However, since FERs below 5% are not objectionable in video quality terms, this FER increase is an acceptable compromise for attaining a higher effective video bit rate.

The effective video bit rate for the realistic adaptive modem with the three sets of switching thresholds is shown in Figure 3.10. The more conservative set of switching thresholds reduces the effective video bit rate but also reduces the transmission FER. The aggressive switching thresholds increase the effective video bit rate, but also increase the transmission FER. Therefore the optimal switching thresholds should be set such that the transmission FER is deemed acceptable in the range of channel SNRs considered. Let us now consider the performance improvements achievable, when employing powerful turbo codecs.

3.6.2 Turbo-coded AQAM Videophone Performance

Let us now demonstrate the additional performance gains that are achievable when a somewhat more complex turbo codec [134] is used in comparison to similar-rate algebraically decoded binary BCH codecs [11]. The generic system parameters of the turbo-coded reconfigurable multi-mode video transceiver are the same as those used in the BCH-coded version summarized in Table 3.2. Turbo-coding schemes are known to perform best in conjunction with square-shaped turbo interleaver arrays and their performance is improved upon extending the associated interleaving depth, since then the two constituent encoders are fed with more independent data. This ensures that the turbo decoder can rely on two quasi-independent data streams in its efforts to make as reliable bit decisions as possible. A turbo interleaver size of 18×18 bits was chosen, requiring 324 bits for filling the interleaver. The required so-called recursive systematic convolutional (RSC) component codes had a coding rate of 1/2 and a constraint length of $K = 3$. After channel coding the transmission burst length became 648 bits, which facilitated the decoding of all AQAM transmission bursts independently. The operational-mode specific turbo transceiver parameter are shown in Table 3.5, which should be compared to the corresponding BCH-coded parameters of Table 3.3. The turbo-coded parameters result in a 10% lower effective throughput bit rate compared to the similar-rate BCH-codecs under error-free conditions. However, Figure 3.11 demonstrates that the PSNR video quality versus channel SNR performance of the turbo-coded AQAM modem becomes better than that of the BCH-coded scenario, when the channel quality degrades. Having highlighted the operation of wideband single-carrier burst-by-burst AQAM modems, let us now consider briefly in the next two sections how the above

Table 3.5: Operational-mode specific turbo-coded transceiver parameters.

Features	Multi-rate System			
Mode	BPSK	4QAM	16QAM	64QAM
Bits/Symbol	1	2	4	6
FEC	Half-Rate Turbo coding with CRC			
Transmission bit rate (kbit/s)	140.4	280.8	561.6	842.5
Unprotected bit rate (kbit/s)	66.3	136.1	275.6	415.2
Effective Video-rate (kbit/s)	60.9	130.4	270.0	409.3
Video fr. rate (Hz)	30			

Figure 3.11: Decoded video quality (PSNR) versus transmission FER (or packet loss ratio) comparison of the realistic adaptive burst-by-burst modems (AQAM) using either BCH or turbo coding. The channel parameters were defined in Table 3.1 [225] Cherriman, Wong, Hanzo, 2000 ©IEEE.

burst-by-burst adaptive principles can be extended to CDMA and Orthogonal Frequency Division Multiplex (OFDM) systems [13, 226].

3.7 Burst-by-Burst Adaptive Joint-Detection CDMA Video Transceiver

3.7.1 Multi-user Detection for CDMA

In the previous chapter a simple conceptual introduction was provided to CDMA, assuming the employment of simple single-user receivers. Then the most recent family of CDMA-based third-generation standards was reviewed. In this chapter we introduce a number of advanced

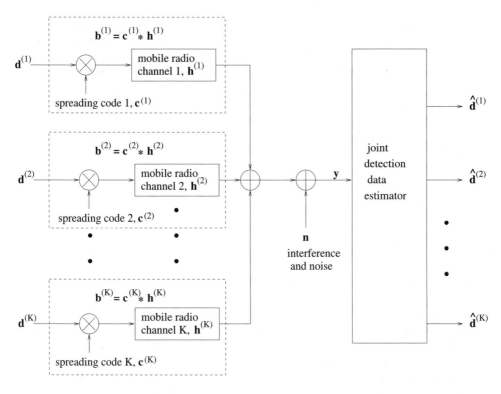

Figure 3.12: System model of a synchronous CDMA system on the UL using joint detection.

near-instantaneously adaptive transceiver concepts, which may find their way into future standards, in order to enhance the performance of the existing systems. We also introduce the concept of multi-user detection in an effort to maintain a near-single-user performance, whilst supporting a multiplicity of users. These adaptive system concepts are discussed in significantly more depth in [94, 192].

The effects of multi-user interference (MAI) are similar to those of the Intersymbol Interference (ISI) inflicted by the multipath propagation channel. More specifically, each user in a K-user system will suffer from MAI due to the other $(K-1)$ users. This MAI can also be viewed as a single user's signal contaminated by the ISI due to $(K-1)$ propagation paths in a multipath channel. Therefore, conventional equalization techniques used to mitigate the effects of ISI can be modified for employment in multi-user detection assisted CDMA systems. The so-called joint detection (JD) receivers constitute a category of multi-user detectors developed for synchronous burst-based CDMA transmissions and they utilize these techniques.

Figure 3.12 depicts the block diagram of a synchronous joint-detection assisted CDMA system model for UL transmissions. There are a total of K users in the system, where the information is transmitted in bursts. Each user transmits N data symbols per burst and the data vector for user k is represented as $\mathbf{d}^{(k)}$. Each data symbol is spread with a user-specific spreading sequence, $\mathbf{c}^{(k)}$, which has a length of Q chips. In the UL, the signal of each user passes through a different mobile channel characterized by its time-varying complex impulse response, $\mathbf{h}^{(k)}$. By sampling at the chip rate of $1/T_c$, the impulse response can be represented by W complex samples. Following the approach of Klein and Baier [227], the received burst can be represented as $\mathbf{y} = \mathbf{A}\mathbf{d} + \mathbf{n}$, where $\mathbf{y}$ is the received vector and consists of the synchronous sum of the transmitted signals of all the K users, corrupted by a noise sequence, $\mathbf{n}$. The matrix $\mathbf{A}$ is referred to as the system matrix and it defines the system's response, representing the effects of MAI and the mobile channels. Each column in the matrix represents the combined impulse response obtained by convolving the spreading sequence of a user with its channel impulse response, $\mathbf{b}^{(k)} = \mathbf{c}^{(k)} * \mathbf{h}^{(k)}$. This is the impulse response experienced by a transmitted data symbol. Upon neglecting the effects of the noise the joint detection formulation is simply based on inverting the system matrix $\mathbf{A}$, in order to recover the data vector constituted by the superimposed transmitted information of all the K CDMA users.

3.7.2   JD-ACDMA Modem Mode Adaptation and Signalling

In mobile communications systems typically power control techniques are used to mitigate the effects of pathloss and slow fading. However, in order to counteract the problem of fast fading and co-channel interference, agile and tight-specification power control algorithms are required [228]. Another technique that can be used to overcome the problems due to fading is adaptive-rate transmission [199, 229], where the information rate is varied according to the quality of the channel.

Different methods of multi-rate transmission have been proposed by Ottosson and Svensson [230]. According to the multi-code method, multiple codes are assigned to a user requiring a higher bit rate [230]. Multiple data rates can also be provided by a multiple processing-gain scheme, where the chip rate is kept constant but the data rates are varied by changing the processing gain of the spreading codes assigned to the users. Performance comparisons for both of these schemes have been carried out by Ottosson and Svensson [230] and Ramakrishna and Holtzman [231], demonstrating that both schemes achieved similar performance. Saquib and Yates [232] and Johansson and Svensson [233] have also investigated the employment of the so-called decorrelating detector and the successive interference cancellation receiver for multi-rate CDMA systems.

Adaptive rate transmission schemes, where the transmission rate is adapted according to the channel quality have also been proposed. Abeta *et al.* [234] have conducted investigations into an adaptive CDMA scheme, where the transmission rate is modified by varying the channel code rate and the processing gain of the CDMA user, employing the carrier to interference and noise ratio (CINR) as the switching metric. In their investigations, the overall packet rate was kept constant by transmitting in shorter bursts, when the transmission bit rate was high and lengthening the burst when the bit rate was low. This resulted in a decrease in interference power, which translated to an increase in system capacity. Hashimoto *et al.* [235] extended this work to show that the proposed system was capable of achieving a higher capacity with a smaller hand-off margin and lower average transmitter power. In these schemes, the conventional Rake receiver was used for the detection of the data symbols. Kim [229] analysed the performance of two different methods of combatting the mobile channel's variations, which were the adaptation of the transmitter power to compensate for channel variations or the switching of the information rate to suit the channel conditions.

Using a Rake receiver, it was demonstrated that rate adaptation provided a higher average information rate than power adaptation for a given average transmit power and a given BER.

In our design example here we also propose to vary the information rate in accordance with the channel quality. However, in comparison to conventional power control techniques— which may disadvantage other users by increasing their transmitted powers in an effort to maintain the quality of their own links—the JD-AQAM scheme employed does not disadvantage other users. This is achieved by "non-destructively" adjusting the modulation mode of the user supported according to the near-instantaneous channel quality experienced. Additionally, burst-by-burst adaptive transceivers are capable of increasing the network capacity, as we will demonstrate in the book. This is because conventional transceivers would drop a call, when the interference levels become excessive. By contrast, adaptive transceivers reconfigure themselves in a more robust coding/modulation mode.

In this section we will quantify the expected video performance of a range of intelligent multi-mode CDMA transceivers, employing JD multi-user reception CDMA techniques at the BS, which are optional in the 3G system proposals due to their high implementational complexity and hence are likely to be employed only in future implementations of the 3G standards. As a potential further future enhancement, we will also invoke the powerful principle of burst-by-burst adaptive JD-CDMA (JD-ACDMA) transmissions, which was discussed in some depth in Section 3.7. Burst-by-burst adaptive transmissions can be readily accommodated by JD-CDMA receivers, as it will be augmented in more detail below. The duplex JD-ACDMA video transceiver used in our system design example operates on the basis of the following philosophy.

- The channel quality estimation is based on evaluating the Mean Squared Error (MSE) at the output of the JD-CDMA multi-user equalizer at the receiver, as suggested for wideband single-carrier Kalman-filtered DFE-based modems by Liew *et al.* in [236].

- The decision concerning the modem mode to be used by the local transmitter for the forthcoming CDMA transmission burst is based on the prediction of the expected channel quality.

- Specifically, if the channel quality can be considered predictable, then the channel quality estimate for the UL can be extracted from the received signal and *the receiver instructs the local transmitter as to what modem mode to use in its next transmission burst. We refer to this regime as open–loop adaptation.* In this case, the transmitter has to explicitly signal the modem modes to the receiver.

- By contrast, if the channel cannot be considered reciprocal, then the channel quality estimation is still performed at the receiver, but *the receiver has to instruct the remote transmitter as to what modem modes have to be used at the transmitter, in order to meet the target integrity requirements of the receiver. We refer to this mode as closed–loop adaptation.*

3.7.3   The JD-ACDMA Video Transceiver

In this JD-CDMA system performance study we transmitted 176×144 pixel Quarter Common Intermediate Format (QCIF) and 128×96 pixel Sub-QCIF (SQCIF) video sequences at 10 frames/s using a reconfigurable Time Division Multiple Access/Code

Division Multiple Access (TDMA / CDMA) transceiver, which can be configured as a 1, 2 or 4 bit/symbol scheme. The H.263 video codec [237] extensively employs variable-length compression techniques and hence achieves a high compression ratio. However, as all entropy- and variable-length coded bit streams, its bits are extremely sensitive to transmission errors.

This error sensitivity was counteracted in our system by invoking the adaptive video packetization and video packet dropping regime of [133], when the channel codec protecting the video stream became incapable of removing all channel errors. Specifically, we refrained from decoding the corrupted video packets in order to prevent error propagation through the reconstructed video frame buffer [133, 237]. Hence—similarly to our AQAM/TDD-based system design example—these corrupted video packets were dropped at both the transmitter and receiver and the reconstructed video frame buffer was not updated, until the next video packet replenishing the specific video frame area was received. This required a low-delay, strongly protected video packet acknowledgement flag, which was superimposed on the transmitted payload packets [133]. As in the system design example of the previous section, the associated video performance degradation was found perceptually unobjectionable for transmission burst error rates below about 5%.

The associated JD-ACDMA video system parameters are summarized in Table 3.7, which will be addressed in more depth during our further discourse. Employing a low spreading factor of 16 allowed us to improve the system's multi-user performance with the aid of joint-detection techniques [95], whilst imposing a realistic implementational complexity. This is because the JD operation is based on inverting the system matrix, which is constructed from the convolution of the channel's impulse response (CIR) and the spreading codes. Hence maintaining a low spreading factor (SF) is critical as to the implementational complexity. We note furthermore that the implementation of the joint detection receivers is independent of the number of bits per symbol associated with the modulation mode used, since the receiver simply inverts the associated system matrix and invokes a decision concerning the received symbol, irrespective of how many bits per symbol were used. *Therefore, joint detection receivers are amenable to amalgamation with the above 1, 2 and 4 bit/symbol CDMA modem, since they do not have to be reconfigured each time the modulation mode is switched.*

In this performance study we used the Pan-European FRAMES proposal [224] as the basis for our CDMA system. The associated transmission burst structure is shown in Figure 3.13, while a range of generic system parameters are summarized in Table 3.6. In our

Figure 3.13: Transmission burst structure of the FMA1 spread speech/data mode 2 of the FRAMES proposal [224].

Table 3.6: Generic system parameters using the FRAMES spread speech/data mode 2 proposal [224].

Parameter	
Multiple access	TDMA/CDMA
Channel type	COST 207 Bad Urban
Number of paths in channel	7
Normalized Doppler frequency	3.7×10^{-5}
CDMA spreading factor	16
Spreading sequence	Random
Tx. Frame duration	4.615 ms
Tx. Slot duration	577 μs
Joint detection CDMA receiver	Whitening matched filter (WMF) or Minimum mean square error block decision feedback equalizer (MMSE-BDFE)
No. of Slots/Frame	8
TDMA slots/Video packet	3
Chip Periods/TDMA slot	1250
Data Symbols/TDMA slot	68
User Data Symbol Rate (kBd)	14.7
System Data Symbol Rate (kBd)	117.9

performance studies we used the COST207 [77] seven-path bad urban (BU) channel model, whose impulse response is portrayed in Figure 3.14.

Again, the remaining generic system parameters are defined in Table 3.6. In our JD-ACDMA design example we investigated the performance of a multi-mode convolutionally coded video system employing joint detection, while supporting two users. The associated convolutional codec parameters are summarized in Table 3.7 along with the operational-mode specific transceiver parameters of the multi-mode JD-ACDMA system. As seen in Table 3.7, when the channel is benign, the unprotected video bit rate will be approximately 26.9 kbit/s in the 16QAM/JD-CDMA mode. However, as the channel quality degrades, the modem will switch to the BPSK mode of operation, where the video bit rate drops to 5 kbit/s and for maintaining a reasonable video quality, the video resolution has to be reduced to SQCIF (128 × 96 pels).

3.7.4 JD-ACDMA Video Transceiver Performance

The burst-by-burst adaptive JD-ACDMA scheme of our design example maximizes the system's throughput expressed in terms of the number of bits per transmitted non-binary symbol by allocating the highest possible number of bits to a symbol based on the receiver's perception concerning the instantaneous channel quality. When the instantaneous channel conditions degrade, the number of bits per symbol (BPS) is reduced in order to maintain the required target transmission burst error rate. Figure 3.15 provides a snap-shot of the JD-ACDMA system's mode switching dynamics, which is based on the fluctuating channel conditions determined by all factors influencing the channel's quality, such as pathloss, fast-fading, slow-fading, dispersion, co-channel interference, etc. The adaptive modem uses the

Figure 3.14: Normalized channel impulse response for the COST 207 [77] seven-path Bad Urban channel.

Table 3.7: Operational-mode specific JD-ACDMA video transceiver parameters used in our design example.

Features	Multi-rate System		
Mode	BPSK	4QAM	16QAM
Bits/Symbol	1	2	4
FEC	Convolutional Coding		
Octal Gen. Pol.	561; 753		
Coding-rate	$R = 1/2$		
Constraint-length	$K = 9$		
Transmitted bits/packet	204	408	816
Total bit rate (kbit/s)	14.7	29.5	58.9
FEC-coded bits/packet	102	204	408
Assigned to FEC-coding (kbit/s)	7.4	14.7	29.5
Error detection per packet	16 bit CRC		
Feedback bits/packet	9		
Video packet size	77	179	383
Packet header bits	8	9	10
Video bits/packet	69	170	373
Unprotected video-rate (kbit/s)	5.0	12.3	26.9
Video framerate (Hz)	10		

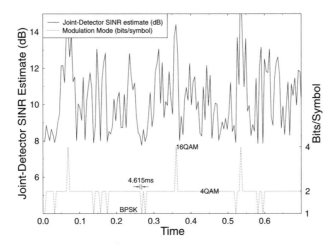

Figure 3.15: Example of modem mode switching in a dynamically reconfigured burst-by-burst modem in operation, where the modulation mode switching is based upon the SINR estimate at the output of the joint-detector over the channel model of Figure 3.14.

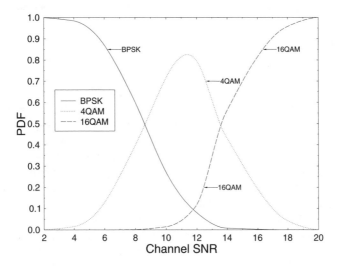

Figure 3.16: PDF of the various adaptive modem modes versus channel SNR over the channel model of Figure 3.14.

SINR estimate at the output of the joint-detector, in order to estimate the instantaneous channel quality, and hence to set the modulation mode. The probability density function (PDF) of the JD-ACDMA scheme using each modulation mode for a particular average channel SNR is portrayed in Figure 3.16. It can be seen at high channel SNRs that the modem predominantly uses the 16QAM/JD-ACDMA modulation mode, while at low channel SNRs the BPSK mode is most prevalent. However, the PDF is widely spread, indicating that often the channel quality is misjudged by the receiver due to unpredictable channel quality fluctuations caused by a high doppler frequency or co-channel interference, etc. Hence in

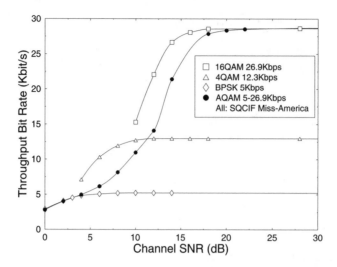

Figure 3.17: Throughput bit rate versus channel SNR comparison of the three fixed modulation modes (BPSK, 4QAM, 16QAM) and the adaptive burst-by-burst modem (AQAM), both supporting two users with the aid of joint detection over the channel model of Figure 3.14.

certain cases BPSK is used under high channel quality conditions or 16QAM is employed under hostile channel conditions.

The advantage of the dynamically reconfigured burst-by-adaptive JD-ACDMA modem over a statically reconfigured system, which would be incapable of near-instantaneous channel quality estimation and modem mode switching is that the video quality is smoothly—rather than abruptly—degraded, as the channel conditions deteriorate and vice versa. By contrast, a less "agile" statically switched or reconfigured multi-mode system results in more visible reductions in video quality, when the modem switches to a more robust modulation mode, as it is demonstrated in Figure 3.17. Explicitly, Figure 3.17 shows the throughput bit rate of the dynamically reconfigured burst-by-burst adaptive modem, compared to the three modes of a less agile, statically switched multi-mode system. The reduction of the fixed modem modes' effective throughput at low SNRs is due to the fact that under such channel conditions an increased fraction of the transmitted packets have to be dropped, reducing the effective throughput, since dropped packets do not contribute towards the system's effective throughput. The figure shows the smooth reduction of the throughput bit rate, as the channel quality deteriorates. The burst-by-burst modem matches the BPSK mode's bit rate at low channel SNRs, and the 16QAM mode's bit rate at high SNRs. In this example the dynamically reconfigured burst-by-burst adaptive modem characterized in the figure perfectly estimates the prevalent channel conditions although in practice the estimate of channel quality is not perfect and it is inherently delayed. Hence our results constitute the best-case performance.

The smoothly varying effective throughput bit rate of the burst-by-burst adaptive modem translates into a smoothly varying video quality, as the channel conditions change. The video quality measured in terms of the average PSNR is shown versus the channel SNR in Figure 3.18 in contrast to that of the individual modem modes. The figure demonstrates that the burst-by-burst adaptive modem provides equal or better video quality over a large

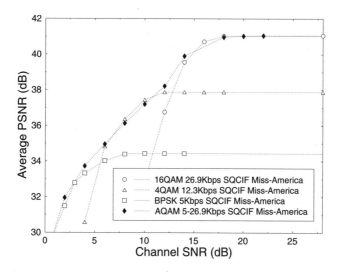

Figure 3.18: Average decoded video quality (PSNR) versus channel SNR comparison of the fixed modulation modes of BPSK, 4QAM and 16QAM, and the burst-by-burst adaptive modem. Both supporting two-users with the aid of joint detection. These results were recorded for the Miss-America video sequence at SQCIF resolution (128 × 96 pels) over the channel model of Figure 3.14.

proportion of the SNR range shown than the individual modes. However, even at channel SNRs, where the adaptive modem has a slightly reduced PSNR, the perceived video quality of the adaptive modem is better, since the video packet loss rate is far lower than that of the fixed modem modes.

3.8 Subband-adaptive OFDM Video Transceivers

In order to demonstrate the benefits of the proposed near-instantaneously adaptive video transceivers also in the context of OFDM schemes [13, 226], in this section we compare the performance of a subband-adaptive OFDM video scheme [192] to that of a fixed modulation mode transceiver under identical propagation conditions, while having the same transmission bit rate. The subband-adaptive modem is capable of achieving a lower BER, since it can disable transmissions over low quality sub-carriers and compensate for the lost throughput by invoking a higher-order modulation mode, than that of the fixed-mode transceiver over the high-quality sub-carriers.

Table 3.8 shows the system parameters for the fixed-mode BPSK and QPSK transceivers, as well as for the corresponding AOFDM transceivers. The system employs constraint length three, half-rate turbo coding, using octal generator polynomials of 5 and 7 as well as random turbo interleavers, where the channel- and turbo-interleaver depth was adjusted for each AOFDM transmission burst, in order to facilitate burst-by-burst or symbol-by-symbol based OFDM demodulation and turbo decoding. Therefore the unprotected bit rate is approximately half the channel coded bit rate. The protected to unprotected video bit rate ratio is not exactly half, since two tailing bits are required to reset the convolutional encoders' memory

Table 3.8: System parameters for the fixed QPSK and BPSK transceivers, as well as for the corresponding subband-adaptive OFDM (AOFDM) transceivers for Wireless Local Area Networks (WLANs).

	BPSK mode	QPSK mode
Packet rate	4687.5 Packets/s	
FFT length	512	
OFDM symbols/packet	3	
OFDM symbol duration	2.6667 μs	
OFDM time frame	80 Timeslots = 213 μs	
Normalized Doppler frequency, f'_d	1.235×10^{-4}	
OFDM symbol normalized Doppler frequency, F_D	7.41×10^{-2}	
FEC coded bits/packet	1536	3072
FEC-coded video bit rate	7.2 Mbps	14.4 Mbps
Unprotected bits/packet	766	1534
Unprotected bit rate	3.6 Mbps	7.2 Mbps
Error detection CRC (bits)	16	16
Feedback error flag bits	9	9
Packet header bits/packet	11	12
Effective video bits/packet	730	1497
Effective video bit rate	3.4 Mbps	7.0 Mbps

to their default state in each transmission burst. In both the BPSK and QPSK modes 16-bit Cyclic Redundancy Checking (CRC) is used for error detection and 9 bits are used to encode the reverse link feedback acknowledgement information by simple repetition coding. The packet acknowledgement flag decoding ensues using majority logic decisions. The packetization [192] requires a small amount of header information added to each transmitted packet, which is 11 and 12 bits per packet for BPSK and QPSK, respectively. The effective or useful video bit rates for the fixed BPSK and QPSK modes are then 3.4 and 7.0 Mbps.

The fixed-mode BPSK and QPSK transceivers are limited to one and two bits per symbol, respectively. By contrast, the proposed AOFDM transceivers operate at the same bit rate as their corresponding fixed modem mode counterparts, although they can vary their modulation mode on a subband by subband basis between 0, 1, 2 and 4 bits per symbol. Zero bits per symbol implies that transmissions are disabled for the subband concerned.

The "micro-adaptive" nature of the subband-adaptive modem is characterized by Figure 3.19, portraying at the top a contour plot of the channel SNR for each subcarrier versus time. This channel SNR fluctuation was recorded here for the short indoor WLAN channel impulse response of Figure 3.20 having a maximum dispersion of about 60 ns, which was referred to as the short Wireless Asynchronous Transfer Mode (WATM) channel in [13].

At the centre and bottom of the figure the modulation mode chosen for each 32-subcarrier subband is shown versus time for the 3.4 and 7.0 Mbps target-rate subband-adaptive modems, respectively. Again, this was recorded for the short WATM channel impulse response of Figure 3.20. It can be seen that when the channel is of high quality—like for example at about frame 1080—the subband-adaptive modem used the same modulation mode as the equivalent

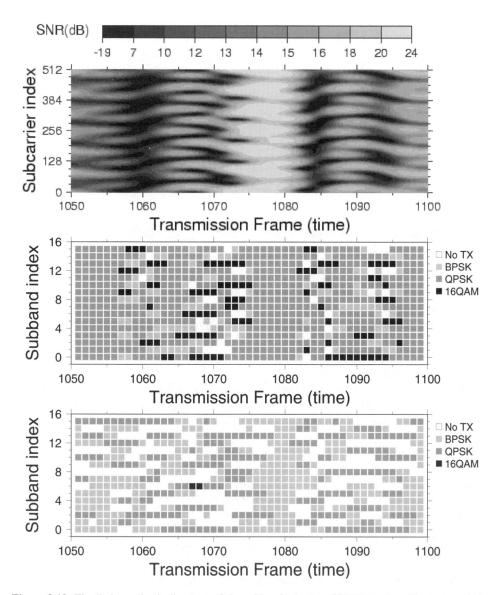

Figure 3.19: The "micro-adaptive" nature of the subband-adaptive OFDM modem. The top graph is a contour plot of the channel SNR for all 512 subcarriers versus time. The bottom two graphs show the modulation modes chosen for all 16 32-subcarrier subbands for the same period of time. The middle graph shows the performance of the 3.4 Mbps subband-adaptive modem, which operates at the same bit rate as a fixed BPSK modem. The bottom graph represents the 7.0 Mbps subband-adaptive modem, which operated at the same bit rate as a fixed QPSK modem. The average channel SNR was 16dB. ©IEEE, 2001, Hanzo, Cherriman, Streit [192].

Figure 3.20: Indoor three-path WATM channel impulse response. ©IEEE, 2001, Hanzo, Webb, Keller [13].

fixed rate modem in all subcarriers. When the channel is hostile—like around frame 1060—the subband-adaptive modem used a lower-order modulation mode in some subbands, than the equivalent fixed mode scheme, or in extreme cases disabled transmission for that subband. In order to compensate for the loss of throughput in this subband a higher-order modulation mode was used in the highest quality subbands.

One video packet is transmitted per OFDM symbol, therefore the video packet loss ratio is the same as the OFDM symbol error ratio. The video packet loss ratio is plotted versus the channel SNR in Figure 3.21. It is shown in the graph that the subband-adaptive transceivers—or synonymously termed as microscopic-adaptive (μAOFDM), in contrast to OFDM symbol-by-symbol adaptive transceivers—have a lower packet loss ratio (PLR) at the same SNR compared to the fixed modulation mode transceiver. Note in Figure 3.21 that the subband-adaptive transceivers can operate at lower channel SNRs than the fixed modem mode transceivers, while maintaining the same required video packet loss ratio. Again, the figure labels the subband-adaptive OFDM transceivers as μAOFDM, implying that the adaptation is not noticeable from the upper layers of the system. A macro-adaption could be applied in addition to the microscopic adaption by switching between different target bit rates on an OFDM symbol-by-symbol basis, as the longer-term channel quality improves and degrades. This issue was further investigated in [192].

The figure shows that when the channel quality is high, the throughput bit rate of the fixed and adaptive transceivers is identical. However, as the channel degrades, the loss of packets due to channel impairments results in a lower throughput bit rate. The lower packet loss ratio of the subband-adaptive transceiver results in a higher throughput bit rate than that of the fixed modulation mode transceiver. Finally, these improved throughput bit rate results translate to the enhanced decoded video quality performance results evaluated in terms of Peak Signal-to-Noise Ratio (PSNR) in Figure 3.22. Again, for high channel SNRs the performance of the fixed and adaptive OFDM transceivers is identical. However, as the channel quality degrades, the video quality of the subband-adaptive transceiver degrades less dramatically than that of the corresponding fixed modulation mode transceiver.

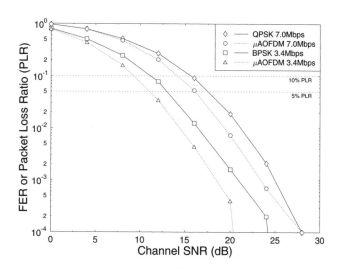

Figure 3.21: Frame Error Rate (FER) or video packet loss ratio (PLR) versus channel SNR for the
BPSK and QPSK fixed modulation mode OFDM transceivers and for the corresponding
subband-adaptive μAOFDM transceiver, operating at identical effective video bit rates,
namely at 3.4 and 7.0 Mbps, over the channel model of Figure 3.20 at a normalized
Doppler frequency of $F_D = 7.41 \times 10^{-2}$. ©IEEE, 2001, Hanzo, Cherriman, Streit [192].

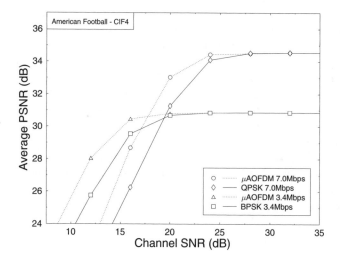

Figure 3.22: Average video quality expressed in PSNR versus channel SNR for the BPSK and
QPSK fixed modulation mode OFDM transceivers and for the corresponding μAOFDM
transceiver operating at identical channel SNRs over the channel model of Figure 3.20 at a
normalized Doppler frequency of $F_D = 7.41 \times 10^{-2}$. ©IEEE, 2001, Hanzo, Cherriman,
Streit [192].

3.9   Summary and Conclusions

In contrast to the statically reconfigured narrow-band multimode video transceivers [192], in this chapter we have advocated BbB-AQAM based wireless transceivers [192]. We justified their service-related benefits in terms of the video quality improvements perceived by the users of such systems. As an example, the channel quality perceived by the channel equalizer or the multi-user equalizer was used for controlling the AQAM modes. When an adaptive packetizer is used in conjunction with the AQAM modem, it continually adjusts the video codec's target bit rate in order to match the instantaneous throughput provided by the adaptive modem.

We have also shown that the delay between the instants of channel estimation and AQAM mode switching has an effect on the performance of the proposed AQAM video transceiver. This performance penalty can be mitigated by reducing the modem mode signaling delay. It was also demonstrated that the system can be tuned to the required FER performance using appropriate AQAM switching thresholds. In harmony with our expectations, we found that the more complex turbo channel codecs were more robust against channel effects than the lower-complexity binary BCH codecs. Finally, the AQAM principles were extended to joint-detection assisted AQAM/CDMA and adaptive OFDM systems, where similar findings were confirmed to those found in the context of unspread AQAM.

It is a natural thought to combine these adaptive transceivers [94, 191–193] with diversity aided Multiple Input, Multiple Output (MIMO) systems and space-time coding [218, 238–241] in a further effort towards mitigating the effects of fading and rendering the channel more Gaussian-like. A vital question in this context is, whether adaptive transceivers retain their performance advantages in conjunction with MIMOs? As expected, no significant joint benefits accrue, since both of these regimes aim at mitigating the effects of fading and once the fading is mitigated sufficiently for it to become near-Gaussian, no further fading counter-measures are necessary. It is worth noting, however that MIMOs have been predominantly studied in the context of narrowband or non-dispersive fading channels or in conjunction with OFDM—a scheme that decomposes a high-rate bit stream into a high number of low-rate bit streams—thereby rendering the dispersive channel non-dispersive for each of the low-rate composite streams.

A further problem, when invoking high-order receiver diversity in an effort to mitigate the effects of fading and hence rendering the wireless channel Gaussian-like is that the receiver complexity increases. It is a more attractive proposition to employ complex, transmit diversity assisted base stations, which allows us to aim for low-complexity terminals. In this context in recent years space-time codecs have found favor and have also been proposed for the IMT2000 system and for multi-user HIPERLAN 2 type systems.

Chapter 4

Intelligent Antenna Arrays and Beamforming

4.1 Introduction

Adaptive beamforming was initially developed in the 1960s for the military applications of sonar and radar, in order to remove unwanted noise and jamming from the output. The related literature of the past 40 years is extremely rich [242–278] and since this book is mainly concerned with the networking aspects of wireless systems, rather than with specific antenna array designs, here we will restrict our discussions on the topic to a rudimentary overview.

The first fully adaptive array was conceived in 1965 by Applebaum [279], which was designed to maximize the Signal-to-Noise Ratio (SNR) at the array's output. An alternative approach to cancelling unwanted interference is the Least Mean Squares (LMS) error algorithm of Widrow [280]. While a simple idea, satisfactory performance can be achieved under specific conditions. Further work on the LMS algorithm, by Frost [281] and Griffiths [282], introduced constraints to ensure that the desired signals were not filtered out along with the unwanted signals. The optimization process takes place as before, but the antenna gain is maintained constant in the desired direction. For stationary signals, both algorithms converge to the optimum Wiener solution [3, 281, 283]. A different technique was proposed in 1969 by Capon [284] using a Minimum-Variance Distortionless Response (MVDR) or the Maximum Likelihood Method (MLM). In 1974, Reed *et al.* demonstrated the power of the Sample-Matrix Inversion (SMI) technique, which determines the adaptive antenna array weights directly [285]. Unlike the algorithms of Applebaum [279] and Widrow [280], which may suffer from slow convergence if the eigenvalue spread of the received sample correlation matrix is relatively large, the performance of the SMI technique is virtually independent of the eigenvalue spread.

In recent years the tight frequency reuse of cellular systems has stimulated renewed research interests in the field [3, 6, 283, 286]. In this book we will attempt to review the recent literature and highlight the most important research issues for UMTS, HiperLAN and WATM applications, while providing some performance results. We commence in Section 4.2

3G, HSPA and FDD versus TDD Networking Second Edition
L. Hanzo, J. S. Blogh and S. Ni © 2008 John Wiley & Sons, Ltd

by reviewing beamforming and its potential benefits, then we provide a generic signal model in Section 4.2.3 and we describe the processes of element and beam space beamforming. In Section 4.3 we highlight a range of adaptive beamforming algorithms and consider the less commonly examined DL scenario in Section 4.3.5. Finally, in Section 4.3.6 we provide some performance results and outline our future work.

4.2 Beamforming

The signals induced in different elements of an antenna array are combined to form a single output of the array. This process of combining the signals from the different elements is known as beamforming. This section describes the basic characteristics of an antenna, the advantages of using beamforming techniques in a mobile radio environment [3, 6], and a generic signal model for use in beamforming calculations. For further details on the associated issues the reader is referred to [3, 6, 8, 279–283, 285–291].

4.2.1 Antenna Array Parameters

Below we provide a few definitions used throughout this report in order to describe antenna systems:

Radiation Pattern. The radiation pattern of an antenna is the relative distribution of the radiated power as a function of direction in space. The radiation pattern of an antenna array is the product of the element pattern and the array factor, both of which are defined below. If $f(\theta, \phi)$ is the radiation pattern of each antenna element and $F(\theta, \phi)$ is the array factor, then the array's radiation pattern, $G(\theta, \phi)$, which is also referred to as the beam pattern, is given by

$$G(\theta, \phi) = f(\theta, \phi)F(\theta, \phi). \tag{4.1}$$

Figure 4.1 gives an example of a stylized antenna element response, an array factor of an 8 element linear array with an element spacing of $\lambda/2$ steered at $0°$ and the radiation pattern, which results from combining the two.

Array Factor. The array factor, $F(\theta, \phi)$, is the far-field radiation pattern of an array of isotropically radiating elements, where θ is the azimuth angle and ϕ is the elevation angle.

Main Lobe. The main lobe of an antenna radiation pattern is the lobe containing the direction of maximum radiated power.

Sidelobes. Sidelobes are lobes of the antenna radiation pattern, which do not constitute the mainlobe. They allow signals to be received in directions other than that of the main lobe and hence they are undesirable, but they are also unavoidable.

Beamwidth. The beamwidth of an antenna is the angular width of the main lobe. The 3 dB beamwidth is the angular width between the points on the main lobe that are 3 dB below the peak of the main lobe. A smaller beamwidth results from an array of a greater aperture size, which is the distance between the two farthest elements of the array.

Antenna Efficiency. Antenna efficiency is the ratio of the total power radiated by the antenna to the total power input to the antenna.

Grating Lobes. When the distance between the antenna array elements, d, exceeds $\lambda/2$, spatial under-sampling of the received radio frequency carrier wave takes place, causing secondary maxima [2, 288], referred to as grating lobes, to appear in the radiation pattern,

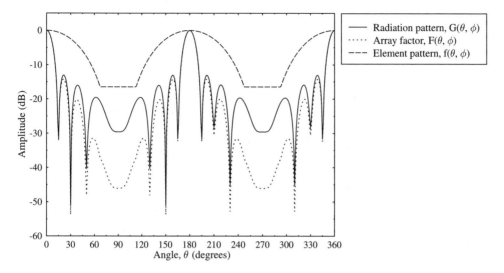

Figure 4.1: The array factor of an eight element linear array with an element spacing of $\lambda/2$ steered at
$0°$, the response of each antenna element and the radiation pattern resulting from combining
the two.

which can be clearly seen in Figure 4.2. The spatial under-sampling results in ambiguities
in the directions of the arriving signals, which manifests itself as copies of the main lobe in
unwanted directions. The grating lobe phenomenon in spatial sampling is analogous to the
well known aliasing effect in temporal sampling [288]. Therefore, the distance, d, between
adjacent sensors in the array must be chosen to be less than or equal to $\lambda/2$, if grating lobes
are to be avoided [288, 292]. However, an inter-element spacing of greater than $\lambda/2$ improves
the spatial resolution of the array [2], i.e. reduces the 3 dB beamwidth as shown in Figure 4.2,
and reduces the correlation between the signals arriving at adjacent antenna elements.

4.2.2 Potential Benefits of Antenna Arrays in Mobile Communications

4.2.2.1 Multiple Beams [6]

The formation of multiple beams, or sectorization, uses multiple antennas at the base station
in order to form beams that cover the whole cell site [292]. For example, three beams, each
with a beamwidth of $120°$ may cover the entire $360°$ as seen in Figure 4.3. The coverage area
of each beam may be regarded as a separate cell, with frequency assignment and handovers
between beams performed in the usual manner [293]. No intelligence is required to locate
a subscriber within a beam and to connect that beam to a radio channel unit. The use of
multiple beams results in a reduction of the co-channel interference. In the UL scenario, the
signal received from the mobile station constitutes interference at only two base stations, and
additionally in only one sector. In the DL, the situation is similar, only now the sectors which
can interfere with the user in the central cell are the images of the interfering sectors on the
UL [19], again, as shown in Figure 4.3.

Figure 4.2: The array factor of an eight element uniform linear array with element spacing of $\lambda/2$ and $3\lambda/2$. The grating lobes associated with the spatial under-sampling-induced secondary maxima of the radiated carrier wave are clearly visible for the case when the element spacing is $3\lambda/2$.

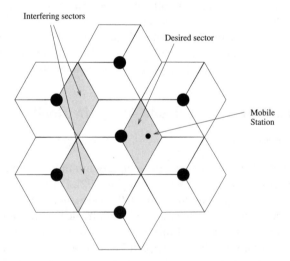

Figure 4.3: An example of sectorization, using three sectors per base station, showing the reduced levels of interference with respect to an omni-directional base station antenna scenario.

Figure 4.4: Switched-diversity combining.

4.2.2.2 Adaptive Beams [6]

The combined antenna array is used to find the location of each mobile, and then beams are formed, in order to cover different mobiles or groups of mobiles [20, 294]. Each beam having its own coverage area may be considered as a co-channel cell, and thus be able to use the same carrier frequency [7, 292]. In conventional sectorization the location of the beams is fixed, while the adaptive system allows the beams to cover specific areas of the cell within which users are located [17]. In intelligent near-future systems the beams may follow the mobiles, which benefit from the concentrated transmission power, with inter-beam handovers occurring as necessary.

4.2.2.3 Null Steering [6, 295]

In contrast to steering beams towards mobiles, null steering creates spatial radiation nulls towards co-channel mobiles [65]. The realization of true nulls or zero response is not possible due to practical considerations, such as the isolation of the radio frequency components. The formation of spatial radiation nulls in the antenna response towards co-channel mobiles reduces the co-channel interference both on the UL and the DL [2, 294].

4.2.2.4 Diversity Schemes [6, 296]

The simplest and most commonly used diversity scheme is *switched diversity*. In this scheme the system switches between antennas, such that only one is in use at any one time [1, 297], as shown in Figure 4.4. The switching criterion is often the loss of received signal level at the antenna being used. The switching may be performed at the Radio Frequency (RF) stage, avoiding the need for a down-converter for each antenna.

Selection diversity is a more sophisticated version of switched diversity, where the system can monitor the signal level on all of the antennas simultaneously, and select the specific branch exhibiting the highest SNR at any given time, thus requiring an RF front-end for each antenna in the system [1], as seen in Figure 4.5.

In a Rayleigh fading environment, the fading at each branch can be assumed to be independent provided that the branches are sufficiently far apart. If each branch has an instantaneous SNR of γ_l, the probability density function of γ_l is given by [3]

$$p(\gamma_l) = \frac{1}{\Gamma} e^{\frac{-\gamma_l}{\Gamma}} \tag{4.2}$$

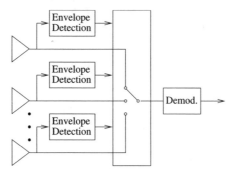

Figure 4.5: Selective-diversity combining.

where Γ denotes the mean SNR at each branch. The probability that a single branch has a SNR less than some threshold γ is given by [3]

$$P[\gamma_l \leq \gamma] = \int_0^\infty p(\gamma_l)d\gamma_l = 1 - e^{-\frac{\gamma}{\Gamma}}. \tag{4.3}$$

Therefore, the probability that all the branches fail to achieve an SNR higher than γ is [3]:

$$P_L(\gamma) = P[\gamma_1, \gamma_2, \ldots, \gamma_L \leq \gamma] = \left(1 - e^{-\frac{\gamma}{\Gamma}}\right)^L, \tag{4.4}$$

from which the probability density function of the fading magnitude in conjunction with selection diversity can be obtained,

$$p_L(\gamma) = \frac{d}{d\gamma}P_L(\gamma) = \frac{L}{\Gamma}\left(1 - e^{-\frac{\gamma}{\Gamma}}\right)^{L-1}e^{-\frac{\gamma}{\Gamma}}, \tag{4.5}$$

leading to the average SNR, $\bar{\gamma}$, of selection diversity assisted Rayleigh fading channels as [3]:

$$\bar{\gamma} = \int_0^\infty \gamma p_L(\gamma)d\gamma = \Gamma\sum_{l=1}^L \frac{1}{l}. \tag{4.6}$$

In maximal ratio combining, which is also often referred to as optimal diversity combining, the signal of each antenna is weighted by its instantaneous SNR. The weighted signals are then combined for forming a single output, as shown in Figure 4.6. It has been shown that the maximal ratio combining technique is optimal, if the diversity branch signals are uncorrelated and follow a Rayleigh distribution [21], provided that the noise has a Gaussian distribution and a zero mean. If each branch has a gain, g_l, the output of the combiner is [3]

$$s_L = \sum_{l=1}^L g_l s_l, \tag{4.7}$$

and if each branch has the noise power, σ_n^2, the total noise power at the output of the combiner is [3]:

$$\sigma_N^2 = \sigma_n^2 \sum_{l=1}^L g_l^2. \tag{4.8}$$

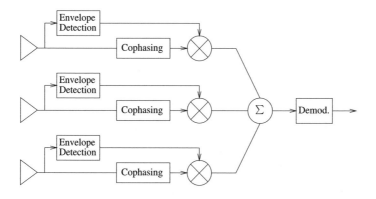

Figure 4.6: Optimal combining.

Therefore, the SNR at the output of the combiner is given by

$$\gamma_L = \frac{s_l^2}{2\sigma_N^2}. \tag{4.9}$$

It can be easily shown that γ_L is maximized, when $g_l = s_l^2/\sigma_n^2$, which is the SNR in each branch. The expansion of Equation 4.9 is thus

$$\gamma_L = \frac{1}{2} \frac{\left(\sum_{l=1}^{L} \frac{s_l^2}{\sigma_n^2} s_l\right)^2}{\sigma_n^2 \sum_{l=1}^{L} \left(\frac{s_l^2}{\sigma_n^2}\right)^2} = \frac{1}{2} \sum_{l=1}^{L} \frac{s_l^2}{\sigma_n^2} = \sum_{l=1}^{L} \gamma_l. \tag{4.10}$$

As γ_L has a chi-squared distribution [3], the probability density function of γ_L is [3]:

$$p(\gamma_L) = \frac{\gamma_L^{L-1} e^{-\frac{\gamma_L}{\Gamma}}}{\Gamma^L (L-1)!}. \tag{4.11}$$

The probability that γ_L is less than the threshold, γ, is [3]

$$P[\gamma_L \le \gamma] = \int_0^{\gamma} p(\gamma_L) d\gamma_L = 1 - e^{-\frac{\gamma}{\Gamma}} \sum_{l=1}^{L} \frac{(\frac{\gamma}{\Gamma})^{l-1}}{(l-1)!}. \tag{4.12}$$

The expectation of Equation 4.12, $\bar{\gamma}_L$, is the average SNR at the output of the combiner:

$$\bar{\gamma}_L = \sum_{l=1}^{L} \bar{\Gamma} = L\Gamma, \tag{4.13}$$

where Γ is the mean SNR at each branch.

Optimal combining processes the signals received from an antenna array such that the contribution from unwanted co-channel sources is reduced, whilst enhancing that of the desired signal. The explicit knowledge of the directions of the interferences is not

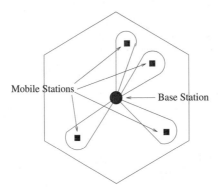

Figure 4.7: A cell layout showing how an antenna array can support many users on the same carrier frequency and timeslot with the advent of spatial filtering or Space Division Multiple Access (SDMA).

necessary, but some characteristics of the desired signal are required in order to protect it from cancellation as if it were an unwanted co-channel source [6]. A popular technique is to use a reference signal, such as a channel sounding sequence, which must be correlated with the desired signal. The scheme then phase-coherently combines all the signals that are correlated with the reference signal, whilst simultaneously cancelling the waveforms that are not correlated with this signal, resulting in the removal of co-channel interferences.

A base station using an optimal combining antenna array may adjust the array weights during the receive cycle, in order to enhance the signal arriving from a desired mobile. A system using the same frequency for receiving and transmitting the signals in different timeslots, such as in the Time Division Duplex (TDD) Digital European Cordless Telephone (DECT) [298, 299] system may be able to use the complex conjugate of these weights during the transmit cycle in order to pre-process the transmit signal and to enhance the signal received at the desired mobile, whilst suppressing this signal at the other mobiles. This process relies on the fact that the weights were adjusted during the receive cycle to reduce co-channel interference, thus placing nulls in the directions of co-channel mobiles [6]. Therefore, by employing the complex conjugate of these weights during the transmit cycle, the same antenna pattern may be produced, resulting in no energy transmitted towards the co-channel mobiles [6].

4.2.2.5 Reduction in Delay Spread and Multipath Fading

Delay spread is caused by multipath propagation, where a desired signal arriving from different directions is delayed due to the different distances travelled [17]. In transmit mode an intelligent antenna is able to focus the energy in the required direction, assisting in reducing the multipath reflections and thus delay spread. In receive mode the antenna array is able to perform optimal combining after delay compensation of the multipath signals incident upon it [1]. Those signals whose delays cannot be compensated for may be cancelled by the formation of nulls in their directions [18].

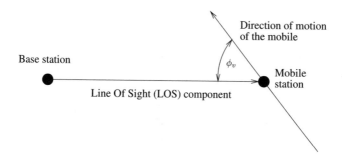

Figure 4.8: Illustration of the Line Of Sight (LOS) component arriving at the mobile from the base station showing the direction of motion of the mobile, ϕ_v.

The directive nature of an antenna array also results in a smaller spread of Doppler frequencies encountered at the mobile [300]. For an omni-directional antenna at both the base station, and at the mobile the Direction-Of-Arrival (DOA) at the mobile is uniformly distributed. Hence the Doppler spectrum is given by Clarke's model [21] as:

$$S_r(f) = \frac{A_o^2}{\pi f_m \sqrt{1 - (f/f_m)^2}}, \qquad |f| < f_m \qquad (4.14)$$

where A_o is the mean power transmitted and $f_m = v/\lambda$ is the maximum Doppler shift, where v is the velocity of the mobile and λ is the carrier wavelength. However, if a directional antenna is used at the base station then the Doppler power spectral density is given by [300]:

$$S_r(f) = \frac{A_o^2}{f_m \sqrt{1 - (f/f_m)^2}}$$
$$\times \left[f_\theta(\phi_v + |\cos^{-1}(f/f_m)|) + f_\theta(\phi_v - |\cos^{-1}(f/f_m)|) \right], \quad |f| < f_m, \qquad (4.15)$$

where ϕ_v, as shown in Figure 4.8, is the direction of motion of the mobile with respect to the direction of the base station from the mobile and $f_\theta()$ is the PDF of the DOA of the multipath components at the mobile, as given by [300]:

$$f_\theta(\theta) = \begin{cases} \dfrac{R^2}{I}, & -\theta_1 < \theta \le \theta_1 \\[2mm] \dfrac{(D\tan(\alpha))^2}{I(\sin(\theta) + \cos(\theta)\tan(\alpha))^2}, & \theta_1 < |\theta| \le \theta_2 \\[2mm] \dfrac{R^2}{I}, & \theta_2 < \theta \le -\theta_2 \end{cases} \qquad (4.16)$$

where

$$I = 2R^2(\pi + \theta_1 - \theta_2) + 4D\sin(\alpha)\sqrt{R^2 - D^2\sin^2(\alpha)}. \qquad (4.17)$$

Furthermore, 2α is the beamwidth of the so-called idealized "flat-top" directional antenna, which has zero gain except over the angular spread of 2α, where the gain is 1, R is the radius

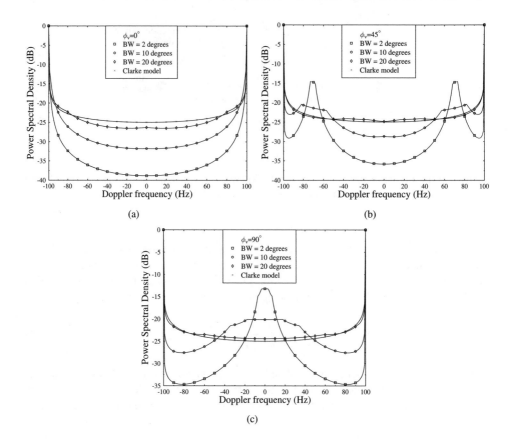

Figure 4.9: Doppler spectra at the mobile for: (a) $\phi_v = 0°$; (b) $\phi_v = 45°$; (c) $\phi_v = 90°$, when using a directional antenna at the base station, and an omnidirectional antenna at the mobile, is compared with Clarke's model. $R = 1$ km, $D = 3$ km, $f_m = 100$ Hz.

of the circular area containing all the scatters and D is the separation distance between the base station and the mobile. Finally, θ_1 and θ_2 are constants calculated using

$$\theta = \cos^{-1}\left[\frac{D}{R}\sin^2(\alpha) \pm \frac{\cos(\alpha)}{R}\sqrt{R^2 - D^2\sin^2(\alpha)}\right].$$

Figure 4.9 shows examples of the Doppler spectra for beamwidths of 2, 10 and 20 degrees for a mobile moving at angles of 0, 45 and 90 degrees with respect to the main LOS component, with a base station to mobile distance of 3 km, where the scatterers are all located within a circle of 1 km radius of the mobile.

4.2.2.6   Reduction in Co-channel Interference

An antenna array allows the implementation of spatial filtering, as shown in Figure 4.7, which may be exploited in both transmitting as well as receiving modes in order to reduce co-channel interferences [1, 2, 14, 15]. When transmitting, the antenna is used to focus the radiated energy in order to form a directive beam in the area, where the receiver is likely to be. This in turn means that there is less interference in the other directions, where the beam is not pointing. The co-channel interference generated in transmit mode may be further reduced by forming beams exhibiting nulls in the directions of other receivers [6, 16]. This scheme deliberately reduces the transmitted energy in the direction of co-channel receivers and hence requires prior knowledge of their positions.

The employment of antenna arrays for reducing co-channel interference in the receive mode has been reported widely [1, 2, 6, 16–18]. It does not require knowledge of the co-channel interference, but must have some information concerning the desired signal, such as the direction of its source, a reference signal, such as a channel sounding sequence, or a signal that is correlated with the desired signal.

4.2.2.7   Capacity Improvement and Spectral Efficiency

The spectral efficiency of a network refers to the amount of traffic a given system with a certain spectral allocation could handle. An increase in the number of users of the mobile communications system without a loss of performance increases the spectral efficiency. Channel capacity refers to the maximum data rate a channel of a given bandwidth can sustain. An improved channel capacity leads to an ability to support more users of a specified data rate, implying a better spectral efficiency. The increased QoS that results from the reduced co-channel interference and reduced multipath fading [18,19] upon using smart antennas may be exchanged for an increased number of users [2, 20].

4.2.2.8   Increase in Transmission Efficiency

An antenna array is directive in nature, having a high gain in the direction where the beam is pointing. This property may be exploited in order to extend the range of the base station, resulting in a larger cell size or may be used to reduce the transmitted power of the mobiles. The employment of a directive antenna allows the base station to receive weaker signals than an omni-directional antenna. This implies that the mobile can transmit at a lower power and its battery life becomes longer, or it would be able to use a smaller battery, resulting in a smaller size and weight, which is important for hand-held mobiles. A corresponding reduction in the power transmitted from the base station allows the use of electronic components having lower power ratings and therefore, lower cost.

4.2.2.9   Reduction in Handovers

When the amount of traffic in a cell exceeds the cell's capacity, cell splitting is often used in order to create new cells [2], each with its own base station and frequency assignment. The reduction in cell size leads to an increase in the number of handovers performed. By using antenna arrays to increase the capacity of a cell [1] the number of handovers required may actually be reduced. Since each beam tracks a mobile [2], no handover is necessary, unless different beams using the same frequency cross each other.

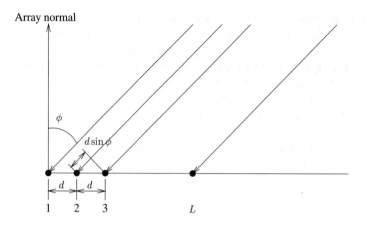

Figure 4.10: Reception by a uniformly spaced linear antenna array.

4.2.3   Signal Model

Consider an array of L omni-directional antenna elements situated in the far field of a sinusoidal point source, as shown in Figure 4.10. Given that the array element separation is d and the plane wavefront is impinging upon the array at an angle of θ with respect to the array normal, the wavefront arrives at the $l + 1^{th}$ element before arriving at the lth element. Again, as seen in Figure 4.10, the extra distance that the wavefront must travel to reach the lth element relative to the $l + 1^{th}$ element is $d \sin \theta$. However, for an arbitrary array of L elements the relative delays, assuming that the point of zero delay is the origin, are given by

$$t_l(\theta) = \frac{x_l \sin \theta + y_l \cos \theta}{c}, \qquad l = 1, \dots, L \qquad (4.18)$$

where c is the speed of wave propagation, i.e. the speed of light, while x_l and y_l are the x and y-coordinates of the lth element with respect to the origin located at $(0,0)$. The extra cosine term is due to the potential y-offset from the x-axis of the array elements which is zero, and thus omitted, from the example shown in Figure 4.10. The signal, $x_{l,i}(t)$, induced in the lth element due to the ith source can be expressed as

$$x_{l,i}(t) = m_i(t)e^{j\omega t_l(\theta)}, \qquad (4.19)$$

with $m_i(t)$ denoting the complex modulating function. This expression is based upon the narrow-band assumption for array signal processing, which assumes that the bandwidth of the signal is sufficiently small, so that the weighting co-efficients maintain a constant phase variation across all of the antenna array elements.

Assuming M directional sources and isotropic background noise, the total signal at the lth element is

$$x_l = \sum_{i=1}^{M} m_i(t)e^{j\omega t_l(\theta)} + n_l(t), \qquad (4.20)$$

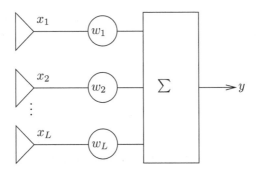

Figure 4.11: A beamformer sums the weighted antenna element signals, yielding the received signal $y(t) = \sum_{l=1}^{L} w_l^* x_l(t)$.

where $n_l(t)$ is a random noise component on the lth antenna array element, which includes background noise and electronic noise. It is assumed to be white noise with a mean of zero and a variance of σ_n^2.

The array factor, $F(\theta)$ which was introduced in Section 4.2.1 may be calculated thus as:

$$F(\theta) = \sum_{l=1}^{L} w_l e^{-j\omega t_l(\theta)}, \qquad (4.21)$$

where w_l is the complex weighting applied to the lth element to steer the antenna beam in the direction of θ_0. The maximum value of $F(\theta)$ will occur when $\theta = \theta_0$, as shown previously in Figure 4.1.

Consider the narrow-band receiving beamformer, shown in Figure 4.11, where signals from each element are multiplied by a complex weight, $w_l, l = 1, \ldots, L$ and summed, in order to form the array output. The array output, $y(t)$ in Figure 4.11, at time t is given by

$$y(t) = \sum_{l=1}^{L} w_l^* x_l(t), \qquad (4.22)$$

where * denotes the complex conjugate, $x_l(t)$ is the signal arriving from the lth element of the array, and w_l is the weight applied to the lth element. Representing the weights of the beamformer of Figure 4.11 as:

$$\underline{w} = [w_1, w_2, \ldots, w_L]^T, \qquad (4.23)$$

and the signals induced in all elements as

$$\underline{x} = [x_1(t), x_2(t), \ldots, x_L(t)]^T, \qquad (4.24)$$

the output of the beamformer receiver in Figure 4.11 becomes

$$y(t) = \underline{w}^T \underline{x}(t), \qquad (4.25)$$

where the superscripts T and H, respectively, denote the transpose and complex conjugate transpose of a vector or matrix.

Let R define the L-by-L correlation matrix of the signal received by the L elements:

$$R = E[\underline{x}(t)\underline{x}^H(t)] = E\left\{\begin{bmatrix} x_1(t) \\ x_2(t) \\ \vdots \\ x_L(t) \end{bmatrix} \begin{bmatrix} x_1^*(t) & x_2^*(t) & \cdots & x_L^*(t) \end{bmatrix}\right\}, \quad (4.26)$$

where the superscript H denotes Hermitian transposition (i.e., transposition combined with complex conjugation).

The correlation matrix R may be expressed in the expanded form:

$$R = \begin{bmatrix} r(0) & r(1) & \cdots & r(L-1) \\ r(-1) & r(0) & \cdots & r(L-2) \\ \vdots & \vdots & \ddots & \vdots \\ r(-L+1) & r(-L+2) & \cdots & r(0) \end{bmatrix}. \quad (4.27)$$

The element $r(0)$ on the main diagonal is always real-valued. For complex-valued data, the remaining elements of R assume complex values. The correlation matrix of a stationary discrete-time stochastic process is Hermitian [288], i.e. $R^H = R$. Alternatively, this may be written as $r(-k) = r^*(k)$, where $r(k)$ is the autocorrelation function of the stochastic process for a lag of k. Therefore, Equation 4.27 may be rewritten as

$$R = \begin{bmatrix} r(0) & r(1) & \cdots & r(L-1) \\ r^*(1) & r(0) & \cdots & r(L-2) \\ \vdots & \vdots & \ddots & \vdots \\ r^*(L-1) & r^*(L-2) & \cdots & r(0) \end{bmatrix}. \quad (4.28)$$

The elements of the matrix, R, denote the correlation between the output signals of the various antenna elements of Figure 4.11. For example, R_{ij} denotes the correlation between the ith and the jth elements of the array. Given that the steering vector associated with the direction θ_i, or the i_{th} source, can be described by an L-dimensional complex vector $\underline{s}_i$ as [283],

$$\underline{s}_i = [\exp(j\omega t_1(\theta_i)), \ldots, \exp(j\omega t_L(\theta_i))]^T, \quad (4.29)$$

where L is the number of elements in the antenna array, and t_i is the time delay taken by a plane wave arriving from the ith source, located in the direction θ_i, and measured from the element at the origin, then the correlation matrix, R, of the array elements' outputs in Figure 4.11 may be expressed as [283]:

$$R = \sum_{i=1}^{M} p_i \underline{s}_i \underline{s}_i^H + \sigma_n^2 I, \quad (4.30)$$

where p_i is the power of the ith source, σ_n^2 is the noise power and I is the identity matrix.

Using matrix notation, the correlation matrix, R, may be expressed in the following form [283, 301]:

$$R = ASA^H + \sigma_n^2 I = U\Lambda U^H, \quad (4.31)$$

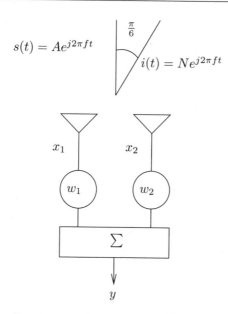

Figure 4.12: Example of a beamforming receiver problem with a wanted signal at 0° and interfering signal at 30° using an array element spacing of $\lambda/2$.

where $S = E[\underline{s}_i\underline{s}_i^H]$ is the covariance matrix of the array elements' outputs in Figure 4.11, $A = [\underline{s}_1, \underline{s}_2, \ldots, \underline{s}_M]$ and is the $L \times M$ matrix of steering vectors. Furthermore, the diagonal matrix $\Lambda = \text{diag}[\lambda_1, \lambda_2, \ldots, \lambda_L]$ is constituted by the real eigenvalues of R, while U contains the corresponding unit-norm eigenvectors of R.

4.2.4 A Beamforming Example

Consider the antenna array shown in Figure 4.12, which consists of two omni-directional antenna elements having a spacing of $\frac{\lambda}{2}$. The desired unmodulated carrier signal, $s(t) = Ae^{j2\pi ft}$, arrives from the angle of $\theta_s=0$ radians. The interfering signal, $i(t) = Ne^{j2\pi ft}$, arrives from the direction of $\theta_i=\frac{\pi}{6}$ radians or 30°. Both signals have the same frequency, f. The signal arriving from each antenna array element is multiplied by a variable complex weight, and the weighted signals are then summed in order to form the array output. The array output due to the desired signal is

$$y_s(t) = Ae^{j2\pi ft}(w_1 + w_2). \tag{4.32}$$

For the array output, $y(t)$ in Figure 4.12, to be the desired signal $s(t)$, the following equation must be satisfied:

$$Ae^{j2\pi ft}(w_1 + w_2) = Ae^{j2\pi ft}, \tag{4.33}$$

which leads to

$$\begin{aligned} \Re[w_1] + \Re[w_2] &= 1 \\ \Im[w_1] + \Im[w_2] &= 0. \end{aligned} \tag{4.34}$$

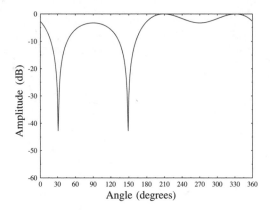

Figure 4.13: The beam pattern produced using Equation 4.21 for a two element array with an element spacing of $\lambda/2$ and element weights of $0.5 \pm j0.5$. The desired signal is at $0°$, the interference is at $30°$, while SNR $= 9.0$ dB and INR $= 9.0$ dB.

The interfering signal arrives at the second array element with a phase lead of $\frac{\pi}{2}$ relative to the first element, since their spacing is $\lambda/2$ and the angle of incidence is $30°$. Therefore, the array output due to the interfering signal is

$$y_i(t) = w_1 N e^{j2\pi ft} + w_2 N e^{j(2\pi ft + \pi/2)}. \tag{4.35}$$

For this to become zero we require that:

$$\begin{aligned} \Re[w_1] - \Re[w_2] = 0 \\ \Im[w_1] + \Im[w_2] = 0. \end{aligned} \tag{4.36}$$

Solving the simultaneous Equations 4.34 and 4.36 yields

$$w_1 = 0.5 - j0.5, w_2 = 0.5 + j0.5. \tag{4.37}$$

The beam pattern obtained using these weights is shown in Figure 4.13. The desired signal at $0°$ is attenuated by about 3 dB, but the unwanted interference at an angle of $30°$ is subjected to an attenuation of more than 30 dB. This example shows how beamforming and the cancellation of unwanted interferences may be accomplished. However, a practical beamformer does not require the information regarding the location, number and nature of the signal sources.

4.2.5   Analog Beamforming

An antenna array consists of a number of antenna elements, the outputs of which are combined via an amplitude and phase control network, in order to form a desired antenna beam [20]. It is possible to perform analog beamforming at the RF stage [20], using phase shifters and amplifiers, however, the high specification required of these devices renders them costly. An alternative solution is to down-convert the RF signal to an Intermediate Frequency (IF) and to perform the beamforming at the IF stage [3]. The disadvantage of this technique

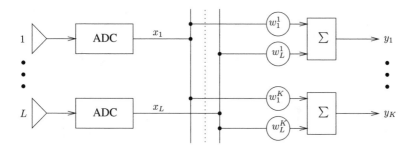

Figure 4.14: An element-space beamformer receiver with L antenna elements capable of forming K beams.

is that each antenna must have its own RF-to-IF receiver. Multiple beamformers must be used to form multiple beams, resulting in the distribution of the signal energy across all the formed beams. The output SNR is thus reduced, when the lower signal energy of the beams is combined with the increased noise injected by the increased number of RF and IF stages.

4.2.6 Digital Beamforming

The philosophy of digital beamforming is similar to that of analog beamforming in that they both adjust the amplitude and phase of the signal arriving from each antenna element, but they use different techniques to reach the same objective. The digitization of the signal received at each antenna element ensures a higher information processing accuracy [295]. The RF signal received at each element is either digitized at RF or down-converted to IF and then digitized using an Analog-to-Digital Convertor (ADC). The digital baseband signals then represent the amplitudes and phases of the signals received at each element of the array [295]. The process of beamforming weights these digital signals, thereby adjusting their amplitudes and phases, such that when added together they form the desired beam [20]. The receivers used in a digital beamforming system need not be as closely matched in phase and amplitude, as in an analog network, since a calibration process can be performed by the controlling software, and any discrepancies can be removed by adjusting the weights appropriately [295].

4.2.7 Element-space Beamforming

The beamforming process described in Sections 4.2.3-4.2.6 is referred to as element-space beamforming, where the digitized data signals, x_l, $l = 1, \ldots, L$, received from the array elements are directly multiplied by a set of weights, $w_l, l = 1, \ldots, L$, in order to form a beam at the desired angle, θ_k. By multiplying the received data signals, $x_1, \ldots, x_L$, by different sets of weights, w_l^k, where $l = 1, \ldots, L$, and $k = 1, \ldots, K$, *it is possible to form beams steered in any direction*, θ_k, where, again $k = 1, \ldots, K$. More explicitly, by multiplying the signal received at each antenna element by a given complex-valued weight, which may be different for each antenna element, the desired signal may be recovered. Each of the beamformers creates an independent beam, at an angle, θ_k, for receiving an arbitrary mobile's signal, by

Figure 4.15: A beam-space beamformer receiver with L antenna elements capable of forming K beams [3].

applying independent weights, $w_l^k, l = 1, \ldots, L, k = 1, \ldots, K$, to the array signals, yielding:

$$y(\theta_k) = \sum_{l=1}^{L} w_l^{k*} x_l, \qquad k = 1, \ldots, K \qquad (4.38)$$

where $y(\theta_k)$ is the output of the beamformer in the direction of source k, $k = 1, \ldots, K$, which is located at the angle θ_k, $x_l(t)$ is a sample from the lth array element and $w_l^k, l = 1, \ldots, L$ represents the weights for forming a beam at angle θ_k. This equation is very similar to Equation 4.22, except for the addition of the superscript k, $k = 1, \ldots, K$ denoting the kth beam.

Figure 4.14 shows an element-space beamformer with L antenna elements, capable of forming K independent beams for receiving K mobiles' signals. Each of the K beams may independently reject sources of interference, whilst receiving the desired signal.

4.2.8 Beam-space Beamforming

In contrast to the method of element-space beamforming, where the signals arriving from each of the L elements are weighted and summed to produce the desired output, the beam-space technique forms multiple fixed beams, using a fixed beamforming network, which may be spatially orthogonal. The output of each beam is then weighted and the resultant signals are combined to produce the desired output [3,283,287,288]. The signals from the beams, which are not used to supply the desired response may be used to cancel unknown interference [288].

Assuming that the outputs from each antenna element are equally weighted and have a uniform phase delay, the response of the array, the array factor $F(\Phi, \alpha)$ in Equation 4.21, produced by an incident plane wave arriving at the antenna array from direction θ, measured with respect to the normal of the antenna array, is given by [288]

$$F(\Phi, \alpha) = \sum_{n=-N}^{N} e^{jn\Phi} e^{-jn\alpha}, \qquad (4.39)$$

where $L = (2N + 1)$ is the total number of elements in the array, $\Phi = \frac{2\pi d}{\lambda} \sin \theta$ is the electrical angle, where d is the inter-elemental distance and α is a constant known as the uniform phase factor. Substituting Φ into Equation 4.39 leads to

$$F(\Phi, \alpha) = \sum_{n=-N}^{N} e^{j\omega t_n(\theta)} e^{-jn\alpha}, \tag{4.40}$$

where $t_n(\theta) = \frac{d \sin \theta}{c}$ and c is the propagation velocity of the received signal. This equation corresponds to Equation 4.21.

For $d = \lambda/2$, we have $\Phi = \pi \sin \theta$ [288]. Summing the geometric series in Equation 4.39, leads to [288]

$$F(\Phi, \alpha) = \frac{\sin[\frac{1}{2}(2N + 1)(\Phi - \alpha)]}{\sin[\frac{1}{2}(\Phi - \alpha)]}. \tag{4.41}$$

By assigning different values to α, the main beam of the antenna may be swept across the range, $-\pi \leq \Phi \leq \pi$. In order to generate an orthogonal set of $2N = L - 1$ beams, the uniform phase factor, α, may be assigned the following values [288]:

$$\alpha = \frac{\pi}{2N + 1} k, \qquad k = \pm 1, \pm 3, \dots, \pm 2N - 1. \tag{4.42}$$

Figure 4.16 illustrates the variations in the magnitude of the array factor, $F(\Phi, \alpha)$, with $-\pi \leq \Phi \leq \pi$ for the case of $2N + 1 = 5$ elements and $\alpha = \pm\pi/5, \pm3\pi/5$. The orthogonal beams generated by the beamforming network represent $2N$ independent directions, one per beam. Depending on the target direction of interest, a particular beam of the set is identified as the main beam and the remainder are viewed as auxiliary beams. From Figure 4.16 it can be seen that each of the auxiliary beams has a null in the direction of the main beam. Because of the fixed nature of these unweighted beams formed by the fixed beamformers of Figure 4.15, individual beam control requires interpolation between beams in order to fine-steer the resultant beam and linear combination of auxiliary beams to create nulls in the direction of interfering sources. Alternatively, beam-space beamforming requires a set of beam-space combiners to generate weighted outputs as shown in Figure 4.15. The Fast Fourier Transform (FFT) block in the diagram generates the orthogonal beams, the process by which this is done is analogous to the performance of an FFT in the time-domain, where it may be viewed as a bank of non-overlapping narrow-band filters whose passbands span the frequency of interest [288]. Hence, the L point FFT generates L spatially orthogonal beams.

4.3   Adaptive Beamforming

An antenna array uses an array of simple antennas, such as omni-directional antennas, and combines the signal induced in these antennas to form the array output. Each antenna forming part of the array is known as an element of the array. The direction where the maximum gain would appear is controlled by adjusting the phase between the different antenna elements. The phase and gain of the signals induced in each array element is adjusted such that the signals due to a source in the direction in which maximum gain is required are added in-phase. An adaptive antenna adjusts these phases and gains, known as weights, so that when the outputs from the antenna elements are combined, the desired output is achieved [6, 289]. The properties of the antenna array may be varied over time in order to optimize the system's performance with respect to different optimization criteria. This

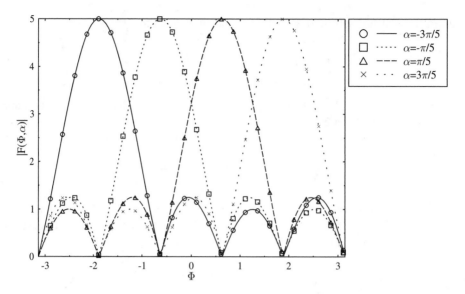

Figure 4.16: The array factor, $F(\Phi, \alpha)$, of a five element antenna array using beam-space beamforming showing the four spatially orthogonal beams that may be generated.

criteria can include maximum power, maximum SNR, minimum interference and maximum Signal to Interference plus Noise Ratio (SINR) [287]. Depending upon the operational environment that the antenna is currently in, it can change its performance metric and control algorithm, in order to provide the best service for the users of the network [302]. For example, conventional beamforming/diversity may be used to give maximum received signal power, while a null steering algorithm results in minimum interference. Finally, maximizing the SINR corresponds to optimum diversity combining. Given these examples and the generic optimization criteria to maximize reliable information flow to users with minimum required resources such as power and bandwidth, it is plausible that using a range of different schemes may be necessary. The term intelligent antenna encompasses the technologies of diversity combining [1, 3, 6, 296, 297, 303], adaptive beamforming [3, 6, 8], optimum combining [3, 6], adaptive matching of the antenna's impedance to the receiver [304, 305], and space division multiple access [6, 8, 306, 307].

An adaptive antenna's parameters are automatically adjusted, in order to obtain an optimal or near-optimal array output. The optimization cost-function and the method used to achieve this state are dependent upon the optimization algorithm chosen. The need for an adaptive solution is obvious, once one considers that interference is seldom constant in either terms of either time or space and a fixed antenna response would be of little, if any, use.

4.3.1 Fixed Beams

The simplest technique of improving the system's performance is to use fixed multiple beams for both reception and transmission at the base station [292]. The strongest beam in the UL will also be used for the DL, since this is deemed to be the beam targeted at the desired user. On the UL, the base station determines the direction of the path on which the strongest component of the desired signal arrives at the base station. On the DL, the base station points a beam in the corresponding direction. Although this simple technique is not optimal, the SINR achievable at the mobile can be improved.

Leth-Espensen *et al.* [20] describe a system of array processing, where an algorithm searches through the 22 fixed beams that may be generated by the antenna array, in order to find the strongest receiver beam of the desired signal. More explicitly, an exhaustive search is performed over nine delay taps and the 22 directions until the tap and direction, which result in the maximum received power are obtained. The estimated Direction of Arrival (DoA) was compared to the actual DoA found using a Global Positioning System (GPS) receiver. When averaging the received signals over 21 GSM transmission bursts ($21 \times 8 \times 576\mu s \approx 100$ ms) the direction estimates occasionally indicated a direction quite different from that of the mobile. This was attributed to the received signal's lack of power due to undergoing a deep fade at that time. Increasing the number of bursts, over which the received signal was averaged, to 104 (≈ 480 ms) gave significantly improved results. The performance of eight element arrays processing either 22 beams or eight beams as well as that of four element arrays processing eight or four beams were compared. The average performance gain of the eight element array using 22 beams over that of a single element was 9.8 dB. For the eight element, eight beam antenna the corresponding improvement was 8.8 dB and for the four element, eight beam array the gain was 8.7 dB. Finally, the gain offered by the four beam, four element array was 5.4 dB.

In a switched beam system [293] a mobile station is located within a specific antenna beam and the antenna is then switched in the required operational mode in order to communicate with the specific user supported by the selected beam. If one considers a cell split into three sectors, each of 120° coverage, the available channels are divided equally amongst the sectors. No intelligence is required to locate a mobile station within a sector and to initiate a call. In the event of the mobile station changing sector a handover is performed. An intelligent antenna system is able to switch from a given beam to a new beam without necessitating a handover, i.e. any of the beams can be assigned to one or more of the transceivers. Therefore, should all the users be located in one sector, then as many users as there are transceivers can be served. In contrast, using a conventionally sectorized base station the transceivers in the empty sectors would not be used, while calls in the high-traffic sectors would be blocked [293].

4.3.2 Temporal Reference Techniques

Temporal reference techniques refer to the design of array processors which optimize the receive antenna array weights, in order to be able to identify a known sequence at the output of the antenna array. This known desired sequence is termed the reference signal, which must be specifically designed so as to be easily identifiable, for example with the aid of a high auto-correlation peak, while being readily distinguishable from or uncorrelated with unwanted interferences and noise sources [3, 280, 290]. For example, in GSM [11] there are eight different channel sounding sequences used for identifying the eight co-channel base stations, therefore, inevitably, co-channel interferers will use identical sounding sequences to those used by the desired mobile user, hence the system may become unable to distinguish between the wanted signal and a co-channel interferer [1]. The spreading codes used in CDMA are

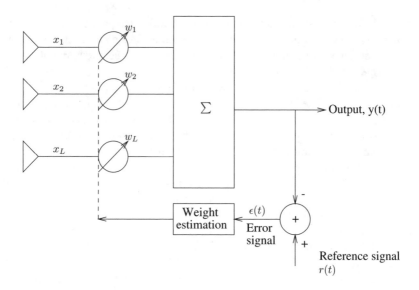

Figure 4.17: The structure of a temporal reference based beamformer with L antenna elements.

inherently unique and they are therefore suitable for use as the user specific sequence. A significant advantage of the temporal reference technique is that, unlike the spatial reference approach, it does not need careful characterization of the antenna array. Effects such as mutual coupling between the antenna array elements are readily handled by the adaptation routine, since the array weights are adjusted automatically, in order to cancel them [1].

Figure 4.17 shows the structure of a temporal reference based beamformer, where the array output is subtracted from the reference signal, $r(t)$ which assists in identifying the desired user, in order to generate the error signal $\epsilon(t) = r(t) - \underline{w}^H \underline{x}(t)$, which is then used to control the weights. The weights are adjusted such that the Mean Squared Error (MSE) between the array output and the reference signal is minimized, where the error is expressed as:

$$\epsilon^2(t) = [r(t) - \underline{w}^H \underline{x}(t)]^2. \tag{4.43}$$

Taking the expected values of both sides of Equation 4.43 we get

$$E[\epsilon^2(t)] = E[r^2(t)] - 2\underline{w}^H \underline{z} + \underline{w}^H R \underline{w}, \tag{4.44}$$

where $\underline{z} = E[\underline{x}(t) r^*(t)]$ is the cross-correlation between the reference signal and the array signal vector $\underline{x}(t)$ and $R = E[\underline{x}(t) \underline{x}^H(t)]$, as defined in Equations 4.26 and 4.27, is the correlation matrix of the array output signals.

The MSE surface is a quadratic function of the complex array weight vector $\underline{w}$ and it is minimized by setting its gradient with respect to $\underline{w}$ equal to zero:

$$\nabla_{\underline{w}}(E[\epsilon^2(t)]) = -2\underline{z} + 2R\underline{w} = 0, \tag{4.45}$$

yielding the well-known Wiener–Hopf equation for the optimal weight vector [3, 280, 283, 287, 288, 290] in the form of:

$$\underline{w}_{opt} = R^{-1}\underline{z}. \tag{4.46}$$

The Minimum Mean Square Error (MMSE) at the output of the array processor, also known as the Wiener filter, using these weights is given by [283]:

$$MMSE = E[|r(t)|^2] - \underline{z}^H R^{-1} \underline{z}. \tag{4.47}$$

In [308] a 16-bit reference signal was used in order to uniquely identify the mobiles. This contribution proposes an adaptive antenna algorithm suitable for GSM and the urban environment, since this is where the highest capacity is generally needed. More specifically, the 16-bit reference signal used in this system is the GSM equalizer's training sequence, which is one of the eight legitimate 16-bit codes exhibiting the highest main-peak to side-peak ratio in its auto-correlation function, which were found by exhaustive computer search of all 2^16 possible sequences. These 16-bit sequences were then extended to 26 bits by quasi-periodically repeating five bits at both ends of the sequence. Neighboring base stations, and hence their mobiles, use a different one from the set of eight codes, as detailed in [11]. The algorithm described in this paper [308] calculates the initial weight vector using just the known training sequence. This weight vector is then applied to all the data in the burst and the result is passed to the GSM channel equalizer in order to detect the unknown bits. The detected bits are then input to the GSM modulator, in order to construct a modulated reference waveform for the entire burst and a new weight vector is calculated. This weight vector is applied to the whole data burst and the result is again passed to the GSM equalizer. Therefore, the SINR is improved for the whole burst, rather than just for the training sequence. In the simulations carried out in [308] the process was repeated for a maximum of 20 iterations or until the same data bits were returned twice. It was found that the typical number of iterations required was three or four. The effect of varying the number of antenna elements was investigated. If the multipath components of the wanted signal are sufficiently delayed, so that they are uncorrelated with the reference signal, they are cancelled. These delayed paths can be exploited, if tapped delay-line filters are used in conjunction with amplitude and phase weighting of the antenna elements. The paper presents results for an eight element linear array with up to three taps.

Barrett and Arnott [1] describe a similar system, in which the modulated training sequence is compared to the signal at the array's output. After the training sequence has been received and the data detection begins, the system switches into decision directed mode, in which the demodulator decisions are remodulated in order to form the reference signal on the basis of the total received burst. Provided that the error rate is adequate (better than 10^{-2}), a reference signal generated by this method would allow the system to track interference changes in the propagation environment. Field trials were conducted for a system using an eight element adaptive antenna. The data received at each antenna was digitized and stored, in order to allow offline processing, enabling the comparison of different processing functions operating on the basis of the same recorded data. The results show a substantial improvement in terms of the demodulated SNR, when compared to that of a single element antenna. The optimum combining was implemented by updating the array weights every transmission burst (every 10 ms), and each update used 100 data snapshots taken from within the burst. The reference signal was obtained using decision directed operation (no training sequence was used) and the weights were updated using the Normalized Least Mean Squares (NLMS) algorithm. The amplitude resolution of the data and weights was eight bits. The results using optimum combining were found to be superior to those obtained using selection diversity.

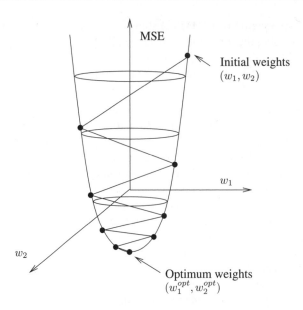

Figure 4.18: An example of the quadratic error surface and the weights of a two element system following the negative direction of the gradient in order to minimize the MSE.

4.3.2.1 Least Mean Squares

The Least Mean Squares (LMS) algorithm is the most common technique used for continuous adaptation [3, 280, 283, 288, 290]. It is based on the steepest-descent method, a well-known optimization technique that recursively computes and updates the weight vector. The algorithm updates the weights at each iteration by estimating the gradient of the quadratic error surface and then changing the weights in the direction opposite to the gradient by a small amount in an attempt to minimize the MSE, as seen in Figure 4.18. The desired response, generated for example by inputting the reference sequence to the modulator is supplied to the algorithm, allowing the estimation error and thus the error surface, to be calculated. The constant that determines the amount by which the weights are adjusted during each iteration is referred to as the step size. When the step size is sufficiently small, the process leads these estimated weights to the near-optimal weights in Figure 4.18, whilst large step sizes allow faster convergence, but exhibit a larger residual MSE due to the non-optimal weights [288].

The updated value of the weight vector at time $n + 1$ is computed using [3, 8, 283, 287–289, 309]:

$$\underline{w}(n + 1) = \underline{w}(n) - \frac{1}{2}\mu\nabla(J(n)), \tag{4.48}$$

where $\underline{w}(n+1)$ denotes the new weights computed at the $(n+1)^{th}$ iteration; μ is the positive step size that controls the rate of convergence and hence determines how close the estimated weights approach the optimal weights and $\nabla(J(n))$ is an estimate of the gradient of the MSE, $J(n)$, where $J(n)$ is given by [283]:

$$J(n) = E[|r(n + 1)|^2] + \underline{w}^H(n)R\underline{w}(n) - 2\underline{w}^H(n)\underline{z}, \tag{4.49}$$

where $r(n + 1)$ is the reference signal at time $n + 1$ and $\underline{z} = E[\underline{x}(t)r^*(t)]$ is the cross-correlation vector between the input vector $\underline{x}(n)$ and the desired response $r(n)$, while the correlation matrix, R, was defined in Equations 4.26 and 4.27.

Differentiating Equation 4.49 with respect to $\underline{w}(n)$ gives:

$$\nabla(J(n)) = 2R\underline{w}(n) - 2\underline{z}. \tag{4.50}$$

Therefore, the instantaneous estimate of the gradient vector becomes:

$$\hat{\nabla}(J(n)) = 2\underline{x}(n)\underline{x}^H(n)\underline{w}(n) - 2\underline{x}(n)r^*(n) \tag{4.51}$$
$$= 2\underline{x}(n)\epsilon^*(n),$$

where $\epsilon^*(\underline{w}(n))$ is the error between the array output and the reference signal, which is formulated as:

$$\epsilon^*(n) = \underline{x}^H(n)\underline{w}(n) - r^*(n). \tag{4.52}$$

The array output in Figure 4.17 is given by:

$$y(n) = \underline{w}^H(n)\underline{x}(n). \tag{4.53}$$

Upon substituting Equation 4.52 in Equation 4.48 the weight adaptation equation becomes:

$$\underline{w}(n + 1) = \underline{w}(n) - \mu\underline{x}(n)\epsilon^*(n). \tag{4.54}$$

Therefore, as Equation 4.52 shows, the estimated gradient, $\hat{\nabla}(J(n))$, is a function of the error, $\epsilon(n)$, between the array output, $y(n)$, and the reference signal, $r(n)$, and the received array signals, $\underline{x}(n)$, after the nth iteration. Convergence is guaranteed only, if [283, 289],

$$0 < \mu < \frac{1}{\lambda_{max}}, \tag{4.55}$$

where λ_{max} is the maximum eigenvalue of R, the correlation matrix of Equations 4.26 and 4.27. Therefore, the eigenvalue spread or ratio of the matrix R controls the rate of convergence [288] according to:

$$\chi(R) = \frac{\lambda_{max}}{\lambda_{min}}, \tag{4.56}$$

where $\chi(R) \geq 1$.

Under these conditions the algorithm is stable and the mean value of the estimated array weights converges to the values of the optimal weights. Within these bounds, the speed of adaptation and also the noise contaminating the weight vector are both determined by the size of μ. Since the trace of R is given by the sum of the diagonal elements of R [288], λ_{max} therefore cannot be greater than the trace of R, that is,

$$\lambda_{max} \leq tr[R] = \sum_{i=1}^{L} \lambda_i \tag{4.57}$$

where L is the number of antenna elements, and λ_i is the ith eigenvalue of R. Hence we have:

$$0 < \mu < \frac{1}{tr[R]}. \tag{4.58}$$

This is a more restrictive bound on μ than Equation 4.55, but it is much easier to apply, because the elements of R and the signal power can generally be more readily estimated than the eigenvalues of R. The efficiency of the LMS algorithm has been shown to approach a theoretical limit for adaptive algorithms, when the eigenvalues of R are equal or nearly equal [310]. When the eigenvalues of the correlation matrix R are widely spread, i.e. $\chi(R) = \frac{\lambda_{max}}{\lambda_{min}} \gg 1$, then, according to Haykin [288], the excess mean-squared error produced by the LMS algorithm with respect to the minimum is determined primarily by the largest eigenvalues [288], and the time taken for the average weight vector to converge is limited by the smallest eigenvalues. However, as the spread of the eigenvalues increases, the highest acceptable value of the stepsize μ required for maintaining stability decreases inevitably, resulting in slower convergence to the optimal weights. Selecting too small a value for μ results in a slow rate of convergence, and in a non-stationary environment may cause the estimated weights to lag behind the evolution of the optimal weights [295], a phenomena known as the weight vector lag. Alternatively, using too high a value for μ allows the vicinity of the solution point to be reached more rapidly, but the weights then wander around a larger region and cause a weight mis-adjustment error, as was demonstrated in Figure 4.18 [311]. This is due to μ being equivalent to the reciprocal of the memory of the system, where a large value of μ uses fewer samples to estimate R, and hence a degraded estimation is performed, resulting in an increase in the average excess mean-squared error after adaptation.

4.3.2.2 Normalized Least Mean Squares Algorithm

In the LMS algorithm, the correction $\mu \underline{x}(n)\epsilon^*(n)$ applied to the weight vector at time $n+1$ in Equation 4.54 is directly proportional to the input vector $\underline{x}(n)$. Therefore, when $\underline{x}(n)$ is large, the LMS algorithm experiences a gradient noise amplification problem [288]. Therefore an algorithm which normalizes the weight vector correction with respect to the squared Euclidean norm of the input vector $\underline{x}(n)$ at time n can be invoked. At the nth iteration the step size is then given by [283, 288]:

$$\mu(n) = \frac{\mu_0}{\underline{x}^H(n)\underline{x}(n)} = \frac{\mu_0}{\|\underline{x}(n)\|^2}, \tag{4.59}$$

where μ_0 is a constant. The normalized LMS algorithm is convergent in the mean-square sense, if $0 < \mu_0 < 2$ [288]. However, if the input vector $\underline{x}(n)$ is small, then numerical problems may arise due to the associated division by a small number. Therefore Equation 4.59 may be modified to:

$$\mu(n) = \frac{\mu_0}{a + \|\underline{x}(n)\|^2}, \tag{4.60}$$

where $a > 0$. Hence, the weight update formula of Equation 4.54 is modified to:

$$\underline{w}(n+1) = \underline{w}(n) + \frac{\mu_0}{a + \|\underline{x}(n)\|^2}\underline{x}(n)\epsilon^*(n). \tag{4.61}$$

4.3.2.3 Sample Matrix Inversion

The Sample Matrix Inversion (SMI) algorithm is a method of directly calculating the antenna array weights based on an estimate of the correlation matrix, $R = E[\underline{x}(t)\underline{x}^H(t)]$ of the adaptive array output samples. The Wiener–Hopf solution for the optimal weights is repeated here from Equation 4.46, for convenience:

$$\underline{w}_{opt} = R^{-1}\underline{z}, \tag{4.62}$$

where $\underline{z} = E[\underline{x}(t)r(t)]$ is the cross-correlation between the reference signal, $r(t)$ and the array output signal, $\underline{x}(t)$. If the signal, noise and interference characteristics are stationary, then the correlation matrix can be evaluated and the optimal solution for the adaptive weights can be computed directly using the above equation, with the aid of matrix inversion. In practice however, due to the non-stationary mobile environments encountered, the adaptive processor must continually update the weight vector, in order to meet the new conditions imposed by the time-varying mobile environment. This need to regularly update the weight vector leads to the requirement of obtaining estimates of R and $\underline{z}$ in a finite observation interval, and thus to obtain a weight vector estimate. This approach is termed block-adaptive, where the statistics are estimated from a temporal block of data and are used in a periodic optimum weight calculation process. In the GSM system [11] it may be possible to use the synchronization/channel sounding sequence in each burst to recompute the antenna array weights for each 4.615 ms burst.

If the cross-correlation vector $\underline{z} = E[\underline{x}(t)\underline{x}^H(t)]$ is assumed to be known, then the optimal weight vector estimate, $\underline{\hat{w}}$ of Equation 4.62, for the situation when $\underline{x}(t)$ contains the reference-signal related desired signal, where $\hat{R}_{xx}$ is the block based estimate of the true correlation of the array's output samples, namely that of R_{xx}, may be determined using

$$\underline{\hat{w}}_1 = \hat{R}_{xx}^{-1}\underline{z}. \tag{4.63}$$

However, in the scenario when the received signal $\underline{x}(t)$ contains either noise of the interfering users' signals rather than the desired signal, the estimate of the correlation matrix R_{xx} is denoted by $\hat{R}_{nn}$, and the optimal antenna weights may be calculated thus according to:

$$\underline{\hat{w}}_2 = \hat{R}_{nn}^{-1}\underline{z}. \tag{4.64}$$

Therefore, the SNR at the output of the combiner seen in Figure 4.17 may be written as [290]:

$$\left(\frac{s}{n}\right)_i = \frac{\underline{\hat{w}}_i^H \underline{s}\underline{s}^H \underline{\hat{w}}_i}{\underline{\hat{w}}_i^H R_{nn} \underline{\hat{w}}_i}, \tag{4.65}$$

where i assumes values of 1 or 2, according to the first of second scenarios above, and $\underline{s}$ denotes the reference-signal related desired signal component of the array output signal vector $\underline{x}$. The SNR $(s/n)_2$ is only defined during those time intervals, when a reference-signal related desired signal is actually present; the weight adjustment is assumed to take place when the desired signal is absent.

The estimate of the sample correlation matrix can be evaluated according to:

$$\hat{R}_{xx} = \frac{1}{N}\sum_{n=1}^{N} \underline{x}(n)\underline{x}^H(n), \tag{4.66}$$

where N is the size of the observation interval expressed in terms of the number of array output samples considered. Again, this approach is termed block-adaptive, where the statistics

Figure 4.19: The expected normalized SNR, $E[\rho_2]$ evaluated from Equation 4.69, for various numbers
of array output samples, in terms of the number of antenna array elements, used to
construct the noise- or interference-only correlation matrix. Simulated results for identical
scenarios are also presented for comparison. The SNR at each antenna array element was
12.0 dB.

are estimated from a temporal block of data and used during the optimum weight calculation
process. Given that each element of the matrix, $\hat{R}_{xx}$, is a random variable, the output SNR
is also a random variable [285, 290]. The maximum achievable SNR at the output of the
combiner seen in Figure 4.17 that may be obtained is:

$$SNR_{opt} = \underline{s}^H R_{nn}^{-1} \underline{s}. \tag{4.67}$$

The actual SNRs obtained using $\hat{\underline{w}}_1$ and $\hat{\underline{w}}_2$ may be normalized as follows [285, 290]:

$$\rho_i = \frac{(s/n)_i}{SNR_{opt}}. \tag{4.68}$$

Reed [285] examined the number of samples, N, required in order to achieve a high-
quality estimate of the noise- or interference-related co-variance matrix, R_{nn}, and derived
the expected value of the normalized SNR at the output of the combiner seen in Figure 4.17,
which was found to be:

$$E[\rho_2] = \frac{N + 2 - L}{N + 1}, \tag{4.69}$$

where L is the number of elements in the antenna array.

The expectation of the normalized SNR in Equation 4.69 employing the antenna weights
calculated on the basis of the noise- or interference-only related co-variance matrix, is plotted
in Figure 4.19 for two, four and eight element antenna arrays. Explicitly, Figure 4.19 suggests
that as long as, N, the number of samples used to estimate the noise- or interference-related

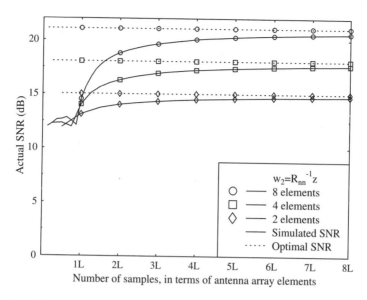

Figure 4.20: The SNR at the output of the array combiner determined by simulation and the optimal SNR according to Equation 4.67 for a varying number of array output samples, in terms of the number of antenna array elements, used to construct the noise- or interference-only correlation matrix. The SNR at each antenna array element was 12.0 dB.

correlation matrix, R_{nn}, is greater than twice the number of antenna elements, the loss in $E[\rho_2]$ due to non-optimal weights is less than 3 dB. The expected values of $E[\rho_2]$ evaluated from Equation 4.69 are compared to values determined using simulations. The simulation based and theoretical SNRs were in good agreement. It is interesting to note that although both the normalized simulated and theoretical SNRs approach unity, implying approaching the optimum SNR in Equation 4.67, however the rate of convergence for both the theoretical and simulated values slows down, as the number of antenna elements used to form the antenna array increases. This is expected, since as the number of antenna array elements increases, so does the optimum SNR that may be obtained according to Equation 4.67, as also seen in Figure 4.20.

Thus far we have assumed the knowledge of the cross-correlation vector $\underline{z}$, which is unrealistic in a practical system. Therefore, the optimal weight vector may be determined with the aid of the estimated cross-correlation vector $\underline{\hat{z}}$ according to:

$$\underline{\hat{w}}_2 = \hat{R}_{xx}^{-1} \underline{\hat{z}}, \tag{4.70}$$

where $\underline{\hat{z}}$ is the sample cross-correlation vector given by

$$\underline{\hat{z}} = \frac{1}{N} \sum_{n=1}^{N} \underline{x}(n) r^*(n), \tag{4.71}$$

and $r(n)$ is the reference signal.

Figure 4.21: The normalized SNR, ρ_i, for various numbers of array output samples, in terms of the number of antenna array elements. Results are shown for $\underline{w}_1 = \hat{R}_{xx}^{-1}\underline{z}$, $\underline{w}_2 = \hat{R}_{nn}^{-1}\underline{z}$, and $\underline{w}_3 = \hat{R}_{xx}^{-1}\hat{\underline{z}}$ for both theory, according to Equations 4.63, 4.64 and 4.70, and simulation. The antenna array consisted of two antenna elements, separated by $\lambda/2$, and the SNR at each of which was 12.0 dB.

The normalized SNR for a two element antenna array was determined by simulation using Equation 4.70, for estimating the optimum antenna array weights, is presented in Figure 4.21. This figure shows that the SNR of the received signal, using the antenna weights determined when the desired signal was present, is significantly lower than when using the weights obtained when the desired user's signal was absent. The simulated SNR, for the case of two antenna elements, when the desired signal was received is significantly higher than that predicted theoretically by Equation 4.68, although this phenomenon does not appear for the four and eight element antenna arrays characterized in Figure 4.22.

The SNR obtained using Equation 4.70 is shown to be comparable to the SNR obtained with the noise- or interference-only correlation matrix, R_{nn}, which is because the estimates $\hat{\underline{z}}$ and $\hat{R}_{xx}$ are highly correlated under strong desired signal conditions, and the errors in each estimate tend to compensate each other, thus yielding an improved weight estimate and faster convergence. Improvement of the transient response through careful selection of the initial weight vector is possible by invoking the following relationship [290]:

$$\hat{\underline{w}}_1 = \left[\frac{1}{N}\left(\sum_{n=1}^{N}\underline{x}(n)\underline{x}^H(n) + \alpha I\right)\right]^{-1}\underline{z} \qquad (4.72)$$

where α is a scalar constant and I is the $N \times N$ identity matrix.

(a)

(b)

Figure 4.22: The normalized SNR, ρ_i, for various numbers of samples, in terms of the (a) four and (b) eight antenna array elements. Results are shown for $\underline{w}_1 = \hat{R}_{xx}^{-1}\underline{z}$, $\underline{w}_2 = \hat{R}_{nn}^{-1}\underline{z}$, and $\underline{w}_3 = \hat{R}_{xx}^{-1}\hat{\underline{z}}$ for both theory, according to Equations 4.63, 4.64 and 4.70, and simulation. The antenna elements were separated by $\lambda/2$. The SNR at each antenna element was 12.0 dB.

The estimate of R may be updated, when new samples arrive from the antenna, according to [283]:

$$\hat{R}(n+1) = \frac{n\hat{R}(n) + \underline{x}(n+1)\underline{x}^H(n+1)}{n+1}, \qquad (4.73)$$

and a new estimate of the weights $\hat{\underline{w}}(n+1)$ at time instant $n+1$ may be made. The expression of the optimal weights in Equation 4.46 requires the inverse of R, and this process of estimating R and then its inverse may be combined to update the inverse of R from the array signal samples, $\underline{x}(n)$, using the Matrix Inversion Lemma [283, 287] which is given in its general form as:

$$(A + XX^H)^{-1} = A^{-1} - \frac{A^{-1}XX^H A^{-1}}{1 + X^H A^{-1}X}, \qquad (4.74)$$

thus leading to:

$$\hat{R}^{-1}(n) = (n+1)\hat{R}^{-1}(n-1) - \frac{(1+\frac{1}{n})\hat{R}^{-1}(n-1)\underline{x}(n)\underline{x}^H(n)\hat{R}^{-1}(n-1)}{n + \underline{x}^H(n)\hat{R}^{-1}(n-1)\underline{x}(n)}, \quad (4.75)$$

with

$$\hat{R}^{-1}(0) = \frac{1}{\epsilon_0}I, \quad \epsilon_0 > 0, \quad (4.76)$$

where I is the $N \times N$ identity matrix. This method of estimating the array weights using the inverse update technique is known as the Recursive Least Squares (RLS) algorithm.

Unlike for the LMS algorithms, the performance of the SMI algorithm is almost independent of the eigenvalue spread of R and it is similar to that of the steepest descent algorithm using a correlation matrix, R, of equal eigenvalues [287]. The matrix estimation in Equation 4.66 is only suitable for use in a stationary environment [287]. In a time varying environment a de-weighted matrix estimate may be more applicable [309], yielding:

$$\hat{R}(n) = \alpha\hat{R}(n-1) + (1-\alpha)\underline{x}(n)\underline{x}^H(n), \quad 0 < \alpha < 1 \quad (4.77)$$

where α is the so-called "forgetting factor".

Hence, Equation 4.75 becomes

$$\hat{R}^{-1}(n) = \alpha^{-1}\hat{R}^{-1}(n-1) - \frac{(1-\alpha)\alpha^{-2}\hat{R}^{-1}(n-1)\underline{x}(n)\underline{x}^H(n)\hat{R}^{-1}(n-1)}{1 + (1-\alpha)\alpha^{-1}\underline{x}(n)\hat{R}^{-1}(n-1)\underline{x}^H(n)}. \quad (4.78)$$

The vector, $\hat{\underline{z}} = E[\underline{x}(n)r^*(n)]$, containing the correlation between the reference signal, $r(n)$, and the array output signals, $\underline{x}(n)$, must also be updated for each block of N received samples according to:

$$\hat{\underline{z}} = \frac{1}{N}\sum_{n=1}^{N}\underline{x}(n)r^*(n). \quad (4.79)$$

If an error term, $e = \hat{\underline{z}} - \underline{z}$, between the estimate of the correlation vector $\underline{z}$ and its actual value is used to represent the errors due to the estimation process, we may write

$$e = \hat{R}\underline{w}_{opt} - \underline{z}. \quad (4.80)$$

Therefore, the weight vector derived using the SMI method is a least squares solution. It can be shown theoretically that the array weights derived by the SMI approach converge more rapidly towards their final values than those generated by the LMS algorithm. However, there are practical difficulties associated with the employment of the SMI algorithm. Specifically, the inversion of the potentially large correlation matrix, R, requires a high complexity. Specifically, the complexity of the matrix inversion is proportional to L^3, where L is the matrix dimensionality, and it is thus very computationally expensive. However, the matrix inversion may be avoided by using the recursive techniques of Equation 4.75.

In [15] Strandell *et al.* investigated the performance of an adaptive antenna system using the SMI adaptation algorithm. The system was integrated into an existing DCS-1800 base station and used the 26-bit equalizer training sequence in each traffic burst as the reference signal. The performance of the adaptive antenna was evaluated in the laboratory initially, so as to avoid multipath propagation. It was shown that the algorithm was capable of suppressing an interferer, when the power of the interferer was within the dynamic range of the Analog-to-Digital Converter (ADC) used to digitize the signals arriving at the antenna array elements. The ADC had an eight-bit resolution giving approximately a 48 dB dynamic range spanning from -32 dBm to -80 dBm. Consequently, below -80 dBm the interferer is buried in the noise and no suppression is possible. Therefore, stronger interferers are suppressed more effectively than weak ones. The adaptive antenna was found to improve the SIR by more than 30 dB in conjunction with an interferer power at -40 dBm and a desired input signal power between -70 dBm and -40 dBm. When either of the signal levels exceeded the dynamic range of the ADC, the SIR improvement was very low, even less than 0 dB in some circumstances.

The performance of the antenna was then evaluated in an open terrain environment, with no obstacles within 500 m of the antenna. It was found that even though there was some array pattern distortion, or angular pointing error in the direction of the main beam, the interfering signal located at an angle of 90° with respect to the desired signal was suppressed by about 25 dB relative to the main beam. The pointing error of the main beam was due to the relatively short, 26-bit training sequence used, leading to a poorly estimated array output correlation matrix, when the desired signal was present in the matrix [312]. A solution to this problem is the positive diagonal loading technique [312], where adding a small value to the diagonal elements of the matrix results in faster weight convergence. In conjunction with a perfectly estimated array output correlation matrix all the noise eigenvalues are identical and equal to the noise variance [15]. In contrast, a poor estimate of the array output correlation matrix gives non-identical eigenvalues, resulting in a distorted array pattern. If the loading value is larger than the noise eigenvalues, but smaller than the eigenvalues of the desired and interfering signal, then the overall noise level is increased, resulting in almost identical noise eigenvalues [313]. The loading value l was chosen so that $l/\sigma^2 \approx 10^2$ [312]. The diagonal loading decreases the SIR, but increases the SNR due to the lower sidelobe levels, leaving the SINR unchanged [312]. The SIR improvement achieved by the adaptive antenna was measured for Direction-Of-Arrival (DOA) separations ranging from 2.5° to 180° at a constant input SIR of 20 dB. The interference suppression capability varied from 31 dB for a 180° angular separation to 26 dB for a 2.5° separation. However, as a consequence of the limited array beamwidth, the SNR gain decreased upon decreasing the DOA separation, reaching a minimum of -10 dB at 5° separation.

4.3.2.4 Recursive Least Squares

The RLS algorithm exploits the matrix inversion lemma defined in Equation 4.74 for updating the antenna array element weights. As the RLS algorithm utilizes information contained in the array's combiner output data as shown by Equations 4.74 and 4.75, extending back to the time when the algorithm was initiated, the rate of convergence is typically an order of magnitude higher than that of the LMS algorithm. This performance improvement, however, is achieved at the expense of a substantial increase in computational complexity.

The correlation matrix, R of the array output, at time n, may be updated thus according to [3, 283, 288]:

$$R(n) = \delta_0 R(n-1) + \underline{x}(n)\underline{x}^H(n), \tag{4.81}$$

where, similarly to Equation 4.77, the "forgetting factor", δ_0, is used to de-emphasize old array output samples. The value $1/(1 - \delta_0)$ is known as the memory of the algorithm, and for example when $\delta_0 = .99$, the memory of the algorithm is approximately 100 samples, while $R(n - 1)$ is the previous value of the correlation matrix, R, at time $n - 1$.

Similarly, the cross-correlation vector between the array output signal and the desired signal may be calculated as:

$$\underline{z}(n) = \delta_0 \underline{z}(n - 1) + \underline{x}(n)r(n). \tag{4.82}$$

Equation 4.46 states how the optimal receive antenna weights may be obtained, which is repeated here for convenience:

$$\underline{w}_{opt} = R^{-1}\underline{z}, \tag{4.83}$$

leading to,

$$
\begin{aligned}
\hat{\underline{w}}(n) &= R^{-1}(n)\underline{z} \\
&= \delta_0 R^{-1}(n)\underline{z}(n - 1) + R^{-1}(n)\underline{x}(n)r^*(n),
\end{aligned} \tag{4.84}
$$

where

$$R^{-1}(n) = \frac{1}{\delta_0}\left[R^{-1}(n - 1) - \frac{R^{-1}(n - 1)\underline{x}(n)\underline{x}^H(n)R^{-1}(n - 1)}{\delta_0 + \underline{x}^H(n)R^{-1}(n - 1)\underline{x}(n)}\right] \tag{4.85}$$

with

$$R^{-1}(0) = \frac{1}{\epsilon_0}I, \qquad \epsilon_0 > 0, \tag{4.86}$$

as in Equation 4.76, when using the SMI algorithm. Therefore, with the aid of:

$$R^{-1}(n) = \frac{1}{\delta_0}\left[R^{-1}(n - 1) - q(n)\underline{x}^H(n)R^{-1}(n - 1)\right], \tag{4.87}$$

where

$$q(n) = \frac{R^{-1}(n - 1)\underline{x}(n)}{\delta_0 + \underline{x}^H(n)R^{-1}(n - 1)\underline{x}(n)} \tag{4.88}$$

we arrive at [3, 288],

$$\underline{w}(n) = \underline{w}(n - 1) + q(n)[r^*(n) - \underline{w}^H(n - 1)\underline{x}(n)], \tag{4.89}$$

where the square-bracketed term represents the error, $e(n) = r^*(n) - y(n)$ between the desired signal and the array output signal after processing. As can be seen from Equation 4.85, the inversion of the correlation matrix, $R(n)$ required by Equation 4.83, has been replaced by the simple update formula of Equation 4.87, requiring scalar division, thus significantly reducing the complexity imposed.

4.3.3 Spatial Reference Techniques

Spatial reference adaptation [1, 3, 8, 279–283, 301] relies on information regarding the direction of arrival of the desired signal and its multipath components. There are numerous different methods for obtaining estimates of the DOA information with the aid of the received antenna array signals [3, 283, 301]. Wave-number estimation techniques [3, 283, 284, 301] are based on the decomposition of the array output correlation matrix, $R = E[\underline{x}(t)\underline{x}^H(t)]$, whose terms consist of estimates of the correlation between the signals at the elements of the antenna array in Figure 4.10. The so-called MUltiple SIgnal Classification (MUSIC) algorithm [3, 283, 301] and the Estimation of Signal Parameters by Rotational Invariance Techniques (ESPRIT) both use this approach [283, 301]. However, these algorithms are not effective for detecting coherent signals [283,301]. The parametric estimation techniques [283, 301] are mainly maximum likelihood estimation (MLE) based algorithms, where the ML estimates of desired parameters, such as the angles of arrival, are the ones for which the likelihood function is maximized. These techniques impose a high computational complexity and also require the antenna array to be accurately calibrated. Again, further information concerning these algorithms may be found in [7, 9, 283, 314].

4.3.3.1 Antenna Calibration

Antenna calibrating procedures [7, 294, 315] can be readily incorporated in a digital beamforming array, facilitating the realization of highly selective antenna patterns exhibiting ultra-low sidelobes. The feature of self-calibration is an advantage, but may indeed also be an essential requirement for a system employing an array of elemental receivers constituted by multiple, cascaded active components [7, 295]. Several techniques are available, such as the injection of precise radio frequency test signals at the receiver front-ends [15, 294], focusing on a source at a known position in the near or far-field, or employment of a known, well defined scatterer of the transmitted signal.

In order to improve the SIR of the signal received by an adaptive antenna array, nulls can be created in the antenna array's radiation pattern in the direction of strong co-channel interferers. However, the depth and angular position of these nulls are very sensitive to phase and amplitude errors within the antenna array [7]. The performance of RF components generally varies over temperature, time and frequency. A study conducted by Tsoulos and Beach [7] found that a temperature variation of 14°C to 27°C resulted in a maximum amplitude variation of ±1.5 dB and a ±180° maximum phase error across the antenna array. Performing a calibration of phase and amplitude mis-matches between the antenna array elements at the time of manufacture would not take into account temperature variations and ageing effects [294]. Reference [7] noted that even under the same room temperature the amplitude and phase mismatches varied from day to day. Therefore, an online calibration procedure is required that can take place, whilst the base station continues to function normally. Only the active components have to be calibrated, the passive components are assumed to be less susceptible to temperature and time. After calibration the amplitude mismatch was limited to ±0.04 dB and the phase mismatch to ±0.4°.

The calibration process of an 8×8 element receiver antenna array developed for the pan-European TSUNAMI (II) SDMA Field Trial was described by Passman and Wixforth in [315]. The aim of the calibration procedure was to reduce the phase error to less than 3° and the amplitude error to less than 0.5 dB. The receive antenna array, as shown in Figure 4.23, consists of ten linearly spaced active subarrays, each of which consists of eight vertically separated single antenna elements. The 1st and 10th subarrays act as dummy elements in an attempt to maintain a consistent mutual coupling between subarrays across the entire array. The provision of circuitry to allow the reception of both vertically and horizontally

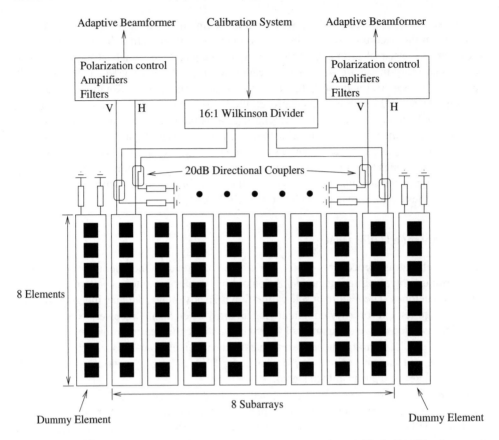

Figure 4.23: Block diagram of the 8×8 element antenna array receiver and in-built calibration system of Passman and Wixforth showing the horizontal and vertical polarization ports [315].

polarized signals at each of the eight subarrays implies that the reception of 16 different polarizations is possible. The calibration of the antenna can be separated into two stages, namely the offline calibration after manufacture and the online calibration performed during operation. The offline calibration measures the characteristics of the passive components in the signal path and assumes that the 16:1 Wilkinson divider and the 20 dB directional couplers are stable over both time and temperature. More specifically, the online calibration procedure uses the Wilkinson divider and the directional couplers to inject a calibration signal into each of the eight signal paths dedicated to horizontal polarization and the eight paths for vertical polarization. The magnitude and phase response of these 16 signal paths is then measured in the baseband in order to characterize the entire antenna system. However, fully characterizing this antenna array receiver at all of the frequencies of interest would generate vast amounts of data, and require an impractical length of time.

Fortunately, it is possible to use a reduced set of measurements [315]. Measurements of the antenna array's forward transfer function, S_{21}, between the central calibration port and the 16 receiver ports for both the vertically and horizontally polarized signals were found to be essential for characterizing the calibration network itself. Phase differences of up to 20° and amplitude variations of 2 dB were measured between two seemingly identical calibration signal paths, despite the symmetrical layout of the Wilkinson divider [315]. Further measurements of S_{21} between each subarray port and all other subarray ports, in order to account for mutual coupling of the subarrays showed coupling levels of below -30 dB between all ports. Thus far, the characterization of the calibration network has required 16 phase and magnitude values, while the mutual coupling between the subarrays necessitated a further $(2 \times 8)^2 = 256$ readings. Additionally, any imbalances between subarrays in the magnitude radiation patterns over all specified azimuth and elevation angles must also be measured, leading to a still significant amount of information that must be processed.

Simmonds and Beach [294] described how an 8 element adaptive antenna array can be calibrated, with no interruption to the network, for both transmission and reception. The aim of the scheme was to achieve a post-calibration accuracy of 3° phase and 0.5 dB magnitude error across the array. The design of the process allows the receive calibration to be performed during the unallocated timeslots within the DCS1800 frame structure. A Continuous Wave (CW) signal is injected simultaneously into each of the receiver antenna array elements via directional couplers and a power divider/combiner. Digital attenuators allow the injected signal strength to be varied over a range of 60 dB in 2 to 3 dB steps. The errors associated with the received signal phase and amplitude are measured in the baseband and the beamformer weights are adjusted appropriately, in order to produce the desired beam pattern. Moreover, the same technique cannot be used for transmitter calibration, since this would result in spurious RF transmission. In the proposed scheme the transmitter would be calibrated in the even timeslots, except for timeslot zero, which is used for the Broadcast Control CHannel (BCCH) in DCS1800 and GSM [294]. Each branch of the antenna array is sampled using the directional couplers and the resulting signals are down-converted to baseband. These 8-bit quantized I and Q samples are compared to the baseband digital beamformer outputs, in order to obtain the correction factor required for each array path.

4.3.4 Blind Adaptation

Blind adaptation [3, 283, 301] of the array weights has several advantages over both the spatial [6, 7, 9, 283, 301, 314] and temporal reference [1, 3, 6, 280, 290, 308] based systems of Sections 4.3.3 and 4.3.2. Temporal reference assisted systems must achieve synchronization and perform demodulation, before weight adaptation can commence, whereas spatial reference aided systems require very strictly calibrated hardware and rely on DoA information. However, the typically large angular spread of the incoming signals in small picocells makes this difficult to attain. In contrast, a blind adaptation scheme [3, 283, 301, 307] does not require training sequences or any information concerning the antenna array's geometry. Dispensing with the reference or training sequence results in potentially increased data rates. For example, a capacity increase of 17% can be achieved in the UL for GSM [307] upon invoking blind joint space-time equalization. However, using for example the so-called constant modulus adaptive algorithm [316] can lead to the capture of interfering signals instead of the wanted signal, an issue argued more explicitly in [317, 318].

4.3.4.1   Constant Modulus Algorithm

The Constant Modulus (CM) algorithm [316] operates on the principle that the amplitude of the receive antenna array output should remain constant, unless the interference causes fluctuations. If the transmitted signal, $s(n)$, has a constant envelope, then the combiner output, $y(n)$ in Figure 4.17, should also have a constant envelope. However, if multipath fading occurs, then the combiner output, $y(n)$, will have a fluctuating envelope. The objective of CM beamforming [3, 283, 301] is to restore the array output to a constant envelope signal, on average. This can be accomplished by adjusting the array weight vector, $\underline{w}$, in such a way, so as to minimize a certain cost function.

In the classic paper by Godard [316], who used the CM property in order to carry out blind channel equalization, the criterion was to minimize the functions $D^{(p)}$, referred to as the dispersion of order p ($p > 0$ integer), defined by

$$D^{(p)}(n) = E\Big[(|y(n)|^p - R_p(n))^2\Big], \tag{4.90}$$

with R_p being real positive constants given by:

$$R_p(n) = \frac{E[|a(n)|^{2p}]}{E[|a(n)|^p]}, \tag{4.91}$$

where $a(n)$ is the transmitted data symbol. The standard cost function of [3, 283, 316]

$$G^{(p)}(n) = E\Big[(|y(n)|^p - |a(n)|^p)^2\Big], \tag{4.92}$$

which is not used in blind array weight adaptation, is independent of the carrier phase but depends on the magnitude of the antenna array's output signal, $|y(n)|$, and that of the transmitted signal, $|a(n)|$. In contrast, the function $D^{(p)}$, used in the CM algorithm is independent of both the carrier phase and the data symbol's magnitude [316].

The most often used practical case is that of $p = 2$, where

$$D^{(2)}(n) = E[(|y(n)|^2 - R_2(n))^2], \tag{4.93}$$

with

$$R_2 = \frac{E[|a(n)|^4]}{E[|a(n)|^2]}, \tag{4.94}$$

and the cost function, $D^2(n)$, is effectively the mean squared error between the magnitude of the antenna array's output signal squared and the constant $R_2(n)$. Hence, again, the main difference between the conventional cost function of Equation 4.92 and that of the constant modulus algorithm in Equation 4.90 is that the constant modulus algorithm does not assume the knowledge of the data sequence's magnitude, $|a(n)|$, it rather attempts to minimize the difference with respect to the constant quantity $R_2(n)$, which is related to the moments of $|a(n)|$ by Equation 4.94. In other words, the CM beamforming algorithm directs the combiner's output to a constant envelope.

In [283] a cost function is given in the form of:

$$J(n) = \frac{1}{2}E[(|y(n)|^2 - y_0^2)^2], \tag{4.95}$$

where y_0 is the desired amplitude in the absence of interference.

The objective is to find a set of values for the array weight vector, $\underline{w}$, that will minimize the given cost function. This may be accomplished using the following equation [283]:

$$\underline{w}(n+1) = \underline{w}(n) - 2\mu(|y(n)|^2 - y_0^2)y(n)\underline{x}(n+1), \tag{4.96}$$

or employing the update formula of [3]:

$$\underline{w}(n+1) = \underline{w}(n) + \mu[R_p(n) - |y(n)|^2]y(n)\underline{x}(n), \tag{4.97}$$

which are used in a steepest descent fashion to update the array weights and are essentially identical, apart from Equation 4.96 using the current sample, $\underline{x}(n+1)$, of the array's output, while Equation 4.97 using the previous sample, $\underline{x}(n)$. These equations are identical to the update regime of Equation 4.54 used in the LMS algorithm, with the only difference being the error term.

There are two conditions, which may lead to a zero-gradient situation, where the algorithm stops adapting. The first is the condition of $|y(n)| = 1$, which represents the desired convergence optimum. The second is $y(n) = 0$, which also forces the gradient to become zero. However, fortunately this is not a practical problem, since the point $y(n) = 0$ is not a stable equilibrium and the system noise moves the weight vector from this zero-gradient point. A further problem in a hostile fading environment is that the beamformer may incorrectly select the interference as the signal to process, so as to maintain a constant modulus, rather than the desired signal.

In [307] a blind array weight adaptation technique was described by Laurila and Bonek, which performs joint space-time equalization, separation and detection of multiple unsynchronized co-channel digital signals. The scheme exploits the facts that the signals are of fixed symbol rate, have a CM and a Finite Alphabet (FA) of symbols. Simulations were conducted for an eight-element Uniform Linear Array (ULA) with an element spacing of $\lambda/2$ [307]. The equalizer order was five. Although the simulation parameters were not optimized, the system gave results demonstrating that comparable BER can be achieved, when compared to reference-assisted adaptation methods.

4.3.5 Adaptive Arrays in the Downlink

Adaptive arrays have been more often studied for receiving UL data at the base station. However, they are equally suitable for transmitting data by the base station in the DL. It is possible to steer a transmitting array in the same way as one used for reception, so as to minimize the DL interference inflicted upon co-channel mobiles. The wide frequency separation between the UL and DL frequency bands used in the Frequency Division Duplexing (FDD) GSM system, for example, results in uncorrelated fading between the up- and the down-link. Therefore, the weights calculated for reception are typically unsuitable for employment in transmit mode. In contrast, in a Time Division Duplexing (TDD) system, such as UTRA [11] it may be possible to re-use the receive mode weights, provided that the location of the mobile has not changed significantly between timeslots, i.e. if the duration of the timeslots is sufficiently short.

In the UL scenario, the receive array can adapt to changes in the propagation medium by observing its own outputs and modifying its own processing, since there is an in-built feedback mechanism, as was shown in Figure 4.17. When used in the transmit mode, an adaptive antenna array at the base station needs an additional feedback signal from the mobile receivers, in order to give the base station a means of measuring its own beam patterns. The array, by directing a mainlobe towards a mobile, could nonetheless produce a spurious fade in the desired signal or inflict interference upon other mobiles.

The scheme proposed by Gerlach and Paulraj [319] uses feedback of the signals received at the mobiles, in order to calculate the transmitter antenna weights to employ. The paper describes a system where data transmission is temporarily halted in for the transmission of probing signals. Each probing signal is sent on an orthogonal channel in the time, frequency or code domain so that the receivers may measure the response of each probing signal. The responses to each of the probing signals at each of the receivers are fed back to the transmitter, allowing the channel responses to be estimated. Simulations were performed, by Gerlach and Paulraj [319], which showed that at a low mobile speed of 2.5 miles per hour (mph) adequate signal separation required a data feedback rate in the order of a few kbit/s, making the approach only viable for static or slow-moving receivers. It is worth noting here that the 3G UTRA system has a total control channel rate of about 10 kbit/s.

Further to this scheme, Gerlach and Paulraj [320] presented a method which reduces the feedback rate by exploiting that as the array's weight vector fluctuates due to the mobile receiver's motion, the weight vector's fluctuations will be confined to a certain subspace of its total vector space. In contrast to the channel weight vector itself, the channel vector's subspace is much more stable during the mobile's motion, and this fact can reduce the required feedback rates. The method is best suited to environments having either a low number of propagation paths, or for several paths approaching the base station from similar angles. This implies that there must be only a few scattering bodies near to the base station. As the mobile receiver moves, its array weight vector varies at the fast fading rate, but the fluctuations are confined to the subspace Ψ_k, where the subscript k denotes the kth mobile, which varies slowly. A beamformer based on this more stable subspace structure, rather than the array weight vector, will need a lower feedback rate. Hence, the subspace structure tends to be more useful when the subspace dimension, $\dim[\Psi_k]$, is small. The subspace dimension, $\dim[\Psi_k]$, will only be small however, if the number of propagation paths is low or if all of the paths have approximately the same angle of departure from the array. The paper derives a subspace beamformer and presents results obtained using simulations. The required feedback rate for a mobile moving at 35 mph was estimated to be 250 bits/s. While this is a best-case estimate, it is significantly reduced in comparison to the rate in [319] and it is also less than the feedback rate used for power control in Qualcomm's IS-95 cellular system [320]. Hence, such a regime could realistically be used in a UTRA-type system.

Martin and Gaspard [292] presented a system based on the Discrete Fourier Transform (DFT) Beamspace technique. Each user's signal was transmitted on the particular DFT beam, which offered the largest mean power level during the UL reception. With a four-element linear array the system provided a 175% radio capacity gain over a conventional base station. An eight-element array resulted in a gain of 200% in radio capacity. However, the DL capacity using this method was not matched to the UL capacity. Similarly enhanced DL capacity was achieved using exact DOA information, where the DL's transmission beam was steered in the direction of the strongest multipath component received at the UL. This provided an estimated 350% increase in radio capacity.

Table 4.1: Eigenvalue spread, $\chi(R) = \frac{\lambda_{max}}{\lambda_{min}}$, of Equation 4.56 evaluated for the array output cross-correlation matrix, R, for different values of SNR and INR.

SNR (dB)	INR (dB)	$\chi(R)$
3.0	3.0	4.4
3.0	9.0	8.3
3.0	27.0	402.2
9.0	3.0	8.3
9.0	9.0	5.4
9.0	27.0	120.6
27.0	3.0	403.3
27.0	9.0	120.6
27.0	27.0	5.8

Monot *et al.* [16] also used a DOA based system. Their prototype implemented the Capon [284] and the MUSIC [3, 283, 301] algorithms using a five element antenna array, and it was reported to have successfully estimated the DOA of the different paths, in an environment consisting of one main path and a set of spatially dispersed other paths.

4.3.6   Adaptive Beamforming Performance Results

The performance of the SMI algorithm of Section 4.3.2.3 using the direct matrix inversion formula of Equation 4.70 and the iterative matrix inversion lemma in Equation 4.75 as well as that of the ULMS and NLMS algorithms of Sections 4.3.2.1 and 4.3.2.2 was compared for identical scenarios. The effects of varying the reference signal lengths and the SNR as well as INR on the level of interference rejection were measured. For the situations exposed to different SNRs and INRs, the eigenvalue spread $\chi(R)$ of Equation 4.56 is summarized in Table 4.1.

The effects of varying the reference signal length, the signal-to-noise, and the interference-to-noise ratios on the interference rejection achieved were evaluated and a complexity analysis was performed. The ability of the various beamforming algorithms to combine multipath signals, whilst rejecting interference was also investigated. The modulation scheme used in the simulations was BPSK. Our associated results are summarized in the forthcoming sections.

4.3.6.1   Two Element Adaptive Antenna Using Sample Matrix Inversion

Recall that the SMI algorithm of Section 4.3.2.3 directly inverts the sample correlation matrix, $\hat{R}_{xx} = E[\underline{x}(t)\underline{x}^H(t)]$, in order to find the optimal antenna element weights according to Equation 4.70. Specifically, we have $\underline{\hat{w}}_3 = \hat{R}_{xx}^{-1}\underline{\hat{z}}$, where $\underline{\hat{z}}$ is the sample cross-correlation vector between the array output vector, $\underline{x}$, and the reference signal, r. The iterative version of this technique, as described in Section 4.3.2.3, forms the inverse of the sample correlation matrix, $\hat{R}_{xx}^{-1}$, based on the received signal samples using Equation 4.75, and iteratively

Figure 4.24: The interference rejection achieved using SMI beamforming upon varying the reference signal lengths for a two element antenna array using an element spacing of $\lambda/2$ at (a) equal SNR and INR and (b) unequal SNR and INR. The source was at $0°$ and the interferer at $30°$, whilst $\epsilon_0 = 0.01$ evaluating $10\,000$ averaged runs over a Gaussian channel.

updates it according to:

$$\hat{R}^{-1}(n) = \hat{R}^{-1}(n-1) - \frac{\hat{R}^{-1}(n-1)\underline{x}(n)\underline{x}^H(n)\hat{R}^{-1}(n-1)}{1 + \underline{x}^H(n)\hat{R}^{-1}(n-1)\underline{x}(n)}, \qquad (4.98)$$

with $\hat{R}^{-1}(0) = \frac{1}{\epsilon_0}I$ where ϵ_0 is a scalar value greater than zero.

The interference rejection achieved using the SMI algorithm as a function of the reference signal length is shown in Figure 4.24(a) for equal values of SNR and INR, i.e. for equal signal and interferer powers. The graph also shows how the interference rejection increases, as the SNR and INR are increased. The performance of the direct inversion method of

Figure 4.25: The interference rejection achieved versus SNR and INR using SMI beamforming upon varying the reference signal lengths for a two element antenna array using an element spacing of $\lambda/2$, at equal SNR and INR. The source was at $0°$ and the interferer at $30°$, whilst $\epsilon_0 = 0.01$ evaluating $10\,000$ averaged runs over a Gaussian channel.

Figure 4.26: The interference rejection achieved using ULMS beamforming upon varying the reference signal lengths, a two element antenna array, using an element spacing of $\lambda/2$, at equal SNR and INR. The source was at $0°$ and the interferer at $30°$, whilst $\mu = 0.00005$ evaluating $10\,000$ averaged runs over a Gaussian channel.

Equation 4.70 and the iterative method of Equation 4.75 was found to be identical using a value of $\epsilon_0 = 0.01$ in Equation 4.76 to initialize the estimate of R^{-1}. For a setting of $\epsilon_0 = 0.3$ the difference between the rejection levels was of the order of 0.01 dB, while a 0.1 dB interference rejection reduction resulted from $\epsilon_0 = 0.9$. As stated earlier in Section 4.3.2.3, an adequate performance can be achieved after processing only $2M$ data samples, where M

Figure 4.27: The interference rejection achieved using ULMS beamforming upon varying the reference signal lengths for a two element antenna array using an element spacing of $\lambda/2$, at equal SNR and INR. The source was at $0°$ and the interferer at $30°$, whilst (a) $\mu = 0.000\,000\,5$ and (b) $\mu = 0.0005$ evaluating $10\,000$ averaged runs over a Gaussian channel.

is the number of sources present, which was two in this case. Figure 4.24(b) shows that the interference rejection is only affected by the INR and appears to be independent of the SNR.

The rate at which the interference rejection increases, as the SNR and INR improve is shown in Figure 4.25. The increased SNR and INR values allow for more accurate estimates of the array output cross-correlation matrix, R, thus resulting in improved interference rejection. The rate of increase of the interference rejection slows down as the SNR and INR increase, since the limit of the estimation accuracy is approached. As expected, the longer reference lengths allow for a better estimate of R and hence exhibit higher interference rejection levels for sufficiently high SNR and INR values. In contrast, for low SNR and INR

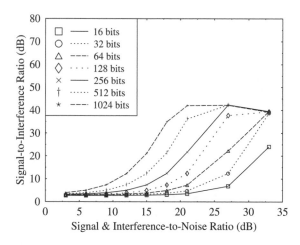

Figure 4.28: The interference rejection achieved using ULMS beamforming upon varying the reference signal lengths, SNR and INR. A two element antenna array was used with an element spacing of $\lambda/2$, at equal SNR and INR. The source was at $0°$ and the interferer at $30°$ while $\mu = 0.000\,05$ evaluating 10 000 averaged runs over a Gaussian channel.

the estimation quality of R is poor, resulting in marginal performance improvements due to extending the reference sequence length.

4.3.6.2   Two Element Adaptive Antenna Using Unconstrained Least Mean Squares

The Unconstrained Least Mean Squares (ULMS) technique [3,280–283] of beamforming was described in more detail in Section 4.3.2.1 but is based around the weight update formula of Equation 4.54, i.e. $\underline{w}(n+1) = \underline{w}(n) - \mu\underline{x}(n)\epsilon^*(n)$, where μ is a constant controlling the rate of convergence and $\epsilon(n)$ is the error between the combiner output, $y(n)$, and the reference signal, $r(n)$. For each array output sample, $x(n)$, the new antenna element weights are calculated, in order to minimize the mean square error between the measured array output and the desired array output.

The performance of the ULMS algorithm of Section 4.3.2.1 was studied using $\mu = 0.000\,000\,5$, $\mu = 0.000\,05$, $\mu = 0.0005$ in Equation 4.54 and varying the prevalent SNR and INR. It was found that convergence was extremely slow using $\mu = 0.000\,000\,5$, and a reasonable level of interference rejection required an SNR and INR of 33.0 dB in conjunction with a reference length of 1024 bits. This shows the dependence of the ULMS algorithm upon the received signal strength, which is evidenced by Figures 4.26, 4.27(a) and 4.27(b). Additionally, Figure 4.29(a) shows that step size is insufficient to allow convergence to an acceptable level of interference rejection regardless of the reference length or the number of iterations. In contrast, using a value of $\mu = 0.000\,05$ in Figure 4.26 results in significantly faster convergence for all SNRs and INRs, where best performance was achieved by the stronger signals. However, the step size is excessive for SNRs and INRs in excess of about 20 dB and the phenomenon of weight jitter can be seen becoming apparent. Figure 4.28 illustrates this further since it can be seen that the interference rejection achieved actually decreases for high SNRs and INRs upon increasing the number of iterations. Again, this

Figure 4.29: The interference rejection achieved using ULMS beamforming upon varying the reference signal lengths, SNR and INR. A two element antenna array was used with an element spacing of $\lambda/2$, at equal SNR and INR. The source was at $0°$ and the interferer at $30°$, whilst (a) $\mu = 0.000\,000\,5$ and (b) $\mu = 0.0005$ evaluating $10\,000$ averaged runs over a Gaussian channel.

phenomenon is due to weight jitter around the optimal solution for high values of SNR as well as INR and it becomes more prevalent for a large step size of 0.0005, which may be seen in Figure 4.29(b). Increasing the step size to 0.0005 results in a levelling off or even a reduction in the interference rejection achieved, as shown in Figure 4.27(b). Therefore, if the step size, μ, is chosen to be small, weak signals associated with low SNRs and INRs limit the convergence speed and may not be of much practical use, while strong signals allow for rapid convergence, as displayed in Figure 4.26. However, if μ is large then the convergence is rapid

Figure 4.30: The interference rejection achieved using NLMS beamforming upon varying the reference signal lengths for a two element antenna array with an element spacing of $\lambda/2$, at equal SNR and INR. The source was at $0°$ and the interferer at $30°$ whilst $\mu_0 = 0.2$ evaluating 10 000 averaged runs over a Gaussian channel.

Figure 4.31: The interference rejection achieved using NLMS beamforming upon varying the reference signal lengths for a two element antenna array with an element spacing of $\lambda/2$, at equal SNR and INR. The source was at $0°$ and the interferer at $30°$ whilst $\mu_0 = 0.5$ evaluating 10 000 averaged runs over a Gaussian channel.

even for weak signals, but the algorithm exhibits weight jitter, resulting in poor performance and potential instability.

4.3.6.3 Two Element Adaptive Antenna Using Normalized Least Mean Squares

The Normalized Least Mean Squares (NLMS) algorithm [283, 288] of Section 4.3.2.2 uses a data dependent step size calculated using Equation 4.59, namely $\mu(n) = \frac{\mu_0}{\|\underline{x}(n)\|^2}$, in order to eliminate the deficiencies of the ULMS method of Section 4.3.2.1. Figure 4.30 characterizes the algorithm's convergence, when $\mu_0 = 0.2$ in Equation 4.59.

Figure 4.32: The interference rejection achieved using NLMS beamforming upon varying the reference signal lengths for a two element antenna array with an element spacing of $\lambda/2$, at equal SNR and INR. The source was at $0°$ and the interferer at $30°$ whilst $\mu_0 = 1.0$ evaluating 10 000 averaged runs over a Gaussian channel.

When compared to the performance of the algorithm using the larger step sizes of $\mu_0 = 0.5$ and $\mu_0 = 1.0$ in Figures 4.31 and 4.32, it can be seen that for a small reference signal length the level of interference rejection is increased in conjunction with the larger step sizes, due to their faster rates of convergence. However, after the final interference rejection level has been reached, the algorithm performs better for smaller step sizes, attaining a higher level of interference rejection at the end of the convergence phase, and significantly lower weight jitter. For example, using $\mu_0 = 1.0$ when the SNR and INR was 6.0 dB, the interference rejection became approximately 15 dB exhibiting a jitter of ± 2.5 dB. In the case of $\mu_0 = 0.2$, the interference rejection was 20 dB exhibiting virtually no jitter effects. The performance difference became even more marked for higher SNR and INR levels.

Figure 4.33 demonstrates how the interference rejection increases, as the SNR and INR improve. When $\mu_0 = 0.2$, the rate of convergence is too slow for the optimal solution to be reached for reference signal lengths of 16 and 32 bits. For a reference signal length of 64 bits, a near optimal solution is obtained at low values of SNR and INR but as the SNR and INR increase, the performance of the algorithm does not improve beyond a certain point. This performance limitation experience for short reference signal lengths is due to the limited estimation quality of the mean of the received signal. Using a larger step size, hence allowing for faster convergence, resulted in shorter reference signal lengths converging to the optimal weights, although the final value of interference rejection reached did not match that of the smaller step sizes.

The performance of the NLMS beamforming algorithm for unequal values of the SNR and the INR is portrayed in Figure 4.34. From the associated subfigures it can be seen that as the INR improves, i.e. as the interference power increases, so does the interference rejection, regardless of the SNR. However, for a given level of interference, better interference rejection is achieved for a higher SNR, although the rate of convergence may be slower, as seen for the case when we have SNR=18.0 dB and the INR=6.0 dB. Faster convergence was observed for higher values of the INR, for a given SNR. However, for a high INR associated with a low SNR, i.e. for example for SNR=6.0 dB and INR=18.0 dB, significant weight jitter occurred,

Figure 4.33: The interference rejection achieved using NLMS beamforming upon varying the reference
signal lengths, and SNR and INR, for a two element antenna array with an element spacing
$\lambda/2$, at equal SNR and INR. The source was at $0°$ and the interferer at $30°$, whilst (a) $\mu_0 =$
0.2; (b) $\mu_0 = 0.5$; and (c) $\mu_0 = 1.0$, evaluating $10\,000$ averaged runs over a Gaussian
channel.

whilst fast convergence was maintained. Therefore, when the power spread of the received
signals is substantial, the NLMS adaptive beamforming algorithm does not perform as well
as the SMI algorithm. In contrast, when the range of input powers is smaller, the algorithm
performs well and for more than six antenna elements, this is achieved at a lower complexity
than that of the SMI algorithm, as will be shown in Section 4.3.6.5.

4.3.6.4 Performance of a Three Element Adaptive Antenna Array

The interference rejection capabilities of a three element uniformly spaced linear adaptive
array were investigated upon increasing the number of interference sources. The purpose
of these experiments was to determine how the array behaved, when the total number of
sources and interferers exceeded the degrees of freedom of the array, which was defined as

Figure 4.34: The interference rejection achieved using NLMS beamforming upon varying the reference signal lengths, and SNR and INR, for a two element antenna array with an element spacing $\lambda/2$, at unequal SNR and INR: (a) SNR = 6.0 dB, (b) SNR = 12.0 dB, (c) SNR = 18.0 dB. The source was at $0°$ and the interferer at $30°$ whilst $\mu_0 = 0.5$ evaluating $10\,000$ averaged runs over a Gaussian channel.

the number of sources and/or interferences that may simultaneously be steered towards or nulled. The source was located at $15°$, interferer 1 was at $-30°$, interferer 2 at $60°$, interferer 3 was located at $80°$ and lastly, interference source 4 at $-70°$. It was assumed that the sources were point sources located in the far-field of the antenna array, benefiting from pure line of sight propagation without multipaths. Figure 4.35 shows the locations of the desired source and the interfering sources graphically. The simulations were carried out in conjunction with a 256-bit reference signal using the SMI and NLMS algorithms.

From the antenna array beam patterns portrayed in Figure 4.36 it can be observed that successful nulling of the interference source was accomplished for all the scenarios considered. A minimum interference rejection of 40 dB was attained for an INR of 9 dB, and when the INR was increased to 21 dB, an even higher rejection was achieved.

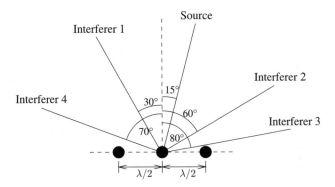

Figure 4.35: Locations of the desired source and the interferers with respect to the three element linear array with $\lambda/2$ element spacing.

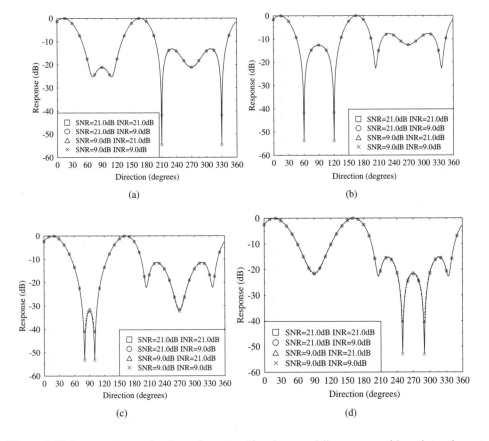

Figure 4.36: Beam patterns of a three element uniformly spaced linear array with an inter-element spacing of $\lambda/2$ with one desired source and one source of interference: (a) source located at $15°$, interference at $-30°$; (b) source located at $15°$, interference at $60°$; (c) source located at $15°$, interference at $80°$; (d) source located at $15°$, interference at $-70°$. The SMI beamforming algorithm was used with a reference length of 256 bits.

(a)

(b)

Figure 4.37: Beam patterns of a three element uniformly spaced linear array having an inter-element
spacing of $\lambda/2$ in conjunction with one desired source and two sources of interference:
(a) source located at 15°, interferers located at $-30°$ and, 80° or $-30°$ and $-70°$;
(b) source located at 15°, interferers located at 60° and 80°, or 60° and $-70°$, or 80°
and $-70°$. The SMI beamforming algorithm was used with a reference length of 256 bits.

Figure 4.38: Beam patterns of a three element uniformly spaced linear array having an inter-element spacing of $\lambda/2$ in conjunction with one desired source and two sources of interference with unequal powers. The SMI beamforming algorithm was used with a reference length of 256 bits. The desired source was at $15°$, interference source 1 was located at $-30°$ and interferer 2 at $60°$.

Figure 4.37 shows the array response for the situation where two interferers are incident upon the antenna array, having equal signal strengths to that of the desired signal. For the cases illustrated in Figure 4.37(a), where one of the sources of interference is at a $-30°$ angle with respect to the array, good rejection of both sources of interference is achieved, whilst maintaining a perfect response in the direction of the desired source. Even for the situation, where the interference sources are located fairly close to each other, i.e. at $-30°$ and $-70°$, strong nulling is maintained. Placing the interferers closer together, at angles of $60°$ and $80°$, resulted in an interference rejection of over 45 dB, albeit exhibiting some beam and null mis-alignment. Spreading the interferers further apart, with each one tending to "end-fire" at opposite ends of the array leads to some beam mis-steering, but nevertheless, maintaining good rejection of the sources of interference. Separating the interferers further so that they were located at $-70°$ and $80°$ yielded significantly poorer results with an average interference rejection of about 25 dB. However, this is still perfectly acceptable and levels significantly higher than this would be unrealizable due to hardware limitations.

From Figure 4.38 it can be seen that, if two sources of interference are present, and one of them is weaker than the other, then the stronger one will be nulled more effectively than the weaker one. The SNR of the desired signal does not appear to affect the interference rejection.

When three sources of interference and one desired signal source are incident upon a three element antenna array, the performance of the array is reduced compared to the situation, when fewer sources impinge upon the array concurrently. In Figure 4.39(a) it can be seen that an interference rejection ratio of at least 15 dB is achieved for all of the interference

(a)

(b)

Figure 4.39: Beam patterns of a three element uniformly spaced linear array having an inter-element spacing of $\lambda/2$ in conjunction with one desired source and three sources of interference: (a) source located at $15°$, interferers located at $-30°$, $60°$ and $80°$ or $-30°$, $60°$ and $-70°$; (b) source located at $15°$, interferers located at $-30°$, $80°$ and $-70°$, or $60°$, $80°$ and $-70°$. The SMI beamforming algorithm was used with a reference length of 256 bits. The SNR and INRs were 21.0 dB.

Figure 4.40: Beam patterns of a three element uniformly spaced linear array having an inter-element spacing of $\lambda/2$ in conjunction with one desired source and three sources of interference: (a) source located at 15°, interferers located at $-30°$, 60° and 80° or $-30°$, 60° and $-70°$; (b) source located at 15°, interferers located at $-30°$, 80° and $-70°$, or 60°, 80° and $-70°$. The SMI beamforming algorithm was used with a reference length of 256 bits. The SNR was 21.0 dB whilst the INRs were 9.0 dB.

(a)

(b)

Figure 4.41: Beam patterns of a three element uniformly spaced linear array having an inter-element spacing of $\lambda/2$ in conjunction with one desired source located at $15°$, and four sources of interference located at $-30°$, $60°$, $80°$ and $-70°$: (a) equal SNR and INR of 21.0 dB; (b) comparison between all SNRs and INRs of 21.0 dB, and all at 21.0 dB except the interferer at $80°$ which has an INR of 9.0 dB. The SMI beamforming algorithm was used with a reference length of 256 bits.

sources simultaneously, where greater than 20 dB rejection ratios are also frequently obtained. The results presented in Figure 4.39(b) are better than those in Figure 4.39(a), exhibiting a minimum interference rejection of 25 dB. Therefore, the interference rejection obtainable when the number of sources equals the number of antenna elements appears to be dependent upon the location of the sources, but on average a good interference rejection performance is observed. Reducing the SNR from 21 dB to 9.0 dB, whilst keeping the INR at 21 dB produced the results depicted in Figure 4.40. The beam patterns in this figure are similar

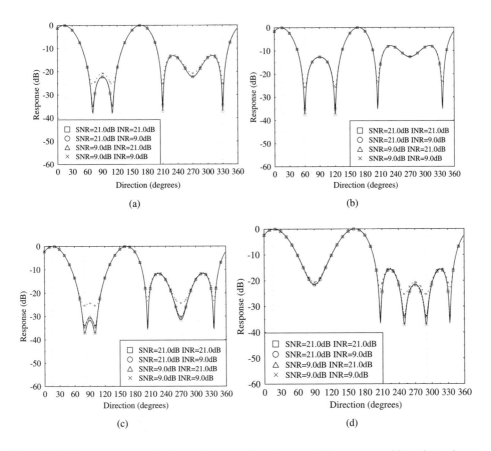

Figure 4.42: Beam patterns of a three element uniformly spaced linear array with an inter-element spacing of $\lambda/2$ with one desired source and one source of interference: (a) source located at $15°$, interference at $-30°$; (b) source located at $15°$, interference at $60°$; (c) source located at $15°$, interference at $80°$; (d) source located at $15°$, interference at $-70°$. The NLMS beamforming algorithm was used with a reference length of 256 bits.

in form to those of Figure 4.39, where the Interference-to-Noise Ratios was 21 dB, but the depths of the nulls are shallower. Although the nulls are less deep, the INRs are not as high, so the resultant SIR should not be any higher. Furthermore, the nulls are generally still more than 15 to 20 dB deep, which should be sufficient for effective interference rejection.

The performance of the three element antenna array when the desired source and the four interfering sources, all exhibiting equal signal power, are incident upon it, is shown in Figure 4.41(a). The antenna array response is virtually identical for the scenario when all the sources have SNRs of 21 dB, to that when the SNRs are equal to 9 dB. The array succeeds in suppressing all of the interference sources by at least 15 dB, where one of the interferers is nulled by more than 40 dB. In the situation when one of the interference sources has an INR of 9 dB, as in Figure 4.41(b), it is nulled less strongly than in the case of an INR of 21 dB. Although the associated null-depth was reduced from 43 dB to 29 dB, due to the associated

Figure 4.43: Beam patterns of a three element uniformly spaced linear array with an inter-element spacing of $\lambda/2$ with one desired source and two sources of interference: (a) source located at $15°$, interferers located at $-30°$ and, $80°$ or $-30°$ and $-70°$; (b) source located at $15°$, interferers located at $60°$ and $80°$, or $60°$ and $-70°$, or $80°$ and $-70°$. The NLMS beamforming algorithm was used with a reference length of 256 bits.

(a)

(b)

Figure 4.44: Beam patterns of a three element uniformly spaced linear array with an inter-element spacing of $\lambda/2$ with one desired source and three sources of interference: (a) source located at $15°$, interferers located at $-30°$, $60°$ and $80°$ or $-30°$, $60°$ and $-70°$; (b) source located at $15°$, interferers located at $-30°$, $80°$ and $-70°$, or $60°$, $80°$ and $-70°$. The NLMS beamforming algorithm was used with a reference length of 256 bits. The SNR and INRs were 21.0 dB.

Figure 4.45: Beam patterns of a three element uniformly spaced linear array with an inter-element spacing of $\lambda/2$ with one desired source and three sources of interference: (a) source located at 15°, interferers located at $-30°$, 60° and 80° or $-30°$, 60° and $-70°$; (b) source located at 15°, interferers located at $-30°$, 80° and $-70°$, or 60°, 80° and $-70°$. The NLMS beamforming algorithm was used with a reference length of 256 bits. The SNR was 21.0 dB whilst the INRs was 9.0 dB.

Figure 4.46: Beam patterns of a three element uniformly spaced linear array with an inter-element spacing of $\lambda/2$ with one desired source located at $15°$, and four sources of interference located at $-30°$, $60°$, $80°$ and $-70°$: (a) equal SNR and INR of 21.0 dB; (b) comparison between all SNRs and INRs of 21.0 dB, and all at 21.0 dB except the interferer at $80°$ which has an INR of 9.0 dB. The NLMS beamforming algorithm was used with a reference length of 256 bits.

$21 - 9 = 12$ dB decrease in the power of the interferer, the SIR only fell by 2 dB to 20 dB. However, the rejection of the other interference sources increased slightly.

The beam patterns obtained for exactly the same scenarios, except using the NLMS beamforming algorithm along with $\mu_0 = 0.5$, are presented in Figures 4.42 to 4.46. From the graphs in Figure 4.42 it can be observed that the nulls formed by the NLMS adaptive beamforming algorithm are not as deep as those of the SMI algorithm. As for the SMI algorithm, the null depths are also shallower when the INRs are lower.

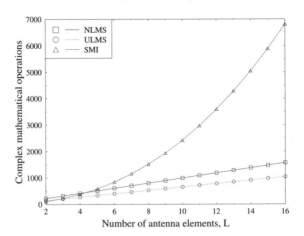

Figure 4.47: The relative complexities of the DMI, ULMS and NLMS beamforming algorithms for a reference signal length of 16 symbols.

In the case of two sources of interference, as shown in Figure 4.43, the algorithm has again successfully nulled the sources, albeit with a lower attenuation than that achieved by the SMI algorithm as may be seen in Figure 4.37. This is, however of purely academic interest, since null depths of 50 dB would be unrealizable. For three interferers, all having the same power as the desired source, this phenomenon persists, as it does when the interference sources are of lower power. The corresponding results for three interferers are portrayed in Figure 4.44 for an SNR and INR value of 21.0 dB. Observe, however, in Figure 4.45(b) that the interference rejection for the source at an angle of 60° is significantly lower at 20 dB than that obtained using the SMI algorithm, which was 27 dB. For deep nulls this difference would have little impact, but at these levels of interference rejection, it may be problematic.

Figure 4.46 shows the beam patterns encountered, when four sources of interference and one desired source are present simultaneously. In conjunction, with the NLMS beamforming algorithm the levels of interference rejection for each interference source are lower than those obtained using the SMI algorithm. Specifically, the associated reductions vary from only 2 dB to 17 dB, having a mean difference of about 8 dB.

4.3.6.5 Complexity Analysis

The relative complexities of the DMI, ULMS and NLMS beamforming algorithms for a reference signal length of 16 symbols are portrayed in Figure 4.47. The direct matrix inversion algorithm requires the average of the cross-correlation matrix, R, which is a square-shaped matrix of size L, where L is the number of antenna elements. In order to calculate each element of the matrix, R, N complex multiplications and $N - 1$ complex additions must be performed, where N is the sample size, in bits, used to generate the cross-correlation matrix, R. Due to the Hermitian nature of the matrix, R, it is only necessary to execute these instructions $L(L + 1)/2$ times, rather than L^2 times, as would be expected.

Therefore, $NL(L + 1)/2$ complex multiplications and $L(L + 1)(N - 1)/2$ complex additions are required for forming the matrix, leading to a total of $L(L + 1)(2N - 1)/2$

complex operations. However, upon assuming that a Multiply-and-ACcumulate (MAC) instruction exists in the implementation, this complexity figure reduces to $NL(L + 1)$. The Hermitian cross-correlation matrix, R, must then be inverted requiring $L^3/2 + L^2$ complex operations [285], rather than the usual L^3 operations required for a non-Hermitian matrix. In order to calculate the correlation between the reference signal and the array output vector requires a further L complex multiplications and $L - 1$ complex additions, reducing to L complex operations assuming a MAC instruction. Then, from the inverted matrix, R^{-1}, and the correlation vector, $\underline{z}$, the weight vector, $\underline{w}$, may be obtained after L^2 complex operations. Therefore, the total complexity of the SMI beamforming algorithm is $L(L + 1)(2N - 1)/2 + L^3/2 + 2L^2 + 2L - 1$ complex operations.

The ULMS adaptive beamformer requires only $2L + 1$ complex multiplications and $2L$ complex additions per iteration, rendering it the least complex algorithm. However, the NLMS technique is more practical, since its performance is less dependent upon the input power. The additional complexity associated with this algorithm is the $L + 1$ complex multiplications required for calculating the current value of μ. Therefore, the final complexity of the NLMS algorithm is equivalent to $3L + 2$ complex multiplications and $3L$ complex additions per bit received. Hence the total number of complex operations required by the NLMS beamforming algorithm is $N(3L + 2 + 3L){=}N(6L + 2)$.

4.4 Summary and Conclusions

In this chapter we commenced in Section 4.2.2 by considering the possible applications of antenna arrays and their related benefits. A signal model was then described in Section 4.2.3 and a rudimentary example of how beamforming operates was presented. Section 4.3 highlighted the process of adaptive beamforming in conjunction with several different temporal reference techniques detailed, along with the approaches of spatial reference techniques and the associated process of antenna array calibration. The challenges that must be overcome before beamforming for the DL becomes feasible were also discussed in Section 4.3.5.

In Section 4.3.6 results were presented showing how the SMI, ULMS and NLMS beamforming algorithms of Sections 4.3.2.3, 4.3.2.1 and 4.3.2.2 behaved for a two element adaptive antenna having varying eigenvalue spread and reference signal length. The SMI algorithm was shown to converge very rapidly, irrespective of the eigenvalue spread, and the level of interference rejection was found to be purely dependent upon the interference power, regardless of the desired signal power. However, in Section 4.3.6.2 the convergence characteristics of the ULMS adaptive beamforming algorithm were shown to be heavily dependent upon both the desired signal power and the interfering signal powers. The NLMS algorithm, in contrast, was found to be far superior in this respect, and considering its significantly lower complexity than that of the SMI technique, offered good performance.

The performance of the SMI and NLMS algorithms was then compared in Section 4.3.6.4 for a three element antenna array with one desired source and between one and four sources of interference. The results obtained in Section 4.3.6.4 further evidenced the better performance of the SMI algorithm, but as was shown in Section 4.3.6.5, this was achieved at a significantly higher complexity, when the number of array elements was higher than four. For a low number of elements, the SMI algorithm was found to have a lower complexity than both the ULMS and the NLMS techniques. However, as the number of antenna elements used in the array increased, the complexity of the SMI technique exponentially increased, while that of the LMS routines increased only linearly. Therefore, for about ten array elements the complexity of the SMI algorithm was about twice that of the NLMS technique.

In the next chapter we consider the performance benefits that may be obtained with the advent of adaptive antenna arrays in a cellular radio network.

Chapter 5

Adaptive Arrays in an FDMA/TDMA Cellular Network

5.1 Introduction

Cellular networks are typically interference limited, with co-channel interference arising from cellular frequency reuse, ultimately limiting the quality and capacity of wireless networks [321, 322]. However, Adaptive Antenna Arrays (AAAs) are capable of exploiting the spatial dimension in order to mitigate this co-channel interference and thus to increase the achievable network capacity [3, 6, 65, 283, 291, 323]. Since an AAA may receive signals with a high gain from one direction, whilst nulling signals arriving from other directions, it is inherently suited to a CCI-limited cellular network. Thus a beam may be formed to communicate with the desired mobile, whilst nulling interfering mobiles [6]. Assuming that each mobile station is uniquely identifiable, it is a relatively simple task to calculate the antenna array's receiver weights, so as to maximize the received SINR. The use of adaptive antenna arrays in a cellular network is an area of intensive research and adaptive antenna array's have been studied widely in the context of both interference rejection and in single-cell situations [1, 15, 18, 302, 308, 309]. More recently, work has been expanded to cover the analysis and performance benefits of using base stations equipped with adaptive antenna arrays across the whole of a cellular network [2, 306, 324].

A further approach to improving the network performance is the employment of Dynamic Channel Allocation (DCA) techniques [325–333], which offer substantially improved call-blocking, packet dropping, and grade-of-service performance in comparison to Fixed Channel Allocation (FCA). A range of so-called distributed DCA algorithms were investigated by Cheng and Chuang [331] where a given physical channel could be invoked anywhere in the network, provided that the associated channel quality was sufficiently high. As compromise schemes, locally optimized distributed DCA algorithms were proposed, for example, by Delli Priscoli *et al.* [334, 335], where the system imposed an exclusion zone for reusing a given physical channel around the locality, where it was already assigned.

3G, HSPA and FDD versus TDD Networking   Second Edition
L. Hanzo, J. S. Blogh and S. Ni   © 2008 John Wiley & Sons, Ltd

In Sections 5.2.1–5.2.3 we briefly consider how an adaptive antenna array may be modeled for employment in a network level simulator, followed by a short overview of a variety of channel allocation schemes in Section 5.3. This section also provides a brief performance summary of the various channel allocation schemes based on our previous work [50, 336], which suggested for the scenarios considered [50, 336] that the Locally Optimized Least Interference Algorithm (LOLIA) provided the best overall compromise in network performance terms. Section 5.4 presents a theoretical analysis of the performance of an adaptive antenna in a cellular network. A summary of several multipath propagation models is given in Section 5.5, with particular emphasis on the Geometrically Based Single-Bounce Statistical Channel Model [337, 338]. The potential methods of cellular network performance evaluation are described in Section 5.3.3.4, as are the parameters of the network simulated in later sections. Simulation results for Fixed Channel Allocation (FCA) and two Dynamic Channel Allocation (DCA) schemes using single element antennas, as well as two- and four-element adaptive antenna arrays for Line-Of-Sight (LOS) scenarios are presented and analyzed in Section 5.6.2.1. Furthermore, simulation-specific details of the multipath model given in Section 5.6.1, with the associated results obtained for the FCA and the LOLIA in the context of two, four and eight element adaptive antenna arrays presented in Section 5.6.2.2. Performance results for a network using power control over a multipath channel in conjunction with two and four element adaptive antenna arrays are provided in Section 5.6.2.3, followed by the description of a network using Adaptive Quadrature Amplitude Modulation (AQAM) in Section 5.6.2.4.

Performance results were also obtained for AQAM and the FCA algorithm as well as the LOLIA, with both two- and four-element adaptive antenna arrays. Results using the "wraparound" technique, described in Section 5.6.1, which removes the cellular edge effects observed at the simulation area perimeter of a "desert-island" scenario, are then presented in Sections 5.6.3.1–5.6.3.4. Finally, a performance summary of the investigated networks is given in Section 5.7.

5.2 Modelling Adaptive Antenna Arrays

The interference rejection capability of an antenna array is determined by both the direction of arrival of the interference and the angle of arrival of the desired signal and therefore ultimately by the angular separation between the two. The direction of arrival and angle of arrival may be used interchangeably throughout our discussions. The number of interferers and their signal strengths also affects the achievable attenuation of each of the interferers. This section attempts to derive a simple relationship between these factors for low-complexity modeling of an adaptive antenna array.

5.2.1 Algebraic Manipulation with Optimal Beamforming

Given that the steering vector associated with the direction θ_i of the ith source can be described by an L-dimensional complex vector $\underline{s}_i$ as [283],

$$\underline{s}_i = [\exp(j\omega t_1(\theta_i)), \ldots, \exp(j\omega t_L(\theta_i))]^T, \tag{5.1}$$

where L is the number of elements in the antenna array, and t_i is the time delay experienced by a plane wave arriving from the ith source direction, θ_i, and measured from the antenna element at the origin. Then the correlation matrix, R, of the steering vector $\underline{s}_i$, may be expressed as [283]:

$$R = \sum_{i=1}^{M} p_i \underline{s}_i \underline{s}_i^H + \sigma_n^2 I, \tag{5.2}$$

where p_i is the power of the ith source, σ_n^2 is the noise power and I is the identity matrix.

Assuming optimal beamforming under the constraint of a unit response in the wanted user's direction, then the weight vector of the AAA is [283]:

$$\underline{w} = \frac{R^{-1}\underline{s}_0}{\underline{s}_0^H R^{-1} \underline{s}_0}. \tag{5.3}$$

The array factor, $F(\theta)$, in the direction θ may be formulated as [65]:

$$F(\theta) = \sum_{l=1}^{L} w_l e^{-j\omega t_l(\theta)}. \tag{5.4}$$

Therefore, given that the desired signal arrives from the direction θ_0, and an interfering signal arrives from the angle θ_1, the corresponding array responses are $F(\theta_0)$ and $F(\theta_1)$, respectively. Hence, the level of interference rejection, $F(\theta_0) - F(\theta_1)$, when one desired signal and one interfering signal are received at a two-element antenna array, may be calculated using Equation 5.4 to be:

$$F(\theta_0) - F(\theta_1) = \frac{(2p_1 + \sigma_n^2)e^{\frac{j\omega\lambda \sin\theta_0}{2c}} - (p_1 + \sigma_n^2)e^{\frac{j\omega\lambda(2\sin\theta_0 - \sin\theta_1)}{2c}} - p_1 e^{\frac{j\omega\lambda \sin\theta_1}{2c}}}{(2p_1 + 2\sigma_n^2)e^{\frac{j\omega\lambda \sin\theta_0}{2c}} - p_1 e^{\frac{j\omega\lambda \sin\theta_1}{2c}} - p_1 e^{\frac{j\omega\lambda(2\sin\theta_0 - \sin\theta_1)}{2c}}}, \tag{5.5}$$

where the terms interference rejection is defined as the difference between the array response in the direction of the desired signal source and that in the directions of the interfering source.

As can be seen from this equation, there is a non-linear relationship between the two angles of arrival and the achievable interference rejection. Furthermore, the achievable interference rejection is independent of the desired signal's received power, p_0, and it is solely dependent upon the power of the interfering signal, p_1. Expanding this technique to either an antenna array having more elements or to catering for more interfering sources, or to multiple incident beams, led to overly complicated expressions which would be too complex to evaluate in real-time. In order to avoid the associated complexity, the quantities required for interference rejection in a given scenario could be stored in lookup tables. However, the size of the table required to store all of the information would be impractical. For example, for the desired source, one dimension would be required for the angle of arrival and then another one for every interference source. Two further table dimensions would be required to store the angle of arrival and interference power. Therefore, the simple situation involving just one interferer, with a received power dynamic range of 40 dB, would require an array of $180 \times 180 \times 40 = 1,296,000$ elements, at an angular resolution of $1°$, and an interferer power resolution of 1 dB. For two interference sources this figure increases to $180 \times 180 \times 40 \times 180 \times 40 = 0.3312 \times 10^9$ elements, which is clearly excessive.

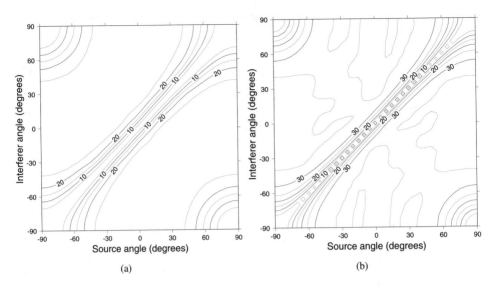

Figure 5.1: Contour plots of interference rejection achieved using a four element antenna array with
an inter-element spacing of $\lambda/2$ using SMI beamforming with a reference signal length of
16 bits: (a) desired signal SNR = 3.0 dB, interference SNR = 3.0 dB; (b) desired signal
SNR = 3.0 dB, interference SNR = 12.0 dB. The angles of arrival of the signals from the
desired source and the interfering source were swept over the range, −90 degrees to +90
degrees.

5.2.2   Using Probability Density Functions

Due to the inherent complexities of performing large-scale network simulations, whilst
invoking the required beamforming operations, we conducted an investigation into the
probability distribution of the interference rejection ratio achieved by an adaptive antenna
array. For our initial studies a two element antenna array with the elements located $\lambda/2$
apart was considered, with one desired source and one interfering source. Therefore, the
average interference rejection achieved in decibels, for a given source-direction and power
as well as interferer-direction and power could be determined. Unfortunately, as it can be
seen from Figure 5.1(a), the achievable interference rejection was not based upon a linear
relationship between the two angles of arrival. Furthermore, Figure 5.1(b) illustrates that the
interference rejection achieved was also related to the power, or the Signal-to-Noise Ratio
(SNR), of the undesired interference source, which was 3 dB or 12 dB. As it was found in
Section 5.2.1, attempting to construct a model or probability density function to cater for these
parameters was not easily achievable. Rather than attempting to find the Probability Density
Function (PDF) relating the two angles of arrival and interference power to the interference
rejection achieved, a brief study was initiated for determining the PDF of the interference
rejection achieved with respect to the angular separation between the desired signal and
interfering signal. Figure 5.2 shows the probability density function of interference rejection
achieved for one interference source and one desired source versus their angular separation.
As this figure shows, the distribution of the interference rejection varies significantly, as the

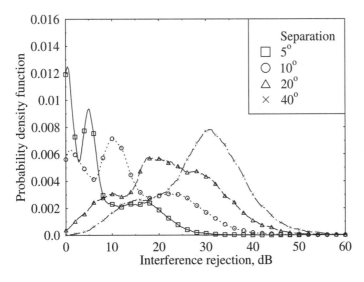

Figure 5.2: The PDF of the interference rejection (dB) achieved for various angular separations of the desired signal and the interfering signal. The angles of arrival of both signals were varied over the range of -90 to $+90$ degrees and were of equal power. The antenna array consisted of two elements separated by $\lambda/2$.

separation between the sources changes. As a consequence of the PDF's dependence on the angular separation encountered, modeling the achievable interference rejection expressed in decibels is an arduous task. Due to the complex nature of the PDF illustrated in Figure 5.2, an analysis of a smaller range of angles of arrival was conducted, in order to construct a piecewise valid model. The results are displayed in Figures 5.3(a) and 5.3(b) for angle of arrival spreads of $\pm 30°$ and $\pm 10°$, respectively. While these PDFs appear to be considerably simpler than that in Figure 5.2, it was not possible to match the PDFs to any commonly known distributions. Additionally, no information was available with regard to the correlation between successive interference rejection values. For these reasons, and due to the difficulties associated with adding multipath, it was decided to cease work on constructing a suitable interference rejection model and instead to implement an actual SMI beamformer within the simulation program as described in the following section.

5.2.3 Sample Matrix Inversion Beamforming

The process of defining a suitable model of an adaptive antenna array was becoming increasingly complex, resulting in the decision to implement an SMI beamformer in the simulation software. The SMI beamforming algorithm of Section 4.3.2.3 was chosen due to its independence from the received signal strengths, as well as due to its fast convergence with the aid of few data samples and for the sake of its good overall performance in terms of its interference rejection capability. The reference signal was chosen to be eight bits in length as a compromise between the quality of the sample correlation matrix, R, and the computational complexity required. Since a cellular network is an interference limited system, the addition

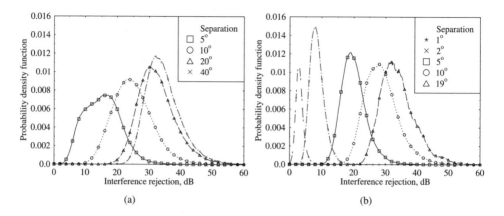

Figure 5.3: The PDF of the interference rejection achieved for the desired signal and the interfering signal angular separations of 5, 10 and 20 degrees: (a) angular spread $= \pm 30°$; (b) angular spread $= \pm 10°$. The desired signal and the interfering signal were of equal power. The antenna array consisted of two elements separated by $\lambda/2$.

of noise to the received signal vector was neglected. A result of this was that occasionally the correlation matrix, R, was non-invertible, which was remedied by diagonally augmenting the matrix with a positive constant as it was suggested in [15, 312, 313]. The addition of multipaths simply required the direction of arrival, and the strength of the multipath rays at the antenna array to be determined before adding these received signal vectors to the total received signal vector of the antenna array. In both the LOS and the multipath scenarios, the transmit/receive channel was assumed to be frequency invariant, thus allowing the same antenna pattern to be used in both the UL and the DL.

5.3   Channel Allocation Techniques

P.J. Cherriman, L. Hanzo[1]

Channel assignment is the process of allocating a finite number of channels to the various base stations and mobile phones in the cellular network. In a system using fixed channel assignment, the channels are assigned to different cells during the network planning stage, and the assignment is rarely altered to reflect changes in traffic levels. A channel is assigned to a mobile at the commencement of the call and the mobile communicates with its base station on this channel until either the call terminates or the mobile leaves the current cell. Dynamic channel allocation, however, assigns a channel that best meets the channel selection criteria, which may be the channel experiencing the minimum interference level, depending upon the cost function used.

With the growth in the number of subscribers to mobile telecommunications systems worldwide and the expected introduction of multimedia services in handheld wireless

---

[1] This section is based on [192].

terminals, a tremendous demand for bandwidth has arisen. Since bandwidth is scarce and becoming increasingly expensive, it must be utilized in an efficient manner.

The main limiting factor in radio spectrum reuse is co-channel interference. In reduced cell-size micro/picocellular architectures, the frequency reuse distance is reduced, thereby increasing the capacity and area spectral efficiency of the system. However, as the channel reuse distance is reduced, the co-channel interference increases. Co-channel interference caused by frequency reuse is the most severe limiting factor of the overall system capacity of mobile radio systems. The most important technique for reducing co-channel interference is power control, an issue, which will be discussed in detail in the context of adaptive modulation during our further discourse. Interference cancellation techniques [339] or adaptive antennas [340–342] can also be used to reduce co-channel interference. However, a simpler and more effective technique used in current systems is employing sectorized antennas [343].

Although handovers are necessary in mobile radio systems, they often cause several problems, and they constitute the major cause of calls being forcibly terminated. As the cell size is decreased, the average sojourn time or cell-crossing time for a user is reduced. This results in an increased number of handovers, requiring more rapid handover completion. In practice a seamless handover is not always possible except when soft-handovers [344] are used in CDMA-based systems. Rapid and numerous handovers require a fast backbone network between the base stations and the mobile switching centers, or they necessitate an increased number of mobile switching centers. Clearly, the handover process is crucial with regard to the perceived Grade of Service (GOS), and a wide range of different complexity techniques have been proposed, for example, by Tekinay and Jabbari [345] and Pollini [346] for the forthcoming future systems. The related issue of timeslot reassignment was investigated by Bernhardt [347].

5.3.1 Overview of Channel Allocation

The purpose of channel allocation algorithms is to exploit the variability of the radio channel propagation characteristics in order to allow increased efficiency radio spectrum utilization, while maintaining required signal quality. The most commonly used signal quality measure is the signal-to-interference ratio (SIR), also known as the carrier-to-interference ratio (CIR). The signal quality measure that we have used previously was the signal-to-interference+noise ratio (SINR). The SINR is approximately equal to the signal-to-noise ratio (SNR) in a noise-limited environment and approximately equal to the SIR in an interference-limited environment.

The radio spectrum is divided into sets of noninterfering physical radio channels, which can be achieved using orthogonal time or frequency slots, orthogonal user signature codes, and so on. The channel allocation algorithm attempts to assign these physical channels to mobiles requesting a channel, such that the required signal quality constraints are met. There are three main techniques for dividing the radio spectrum into radio channels. The first is frequency division (FD), in which the radio spectrum is divided into several nonoverlapping frequency bands. However, in practice the spectral spillage from one frequency band to another causes adjacent channel interference, which can be reduced by introducing frequency guard bands. However, these guard bands waste radio spectrum, and hence there is a compromise between adjacent channel interference and frequency band-packing efficiency.

Tighter filtering can help reduce adjacent channel interference, allowing the guard bands to be reduced.

The second technique is time division (TD), in which the radio spectrum is divided into disjunct timeperiods, which are usually termed timeslots. However, using straight-forward rectangular windowing of the time-domain signal corresponds to convolving the signal spectrum with a frequency-domain sinc-function, resulting in Gibbs-oscillation. Hence, in practical systems a smooth time-domain ramp-up and ramp-down function associated with a time-domain guard period is employed. Therefore, there is a trade-off between complex synchronization, time-domain guard periods, and adjacent channel interference.

The third technique for dividing the radio spectrum into channels is code division (CD). Code division multiple access (CDMA) [66–68, 348] has been used in military applications, in the IS-95 mobile radio system [349], and in the recently standardized Universal Mobile Telecommunications System (UMTS) [348, 350]. In code division, the physical channels are created by encoding different users with different user signature sequences.

In most systems a combination of these techniques is used. For example, the Pan-European GSM system [55] uses frequency division duplexing for up- and down-link transmissions, while accommodating eight TDMA users per carrier. In this chapter, the term "channel" typically implies a physical channel, constituted by a timeslot of a given carrier frequency.

A wide variety of channel allocation algorithms have been suggested for mobile radio systems. The majority of these techniques can be classified into one of three main classes: fixed channel allocation (FCA), dynamic channel allocation (DCA), and hybrid channel allocation (HCA). Hybrid channel allocation is constituted by a combination of fixed and dynamic channel allocation, which is designed to amalgamate the best features of both, in order to achieve better performance or efficiency than DCA or FCA can provide. Several channel allocation schemes and the associated trade-offs in terms of performance and complexity are discussed in detail in the excellent overview papers of Katzela and Naghshineh [351] and those by Jabbari and Tekinay et al. [352, 353]. Figure 5.4 portrays the family tree for the main types of channel allocation algorithms, where the acronyms are introduced during our further discourse. Zander [354] investigated the requirements and limitations of radio resource management in general for future wireless networks. Everitt [355] compared various fixed and dynamic channel assignment techniques and investigated the effect of handovers in the context of CDMA-based systems.

5.3.1.1 Fixed Channel Allocation

In fixed channel allocation (FCA), the available radio spectrum is divided into sets of frequencies. One or more of these sets is then assigned to each base station on a semipermanent basis. The minimum distance between two base stations, they have been assigned the same set of frequencies is referred to as the frequency reuse distance. This distance is chosen such that the co-channel interference is within acceptable limits, when interferers are at least the reuse distance away from each other. The assignment of frequency sets to base stations is based on a predefined reuse pattern. The group of cells that contain one of each of the frequency sets is referred to as the frequency reuse cluster. The grade of frequency reuse is usually characterized in terms of the number of cells in the frequency reuse cluster. The lower the number of cells in a reuse cluster, the more bandwidth-efficient the frequency reuse

Simple FCA	Page 222	Static borrowing	Page 225
Simple borrowing	Page 225	Hybrid borrowing	Page 225
Flexible	Page 226	Hybrid	Page 230
Centrally controlled DCA	Page 228	Distributed DCA	Page 228
Locally distributed DCA	Page 229		

Figure 5.4: Family tree of channel allocation algorithms.

pattern and the higher the so-called area spectral efficiency, since this implies partitioning the available total bandwidth in a lower number of frequency subsets used in the different cells, thereby supporting more users across a given cell area. However, small reuse clusters exhibit increased co-channel interference, which has to be tolerated by the transceiver.

In FCA, the assignment of frequencies to cells is considered semipermanent. However, the assignment can be modified in order to accommodate teletraffic demand changes. Although FCA schemes are very simple, modifying them to adapt to changing traffic conditions or user distributions can be problematic. Hence, FCA schemes have to be designed carefully, in order to remain adaptable and scalable, as the number of mobile subscribers increases. In this context, adaptability implies the ability to rearrange the network to provide increased capacity in a particular area on a long- or short-term basis, where scalability refers to the ability of easily increasing capacity across the whole network via tighter frequency reuse. For example, Dahlin *et al.* [356] suggested a reuse pattern structure for the GSM system that can be scaled to meet increased capacity requirements, as the number of subscribers increases. This is discussed in more detail in the overview paper by Madfors *et al.* [357]. Each measure invoked, in order to further increase the network capacity, increases the system's complexity and hence becomes expensive. Furthermore, such systems cannot be easily modified to provide increased capacity in the specific area of a traffic hot-spot on a short-term basis.

A commonly invoked reuse cluster/pattern is the seven-cell reuse cluster, providing coverage over regular hexagonal shaped cells, which is shown in Figure 5.5. Each cell in the seven-cell reuse cluster has six first-tier co-channel interfering cells at a distance D, the reuse distance. By exploiting the simple hexagonal geometry seen in Figure 5.5 it can be shown that for the seven-cell cluster the reuse distance, D, is 4.58 times the cell

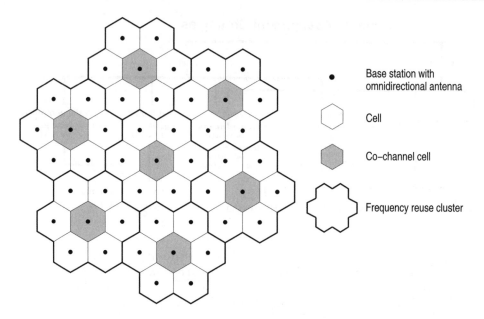

Figure 5.5: A commonly employed frequency reuse pattern for fixed channel assignment (FCA) algorithms. The frequency spectrum divided in seven frequency sets, one set assigned to each cell, yielding a seven-cell reuse cluster. Omnidirectional antennas were used, and the shaded cells represent cells assigned the same frequency set.

radius r [192]. This reuse pattern supports the same number of channels at each cell site, and hence the same system capacity. Therefore, the teletraffic capacity is distributed uniformly across all the cells. Since traffic distributions usually are not uniform in practice, such a system can lead to inefficiencies. For example, under nonuniform traffic loading, some cells may have no spare capacity; hence, new calls in these cells are blocked. However, nearby cells may have spare capacity.

Several studies have suggested techniques to find the optimal reuse pattern for particular traffic and users distributions, as exemplified by the work of Safak [358], on optimal frequency reuse with interference. While such contributions are useful, a practical system would need to modify the whole network configuration every time the traffic or user distributions changed significantly. Therefore, suboptimal but adaptable and scalable solutions are more desirable for practical implementations. When the traffic distribution changes, an alternative technique to modifying the reuse pattern is referred to as channel borrowing, which is the subject of the next section.

5.3.1.1.1   Channel Borrowing.   In channel-borrowing schemes, a cell that has a call setup request but no available channels (which is termed an acceptor cell), can borrow free channels from neighboring cells referred to as donor cells in order to accommodate new calls, which would otherwise have been blocked. A channel can be borrowed only if its use will not interfere with existing ongoing calls. When a channel is borrowed, several cells are then prohibited from using the borrowed channel because it would cause interference. The process of prohibiting the use of borrowed channels is referred to as channel locking [359]. The various channel-borrowing algorithms differ in the way the free channel is chosen from a donor cell to be borrowed by an acceptor cell.

There are three main types of channel-borrowing algorithms: static, simple, and hybrid borrowing; a good overview of these algorithms can be found in [351–353]. Static borrowing could be described as a fixed channel re-allocation strategy rather than channel borrowing. In static borrowing, channels are reassigned from lightly loaded cells to heavily loaded cells, which are at distances in excess of the reuse distance. This reassignment is semipermanent and can be done based on measured or predicted changes in traffic. The other two types of channel borrowing (simple and hybrid) are different from static borrowing in that borrowed channels are returned when the call using the channels ends or is handed off to another base station. Therefore, the simple and hybrid channel borrowing schemes use short-term borrowing in order to cope with traffic excesses.

Simple channel-borrowing schemes allow any of the channels in a donor cell to be lent to an acceptor cell. Hybrid channel borrowing schemes split the channels assigned to each cell into two subsets. One subset of channels cannot be lent to other cells; hence, these are referred to as standard or local channels. The other subset can be lent to other cells, and so they are termed nonstandard or borrowable channels.

Simple borrowing [328, 352, 360] can reduce new call blocking, but it can cause increased interference in other cells; it can also prevent handovers of future calls in these cells. Experiments have shown that simple channel-borrowing algorithms perform better than static fixed channel allocation under light- and moderate traffic loads. However, at high traffic loads the borrowing of channels leads to channel locking, which reduces the channel utilization and therefore results in an increase in new call blocking and in failed handovers. The various simple channel-borrowing algorithms differ in terms of flexibility, complexity and their reduction of channel locking. Some algorithms [328, 360] pick the channel to borrow, while taking into account the associated "cost" in terms of channel locking for each candidate channel. Other algorithms [360] invoke channel reassignment in order to reduce channel locking. The innovative technique used by Jiang and Rappaport [359] to reduce channel locking is to limit the transmission power of borrowed channels.

Hybrid channel borrowing [351, 352] is a hybrid of simple channel borrowing and static fixed channel allocation. By dividing the channels at each base station into two subsets, and only allowing channels of one of the subsets to be borrowed, the chance of channel locking or failed handovers can be mitigated under high traffic loads. A range of algorithms is discussed in the literature, each having different objectives in terms of improving performance in a particular area of operation. Some algorithms [361] have the ratio of channels in each subset assigned a priori, while others dynamically adapt the size of the subsets based on traffic measurements or predictions [362]. The algorithm may also check whether the candidate borrowed channel is free in the co-channel cells [363]. A common technique [360, 364] is to reassign calls using a borrowed channel to another borrowed channel in order to reduce channel locking. A better policy is to reassign a call currently using a borrowed channel to a local channel, thereby returning the borrowed channel to the donor cell. Another procedure [361, 363] to reduce channel locking is to estimate the direction of movement of the mobile in an attempt to reduce future channel locking and interference. A simple technique [365] is to subdivide cells into sectors and only allow borrowed channels to be used in particular sectors of the acceptor cell, thereby reducing channel locking.

5.3.1.1.2   Flexible Channel Allocation.   Flexible channel allocation schemes [351, 352, 366] are similar to hybrid channel allocation schemes (which are described in Section 5.3.1.3) in that they divide the available channels into fixed and dynamic allocation subsets. However, flexible channel allocation is similar to a fixed channel allocation strategy, such as that used in static channel borrowing. In flexible channel allocation, the fixed channel set is assigned to cells in the same way as in fixed and hybrid channel allocation. The dynamic or flexible channels can be assigned to cells depending on traffic measurements or predictions. The difference between so-called hybrid and flexible channel allocation schemes is that in hybrid channel allocation the dynamic channels are assigned to cells only for the duration of the call. In flexible channel allocation the dynamic channels are assigned to cells, when the blocking probability in these cells becomes intolerable. Flexible channel allocation requires much more centralized control than hybrid channel allocation.

5.3.1.2   Dynamic Channel Allocation

Although fixed channel allocation schemes are common in most existing cellular radio systems, the cost of increasing their teletraffic capacity can become high. In theory, the use of dynamic channel allocation allows the employment of all carrier frequencies in every cell, thereby ensuring much higher capacity, provided the transceiver-specific interference constraints can be met. Therefore, it is feasible to design a mobile radio system, which configures itself to meet the required capacity demands as and when they arise. However, in practice there are many complications, which make this simplistic view hard to implement in practice. Dynamic channel allocation is used, for example, in the Digital European Cordless Telephone (DECT) standard [298, 299, 367–369]. Law and Lopes [370] used the DECT system to compare the performance of two distributed DCA algorithms. However, DECT is a low-capacity system, where the timeslot utilization is expected to be comparatively low. For low slot utilization DCA is ideally suited. Dynamic channel allocation becomes more difficult to use in large-cell systems, which have higher channel utilization. Salgado-Galicia *et al.* [371] discussed the practical problems that may be encountered in designing a DCA-based mobile radio system.

Even though much research has been carried out into channel allocation algorithms, particularly dynamic channel allocation, many unknowns remain. For example, the trade-offs and range of achievable capacity gains are not clearly understood. Furthermore, it is not known how to combine even two simple algorithms in order to produce a hybrid that has the best features of both. One reason that the issues of dynamic channel allocation are not well understood is the computational complexity encountered in investigating such algorithms. In addition, the algorithms have to be compared to others in a variety of scenarios. Furthermore, changing one algorithmic parameter in order to improve the performance in one respect usually has some effect on another aspect of the algorithm's performance, due to the parameters highly interrelated nature. This is particularly true, since experience showed that some handover algorithms are better suited for employment in certain dynamic channel allocation algorithms [345]. Therefore the various channel allocation algorithms have to be compared in conjunction with a variety of handover algorithms in order to ensure that the performance is not degraded significantly by a partially incompatible handover algorithm. The large number of parameters and the associated high computational complexity of implementing channel allocation algorithms complicate study of the trade-offs of the various algorithms.

Again, in dynamic channel allocation, typically all channels can be used at any base station as long as they satisfy the associated quality requirements. Channels are then allocated from this pool as and when they are required. This solution provides maximum flexibility and adaptability at the cost of higher system complexity. The various dynamic channel allocation algorithms have to balance allocating new channels to users against the potential co-channel interference they could inflict upon users already in the system. Dynamic channel allocation is better suited to microcellular systems [372] because it can handle the more nonuniform traffic distributions, the increased handover requests, and the more variable co-channel interference better than fixed channel allocation due to its higher flexibility. The physical implementation of DCA is more complex than that of FCA. However, with DCA the complex and labor-intensive task of frequency planning is no longer required.

The majority of DCA algorithms choose the channel to be used based on received signal quality measurements. This information is then used to decide which channel to allocate or whether to allocate a channel at all. It is sometimes better not to allocate a channel if it is likely to inflict severe interference on another user, forcibly terminating existing calls or preventing the setup of other new calls. Ideally, the channel quality measurements should be made at both the mobile and base station. If measurements are made only at the mobile or only at the base station, the channel allocation is partially blind [329]. Channel allocation decisions that are based on blind channel measurements can in some circumstances cause severe interference, leading to the possible termination of the new call as well as curtailing another user's call, who is using the same channel. If measurements are made at both the mobile and the base station, then the measurements need to be compared, requiring additional signaling, which increases the call setup time. The call setup time is longer in DCA algorithms than in FCA due to the time required to make measurements and to compare them. This can be a problem, when, for example, a handover is urgently required.

Probably the simplest dynamic channel allocation algorithm is to allocate the least interfered channel available to users requesting a channel. By measuring the received power within unused channels, effectively the noise plus interference on that channel can be measured. By allocating the least interfered channel, the new channel is not likely to encounter interference, and, due to semireciprocity, it is not likely to cause too much interference to channels already allocated. This works well for lightly loaded systems. However, this algorithm's performance is seriously impaired in high-load scenarios, where FCA would work better. However, the above is a very simple dynamic channel allocation algorithm. In Sections 5.3.4 and 5.4 we will demonstrate that it is possible to achieve a better performance and efficiency than that of FCA even at high traffic loads, when using certain channel allocation algorithms. For these reasons, some channel allocation algorithms use a combination of FCA and DCA to achieve better performance than simple DCA, and better reuse efficiency than FCA. These algorithms are classified as hybrid channel allocation (HCA) algorithms.

The difference between the various dynamic channel allocation algorithms is, essentially, how the allocated channel is chosen. All the algorithms assign a so-called cost to allocating each of the possible candidate channels, and the one with the lowest cost is allocated. The difference between the algorithms is how the "cost" is calculated using the cost function. The cost function can be calculated on the basis of one or more of the following aspects: future call blocking probability; usage frequency of the channel; distance to where the channel is already being used, that is, the actual reuse distance; channel occupancy distribution; radio signal quality measurements; and so on. Some algorithms may give better performance than others, but only in certain conditions. Most DCA algorithms' objectives can be classified into two types, where most of them attempt to reduce interference, while others try to maximize channel utilization in order to achieve spectral compactness.

There are three main types of DCA algorithms, namely:

- centrally controlled algorithms;

- distributed algorithms;

- locally distributed algorithms (hybrid).

5.3.1.2.1   Centrally Controlled DCA Algorithms.   Centrally controlled DCA algorithms are also often referred to as centrally located or centralized DCA algorithms. These algorithms use interference measurements that are made by the mobiles and base stations that are then passed to a central controller, which in most cases would be a mobile switching center. The algorithm that determines the channel allocation is located at the central controller, and it decides on the allocation of channels based on the interference measurements provided by all the base stations and mobiles under its control. These algorithms provide very good performance even at high traffic loads. However, they are complex to implement and require a fast backbone network between the base stations and the central controller. The central controller can become a "bottleneck" and increase the call setup time, which may be critical for "emergency" handovers.

Centralized algorithms [361, 363, 373–375] have been researched actively for over twenty years. One of the simplest is referred to as the First Available (FA) [373, 376] algorithm, which allocates the first channel found that is not reused within a given preset reuse distance. The Locally Optimized Dynamic Assignment (LODA) [361, 363] algorithm bases its allocation decisions on the future blocking probability in the vicinity of the cell. Some algorithms exploit the amount of channel usage to make allocation decisions. The RING algorithm [351, 375], for example, allocates the most often used channel within the cells, which are approximately at the reuse distance, and the terminology RING is justified by the fact that these cells effectively form a ring. There are also several algorithms, which attempt to optimize the reuse distance constraint. The Mean Square (MSQ) algorithm [376] attempts to minimize the mean square distance between cells using the same channel while maintaining the required signal quality. The Nearest Neighbour (NN) and Nearest Neighbour plus One (NN+1) algorithms [373, 376] pick a channel used by the nearest cell, which is at least at a protection distance amounting to the reuse distance (or reuse distance plus one cell radius for NN+1). Other algorithms [375] use channel reassignments to maintain the reuse distance constraint. Recall again that these algorithms were summarized in Figure 5.4.

5.3.1.2.2   Distributed DCA Algorithms.   In contrast to centrally controlled algorithms, distributed algorithms are the least complex DCA techniques, in which the same algorithm is used by each mobile or base station in order to determine the best channel for setting up a call. Each mobile and/or base station makes channel allocation decisions independently using the same algorithm—hence the name distributed algorithms. The algorithmic decisions are usually based on the interference measurements made by the mobile or the base station. These algorithms are easy to implement, and they perform well for low-slot occupancy systems. However, in high-load systems their performance is degraded. Distributed algorithms require less signaling than centralized algorithms. However, the allocation is generally suboptimal owing to their locally based decisions. One real advantage of distributed algorithms is that base stations can easily be added, moved, or removed because the system automatically reorganizes and reconfigures itself. However, the cost of this flexibility is that the local decision making generally leads to a suboptimal channel allocation solution and to a higher probability of interference in neighboring cells. Furthermore, generally distributed algorithms are based on signal strength measurements and estimates of interference. However, these interference estimates can sometimes be poor, which can lead to bad channel allocation decisions. When a new allocation is made, the co-channel interference it inflicts may lead to an ongoing call to experience low-service quality, often termed a service interruption. If a service interruption leads to the ongoing call being terminated prematurely, this is referred to deadlock [351]. Successive service interruptions are termed as instability. A further problem with distributed algorithms is that the same channel can be allocated at the same time to two or more different users in adjacent cells. However, when the mobiles attempt to use the channel, they may find the quality unacceptably low. Therefore, distributed algorithms have to be able to check the quality of an allocation, before it is made permanent, which increases the call setup time further.

Chuang et al. [331] investigated the performance of several distributed DCA algorithms, arguing that under certain conditions these techniques can converge to a local minimum of the total interference averaged over the network. Grandhi et al. [377] and Chuang et al. [330] also evaluated the performance of combining dynamic channel allocation with transmission power control.

Examples of distributed algorithms are the Sequential Channel Search (SCS) and the least interference algorithm (LIA). The SCS algorithm [378] searches the available channels in a predetermined order, picking the first channel found, which meets the interference constraints. The LIA algorithm, alluded to earlier, picks the channel with the lowest measured interference that is available. One of the most complex distributed algorithms is the Channel segregation technique [379], which is a fully distributed, autonomous, self-organizing assignment scheme. Each cell maintains a measure of the relative frequency of channel usage for each channel. This probability-based measure is modified every time an attempt to access a specific channel is made. The channel assigned to the new call is the one with the highest probability of being or having been idle. The algorithm has been shown to reduce blocking and adapt to traffic changes. Although the channel allocation may rapidly converge to a near-optimal solution, it may take a long time to reach a globally optimal solution. As before, for the family tree of these techniques, please refer to Figure 5.4.

5.3.1.2.3 Locally Distributed DCA Algorithms. The third and final class of DCA algorithms are the locally distributed algorithms, which constitute a hybrid of distributed and centralized algorithms. These algorithms provide the greatest number of performance benefits of the centralized algorithms at a much lower complexity. Examples of locally distributed DCA algorithms are those proposed by Delli Priscoli et al. [334, 335] as an evolution of the Pan-European GSM system [55]. Locally distributed DCA algorithms use information from nearby base stations to augment their local channel quality information in order to

Table 5.1: FCA and DCA features.

Fixed Channel Allocation (FCA)	Dynamic Channel Allocation (DCA)
• Better under heavy traffic loads	• Better under light/moderate traffic loads
• Low call setup delay	• Moderate to high call setup delay
• Suited to large-cell environment	• Suited to microcellular environment
• Low flexibility in channel assignment	• Highly flexible channel assignment
• Sensitive to time and spatial changes in traffic load	• Insensitive to time and spatial changes in traffic load
• Low computational complexity	• High computational complexity
• Labor-intensive and complex frequency planning	• No frequency planning required
• Radio equipment only covers channels assigned to cell	• Radio equipment may have to cover all possible channels available
• Low signaling load	• High signaling load
• Centralized control	• Control dependent on the specific scheme from centralized to fully distributed
• Low implementational complexity	• Medium to high implementational complexity
• Increasing system capacity is expensive and time-consuming	• Simple and quick to increase system capacity

make a more informed channel allocation decision. Most of the locally distributed algorithms maintain an Augmented Channel Occupancy (ACO) matrix [332]. This matrix contains the channel occupancy for the local and surrounding base stations from which information is received. After every channel allocation, the information to update the ACO matrices is sent to the nearby base stations. This signaling requires a fast backbone network, but it is far less complex than the signaling required for the centralized algorithms.

The Local Packing Dynamic Distributed Channel Assignment (LP-DDCA) algorithm, proposed in [332], maintains an ACO matrix for every base station for all surrounding cells within the co-channel interference distance or reuse distance from the base station. The LP-DDCA algorithm assigns the first channel available that is not used by the surrounding base stations, whose information is contained in the ACO matrix. There are several algorithms similar to this one, including those by Del Re *et al.* [380], and the Locally Optimized Least/Most Interference Algorithms (LOLIA/LOMIA) that we will use in Section 5.3.3.3 in the context of our performance comparisons.

An overview of the main differences between fixed and dynamic channel allocation is shown in Table 5.1; exploration of its detailed contents is left to the reader. However, this table does not show the increase in spectral efficiency and channel utilization that becomes possible with dynamic schemes, as will be demonstrated during our performance comparisons.

5.3.1.3 Hybrid Channel Allocation

Hybrid channel allocation schemes constitute a compromise between fixed and dynamic channel allocation schemes. They have been suggested in order to combine the benefits of

DCA at low and medium traffic loads with the more stable performance of FCA at high traffic loads. Furthermore, hybrid schemes have been proposed as possible extensions to the fixed channel allocation used in second-generation mobile radio systems. In hybrid channel allocation schemes, the channels are divided into fixed and dynamic subsets. The fixed channels are assigned to the cells, as would be done for fixed channel allocation, and they are the preferred choice for channel allocation. When a cell exhausts all its fixed channels, it attempts to allocate a dynamically assigned channel from the central pool of channels. The algorithm used to pick the dynamically allocated channel depends on the hybrid scheme, but it can be any arbitrary DCA algorithm. The ratio of fixed and dynamic channels could be fixed [381] or varied dynamically, depending on the traffic load. At high loads, best performance is achieved, when the hybrid scheme behaves like FCA, by having none or a limited number of dynamically allocated channels [381,382]. Some hybrid channel allocation algorithms reallocate fixed channels, which become free to calls using dynamic channels in order to free up the dynamic channels. This technique is known as channel reordering [375].

5.3.1.4 The Effect of Handovers

A handover or handoff event occurs when the quality of the channel being used degrades, and hence the call is switched to a newly allocated channel. If the new channel belongs to the same base station, then this is called an intra-cell handover. If the new channel belongs to a different base station, it is referred to as an inter-cell handover. Generally intra-cell handovers occur when the channel quality degrades due to interference or when the channel allocation algorithm decides that a channel reallocation will help increase the system's performance and capacity. Inter-cell handovers occur mainly because the mobile moves outside the cell area; hence, the signal strength degrades, requiring a handover to a nearer base station.

Handovers have a substantial effect on the performance of channel allocation algorithms. At high traffic loads, the majority of forced call terminations are due to the lack of channels available for handover rather than to interference. This can be a particular problem in microcellular systems, where the rate of handovers is significantly higher than that in normal cellular systems.

There are several known solutions to reduce the performance penalty caused by handovers. One of the simplest solutions is to reserve some channels exclusively for handovers, commonly referred to as cutoff priority [345, 383, 384] or guard channel [385] schemes. However, this solution reduces the maximum amount of carried traffic or system capacity and hence yields increased new call blocking. The guard or handover channels do not need to be permanently assigned to cells; they are invoked from an "emergency pool."

Algorithms that give higher priority to requests for handovers than to new calls are called Handover prioritization schemes. Guard channel schemes are therefore a type of handover prioritization arrangement. Another type of handover prioritization is constituted by handover queuing schemes [351,352,383,384]. Normally, when an allocation request for handoff is rejected, the call is forcibly terminated. By allowing handover allocation requests to be queued temporarily, the forced termination probability can be reduced. The simplest handover queuing schemes use a First-In First-Out (FIFO) queuing regime [384]. Tekinay et al. [345] have suggested a nonpreemptive priority handover queuing scheme in which handover requests in the queue that are the most urgent ones are served first.

A further alternative to help reduce the probability of handover failure is to allow allocation requests for new calls to be queued [385]. New call allocation requests can be queued more readily than handovers because they are less sensitive to delay. Handover queuing reduces the forced termination probability owing to handover failures but increases the new call blocking probability. New call queuing reduces the new call blocking probability and also increases the carried teletraffic. This is because the new calls are not immediately blocked but queued, and in most cases they receive an allocation later.

5.3.1.5 The Effect of Transmission Power Control

Transmission power control is an effective way of reducing co-channel interference while also reducing the power consumption of the mobile handset. Jointly optimizing transmission power control with the channel allocation decisions is promising in terms of increasing spectral efficiency. However, little research has been done into this area, apart from a contribution by Chuang and Sollenberger [330] showing the potential benefits. Transmission power control, like channel allocation, can be implemented in a centralized [386, 387] or distributed [388] manner.

An alternative fixed channel allocation strategy, referred to as Reuse partitioning [351], relies on transmission power control. In reuse partitioning, a cell is divided into two or more concentric subcells or zones. If a channel is used in the inner zone with transmission power control, the interference is reduced due to the reduced transmission power. Therefore, the interference from channels used in the inner zones is less than that by those channels, used in the outer zones. Channels used in the inner zones can thus be reused at much shorter distances than those utilized in the outer zones.

By combining transmission power control with dynamic channel allocation, the additional performance gains of reuse partitioning can be achieved. Using reuse partitioning with DCA is far simpler to implement than using FCA, since the system is self-configuring and does not require network reuse pattern planning.

5.3.2 Simulation of the Channel Allocation Algorithms

In this section, we highlight how we simulated the various channel allocation algorithms we investigated. Section 5.3.2.1 describes the simulation program, "Netsim", which was developed to simulate the performance of the channel allocation algorithms. The channel allocation algorithms that we simulated are described in detail in Section 5.3.3. In Section 5.3.3.4, we describe the performance metrics we have used to compare the performance of the channel allocation algorithms. Finally, in Section 5.3.3.5, we describe the model used to generate the nonuniform traffic distributions we used in our simulations.

5.3.2.1 The Mobile Radio Network Simulator, "Netsim"

In order to characterize the performance of the various channel allocation algorithms, we simulated a mobile radio network. The simulator program we developed is referred to as Netsim. The simulated base stations can be placed in a regular pattern or at arbitrary positions within the simulation area. Mobiles are distributed randomly across the simulation area. Each mobile can have different characteristics, such as a particular mobility model or velocity.

Netsim

Calls Attempted =   0038
Timeslots Blocked =   0000
Timeslots Allocated =   0083
Calls Started =   0038
Calls Completed =   0025
Calls In Progress =   0013
Calls Dropped =   0000

Intra Handovers:
              Attempted = 0000
              Succeeded = 0000
              Failures =   0000
              Dropped =   0000

Inter Handovers:
              Attempted = 0045
              Succeeded = 0045
              Failures =   0000
              Dropped =   0000

Total Handovers:
              Attempted = 0045
              Succeeded = 0045
              Failures =   0000
              Dropped =   0000

Call Dropping :
              Total =   0000
              Handover =0000
              Normal =   0000

Frame = 00012000 / 00012000
Percentage Complete = 100.00%

Nonactive Mobile   **Active Mobile**   **Base station**

Radius of cell area   **Statistics**

Figure 5.6: Screenshot of the Netsim program, showing 100 users in a 49-cell simulation. Each base station is located at the center of each cell, and the large circles represent the radius of the cell area. The connection between an active mobile and a base station is represented by a line.

A screenshot from the simulator is shown in Figure 5.6. The figure shows a 49-base station simulation, where the cell areas are represented by circles. The mobiles are shown as small squares, and when they become active, they change color on the video screen. The connection between an active mobile and a base station is represented by a line linking the base station and the mobile. The simulator has the following features:

New Call Queuing Channel allocation requests for new calls are queued if they cannot immediately be served [385]. The new call request is blocked if its request cannot be served within a preset timeout period, referred to as the Maximum new-call queue time.

Handover Prioritization Channel allocation requests for handovers are given priority over new calls, supporting Handover Prioritization [351].

Handover Urgency Prioritization Channel allocation requests for handovers are processed by each base station, so that the more urgent handovers are served first [345].

Handover Hysteresis A call will not be handed over to another base station or channel unless the new channel has a signal quality better than the current channel by at least the preset handover hysteresis level. The only exception is when the current channel quality is below the signal quality level required to maintain the call and the new channel is above this quality level, but the difference between the quality of the new and current channel is less than the hysteresis threshold.

Channel Models The simulator models each propagation channel using one of several pathloss models and a shadow fading model. The shadow fading model can be turned off if necessary.

Call Generation Model Each mobile's activity is described by how much of the time the mobile is active (i.e., making a call). The activity of each mobile is controlled by two parameters, average call duration and average intercall time. The average call duration is the long-term mean of the length of all the calls made by the mobile. The duration of all the calls made by the mobile is Poisson-distributed [330, 389]. The average intercall time is the long-term mean duration of time between calls being made. Similarly to the call durations, the time between calls is also Poisson distributed [330, 389].

Edge effects The cells at the edge of the simulation area behave differently from cells near the center of the simulation area. This is because the cells near the edge have fewer neighboring cells and hence less interference. Therefore, in order to reduce the effect of these edge cells, the statistical results can be gathered only from the cells near the center of the grid (i.e., from the active cells). Furthermore, when a mobile reaches the edge of the simulation area, it is randomly repositioned somewhere else in the simulation area. In order that this does not cause handover problems, active mobiles reaching the edge of the simulation area finish their calls before they are repositioned.

Extensive Statistical data gathering The Netsim simulator stores a large range of statistics from each simulation. For example, the probability density function of the number of simultaneous calls at each base station is stored. Furthermore, the simulation area can be divided into a fine grid, the resolution of which depends on the required accuracy of the statistical evaluation aimed for. Statistics can be gathered separately for each grid square, allowing coverage maps of the simulation area to be generated.

Warmup period When the simulation is first begun, the number of active calls is far below the normal level. There is a latency, before the number of active calls is built up to the correct level, owing to the nature of the Poisson distributed call generation models [330, 389]. Therefore, in order not to bias the results, simulations are conducted for a sufficiently long period of time before the simulation statistics can be gathered. This period of time is referred to as a warmup period.

The Netsim simulator is a network layer-based framework employing a simple physical layer model in order to reduce the complexity of the simulations, which is described in the next section.

5.3.2.1.1 Physical Layer Model. The physical layer, that is, the modulator and de-modulator,are modeled using two parameters, Outage SINR and Reallocation SINR. The Reallocation SINR threshold is always set above the Outage SINR threshold. When the signal quality measured in terms of the signal-to-interference+noise ratio (SINR) (defined in Equation 5.14 drops below the reallocation SINR level, the mobile requests a new channel to hand over to. This handover request can be asking for another channel from the same base station to which the mobile is currently connected and is called an intra-cell handover. Alternatively, the handover can be initiated to a channel from a different base station and is called an inter-cell handover.

If, while waiting for a reallocation handover, the signal quality drops further, below the so-called Outage SINR threshold, the signal is deemed to be lost for that time period. This is referred to as an outage. If a channel is in outage for several consecutive time periods, then the call is forcibly terminated. The parameter termed the Maximum Consecutive Outage reflects the number of consecutive outages that need to occur to cause a call to be forcibly terminated.

The Reallocation SINR threshold should be set at the average SINR required to maintain marginal signal quality. The Outage SINR threshold should be set as the SINR, below which the demodulated signal cannot be decoded error free. This twin-threshold physical layer model is similar to those described by Tekinay et al. [352] and by Katzela et al. [351]. The difference is that our model is based on SINR thresholds instead of received power thresholds used in these references. Since the computational complexity would be too high to simulate fast Rayleigh fading in a network-layer simulation, the SINR threshold of the physical layer model should include a margin to emulate the effects of fast fading, thereby increasing the required outage level.

The simulator calculates the probability of outage as the proportion of time in which a channel was below the Outage SINR threshold (i.e., in outage). The simulator can also calculate the low signal quality probability, as the proportion of time a channel is below the Reallocation SINR threshold.

The next section describes the model used to simulate shadow fading of the radio channels.

5.3.2.1.2 Shadow Fading Model. The channel model used by the Netsim simulator is fairly simple in order to reduce the computational complexity of the simulations. The channel can be modeled using a variety of pathloss models and an optional shadow fading model. This section is concerned with the shadow fading model. Network simulations are particularly complex, since all the possible interfering channels may need to be modeled, that is, from each transmitter to every receiver tuned to the same carrier frequency at the same time.

Shadow fading can be modeled using a correlated signal, which is log-normally dis-tributed [79]. In our previous chapters, shadow fading was modeled by using precalculated shadow fading signal envelopes. However, because of the high number of interfering channels, where the channels should be uncorrelated, a large number of precalculated shadow fading envelopes would be needed. This is impractical because of the associated high storage requirements, and the increased simulation time resulting from storage access delays.

We decided to invoke a method originally used to generate Rayleigh fading rather than shadow fading in order to produce the correlated log-normally distributed shadow fading envelope required. Jakes' method [74] was originally proposed to produce Rayleigh-distributed correlated signal envelope and phase. Jakes' technique is also often called the sum of sinusoids method, which uses the summation of several low-frequency sinusoids with regularly spaced phase differences in order to produce the desired signal. A signal, $r(t)$, exhibiting Rayleigh-distributed envelope or magnitude fluctuations can be produced from the complex summation of two independent Gaussian random variables, which is formulated as:

$$r(t) = X_1 + jX_2. \tag{5.6}$$

Jakes' method produces the required pair of correlated independent Gaussian distributed random variables, X_1, X_2, which are approximated by $x_1(t)$ and $x_2(t)$, given by:

$$x_1(t) = 2 \left[\sum_{n=1}^{N_o} \cos\left(\beta_n\right) \cos\left(\omega_n t\right) \right] + \sqrt{2} \cos(a) \cos\left(\omega_m t\right) \tag{5.7}$$

$$x_2(t) = 2 \left[\sum_{n=1}^{N_o} \sin\left(\beta_n\right) \cos\left(\omega_n t\right) \right] + \sqrt{2} \sin(a) \cos\left(\omega_m t\right) \tag{5.8}$$

$$\beta_n = \frac{n\pi}{(N_o + 1)} \tag{5.9}$$

$$N = 2\left(2N_o + 1\right) \tag{5.10}$$

$$\omega_n = \omega_m \cos\left(\frac{2\pi n}{N}\right) \tag{5.11}$$

$$\omega_m = 2\pi f_d, \tag{5.12}$$

where the functions $x_1(t)$ and $x_2(t)$ produce the in-phase and quadrature components of the Rayleigh-fading signal, $r(t)$. Both the in-phase and quadrature components are the sum of $(N_o + 1)$ oscillators, yielding the sum of sinusoids. The maximum Doppler frequency (f_d) sets the highest oscillator's frequency (ω_m), the phase of which is set by a. The remaining N_o oscillators have frequencies of less than ω_m set by ω_n, the phase of which is set by β_n. Therefore, $x_1(t)$ and $x_2(t)$ are functions of t, with parameters f_d and N_o.

Either one of the variables $x_1(t)$ or $x_2(t)$ can be used to produce the log-normally distributed shadow fading envelope $s(t)$, given by:

$$s(t) = 10^{[x_1(t)/10]} \quad \text{or} \quad s(t) = 10^{[x_2(t)/10]}. \tag{5.13}$$

In the next sections, we describe the investigated algorithms in detail.

5.3.3　Overview of Channel Allocation Algorithms

In this section, we describe the channel allocation algorithms that we have investigated in order to identify the most attractive performance trade-offs. Our simulations have concentrated on dynamic channel allocation (DCA) algorithms (Section 5.3.1.2). However, we have also performed experiments using a basic fixed channel allocation (FCA) algorithm (Section 5.3.1.1) as a benchmarker.

We investigated two classes of dynamic channel allocation (DCA) algorithms, namely, distributed and locally distributed algorithms, described previously in Sections 5.3.1.2.2 and 5.3.1.2.3. We studied four distributed DCA algorithms, which are characterized in

Section 5.3.3.2, while Section 5.3.3.3 portrays the two locally distributed DCA algorithms that we investigated. In the next section, we introduce the fixed channel allocation algorithm employed.

5.3.3.1 Fixed Channel Allocation Algorithm

In order to benchmark our dynamic channel assignment (DCA) algorithms, a fixed channel allocation (FCA) scheme was required. We decided to employ a basic fixed channel assignment algorithm, which uses omnidirectional antennas and a reuse cluster size of seven cells. This structure is commonly used to provide coverage over a grid of regular hexagonally shaped cells. The frequency spectrum was divided into seven frequency sets, and one set was assigned to each cell.

Figure 5.5 shows such a reuse structure, where the shaded cells represent cells assigned the same set of carrier frequencies. The figure shows the center cell and its six first-tier interfering cells. This fixed channel allocation reuse structure provides uniform capacity across all cells, since each cell site has the same number of carrier frequencies. In the next section we describe the distributed DCA algorithms investigated.

5.3.3.2 Distributed Dynamic Channel Allocation Algorithms

In this section we highlight four well-known distributed DCA algorithms that we have studied comparatively. The most plausible technique is the Least Interference Algorithm (LIA) [331], which allocates the channel suffering from the least received instantaneous interference power; hence, it attempts to minimize the total interference within the system. More specifically, this algorithm minimizes the interference at low traffic loads but increases it at high loads. This is because at high loads the LIA algorithm will still attempt to allocate a channel to a new call, even when all the slots have a high level of interference. Again, this increases the total interference load of the system.

The second distributed DCA algorithm we studied is a refinement of the LIA algorithm, which is referred to as the Least interference below Threshold Algorithm (LTA) [331]. This algorithm attempts to reduce the interference caused by the LIA algorithm at high loads by blocking calls from using those channels, where the interference measured is deemed excessive for the transceiver to sustain adequate communications quality. The algorithm allocates the least interfered channel, whose interference is below a preset maximum tolerable interference threshold. Therefore, the LTA algorithm attempts to minimize the overall interference in the system, while maintaining the quality of each call above the minimum acceptable level.

The third algorithm we investigated attempts to utilize the frequency spectrum more efficiently while maintaining acceptable call quality. This algorithm works in a similar way to the LTA algorithm, and it is termed the Highest (or Most) interference below Threshold Algorithm (HTA or MTA) [331]. Since its goal is not to reduce the interference, but to maximize the spectral efficiency, it allocates the most interfered channel, whose interference is below the maximum tolerable interference threshold. The interference threshold is determined by the transceiver's interference resilience.

The final distributed DCA algorithm can be characterized as the Lowest Frequency below Threshold Algorithm (LFA) [331]. This algorithm is a derivative of the LTA algorithm, the

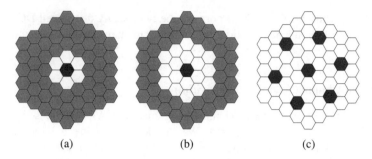

(a)   (b)   (c)

Figure 5.7: The nearest neighbor constraint for (a) $n = 7$ and (b) $n = 19$ for the locally optimized algorithms, LOLIA and LOMIA, compared to a (c) seven-cell reuse cluster for FCA.

difference being that the LFA algorithm attempts to reduce the number of carrier frequencies being used concurrently. This has the advantage that, statistically speaking, fewer transceivers may then be required at each base station. The algorithm allocates the least interfered channel below the maximum tolerable interference threshold, while also attempting to reduce the number of carrier frequencies used. Therefore, no new carrier frequency is invoked from the set of carriers, unless all the available timeslots on the currently used carrier frequencies are considered too interfered. In the next section, we describe the two locally distributed DCA algorithms, whose performance we have compared to the above algorithms using simulations.

5.3.3.3 Locally Distributed Dynamic Channel Allocation Algorithms

We have investigated the performance of two locally distributed dynamic channel allocation algorithms, both of which are quite similar. The Locally Optimized Least Interference Algorithm (LOLIA) attempts to reduce the overall interference in a system, like the LIA and LTA algorithms, while the Locally Optimized Most Interference Algorithm (LOMIA) attempts to increase the spectral efficiency in a similar way to the HTA algorithm.

Specifically, the locally distributed DCA algorithms constitute a hybrid of distributed and centralized channel allocation decisions. They exploit the information provided by neighboring base stations in order to improve the channel allocation decisions, which constitute the centrally controlled part of the distributed/centralized hybrid solution. Their complexity is therefore somewhere between that required for centralized and distributed algorithms.

The LOLIA algorithm carries out its channel allocation decisions in the same way as the distributed LIA algorithm. However, it will not allocate a channel, if it is used in the nearest "n", neighboring cells by another subscriber. Therefore, the nearby base stations exchange information concerning the channels that are currently being used. This requires a fast backbone network but does not rely on central control. The overall level of interference in the system can be reduced by increasing the number of cells, which are classed as neighboring cells. However, the larger "n". the more calls are blocked, since there will be fewer available channels, which are not being used in the nearest "n" base stations. Figure 5.7 shows the arrangement of neighboring cells for $n = 7$ and $n = 19$. The "n" parameter of the algorithm effectively imposes a minimum reuse distance constraint on the algorithm.

The second locally distributed DCA algorithm we consider is similar to LOLIA, but it is based on the HTA and not the LIA distributed algorithm. The LOMIA algorithm picks the most interfered channel, provided that this channel is not used in the nearest "n" neighboring cells. The LOLIA and LOMIA algorithms are similar to those proposed by De Re *et al.* [380] and ChihLin *et al.* [332].

Having described the algorithms that we have simulated in order to identify the performance trade-offs of the various channel allocation algorithms, in the next section we describe the metrics used to compare the performance of the various algorithms.

5.3.3.4 Performance Metrics

Several performance metrics can be used to quantify the performance or QoS provided by a particular channel allocation algorithm. The five performance metrics defined below have been widely used in the literature [331], and we also opted for their employment:

- new call blocking probability, P_B;

- call dropping or forced termination probability, P_D or P_{FT};

- probability of low-quality connection, P_{low};

- probability of outage, P_{out};

- grade of Service, GOS.

The new call blocking probability, P_B, is defined as the probability that a new call is denied access to the network. This may be the case because there are no available channels or the channel allocation algorithm decided that to allow the new call to access any of the available channels would cause increased interference, which might lead to loss of the new call or calls in progress. Ideally, a low call blocking probability is desired. However, it is even more undesirable when calls in progress are lost, and this is where the second performance metric, namely, P_{FT} is useful.

The call dropping probability, P_D, also widely known as the forced termination probability, P_{FT}, is the probability that a call is forced to terminate prematurely. This can be caused by excessive interference. However, generally when a channel becomes excessively interfered with, the mobile or base station will request a new channel. If no channels are available and the quality of the call degrades significantly because of interference or low signal strength, then the call may be forcibly terminated. Calls can also be forcibly terminated when a mobile moves across a cell boundary into a heavily loaded cell. If there are no available channels in the new cell to hand over to, then the call may be lost prematurely. Since premature call termination is annoying to mobile subscribers, the channel allocation algorithm should attempt to keep the call dropping probability low.

The third performance metric we have used is the probability of a low-quality connection or access, P_{low}. This is the probability that either the UL or DL signal quality is below the level required by the specific transceiver to maintain a good-quality connection. A low-quality access could be due to low signal strength or high interference, which is defined as:

$$P_{low} = P\{SINR_{UL} < SINR_{req} \quad \text{or} \quad SINR_{DL} < SINR_{req}\}$$
$$= P\{min(SINR_{UL}, SINR_{DL}) < SINR_{req}\}. \tag{5.14}$$

This metric allows different channel allocation algorithms, which may have similar call dropping and blocking probability to be compared, in order to identify which is better, when calls are in progress. The quantity $SINR_{req}$ is the required reallocation SINR threshold described in Section 5.3.2.1.1. The probability of outage is similar to the probability of low communications quality metric (P_{low}), which was defined in Equation 5.14, except in this case the quantity $SINR_{req}$ is the required SINR value, below which the call is deemed to be in outage, as described in Section 5.3.2.1.1.

The final metric we have used to evaluate the performance of various channel allocation algorithms is the grade of service (GOS). The definition we have used is that proposed by Cheng and Chuang [331] which is stated as follows:

$$\begin{aligned} GOS &= P\{\text{unsuccessful or low-quality call accesses}\} \\ &= P\{\text{call is blocked}\} + P\{\text{call is admitted}\} \\ &\quad \times P\{\text{low signal quality and call is admitted}\} \\ &= P_B + (1 - P_B)P_{low}. \end{aligned} \qquad (5.15)$$

The grade of service is the probability of unsuccessful network access (blocking, P_B) or low-quality access, when a call is admitted into the system (P_{low}). This performance metric is a hybrid of the new call blocking probability (P_B) and the low-quality access probability (P_{low}), when calls are not blocked and it is therefore an important performance metric. Now that we have described the algorithms and the metrics used to compare their performance, the next section describes the model used to generate nonuniform traffic distributions.

5.3.3.5 Nonuniform Traffic Model

Generally, investigations using fixed channel allocation assume a uniform traffic distribution and therefore a uniform carrier frequency allocation per base station. In practice some base stations have more channels, where demand is expected to be increased, for example, at airports and railway stations. However, fixed channel allocation cannot cope with unexpected traffic demand peaks [390], which are sometimes referred to as traffic "hot spots" [365]. Dynamic channel allocation algorithms are better equipped to cope with these unexpected traffic demands, since a DCA system is effectively self-adapting. Furthermore, DCA schemes typically have more potential channels available at each base station. This is an area in which DCA algorithms have a clear advantage over FCA.

Therefore we defined a model to generate a sudden unexpected traffic "hot spot" in order to measure the performance benefits that DCA algorithms provide over FCA. The model we developed is very simple and causes an increase in teletraffic in the cells affected. The model simply limits the maximum velocity of mobile terminals within a particular geographical area. Mobile users can still enter and leave a "hot spot" cell. However, since the users slow down as they enter the cell, the average cell crossing time is increased. This leads to a higher mobile terminal density in the cell, which in turn leads to increased generated teletraffic.

As an example, we refer to Figure 5.8, in which the speed of mobiles in the gray cells is not limited by the model. For our simulations, however, the mobiles all travel at 30 mph. Upon roaming and entering the white cells, these mobiles reduced their speed to 20 mph. The white cells could represent the outskirts of a city. Upon entering the black cell, which could represent a city center, the speed of mobiles is again reduced to 9 mph.

Figure 5.8: Nonuniform traffic conditions exhibiting a traffic "hot spot" in the central cell (black), and a "warm spot" (white) surrounding it. Mobiles in the gray cells move at the standard speed of 13.4 m/s (30 mph). Mobiles in the white ("warm-spot cells") can move at a speed of 9 m/s (20 mph). Mobiles in the black "hot-spot cell" are limited to a speed of 4 m/s (9 mph).

In order to compare our network performance results attained by fixed and various dynamic channel allocation algorithms, with and without adaptive antenna arrays at the base station, it was necessary to consider more than one performance metric. For example, an algorithm may perform very well in one respect, yet have poor performance when measured using an alternative metric. Therefore, it was decided to invoke two different scenarios:

- A *conservative scenario*, where the maximum acceptable value for the call blocking probability, P_B, is 3%, for the call dropping probability, P_{FT}, is 1%, for P_{low} is 1%, and for the GOS is 4%.

- A *lenient scenario*, where the maximum acceptable value for the call blocking probability, P_B, is 5%, for the call dropping probability, P_{FT}, is 1%, for P_{low} is 2%, and for the GOS is 6%.

It must be noted that the maximum allowable GOS does not have to obey Equation 5.15 for the given values of P_b and P_{low}, since they may be traded off against each other. Hence the GOS may be interpreted as a form of "user satisfaction". As a consequence, for example in the lenient scenario the GOS is 6%, rather than the expected 7%, since it may be unacceptable for the user to simultaneously tolerate both a P_b of 5% and at the same time a low-quality link probability of P_{low}=2%. Therefore the required "user satisfaction" may be maintained with the proviso of satisfying any acceptable combination of P_b and P_{low} values, as long as their sum remains below the required GOS level.

The next section presents a summary of the results obtained for the previously described channel allocation algorithms.

5.3.4 DCA Performance without Adaptive Arrays

In our previous work [50, 192, 336, 391] a comparative study of a range of DCA algorithms was conducted and it was found that the algorithm which provided the best overall compromise in terms of the desired performance measures was the Locally Optimized Least Interference Algorithm (LOLIA). The results in Table 5.2 indicate the achievable network capacities, without AAAs and without shadow fading, for various DCA algorithms and for the FCA algorithm. Hence, our further investigations presented here we focus our attention on the

Table 5.2: Maximum number of mobile users that can be supported by the various DCA algorithms [192, 336].

Algorithm	Number of users supported by network	
	Conservative $P_{FT} = 1\%, P_{low} = 1\%$ GOS=4%, $P_B = 3\%$	Lenient $P_{FT} = 1\%, P_{low} = 2\%$ GOS=6%, $P_B = 5\%$
FCA	820	1120
HTA	1435	1520
LFA	1555	1705
LOMIA ($n = 19$)	1505	2040
LTA	1815	1830
LIA	1820	1820
LOLIA ($n = 7$)	1860	2115
LOLIA ($n = 19$)	1935	2005

LOLIA by combining it with adaptive beamforming and other network capacity enhancement techniques.

5.4   Employing Adaptive Antenna Arrays

Here, a study into the usage of an adaptive antenna array in a cellular network is conducted. A theoretical analysis of such a system is performed and the results are presented for later comparison with simulated results. To simplify this process the following assumptions were made:

- There is a uniform distribution of users in each cell.

- There is a blocking probability of P_B in all cells.

- The omni-directional base station antenna has an ideal beam pattern, giving a uniform circular coverage.

- The adaptive base station antenna array can generate m ideal beams, each with a gain of 1.0 over a beamwidth of $\Delta\theta = 2\pi/m$ radians, and a gain of 0.0 over the remaining angular sector, as shown in Figure 5.9

The blocking probability, P_B, is the fraction of attempted calls that cannot be allocated a channel. If the traffic intensity offered is a Erlangs, then the actual traffic carried is $a(1 - P_B)$ Erlangs. The Erlang is a measure of offered tele-traffic, which indicates the quantity of traffic on a channel or group of channels per unit time. This gives a channel usage efficiency of [2]:

$$\eta = \frac{a(1 - P_B)}{N}, \quad (5.16)$$

where N is the total number of channels allocated per cell.

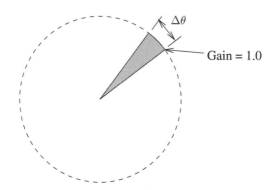

Figure 5.9: Beam pattern of an ideal beamformer with beamwidth $\Delta\theta$.

It was also assumed that the main beam formed by the adaptive antenna was centred about the angle of arrival of the desired mobile's signal and that the mobile was tracked with no error. Additionally, all interfering sources outside the main beam were assumed to be nulled successfully. The ideal beamformer model used has a single mainlobe with a unity-gain beamwidth of $\Delta\theta$ and sidelobes of zero gain, as shown in Figure 5.9. When the desired signal's power, S, does not exceed the co-channel interference power, I, by the required protection ratio, γ. In this situation an "outage" will occur, i.e. we fail to achieve satisfactory reception at the mobile in the presence of interference with the probability of [2, 392–394]:

$$P(\text{outage}) = P(S \leq \gamma I) = P(S/I \leq \gamma) = P(SIR \leq \gamma), \qquad (5.17)$$

where SIR is the signal-to-interference ratio. In other words, $P(\text{outage})$ is the probability of the power of the signal being insufficient to provide reliable communications due to the interference in the channel. Considering only the propagation pathloss, but no fast- and shadow-fading, we have $SIR = S/I = d_i^2/d_w^2 \leq \gamma$, hence for a given interference protection ratio, a locus defined by $d_i/d_w = \sqrt{\gamma}$ can be drawn, as in Figure 5.10. This defines a region, where the signal-to-interference ratio necessary for reliable DL (DL) communications is maintained, and a region where interference occurs.

In a cellular network employing base station (BS) adaptive antenna arrays, the occurrence of co-channel interference is a statistical phenomenon dependent upon the number of co-channel interferers and on the positions of these interferers in the co-channel cells. In general the UL (UL) and DL (DL) interference calculations are different and hence they have to be considered separately. The total probability of co-channel interference-induced outage can be evaluated by [2, 324, 394]:

$$P(\text{outage}) = P(SIR \leq \gamma) = \sum_{n=1}^{N} P(SIR \leq \gamma|n)P(n), \qquad (5.18)$$

where N is the total number of co-channel interferers, usually restricted to the first tier of interferers, shown in white in Figure 5.7(a), i.e. to six, $P(SIR \leq \gamma|n)$ is the conditional probability of co-channel interference, $P(SIR \leq \gamma)$ given n interferers. Furthermore,

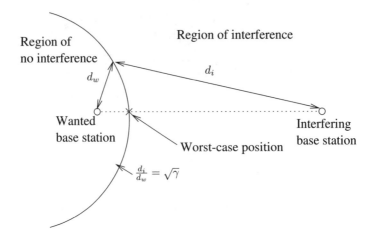

Figure 5.10: Contour defining interference regions in a DL scenario using omnidirectional antennas.

$P(n)$ is the probability that there are n active interfering co-channel cells. Therefore, if the activation of channels is assumed to be independent and identically distributed, $P(n)$ has the form of a binomial PDF [2, 306, 394]:

$$P(n) = \binom{6}{n} p^n (1-p)^{6-n},
\tag{5.19}$$

where p is the probability of finding one interfering co-channel active. The probability p that a single co-channel BS has an active DL co-channel interferer, given that the wanted mobile has been assigned that DL channel already, is [2]:

$$p = \frac{\text{number of active channels}}{\text{total number of channels}} = \frac{a(1-P_B)}{N} = \eta.
\tag{5.20}$$

Therefore, the probability that n co-channel interfering BSs are using the same DL channel as the wanted mobile for its reception becomes:

$$P(n) = \binom{6}{n} \eta^n (1-\eta)^{6-n}.
\tag{5.21}$$

Hence, from Equations (5.18) and (5.21) we have:

$$P(\text{outage}) = P(SIR \leq \gamma) = \sum_{n=1}^{N} P(SIR \leq \gamma|n) \binom{6}{n} \eta^n (1-\eta)^{6-n}.
\tag{5.22}$$

In conjunction with an omnidirectional BS antenna, the probability of an active DL co-channel interferer was given by η, the channel usage efficiency. For an adaptive BS antenna, forming m beams per cell, there will always be six DL beams targeted at the wanted mobiles from the six co-channel base stations. Therefore, for an adaptive base station antenna [2, 306] we have:

$$p = \begin{pmatrix} \text{probability that the beam pointing at the desired} \\ \text{mobile also contains an interfering mobile} \end{pmatrix}$$

$$= \frac{\text{number of active channels in beam}}{\text{total number of channels}} \tag{5.23}$$

$$= \frac{a(1 - P_B)/m}{N} = \frac{\eta}{m}.$$

Hence, for an adaptive BS antenna array with m beams per BS we have:

$$P(n) = \binom{6}{n} \left(\frac{\eta}{m} \right)^n \left(1 - \frac{\eta}{m} \right)^{6-n}, \tag{5.24}$$

leading to the overall outage probability for a BS adaptive antenna array in the form of:

$$P(\text{outage}) = P(SIR \le \gamma) = \sum_{n=1}^{N} P(SIR \le \gamma|n) \binom{6}{n} \left(\frac{\eta}{m} \right)^n \left(1 - \frac{\eta}{m} \right)^{6-n} \tag{5.25}$$

where $P(SIR \le \gamma|n)$ is the conditional outage probability, which is dependent on the mean received signal power and the mean received interference power.

5.5   Multipath Propagation Environments

In Section 5.2 various situations were investigated where only a direct LOS link existed between the base station and the mobile handset. However, in a real environment, a phenomenon known as multipath scattering takes place, which results in the presence of numerous signal components, or multipath components, at the receiver. This is due to reflections, diffractions and signal scattering, caused by objects in the path between the transmitter and the receiver. A simple figure showing an example of the multipath propagation channel is shown in Figure 5.11. Each signal component experiences a different path attenuation and phase rotation, which determines the received signal's amplitude, carrier phase shift, time delay, angle of arrival and Doppler shift [21]. In general, each of these components will be time-varying. We note here that the various UL and DL scenarios will be considered in more depth in Figure 5.22 during our further discourse.

Figure 5.11 shows the multipath environment that may be found on the UL and DL in a macrocellular environment. It is usually assumed that the scatterers surrounding the mobile station are at about the same height as or are higher than the mobile. This implies that the received signal at the mobile antenna arrives from all directions after bouncing from the surrounding scatterers, as illustrated in Figure 5.11. Under these conditions it is assumed that the DL Direction-Of-Arrival (DOA) at the mobile is uniformly distributed over $[0, 2\pi]$ [21, 337]. However, the UL DOA of the received signal at the base station is quite different. In a macrocellular environment, the base station is typically positioned higher than the surrounding scatterers. Hence, the received signals at the base station result predominantly from the scattering process in the vicinity of the mobile station, as it may be seen in Figure 5.11. The UL multipath components are restricted to a smaller angular

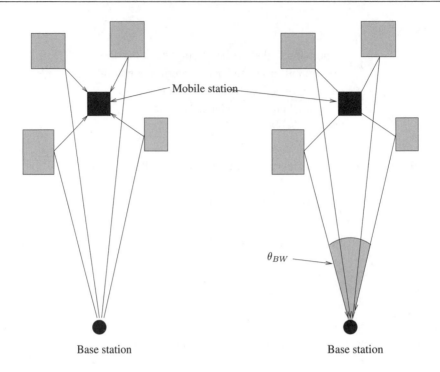

Figure 5.11: Macrocellular UL and DL multipath scattering scenarios.

region, θ_{BW}, and hence the distribution of the UL DOA is no longer uniform over $[0, 2\pi]$. Many different models have been developed for use in different applications. Below a brief description of some of the models follows, but for a more detailed exposition the reader is referred for example, to Ertel *et al.* [337]. The macrocellular models are all based around the same principle of placing a number of scatterers near the mobile station in a given pattern, obeying a geographic probability distribution. In Lee's model, the scatterers are evenly spaced on a circular ring about the mobile, as shown in Figure 5.12. Assuming that the N scatterers are uniformly spaced on the circle having a radius R and orientated such that a scatterer is located along the LOS path, the discrete DOAs are [337]:

$$\theta_i \approx \frac{R}{D} \sin\left(\frac{2\pi}{N}i\right), \qquad i = 0, 1, \ldots, N - 1. \tag{5.26}$$

However, the model was originally designed simply for providing information regarding the signal correlations of the multipath components and when used to provide DOA and Time-Of-Arrival (TOA) information, the simulated results are not consistent with measurements [337].

A model similar to Lee's, known as the discrete uniform distribution, evenly spaces N scatterers within a narrow beamwidth centred about the LOS to the mobile, as shown in Figure 5.13. According to [337], the discrete possible DOAs, assuming that N is odd, are given by:

$$\theta_i = \frac{1}{N - 1}\theta_{BW}i, \qquad i = -\frac{N - 1}{2}, \ldots, \frac{N - 1}{2}. \tag{5.27}$$

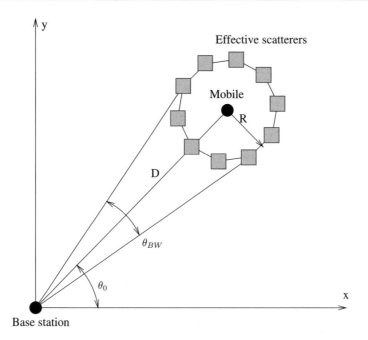

Figure 5.12: Lee's model for multipath scattering using N scatterers in a circle of radius R around the mobile station.

The Geometrically Based Single-Bounce (GBSB) Statistical Channel Models are defined by a spatial scatterer density function. This model involves randomly placing scatterers in the scatterer region according to the spatial scatterer density function. From the location of each of the scatterers, the DOA, TOA, and signal amplitude can be determined. The Geometrically Based Single-Bounce Circular Model (GBSBCM) is shown in Figure 5.14, which was found to be suitable for macrocellular modeling, since it assumes that all the scatterers lie within the radius R about the mobile and $R < D$ [337]. An alternative spatial distribution of the scatterers, known as the Geometrically Based Single-Bounce Elliptical Model (GBSBEM) [337, 338], assumes that the scatterers are uniformly distributed within an ellipse, as shown in Figure 5.15, where the base station and the mobile station are the foci of the ellipse, and the parameters a_m and b_m are the semi-major and semi-minor axis values, which may be calculated as [337, 338]:

$$a_m = \frac{c\tau_m}{2}, \qquad b_m = \frac{1}{2}\sqrt{c^2\tau_m^2 - D^2}, \qquad (5.28)$$

where τ_m is the maximum time of arrival to be considered, D is the distance between the transmitter and the receiver and c is the velocity of light in free space. This model was proposed for microcellular environments [338], where the antenna heights are relatively low, and therefore, multipath scattering near the base station is equally likely, as scattering near the mobile station [338].

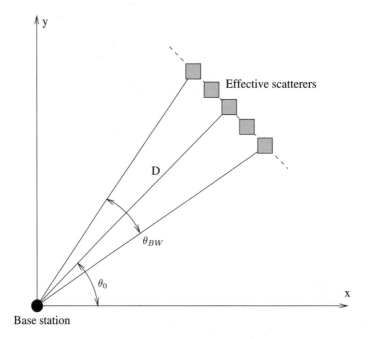

Figure 5.13: The Discrete Uniform Distribution model for multipath scattering using a line of N scatterers centred about the line of sight to the mobile.

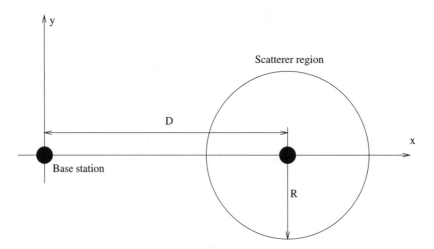

Figure 5.14: The Geometrically Based Single-Bounce Circular Model (GBSBCM), which is suitable for use as a macrocellular model, showing the region in which the scatterers are located.

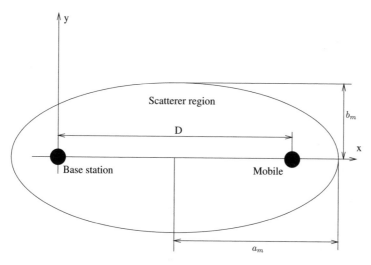

Figure 5.15: The Geometrically Based Single-Bounce Elliptical Model (GBSBEM), which is suitable for use as a microcellular model, showing the region in which the scatterers are located.

The GBSBEM may be used to generate the path time delay, τ_i, the angle of arrival, ϕ_i, the direction of departure, Φ_i, the power of the multipath component, P_i, and the phase angle, α_i. However, here we are only concerned with the angle of arrival information at the base station. The Cumulative Density Function (CDF) of the angle of arrival, ϕ_i, conditioned on the normalized multipath delay, $r_i = c\tau_i/D = \tau_i/\tau_0$, is given as [338]:

$$F_{\phi|r}(\phi_i|r_i) = \begin{cases} \frac{1}{2\pi}\cos^{-1}\left(\frac{1-r_i\cos\phi_i}{r_i-\cos\phi_i}\right) - \frac{\sqrt{r_i^2-1}\sin(-\phi_i)(1-r_i\cos\phi_i)}{2\pi(2r_i^2-1)(r_i-\cos\phi_i)^2} & -\pi \leq \phi_i \leq 0 \\ 1 - \frac{1}{2\pi}\cos^{-1}\left(\frac{1-r_i\cos\phi_i}{r_i-\cos\phi_i}\right) + \frac{\sqrt{r_i^2-1}\sin(\phi_i)(1-r_i\cos\phi_i)}{2\pi(2r_i^2-1)(r_i-\cos\phi_i)^2} & 0 \leq \phi_i \leq \pi. \end{cases}$$

$$(5.29)$$

The conditional probability density function of ϕ_i, may be found by differentiating Equation 5.29 with respect to Φ leading to:

$$f_{\phi|r}(\phi|r_i) = \frac{(r_i^2-1)^{3/2}(r_i^2-2r_i\cos\phi+1)}{\pi(2r_i^2-1)(r_i-\cos\phi)^3} \qquad -\pi \leq \phi \leq \pi, \qquad (5.30)$$

which is plotted in Figure 5.16 for various values of the normalized multipath delay, r_i. From this figure it can be seen that as the normalized multipath delay increases, the distribution of the angles-of-arrival tends to the uniform distribution, since the longer the delays, the greater the distance travelled, which results in a wider range of angles-of-arrival. In contrast, a small value of r_i concentrates the multipath components around the angle-of-arrival of the direct path component.

In simulating multipath component parameters, it is necessary to generate samples of random variables from specified distributions. The normalized path delay, r_i, of the ith multipath component, may be calculated thus as [338]:

$$r_i = \sqrt{\frac{1}{2} + \frac{1}{2}\sqrt{1+4\beta^2 x_i^2}}, \qquad (5.31)$$

Figure 5.16: Probability density function of angle-of-arrival conditioned on the normalized multipath delay, r_i, for various values of r_i, evaluated from Equation 5.30.

Table 5.3: Selection criteria for choosing r_m, the maximum normalized path delay [337, 338].

Criteria	Expression
Maximum path delay, τ_m	$r_m = \tau_m/\tau_0$
Fixed threshold, T (in dB), with pathloss exponent n	$r_m = 10^{(T-L_r)/10n}$
Fixed delay spread, σ_τ	$r_m = 3.24(\sigma_\tau/\tau_0) + 1$
Maximum excess delay, σ_e	$r_m = (\tau_0 + \tau_e)/\tau_0$

where x_i is a uniformly distributed random variable, denoted by $U(0, 1)$, ranging from 0 to 1 and $\beta = r_m\sqrt{r_m^2 - 1}$ depends on the maximum value of the normalized path delay, r_m. The maximum normalized path delay, r_m, may be determined by the four different selection criteria summarized in Table 5.3 [338].

Again, using large values of r_m results in a near-uniform distribution of the angles of arrival, whereas small values of r_m gives low-delay multipath components clustered in angle of arrival about the direct LOS path component.

From normalized path delay r_i and y_i, a uniformly distributed random variable, again formulated as $U(0, 1)$, over 0 to 1, it is now possible to determine the angle-of-arrival of the ith multipath component by solving $y_i = F_{\phi|r}(\phi_i|r_i)$ for ϕ_i, where $F_{\phi|r}(\phi_i|r_i)$ is defined in Equation 5.29.

The corresponding Cumulative Density Function (CDF) is a smooth and monotonic function of the angle-of-arrival, as illustrated in Figure 5.17. The figure shows that, if the

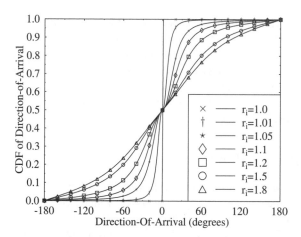

Figure 5.17: Cumulative density function of the angle-of-arrival conditioned on the normalized multipath delay, r_i, for various values of r_i.

normalized path delay, $r_i = 1$, then the angle-of-arrival is $0°$, and that as r_i increases, so does the spread of values of the angle-of-arrival.

Therefore, to summarize, the process of generating the angles-of-arrival obeying the required distribution the following sequence of operations must be performed:

- Determine r_m for the scenario under consideration.

- Calculate $\beta = r_m \sqrt{r_m^2 - 1}$.

- Generate $x_i = U(0, 1)$.

- Calculate $r_i = \sqrt{\frac{1}{2} + \frac{1}{2} \sqrt{1 + 4\beta^2 x_i^2}}$.

- Generate $y_i = U(0, 1)$.

- Solve Equation 5.29 for ϕ_i, given y_i using numerical methods.

5.6 Network Performance Results

Section 5.6.1 describes the processes involved in the simulator used to obtain the network performance results, such as the adaptive beamforming techniques, new call generation and handover queues as well as the multipath propagation model. Section 5.6.2.1 presents our simulation results obtained for the FCA and LOLIA DCA algorithms with a single element antenna, as well as two and four element adaptive antenna arrays, assuming a LOS propagation environment. Further results are presented in Section 5.6.2.2 which were obtained using the multipath channel of Section 5.6.1, using two, four and eight element adaptive antenna arrays. Section 5.6.2.3 characterizes the network performance of using two and four element antenna arrays, in the multipath propagation environment, in conjunction with power control. This is further expanded upon in Section 5.6.2.4, where power control assisted Adaptive Quadrature Amplitude Modulation (AQAM) is employed.

Sections 5.6.3.1–5.6.3.4 present our results for similar scenarios generated using the "wrap-around" rather than the "desert-island" technique, which eliminates the edge effects associated with the reduced interference levels encountered at the boundary of the simulation area. This process is described in Section 5.6.1. Finally, Section 5.6.2.7 provides a summary of the results obtained in this section.

5.6.1   System Simulation Parameters

The performance of the various channel allocation algorithms was investigated in a GSM-like [55] microcellular system, the parameters of which are defined in Table 5.4. The propagation environment was modeled using the power pathloss model having a pathloss exponent of -3.5. The mobile and base station transmit powers were fixed at 10 dBm (10 mW) for the simulations using no power control. The mobile and base station transmit powers were restricted to the range of -20 dBm to $+10$ dBm for the power control assisted and adaptive modulation based simulations. The number of carrier frequencies in the whole system was limited to seven, each supporting eight timeslots, in order to maintain an acceptable computational load. This implied that the DCA system employing seven carrier frequencies in conjunction with eight timeslots, as in GSM for example, was potentially capable of handling a maximum of $7 \times 8 = 56$ (or $12 \times 8 = 96$) instantaneous calls at one base station, provided that their quality was adequate. If a channel allocation request for a new call could not be satisfied immediately, it was queued for a duration of up to 5 s, after which time, if not satisfied, it was classed as blocked. It was assumed that the network was synchronous from cell to cell, thus channels on different timeslots of the same frequency were orthogonal in the time-domain and hence did not interfere with each other. The GSM-like system used a channel bandwidth of 200 kHz, but instead of the Gaussian Minimum Shift Keying (GMSK) [11] based modulation scheme, 4-QAM was employed for the sake of increasing the achievable bandwidth efficiency from 1.35 bps/Hz to 1.64 bps/Hz. Hence, the achievable bit rate was 200 kHz $\times$ 1.64 bps/Hz $= 328$ kbps. When dividing this bit rate amongst the eight users supported by the eight timeslots, the channel rate of the users—when for the sake of a simple argument neglecting transmission overheads, such as the equalizer training sequences, tailing sequences, guard periods and channel coding—became $328/8 = 41$ kbps. The call arrivals were Poisson distributed, and hence the call duration and inter-call periods were exponentially distributed [330, 389] with the mean values shown in Table 5.4.

The physical layer was modeled using two parameters, namely the "Outage SINR" and "Reallocation SINR", defined as the average Signal-to-Interference+Noise Ratio (SINR) required by a transceiver in order to satisfy the FER requirements over a narrowband Rayleigh fading channel. More specifically, Pilot Symbol Assisted (PSA) 4-QAM transmitting 2 bits per symbol was assumed, which had an outage SINR of 17 dB and a reallocation SINR of 21 dB [12, 13]. When the signal quality, expressed in terms of the SINR, drops below the "Reallocation SINR", a low quality access is encountered, and the mobile requests a new physical channel to handover to, thus initiating an intra- or inter-cell handover. If, while waiting for a reallocation handover, the signal quality drops further, below the so-called "Outage SINR", defined as the SINR required to maintain a 10% FER, then an outage is

Table 5.4: Network simulation parameters.

Parameter	Value	Parameter	Value
Noisefloor	-104 dBm	Multiple Access	F/TDMA
Frame duration	0.4615 ms	Cell radius	218 m
BS transmit power	10 dBm	MS transmit power	10 dBm
BS power control	No	MS power control	No
Number of base stations	49	Handover hysteresis	2 dB
Outage SINR threshold	17 dB	Re-alloc. SINR threshold	21 dB
Modulation scheme	4-QAM	Pathloss exponent	-3.5
Number of timeslots	8	Number of carriers	7
Average inter-call-time	300 s	Max new-call queue-time	5 s
Average call duration	60 s	Ref. signal modulation	BPSK
Beamforming algorithm	SMI	Reference signal length	8 bits
MS speed	30 mph	No. of antenna elements	2, 4 & 8
Pathloss at 1 m ref. point	0 dB	Shadow fading	No
Geometry of antenna array	Linear	Array element spacing	$\lambda/2$
Channel/carrier bandwidth	200 kHz		

encountered. A prolonged outage leads to the call being dropped or forcibly terminated. Since a user typically views a dropped call as less desirable than a blocked call, a Handover Queueing System (HQS) was employed. By forming a queue of the handover requests, which have a higher priority during contention for network resources than new calls, it is possible to reduce the number of dropped calls at the expense of a higher blocked call probability. A further advantage of the HQS is that a time window is formed, during which the handover may take place, enabling the user to wait, if necessary, for a slot to become free, thus increasing its chances of a successful handover. This twin-threshold physical layer model is similar to those described by Tekinay and Jabbari [352] and Katzela and Naghshineh [351]. However, the model described here is based on SINR thresholds, rather than on the received power thresholds of Tekinay and Naghshineh [352] and Katzela and Naghshineh [351]. A further metric, namely the low signal quality probability, is calculated as the proportion of time that the SINR is below the "Reallocation SINR" threshold.

Again, the "Outage SINR" and "Reallocation SINR" threshold were determined, with the aid of independent bit-level simulations, for BPSK, QPSK/4-QAM and 16-QAM [12, 13], conducted in a Rayleigh fading environment using approximately half-rate Bose-Chaudhuri-Hocquenghem (BCH) codes, which employed bit interleaving over the different number of bits per transmission frame conveyed by the different modem modes [395]. Thus, the "Reallocation SINR" threshold was determined to be the average SINR required by the specific transceiver employed for maintaining a 5% transmission FER. This SINR value is transceiver dependent and in general can be reduced at the cost of increased transceiver complexity and power consumption. The loss of a maximum of 5% of the speech or video frames can be considered a worst-case scenario for modern "wireless-oriented", i.e. error-resilient source codecs. Therefore, by setting the reallocation threshold at this level, the

Table 5.5: The "Reallocation SINRs" and "Outage SINRs" used in the handover process, found by bit-level simulations for BPSK, QPSK/4-QAM, and 16-QAM modems. The "Reallocation SINR" is the SINR, below which a channel reallocation will be requested, while the "Outage SINR" is the SINR, below which a service outage is declared. Successive service outages render the call to be forcibly terminated.

Modulation scheme	Reallocation SINR threshold (dB) for 5% FER	Outage SINR threshold (dB) for 10% FER
BPSK	17	13
4-QAM	21	17
16-QAM	27	24

system requested handovers to new channels, before the speech or video quality degradation due to excessive FERs became objectionable.

The "Outage SINR" threshold defines the SINR, below which the system declares that the radio channel has degraded to such a level as to cause a service outage. If the radio channel continues to be in outage, then the call is forcibly terminated. The "Outage SINR" threshold was determined by bit-level or physical layer simulations to be the average SINR required for maintaining a 10% FER. Therefore, if the radio channel degrades such that at least 10% of the speech or video frames were lost for some period of time, then the call would be forcibly terminated. The corresponding SINR thresholds based on bit-level simulations of BPSK, QPSK/4-QAM and 16-QAM modems are shown in Table 5.5.

The mobiles were capable of moving freely, at a speed of 30 mph, in a fixed random direction, selected at the start of the simulation from a uniform distribution, within the simulation area of 49 traffic cells, each having a radius of 218 m. Two different types of simulation area were invoked, the classical "desert island" type and the "wraparound" type. The "desert-island" or "urban, sub-urban, rural" environment neglects the interference emanating from the cells surrounding the outside of the simulation area. In other words, the traffic cells at the centre of the simulation area are surrounded by interfering cells and thus are subjected to the highest levels of interference. However, the cells at the edges of the simulation area are not surrounded by interfering cells and hence are subjected to a lower level of interference. This can be likened to an "urban, sub-urban, rural" environment, where the centre cells represent the urban environment and the outer cells are considered to be low traffic-density rural cells in nature. However, this can lead to optimistic results, and hence often a "wraparound" simulation area is used [396, 397]. In order to facilitate the employment of an infinite plane of simulation area, a tessellating rhombic simulation area was used. Hence, the simulation area was replicated around itself, or tiled to form a larger, or effectively infinite, simulation area. More explicitly, mobile stations and their signals were "wrapped around" from one side of the network to the other [396, 397]. Hence, for example, a mobile station in call, which leaves the network at its edge, re-enters the network at the opposite side, whilst inflicting Co-Channel Interference (CCI) to *all* users, which may be positioned at any location in the network. Figure 5.18 depicts this scenario graphically.

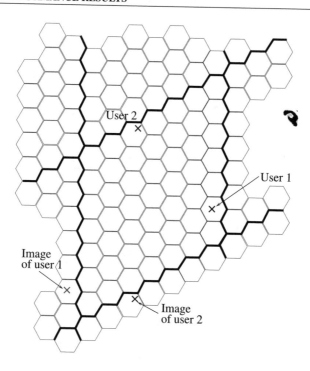

Figure 5.18: The 7×7 rhombic simulation area showing a user and its "wrapped" image.

The receiver antenna array weights were calculated using the Sample Matrix Inversion (SMI) algorithm [283, 285, 290], which determines the value of the AAA weights, such that they are optimized with respect to the received SINR [290]. In order to calculate the receiver antenna weights using the SMI algorithm, an eight-symbol long BPSK reference signal was assigned to the desired mobile. The remaining seven orthogonal eight-symbol duration BPSK reference symbols were then assigned to the interfering mobiles. However, any of these seven codes were allocated to more than one mobile, if the number of interferers was higher than seven. Thus, the desired mobile was uniquely identifiable, with the aid of its reference signal and the receiver antenna weights were optimized for obtaining the maximum received SINR, as detailed in Section 4.3.2.3. The calculation of the receiver antenna array weights was performed on a transmission frame-by-frame basis, leading to updated UL and DL SINRs every transmission frame.

The base station's receiver antenna weights calculated for UL reception may not be suitable for the DL transmission due to the generally uncorrelated UL and DL channels of Frequency Division Duplexed (FDD) systems. However, forming a feedback loop from the mobile to the base stations for conveying the mobile's received reference signal and thus effectively conveying the quality of the mobile's received reference signals for use in an iterative adaptive beamforming algorithm, would allow the base station to use the DL weights as proposed in [319, 320]. In a Time Division Duplexed (TDD) system having a sufficiently short dwell time, the AAA weights calculated for UL reception can also be used for DL transmission, since the propagation channel does not vary significantly between the

Figure 5.19: Screenshot of the simulation software, "Netsim", for a 7-cell, 5-user simulation, showing
the identical UL and DL, or receive and transmit, beam patterns generated by the adaptive
antenna arrays using 2 elements. The squares represent the mobiles, with the large black
circles denoting the base stations. The black lines from the base stations, passing through
the squares, show the array gain in the desired direction. While the half-tone grey lines
point in the direction of interfering sources, where the length of the lines indicates the
antenna gain in that direction.

UL and DL timeslots [6]. However, the system considered here is an FDD based network,
and hence the assumption of channel predictability should therefore give an upper limit to the
performance gains that may be achieved using an adaptive array. From now on we assume
that the base station's receive and transmit, in other words the UL and DL beam patterns are
identical.

An example of the adaptive antenna array beam patterns generated by two element
adaptive antenna arrays is shown in Figure 5.19. In this figure the mobiles are denoted by
the use of small squares, while the base stations are represented by black filled circles. The
solid black lines from the base stations to the users show the direction that the antenna array
is steered in, and the gain in that direction. The half-tone grey lines pointing towards the
mobiles represent the interfering signals, where the length of these lines is proportional to
the gain of the antenna array in that direction. As it can be seen from the figure, the main

Figure 5.20: Screenshot of the simulation software, "Netsim", for a 7-cell, 5-user simulation, showing the identical UL and DL, or receive and transmit, beam patterns generated by the adaptive antenna arrays using 4 elements. The squares represent the mobiles, with the large black circles denoting the base stations. The black lines from the base stations, passing through the squares, show the array gain in the desired direction. While the half-tone grey lines point in the direction of interfering sources, where the length of the lines indicates the antenna gain in that direction.

beamwidth is large and, although there is some beneficial interference nulling, its extent is limited. For the four element adaptive antenna array, as in Figure 5.20, the beams in the direction of the desired users are significantly narrower, and hence the interference sources are nulled much more strongly, as indicated by the shortened half-tone lines. We can observe in both Figures 5.23 and 5.24 that the antenna array beam patterns formed are symmetrical in the y-axis, as a direct consequence of the linear array geometry with the antenna array elements located on the y-axis. Using an alternative array geometry, such as a square or circle shaped one, would prevent this beam pattern symmetry from occuring and thus could potentially improve the achievable performance.

Both a purely LOS propagation environment and a multipath propagation environment were considered. This multipath environment consisted of the LOS ray and two additional rays, each having a third of the power of the LOS ray. The angles-of-arrival at the base

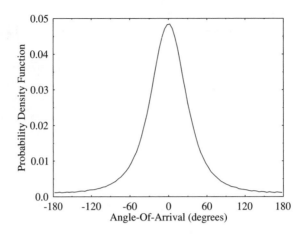

Figure 5.21: Probability density function of the angle-of-arrival of the UL multipath rays, centred about the angle-of-arrival of the line-of-sight path. Furthermore, $r_m = 1.36874$ and 1 000 000 trials were used.

station were determined using the Geometrically Based Single-Bounce Elliptical Model (GBSBEM) of Section 5.5 [337, 338], with its parameters chosen such that the multipath rays had one-third of the received power of the direct ray. The multipath received power criteria of Table 5.3 was used to determine the value of r_m to be used in the GBSBEM. Specifically, we opted for $r_m = 10^{(T-L_r)/10n}$, where T is the received power value in dB, L_r is the reflection loss and n is the pathloss exponent. Furthermore, T was set to 4.8 dB with L_r equal to zero in conjunction with a pathloss exponent of 3.5, in order to achieve the desired received signal power of one-third that of the LOS ray. Hence, using the formulae of Section 5.5, $r_m = 10^{(4.8-0.0)/35} = 1.36874$, leading to, $\beta = r_m \sqrt{r_m^2 - 1} = 1.2792$. Since $r_i = \sqrt{\frac{1}{2} + \frac{1}{2}\sqrt{1 + 4\beta^2 x_i^2}}$ where x_i is a uniformly distributed random variable over $[0, 1]$, r_i varies from 1.0 to 1.36874. The PDF of the angle-of-arrival for r_m=1.36874 is shown in Figure 5.21, which was generated using the GBSBEM algorithm of Section 5.5 for 100 000 trials. It was assumed that all of these multipath rays arrived with zero time delay relative to the LOS path, or that a space-time equalizer [18, 65] was employed, thus making full use of the additional received signal energy. However, the numerous extra desired and interfering signals incident upon the antenna array rapidly consume the finite degrees of freedom of the antenna array, limiting its ability to fully cancel each source of interference.

The addition of multipath rays, for both the desired signal and the interference sources, results in many more received UL signals impinging upon the antenna array at the base station. A result of the increased number of received UL signals is that the limited degrees of freedom of the base station's adaptive antenna array are exhausted, resulting in reduced nulling of the interference sources. A solution to this limitation is to increase the number of antenna elements in the base station's adaptive array, although this has the side effect of raising the cost and complexity of the array. In a macro-cellular system it may be possible to neglect multipath rays arriving at the base station from interfering sources since the majority of the scatterers are located close to the mobile station [21]. In contrast, in a micro-cellular system the scatterers are located in both the region of the reduced-elevation base station and

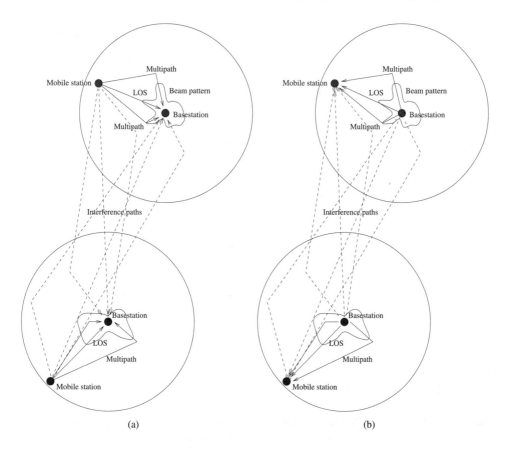

(a)   (b)

Figure 5.22: The multipath environments of (a) the UL and (b) the DL, showing the multipath components of the desired signals, the line-of-sight interference and the associated base station antenna array beam patterns.

that of the mobile, and hence multipath propagation must be considered. Figure 5.22 shows the simulated environment for both the UL and the DL, with the multipath components of the desired signal and interference signals clearly illustrated, where the UL and DL are assumed to be reciprocal. When the DCA algorithm is "listening", in order to determine the best channel to be selected, only the LOS signals are considered, while the multipath signals are neglected. However, at all other times the multipath signals are used in the calculation of the received signal and interference levels.

Figures 5.23, 5.24 and 5.25 show examples of the beam patterns obtained for two, four and eight element adaptive antenna arrays in the presence of multipath propagation. For the two element antenna array, as illustrated in Figure 5.23, the beamwidth of the antenna array is large, thus limiting its efficiency in nulling the sources of interference. Nonetheless, it can be seen that the arrays are attempting to steer towards the desired signals, and away from the sources of interference. Here the desired signals are represented by the three black lines, where the black line passing through a square is the direct ray, while the remaining two

Figure 5.23: Screenshot of the simulation software, "Netsim", for a 7-cell, 5-user scenario, showing the beam patterns generated by the adaptive antenna arrays using 2 elements. The squares represent the users, with the large black circles denoting the base stations. The black lines from the base stations, passing through the squares, show the array gain in the desired direction, the black lines not passing through the squares are the desired user's multipath rays. The dark grey lines are the LOS interference paths, while the interferer's multipath components are illustrated by the light grey lines, where the length of the lines is proportional to the corresponding antenna gains in their directions.

black lines indicate the multipath rays arriving from the desired user. Observe in the figures that most of these lines end on the unity-gain circles, implying that they are received with a unity gain by the base station. Furthermore, the dark grey lines indicate the LOS paths from the interference sources, while the corresponding two multipath rays of the interferers are denoted by the light grey lines. The beam pattern of base station "1" is a good example of how the array is steering towards the desired signal paths, and away from the interference. For base station "5", at the bottom of Figure 5.23, the small angular separation between the arriving signals, and the end-fire location of these sources, makes rejection of the interference harder to accomplish. The use of a four element antenna array, depicted in Figure 5.24, results in more successful nulling of the interference sources, but again, for base station "5" at the bottom, the similar angular location of the desired and interfering sources results in poor interference cancellation performance. Figure 5.25 shows that an eight element adaptive

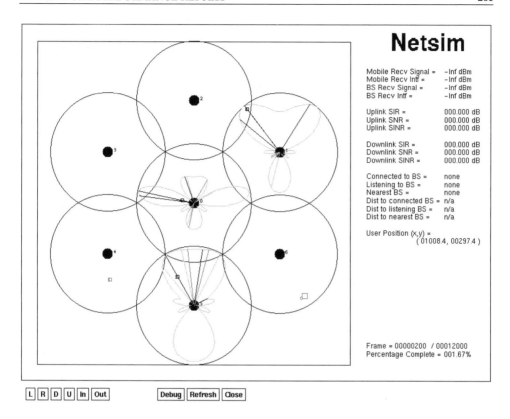

Figure 5.24: Screenshot of the simulation software, "Netsim", for a 7-cell, 5-user scenario, showing the beam patterns generated by the adaptive antenna arrays using 4 elements. The squares represent the users, with the large black circles denoting the base stations. The black lines from the base stations, passing through the squares, show the array gain in the desired direction, the black lines not passing through the squares are the desired user's multipath rays. The dark grey lines are the LOS interference paths, while the interferer's multipath components are illustrated by the light grey lines, where the length of the lines is proportional to the corresponding antenna gains in their directions.

antenna array performs well in most cases, nulling the sources of interference strongly, whilst efficiently steering towards the desired signals. Using an alternative layout of the antenna elements, rather than the uniform linear array, should minimize the possibility of a situation, similar to that of base station "5", where all the sources are located at end-fire, which is the area of poorest performance of the array.

Having described the simulation parameters, in the next section we present our simulation results, quantifying the amount of traffic that can be carried by each of the simulated networks, whilst maintaining the required network quality.

5.6.2   Non-wraparound Network Performance Results

The results presented in this section were obtained for the "desert-island" or "urban, sub-urban, rural" scenario, i.e. with the highest levels of interference present at the centre of the

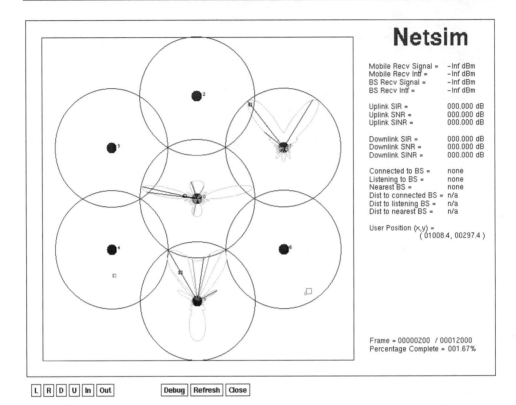

Figure 5.25: Screenshot of the simulation software, "Netsim", for a 7-cell, 5-user scenario, showing the beam patterns generated by the adaptive antenna arrays using 8 elements. The squares represent the users, with the large black circles denoting the base stations. The black lines from the base stations, passing through the squares, show the array gain in the desired direction, the black lines not passing through the squares are the desired user's multipath rays. The dark grey lines are the LOS interference paths, while the interferer's multipath components are illustrated by the light grey lines, where the length of the lines is proportional to the corresponding antenna gains in their directions.

simulation area. Results were obtained for single, two and four element antenna arrays over an LOS channel for both the FCA algorithm and the LOLIA with exclusion zones of 7 and 19 cells. This work was then extended to provide network capacity estimates for non-LOS or multipath channels using adaptive antenna arrays comprising two, four and eight elements. Power control and adaptive modulation techniques were also employed for increasing the network capacity further.

5.6.2.1   Performance Results over a LOS Channel

Figure 5.26 shows the new call blocking probability for a variety of uniform traffic loads, measured in terms of the mean normalized carried traffic, with units of Erlangs/km^2/MHz. The figure shows that for a given traffic load, both FCA and the LOLIA, using an exclusion

Figure 5.26: New call blocking probability performance versus mean carried traffic, for comparison of the LOLIA, with 7 and 19 "local" base stations, and of FCA using a 7-cell reuse cluster, under a uniform geographic traffic distribution, for a single antenna element as well as for two and four element antenna arrays with beamforming in an LOS environment. See Figure 5.32 for the corresponding multipath results.

zone of $n = 19$ maintained a fairly similar probability of new call blocking, regardless of the number of elements in the antenna array. In the case of the FCA algorithm, this was due to the limited number of frequency/timeslot combinations available as a direct result of the fixed nature of the network. However, for the LOLIA having an exclusion zone of 19 cells, the lack of frequency/timeslot combinations was due to the large exclusion zone. Thus, using the smaller exclusion zone of 7 cells led to a significantly reduced new call blocking probability. The figure also shows that, since the new call blocking probability of the LOLIA using $n = 7$ was reduced, thanks to the adaptive antenna arrays, the new call blocking performance was interference limited. This contrasts with the FCA algorithm and the LOLIA using $n = 19$, whose new call blocking performance was limited by the availability of frequency/timeslot combinations. It is interesting to note that, in terms of its new call blocking probability, the FCA algorithm performed better using only one antenna element as a result of its significantly increased call dropping probability, which freed up network resources, thus enabling more new calls to start.

The call dropping probability of the FCA algorithm, and that of the LOLIAs is depicted in Figure 5.27 for one, two and four element antenna arrays, when subjected to varying uniform traffic loads. The FCA algorithm suffered from the highest call dropping probability of the three channel allocation schemes. In conjunction with a four element adaptive antenna array it is similar to the LOLIA using $n = 7$ and a single antenna element for teletraffic loads higher than 10 Erlang/km^2/MHz. For teletraffic levels below this point, the FCA algorithm offered superior performance due to the call dropping probability "floor" experienced by the

Figure 5.27: Call dropping probability performance versus mean carried traffic, for comparison of the LOLIA, with 7 and 19 "local" base stations, and of FCA using a 7-cell reuse cluster, under a uniform geographic traffic distribution, for a single antenna element as well as for two and four element antenna arrays with beamforming in an LOS environment. See Figure 5.33 for the corresponding multipath results.

LOLIA using $n = 7$. The large exclusion zone of the LOLIA using $n = 19$ resulted in a very low probability of forced termination until the system approached its maximum capacity of around 12 Erlang/km^2/MHz, where the dropping probability increased rapidly. However, the performance of the LOLIA with $n = 19$ still exceeded that of both the FCA algorithm and the LOLIA with $n = 7$ due to the low levels of co-channel interference resulting from the high frequency re-use distance associated with the large exclusion zone.

Figure 5.28 shows the probability of low quality access versus various uniform traffic loads. The figure shows our results for the FCA algorithm and the LOLIA for nearest base station constraints of 7 and 19 cells. Again, the LOLIA with $n = 19$ offered the best performance at the lower traffic levels, but the low-quality access probability increased the most rapidly as the traffic load increased. For a given traffic load the LOLIA using $n = 19$ provided the lowest probability of a low quality access. This resulted from the low level of co-channel interference of the network and the interference rejection capabilities of the adaptive antenna arrays. The figure shows that all of the channel allocation schemes benefited from the use of the adaptive antenna arrays.

Figure 5.29 shows the Grade-Of-Service (GOS) for a range of uniform teletraffic loads. The figure shows results for the FCA algorithm and the LOLIAs with nearest base station constraints of 7 and 19 cells, for cases of a single antenna element as well as for two and four element adaptive antenna arrays. The grade of service is better, i.e. lower, for larger exclusion zone size when the traffic load is low, which is reversed for high traffic loads. This is mainly attributable to the higher call blocking probability of the larger exclusion zone of 19 cells,

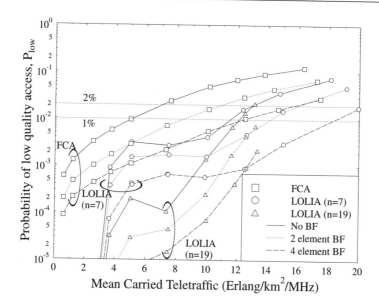

Figure 5.28: Probability of low quality access versus mean carried traffic, for comparison of the LOLIA, with 7 and 19 "local" base stations, and of FCA using a 7-cell reuse cluster, under a uniform geographic traffic distribution, for a single antenna element as well as for two and four element antenna arrays with beamforming in an LOS environment. See Figure 5.34 for the corresponding multipath results.

particularly in the region of the highest traffic loads. The GOS for the FCA scheme follows the probability of a blocked call and the dropping probability trends by increasing smoothly and monotonically with the traffic load.

The effect of beamforming on the number of handovers performed can be seen in Figure 5.30. The performance of the LOLIAs was barely altered by the use of beamforming, with both performing the lowest number of handovers per call. At the highest teletraffic loads it can be seen that the LOLIA using an exclusion zone of 7 base stations benefited slightly from the use of the adaptive antenna arrays. In contrast, the number of handovers performed by the FCA algorithm was reduced significantly as a benefit of using adaptive antennas with a maximum reduction in the mean number of handovers performed per call of 69% for two elements, and of 86% for four elements. This translates into a significantly reduced load for the network, since it has to manage far less handovers, therefore reducing the complexity of the network infrastructure. As the network load exceeded about 12 Erlangs/km^2/MHz, the mean number of handovers performed per call dropped due to the excessive call dropping probability, since calls were being dropped before they could handover, thus reducing the number of handovers.

Figure 5.31 portrays the mean carried teletraffic versus the number of mobiles in the simulated system. The figure shows that at low traffic loads both FCA and the LOLIA carry virtually identical amounts of traffic. However, as the mobile density, and hence the traffic load, is increased, the LOLIA with the nearest base station limit of 19 reaches its maximum traffic load and cannot carry further traffic. In other words, the employment of

Figure 5.29: Grade-Of-Service (GOS) performance versus mean carried traffic, for comparison of the LOLIA, with 7 and 19 "local" base stations, and of FCA using a 7-cell reuse cluster, under a uniform geographic traffic distribution, for a single antenna element as well as for two and four element antenna arrays with beamforming in an LOS environment. See Figure 5.35 for the corresponding multipath results.

adaptive antennas does not enable the network to carry more traffic, since the performance of the network is resource limited, not interference limited. The limiting factor is effectively the high frequency reuse distance of the LOLIA in conjunction with $n = 19$, since the associated low level of interference cannot be substantially further reduced by the adaptive arrays. Hence in Figure 5.31 increasing the number of antenna elements in the adaptive array does not support a substantially increased teletraffic capacity in terms of the number of users supported, since the number of available frequency/timeslot combinations is limited, as indicated by the flattening performance curves.

By contrast, for FCA and the LOLIA in conjunction with $n = 7$, the advantage of using adaptive antennas can be explicitly seen from the figure. Specifically, the FCA and the LOLIA in conjunction with $n = 7$, enable a higher level of traffic to be carried, at a higher quality than a system without adaptive antenna arrays. The performance gain attained by the LOLIA, over the FCA algorithm, is also shown in Figure 5.31, which illustrates the increase in carried traffic as a result of the dynamic configurability of DCA schemes.

It can be seen from Table 5.6 that for all of the channel allocation schemes, the use of adaptive antenna arrays at the receiver resulted in increased carried teletraffic, hence supporting a higher number of simultaneous users. The FCA algorithm benefited most from the use of adaptive antennas with a 67% increase in the number of users supported when using two antenna elements and a 144% rise in the carried traffic, when using an adaptive array with four elements. The LOLIA associated with $n = 7$, supported a higher number of users than FCA although the capacity increases obtained through the use of adaptive antenna arrays

Figure 5.30: The mean number of handovers per call versus mean carried traffic, for comparison of the LOLIA, with 7 and 19 "local" base stations, and of FCA using a 7-cell reuse cluster, under a uniform geographic traffic distribution, for a single antenna element as well as for two and four element antenna arrays with beamforming in an LOS environment. See Figure 5.36 for the corresponding multipath results.

Table 5.6: Maximum mean carried traffic, and the maximum number of mobile users that can be supported by each configuration whilst meeting the preset quality constraints of Section 5.3.3.4. The carried traffic is expressed in terms of normalized Erlangs (Erlang/km^2/MHz), for the network described in Table 5.4 in an LOS environment.

	Conservative			Lenient		
	$P_{FT} = 1\%, P_{low} = 1\%$			$P_{FT} = 1\%, P_{low} = 2\%$		
	$GOS = 4\%, P_B = 3\%$			$GOS = 6\%, P_B = 5\%$		
Algorithm	Users	Traffic	Limiting factor	Users	Traffic	Limiting factor
FCA, 1 element (el.)	815	5.10	P_{low}	1115	7.05	P_{low}
FCA, 2 elements	1360	8.45	P_{low}	1755	11.00	P_{low}
FCA, 4 elements	1985	12.40	P_{low}	2710	15.75	P_{low}
LOLIA ($n = 7$), 1 el.	1855	11.50	P_{low}	2110	13.00	P_{low}
LOLIA ($n = 7$), 2 el.	2260	14.15	P_{low}	2600	16.00	P_{low}
LOLIA ($n = 7$), 4 el.	2935	18.30	P_{low}	>3200	>20.00	P_{low}
LOLIA ($n = 19$), 1 el.	1935	11.35	P_B	2010	11.65	P_B
LOLIA ($n = 19$), 2 el.	1940	11.35	P_B	2045	11.70	P_B
LOLIA ($n = 19$), 4 el.	1960	11.65	P_B	2090	12.00	P_B

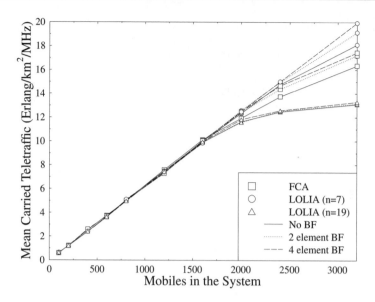

Figure 5.31: Mean traffic carried versus the number of mobiles in the system, for comparison of the LOLIA, with 7 and 19 "local" base stations, and of FCA using a 7-cell reuse cluster, under a uniform geographic traffic distribution, for a single antenna element as well as for two and four element antenna arrays with beamforming in an LOS environment. See Figure 5.37 for the corresponding multipath results.

were more limited. Specifically, a two element array carried an extra 22% of users and with the aid of four elements it supported 58% more users. Using a channel exclusion zone of 19 base stations gave a slight performance advantage over the 7-cell variant for the conservative scenario of Section 5.3.3.4, but only without adaptive antennas. Employing adaptive antennas had little effect on the number of users supported by the network using the LOLIA with $n = 19$, increasing the traffic carried by only a small margin. The corresponding multipath results are summarized in Table 5.7 with network configurations common between the two highlighted in bold.

5.6.2.2 Performance Results over a Multipath Channel

Following our previous simulations, where a purely LOS environment existed between the mobiles and their base stations, this section presents our performance results for the multipath environment described in Section 5.6.1, using two, four and eight element adaptive antenna arrays.

Comparing the blocking probabilities of the multipath environment, in Figure 5.32, with those of the LOS environment, which were portrayed in Figure 5.26, reveals that the FCA algorithm and both the LOLIAs behaved similarly in both propagation environments. Again, only the LOLIA with an exclusion zone of 7 base stations benefited from the use of the adaptive antenna arrays in terms of the new call blocking probability.

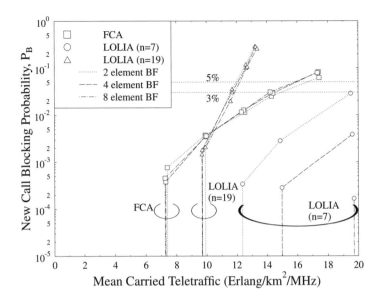

Figure 5.32: New call blocking probability performance versus mean carried traffic, for comparison of the LOLIA, with 7 and 19 "local" base stations, and of FCA using a 7-cell reuse cluster, under a uniform geographic traffic distribution, for two, four and eight element antenna arrays with beamforming in a *multipath environment*. See Figure 5.26 for the corresponding LOS results.

In Figure 5.33 the probability of a dropped call in a multipath environment is presented which, for the FCA algorithm, was similar under the multipath propagation conditions to that of the LOS scenario in Figure 5.27. The LOLIA using an exclusion zone of 7 base stations also exhibited call dropping probabilities close to those observed in the LOS scenario, when using a two element adaptive antenna array. In conjunction with a four element antenna array the performance was slightly degraded in the multipath scenario, but using the eight element antenna array resulted in superior performance to that of the four element array in the LOS environment. There was a slight call dropping performance improvement for the LOLIA using $n = 19$.

The probability of low quality access is depicted in Figure 5.34. The FCA algorithm did not perform as well, with respect to the probability of a low quality access, in the multipath propagation environment, when compared to the LOS case of Figure 5.28. The same was true of the LOLIA using $n = 7$ at higher traffic levels, although, at lower levels of traffic the performance in the multipath case was superior. At low levels of traffic the average level of interference was relatively low, and hence the extra signal power received in the multipath environment resulted in a reduced chance of a low quality access occurring. However, at higher levels of teletraffic, the background interference level was higher than in the LOS scenario of Figure 5.28, and hence the extra received power had a less beneficial impact, in fact the multipath components created additional interference. The LOLIA using an exclusion zone of 19 base stations and an adaptive antenna array of two elements performed better in the multipath case. However, in conjunction with four elements it offered a superior performance

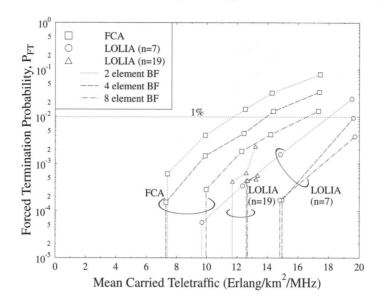

Figure 5.33: Call dropping probability performance versus mean carried traffic, for comparison of the LOLIA, with 7 and 19 "local" base stations, and of FCA using a 7-cell reuse cluster, under a uniform geographic traffic distribution, for two, four and eight element antenna arrays with beamforming in a *multipath environment*. See Figure 5.27 for the corresponding LOS results.

in the LOS scenario of Figure 5.28. Overall, the improvement in the probability of low quality access through increasing the number of adaptive antenna array elements, was reduced in the multipath propagation environment, since the added interference power outweighed the increased received signal power. This ultimately reduced the prevalent SINR even when using adaptive antenna arrays.

As expected on the basis of Equation 5.15, the FCA algorithm and the LOLIA with $n = 19$, offered a similar GOS performance for both the LOS scenario of Figure 5.29 and for the multipath environment. Figure 5.35 also shows that the GOS of the FCA algorithm using a given number of antenna elements is inferior to the GOS of the LOS propagation environment characterized in Figure 5.29, as for the probability of low quality access seen in Figures 5.28 and 5.34. At network loads of less than about 13 Erlang/km^2/MHz, the GOS of the LOLIA with $n = 7$ was superior to that of the LOS environment in Figure 5.29, however, above this carried traffic value the performance was worse.

Figure 5.36 demonstrates the significant impact that adaptive antennas have on the mean number of handovers per call for the FCA algorithm in a multipath environment. As in the LOS propagation environment characterized in Figure 5.30, more handovers per call were initiated when using FCA system employing two or four element antenna arrays, than for either of the LOLIAs using a single antenna element. Furthermore, a higher number of handovers was required in the multipath environment than in the LOS scenario of Figure 5.30, for a given antenna array configuration. The LOLIA schemes performed much fewer handovers than FCA, irrespective of the propagation environment, and generally did

Figure 5.34: Probability of low quality access versus mean carried traffic, for comparison of the LOLIA, with 7 and 19 "local" base stations, and of FCA using a 7-cell reuse cluster, under a uniform geographic traffic distribution, for two, four and eight element antenna arrays with beamforming in a *multipath environment*. See Figure 5.28 for the corresponding LOS results.

not appear to benefit from the employment of adaptive antennas in terms of the required handovers per call.

As it can be seen in Figure 5.37 for the adaptive array, the mean levels of carried teletraffic against the number of mobiles in the system followed a near-linear trend, with the capacity of the LOLIA 19 system rolling off above 2000 users, as for the LOS scenario in Figure 5.31. Above this number of users, very little extra teletraffic was carried, with corresponding several orders of magnitude increases of the blocking, dropping and low quality access probabilities as well as that of the GOS measure. For the channel allocation algorithms operating in a multipath rather than LOS environment, increasing the number of antenna elements did not significantly increase the levels of traffic carried, although the network performance improved in other respects, such as for example the call dropping probability.

Table 5.7 presents similar results to Table 5.6, but for a multipath environment. From this table it can be seen that LOLIA 19 actually performed slightly better in the multipath scenario, than in a LOS situation. This was due to the large reuse distance of the system, resulting in the sum of the three desired multipath signals versus the sum of the interfering signals being higher than the ratio of the LOS desired signal power to the LOS interference power. The LOLIA 7 algorithm, however, did not generally benefit from the multipath environment, since the smaller reuse distance resulted in numerous sources of relatively strong interference, all requiring cancellation. Therefore, as the number of antenna elements increased, so should the number of users supported by the network, as a result of the increased number of degrees of freedom, and therefore, the increased number of sources that may be nulled. The results

Figure 5.35: Grade-Of-Service (GOS) performance versus mean carried traffic, for comparison of the LOLIA, with 7 and 19 "local" base stations, and of FCA using a 7-cell reuse cluster, under a uniform geographic traffic distribution, for two, four and eight element antenna arrays with beamforming in a *multipath environment*. See Figure 5.29 for the corresponding LOS results.

support this expectation with a 17% gain in the number of users, when upgrading the system from two element to four element arrays, and a further 15% improvement in the number of supported users with the aid of eight element antenna arrays instead of the four element arrays. As for the LOS results, the FCA algorithm, again, benefited the most in terms of the number of users supported by the network from the employment of adaptive antenna arrays. The number of users increased by 35%, when doubling the number of antenna elements from two to four, and on doubling from four to eight delivered a further 29% user capacity improvement.

5.6.2.3 Performance over a Multipath Channel using Power Control

This section builds on the results obtained in the previous section for a multipath propagation environment. Simulations were conducted for a standard 7-cell FCA scheme and a LOLIA-assisted system using $n = 7$, both invoking power control. The power control algorithm implemented attempted to independently adjust the mobile and base station transmit powers, such that the UL and DL SINRs were within a given target SINR window. The use of a target window avoided constantly increasing and decreasing the transmission powers, which could lead to potential power control instabilities within the network. Furthermore, using a range of possible transmission powers is analogous to accounting for an inherent power control error plus slow fading phenomenon. The "Target SINR" given in Table 5.8 is the SINR to be maintained by the power control algorithm. The immediate effect of power control on

Figure 5.36: Mean number of handovers per call versus mean carried traffic, for comparison of the LOLIA, with 7 and 19 "local" base stations, and of FCA using a 7-cell reuse cluster, under a uniform geographic traffic distribution, for two, four and eight element antenna arrays with beamforming in a *multipath environment*. See Figure 5.30 for the corresponding LOS results.

the SINR versus the mobile's distance from the base station can be seen in Figure 5.38. This figure shows that power control attempts to maintain a constant SINR, sufficiently high for reliable communications across the network, rather than allowing for unnecessarily high SINRs near the base station and providing insufficient levels of SINR far from the base stations, evident for a cordless telephone type network using no power control. It was found that in conjunction with 4-QAM using a target SINR of 27 dB was most suitable, when using the FCA algorithm. However, the LOLIA required a higher target SINR of 31 dB in order to obtain satisfactory call dropping performance, as a result of its dynamic nature causing the interference levels to vary more rapidly than for the FCA algorithm. In other words, the LOLIA required a higher SINR "headroom" above the re-allocation SINR threshold.

Figure 5.39 shows the new call blocking probability versus the mean normalized carried traffic, expressed in terms of Erlangs/km²/MHz. The figure shows that the blocking performance of the FCA algorithm is limited by the availability of frequency/timeslot combinations, and hence the addition of power control does not improve the new call blocking performance. However, the blocking performance of the LOLIA is not dominated by the availability of frequency/timeslot combinations and hence it can be seen to benefit significantly from using power control.

From Figure 5.40 it can be seen that the Power Control (PC) algorithm substantially improved the call dropping probability of the FCA algorithm in comparison to the scenario without PC in Figure 5.32. Specifically, at the highest traffic loads, the PC-assisted perfor-mance matched that without power control but using antenna arrays with twice the number

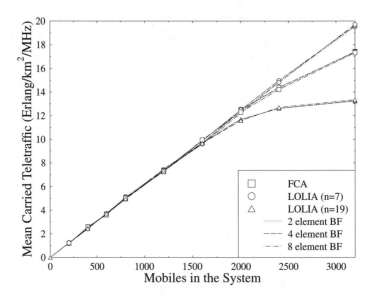

Figure 5.37: Mean traffic carried versus the number of mobiles in the system, for comparison of the LOLIA, with 7 and 19 "local" base stations, and of FCA using a 7-cell reuse cluster, under a uniform geographic traffic distribution, for two, four and eight element adaptive antenna arrays in a *multipath environment*. See Figure 5.31 for the corresponding LOS results.

Table 5.7: Maximum mean carried traffic, and maximum number of mobile users that can be supported by each configuration, whilst meeting the preset quality constraints of Section 5.3.3.4. The carried traffic is expressed in terms of normalized Erlangs (Erlang/km^2/MHz), for the network described in Table 5.4 in a *multipath environment*. The corresponding LOS results are summarized in Table 5.6 with network configurations common between the two highlighted in bold.

	Conservative			Lenient		
	$P_{FT} = 1\%, P_{low} = 1\%$			$P_{FT} = 1\%, P_{low} = 2\%$		
	$GOS = 4\%, P_B = 3\%$			$GOS = 6\%, P_B = 5\%$		
Algorithm	Users	Traffic	Limiting factor	Users	Traffic	Limiting factor
FCA, 2 elements (el.)	1315	8.10	P_{low}	1660	10.30	P_{low}
FCA, 4 elements	1790	11.10	P_{low}	2240	13.60	P_{low}
FCA, 8 elements	2400	14.20	P_B	2780	15.70	GOS
LOLIA ($n = 7$), 2 el.	2310	14.30	P_{low}	2610	16.10	P_{low}
LOLIA ($n = 7$), 4 el.	2735	16.90	P_{low}	3035	18.65	P_{low}
LOLIA ($n = 7$), 8 el.	3155	19.45	P_{low}	>3200	>20.00	P_{low}
LOLIA ($n = 19$), 2 el.	1970	11.55	P_B	2110	11.95	P_B
LOLIA ($n = 19$), 4 el.	1990	11.65	P_B	2155	12.05	P_B
LOLIA ($n = 19$), 8 el.	2095	11.85	P_B	2220	12.20	P_B

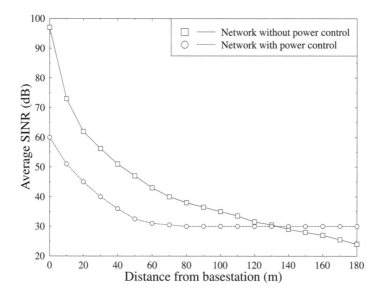

Figure 5.38: Signal-to-Interference plus Noise Ratio (SINR) versus mobile station distance measured from the base station, for networks with and without power control. The unnecessarily high SINR near the base station was a consequence of the base station's inability to power down below the minimum transmit power of −20 dBm, when the mobile station was within a distance of about 60 m from the base station.

Table 5.8: Simulation parameters for the FCA, and DCA-assisted networks using power control.

Parameter	Value	Parameter	Value
Noisefloor	−104 dBm	Multiple Access	F/TDMA
Frame duration	0.4615 ms	Cell radius	218 m
Max. BS transmit power	10 dBm	Maximum MS transmit power	10 dBm
Min. BS transmit power	−20 dBm	Minimum MS transmit power	−20 dBm
Power control stepsize	1 dB	Power control hysteresis	3 dB
Number of base stations	49	Handover hysteresis	2 dB
Outage SINR threshold	17 dB	Power control FCA target SINR	27 dB
Re-alloc. SINR threshold	21 dB	Pow. cont. LOLIA7 target SINR	31 dB
Number of timeslots	8	Number of carriers	7
Average inter-call-time	300 s	Max new-call queue-time	5 s
Average call length	60 s	Reference signal modulation	BPSK
Beamforming algorithm	SMI	Reference signal length	8 bits
MS speed	13.4 m/s	Number of antenna elements	2 &4
Pathloss at 1 m ref. point	0 dB	Pathloss exponent	−3.5
Geometry of antenna array	Linear	Array element spacing	$\lambda/2$
Modulation scheme	4-QAM	Channel/carrier bandwidth	200 kHz

Figure 5.39: New call blocking performance versus mean carried traffic, for comparison of the LOLIA, with 7 "local" base stations, and of FCA using a 7-cell reuse cluster, under a uniform geographic traffic distribution, with and without power control, for two and four element antenna arrays with beamforming in a multipath environment.

of antenna elements. At lower levels of traffic, the performance improvement obtained with the aid of power control was even higher, with the two element array results approaching those of the eight element array without power control. However, below approximately 10 Erlang/km^2/MHz a forced termination probability performance plateau was reached as a result of the power control algorithm limiting the maximum SINR. In contrast, when no power control is used and there are few users, the average SINR is very high and consequently fewer calls are dropped.

The performance gain of the LOLIA using power control is lower than that of the FCA algorithm, but still significant, since its performance is about halfway between that of the LOLIA without power control and using the same number of antenna elements, and that with twice the number of antenna elements.

The probability of low quality access of the PC-assisted scenario is shown in Figure 5.41. The corresponding curves for using no PC were plotted in Figure 5.33. The power controlled variant of the FCA algorithm offered a significantly reduced probability of low quality access for a given number of antenna elements. In fact, the probability of low quality access, when using power control and a two element adaptive antenna array, was lower than that when using a four element array without power control. The LOLIA also benefited to the same extent, with the probability of low quality access when using the power control algorithm equalling that obtained with the aid of twice the number of antenna elements and no power control.

The GOS illustrated in Figure 5.42 is related to the probability of low quality access by Equation 5.15, hence the close resemblance to Figure 5.41. However, it can be seen that the

Figure 5.40: Call dropping performance versus mean carried traffic, for comparison of the LOLIA, with 7 "local" base stations, and of FCA using a 7-cell reuse cluster, under a uniform geographic traffic distribution, with and without power control, for two and four element antenna arrays with beamforming in a multipath environment.

performance difference of the FCA algorithm using two and four element antenna arrays diminished as a result of their similar new call blocking performances, which dominate the GOS metric of Equation 5.15.

Figure 5.43 shows the mean number of handovers performed per call versus the mean carried teletraffic. From this figure it can be seen that the performance of the FCA algorithm was improved significantly as a result of using the power control algorithm. However, the FCA algorithm still required significantly more handovers per call for maintaining the desired call quality than the equivalent LOLIA based network without power control.

From the mean transmission power results of Figure 5.44 it can be seen that, as expected, the mean transmission power increased as the amount of teletraffic carried increased due to the higher levels of interference to be overcome. At high traffic loads the difference between the mean transmission powers of the mobile stations and the base stations, became more significant for the FCA algorithm. This resulted from the DL interfering base stations being, on average, farther away from the served mobile, than the interfering mobiles were from the serving base station on the UL. This was further exacerbated by the omni-directional nature of the mobiles' antennas and the directional nature of the antennas at the base stations. The LOLIA using a 7-cell exclusion zone required a higher mean transmission power than the FCA algorithm, which was attributed to the higher target SINR required by the LOLIA for maintaining an acceptable call dropping performance. When compared to the fixed transmission power of 10 dBm for an identical network operating without power control,

Figure 5.41: Probability of low quality access per call versus mean carried traffic, for comparison of the LOLIA, with 7 "local" base stations, and of FCA using a 7-cell reuse cluster, under a uniform geographic traffic distribution, with and without power control, for 2 and 4 element antenna arrays with beamforming in a multipath environment.

the reductions in transmitted power are significant, with a minimum average transmit power reduction of 6 dB, which substantially extends the mobile stations' battery lives.

Table 5.9 presents the summary of our results obtained for a network using power control in a multipath environment. The table shows that the use of power control has increased the number of users that may be serviced according to the required network performance criteria. The number of users supported by the network using the FCA algorithm increased by 28% to 70%, with a mean of 54% over the conservative and lenient scenarios. The capacity gains obtained with the aid of power control in a network using the LOLIA 7, however, were lower, namely between 9% and 15%, with a mean of almost 13% for both the conservative and lenient scenarios. Whilst the LOLIA 7 capacity gains are fairly modest, the overall call quality of the channel allocation techniques has improved for a given level of traffic, when compared to an identical network without power control.

5.6.2.4 Transmission over a Multipath Channel using Power Control and Adaptive Modulation

The idea behind adaptive modulation is to select a modulation mode according to the instantaneous radio channel quality [12, 13]. Thus, if the channel quality exhibits a high instantaneous SINR, then a high order modulation mode may be employed, enabling the exploitation of the temporarily high channel capacity. In contrast, if the channel has a low instantaneous SINR, using a high-order modulation mode would result in an unacceptable Frame Error Ratio (FER), and hence a more robust, but lower throughput modulation mode

Figure 5.42: Grade-Of-Service (GOS) performance versus mean carried traffic, for comparison of the LOLIA, with 7 "local" base stations, and of FCA using a 7-cell reuse cluster, under a uniform geographic traffic distribution, with and without power control, for two and four element antenna arrays with beamforming in a multipath environment.

would be invoked. Hence, adaptive modulation not only combats the effects of a poor quality channel, but also attempts to maximize the throughput, whilst maintaining a given target FER. Thus, there is a trade-off between the mean FER and the data throughput, which is governed by the modem mode switching thresholds. These switching thresholds define the SINRs, at which the channel is considered unsuitable for a given modulation mode, where an alternative AQAM mode must be invoked.

The power control algorithm invoked attempted to independently adjust the mobile and base station powers, such that the UL and DL SINRs were within a given target SINR window. The employment of a target window avoided constantly increasing and decreasing the transmission powers, which could lead to potential power control instabilities within the network. Furthermore, the affect of a range of different possible transmission powers is analogous to an inherent power control error plus slow fading envelope.

The combination of power control with adaptive modulation leads to several performance trade-offs, which must be considered when designing the power control and modulation mode switching algorithm. For example, the transmitted power could be minimized, which would result in either a high FER and a high throughput, or a low BER and a low throughput. Alternatively, the FER could be lowered even while maintaining a high throughput, when tolerating high transmission powers.

The power control and modulation mode switching algorithm invoked in our simulations attempted to minimize the transmitted power, whilst maintaining a high throughput with a

Figure 5.43: Mean number of handovers per call versus mean carried traffic, for comparison of the LOLIA, with 7 "local" base stations, and of FCA using a 7-cell reuse cluster, under a uniform geographic traffic distribution, with and without power control, for two and four element antenna arrays with beamforming in a multipath environment.

Figure 5.44: Mean transmission power versus mean carried traffic, of the LOLIA, with 7 "local" base stations, under a uniform geographic traffic distribution, with power control, for two and four element antenna arrays with beamforming in a multipath environment.

Table 5.9: Maximum mean carried traffic, and maximum number of mobile users that can be supported by each configuration whilst meeting the preset quality constraints of Section 5.3.3.4. The carried traffic is expressed in terms of normalized Erlangs (Erlang/km^2/MHz) for the network described in Table 5.8 both with and without power control in a *multipath environment*. The figures in bold indicate common network configurations to both the results without power and those with.

Algorithm	Conservative $P_{FT} = 1\%, P_{low} = 1\%$ $GOS = 4\%, P_B = 3\%$			Lenient $P_{FT} = 1\%, P_{low} = 2\%$ $GOS = 6\%, P_B = 5\%$		
	Users	Traffic	Limiting factor	Users	Traffic	Limiting factor
Without power cont.						
FCA, 2 elements (el.)	1315	8.10	P_{low}	1660	10.30	P_{low}
FCA, 4 elements	1790	11.10	P_{low}	2240	13.60	P_{low}
FCA, 8 elements	2400	14.20	P_B	2780	15.70	GOS
LOLIA ($n = 7$), 2 el.	2310	14.30	P_{low}	2610	16.10	P_{low}
LOLIA ($n = 7$), 4 el.	2735	16.90	P_{low}	3035	18.65	P_{low}
LOLIA ($n = 7$), 8 el.	3155	19.45	P_{low}	>3200	>20.00	P_{low}
With power cont.						
FCA, 2 elements	2260	13.30	P_{low}	2455	14.25	P_{FT}
FCA, 4 elements	2510	14.45	P_B	2870	15.95	P_B
LOLIA ($n = 7$), 2 el.	2665	16.30	P_{low}	2935	17.80	P_{low}
LOLIA ($n = 7$), 4 el.	3125	19.08	P_{low}	3295	20.42	P_{FT}

less than 5% target FER. The pseudo-code of the proposed algorithm is described in the next section.

5.6.2.5 Power Control and Adaptive Modulation Algorithm

```
determine lowest SINR out of UL and DL SINRs
if in 16-QAM mode
  if lowest SINR < 16-QAM drop SINR
    drop to 4-QAM mode
  else if lowest SINR < 16-QAM reallocation SINR
    if at maximum transmit power
      revert to 4-QAM
    else
      increase transmit power
  else if lowest SINR < 16-QAM target SINR
    if not at maximum power
      increase transmit power
  else if lowest SINR > 16-QAM target SINR+hysteresis
    decrease transmit power
```

```
else if in 4-QAM mode
  if lowest SINR < 4-QAM drop SINR
    drop to BPSK mode
  else if lowest SINR < 4-QAM reallocation SINR
    if at maximum transmit power
      revert to BPSK
    else
      increase transmit power
  else if lowest SINR < 4-QAM target SINR
    if not at maximum power
      increase transmit power
  else if lowest SINR > 16-QAM target SINR+hysteresis
    change to 16-QAM mode
  else if lowest SINR > 4-QAM target SINR+hysteresis
    if at maximum transmit power
      reduce transmit power
    else
      if lowest SINR > 16-QAM drop SINR
        change to 16-QAM
      else
        decrease transmit power
else if in BPSK
  if lowest SINR < BPSK drop SINR
    outage occurs
  else if lowest SINR < BPSK reallocation SINR
    if not at maximum transmit power
      increase transmit power
  else if lowest SINR > 4-QAM target SINR+hysteresis
    change to 4-QAM
  else if lowest SINR > BPSK target_hysteresis
    if at maximum transmit power
      reduce transmit power
    else
      change to 4-QAM
```

Figure 5.45 shows the flowchart of the AQAM and power control decision tree, when in the 4-QAM mode. The first step in the process is to determine the lower of the UL and the DL SINRs. The next step is to determine whether the BPSK modulation mode should be selected. When in the BPSK mode, outages may occur due to an insufficiently high SINR level and after a given number of BPSK outages the call is dropped. The conditions for this to occur are that either the lower SINR is below the 4-QAM call dropping threshold or that it is below the 4-QAM call reallocation threshold and currently the maximum possible transmission power is used. If the lower SINR is below the 4-QAM call reallocation threshold, or the SINR is below the 4-QAM target SINR, and the maximum transmission power has not been reached, then the transmit power is increased. However, if the SINR is below the 4-QAM target SINR and the maximum possible transmit power is currently used, then the modem remains in the 4-QAM

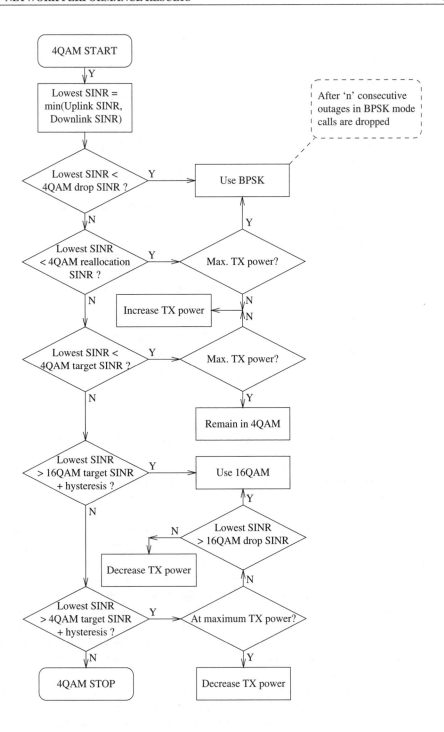

Figure 5.45: The AQAM and power control decision tree for the 4-QAM mode.

mode. The 16-QAM mode is chosen, if the SINR is higher than the 16-QAM target SINR, plus the associated hysteresis. Alternatively, the 16-QAM mode is invoked if the SINR is higher than the 4-QAM target SINR plus the hysteresis, furthermore the transmission power required to obtain this SINR is lower than the maximum transmit power, and the SINR is higher than the 16-QAM call dropping SINR. However, if the SINR is below the 16-QAM call dropping SINR or the maximum transmission power is in use, then the transmit power is reduced in an effort to keep the SINR in the 4-QAM mode's target SINR window. The improved SINR achieved using adaptive antenna arrays at the base station facilitates a higher mean network data throughput.

The FER was evaluated for approximately half-rate Bose-Chaudhuri-Hocquenghem (BCH) codes, which employed interleaving over the different number of bits conveyed by the different modem modes within a transmission frame [133]. The "Reallocation SINR" and the "Outage SINR" are defined as the average SINRs necessary for satisfying the 5% and 10% maximum FER constraints, respectively, using a given modulation mode such as BPSK, 4-QAM, or 16-QAM. The "Target SINR" was chosen so as to maximize the network capacity and represents an FER of approximately 2%.

The calculation of the receive antenna arrays weights was performed on a transmission frame-by-frame basis, leading to updated UL and DL SINRs every transmission frame. These SINR values were then used for selecting the modulation mode and transmission power to be employed, and for determining whether any channel re-allocation was necessary. Hence, frame-by-frame adaptive modulation, power control and dynamic channel allocation was jointly performed.

The system parameters for the network are defined in Table 5.10 and our performance results are provided in the next section.

5.6.2.6 Performance of PC-assisted, AQAM-aided Dynamic Channel Allocation

This section presents the simulation results obtained for a network using burst-by-burst adaptive modulation in order to improve the network's performance. Simulations were conducted for both a standard 7-cell FCA scheme and for the LOLIA using $n = 7$. The benchmark results obtained for a 4-QAM based network using power control were included for comparison purposes. Due to the enhanced network performance resulting from the employment of AQAM, a further constraint of a minimum throughput of 2 bits/symbol was invoked. This ensured a fair comparison with the fixed 4-QAM based network.

Figure 5.46 shows the new call blocking probability versus the mean normalized carried traffic. From this figure it can be seen that in conjunction with the LOLIA there are no blocked calls, except for the highest levels of traffic. In contrast, the performance of the FCA algorithm was degraded by using AQAM. This was the result of the limited availability of frequency/timeslot combinations restricting the achievable performance gain, since the reduced call dropping probability encouraged the prolonged utilization of the limited resources. This however, prevented new call setups.

The corresponding call dropping probability is depicted in Figure 5.47, which shows that when invoking adaptive modulation, the FCA algorithm performs better than the LOLIA below a traffic load of about 14 Erlangs/km²/MHz. Both channel allocation algorithms consistently offered a lower call dropping probability, when employing AQAM compared to when using the fixed-mode 4-QAM modulation scheme. This reduction in the call dropping

Table 5.10: Simulation parameters for the AQAM based network using power control.

Parameter	Value	Parameter	Value
Noisefloor	−104 dBm	Multiple Access	TDMA
Frame length	0.4615 ms	Cell radius	218 m
Min. BS transmit power	−20 dBm	Min. MS transmit power	−20 dBm
Max. BS transmit power	10 dBm	Max. MS transmit power	10 dBm
Power control stepsize	1 dB	Power control hysteresis	3 dB
BPSK outage SINR	13 dB	BPSK reallocation SINR	17 dB
BPSK target SINR	21 dB	4-QAM outage SINR	17 dB
4-QAM reallocation SINR	21 dB	4-QAM target SINR	27 dB
16-QAM outage SINR	24 dB	16-QAM reallocation SINR	27 dB
16-QAM target SINR	32 dB	Pathloss exponent	−3.5
Number of base stations	49	Handover hysteresis	2 dB
Number of timeslots/carrier	8	Number of carriers	7
Average inter-call-time	300 s	Max new-call queue-time	5 s
Average call length	60 s	Ref. signal modulation	BPSK
Beamforming algorithm	SMI	Reference signal length	8 bits
MS speed	30 mph	No. of antenna elements	2 & 4
Pathloss at 1 m ref. point	0 dB	Shadow fading	No
Geometry of ant. array	Linear	Array element spacing	$\lambda/2$
Channel/carrier bandwidth	200 kHz		

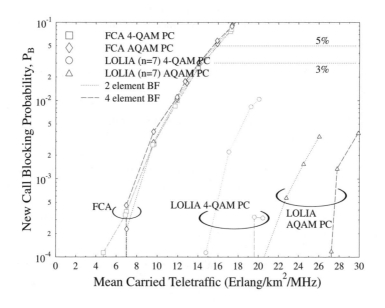

Figure 5.46: New call blocking probability versus mean carried traffic of the LOLIA, with 7 "local" base stations, and of FCA employing a 7-cell reuse cluster, for 2 and 4 element antenna arrays, with and without AQAM.

Figure 5.47: Call dropping or forced termination performance versus mean carried traffic of the LOLIA, with 7 "local" base stations, and of FCA employing a 7-cell reuse cluster, for two and four element antenna arrays, with and without AQAM.

rate using adaptive modulation was brought about by the inherent ability of the AQAM scheme to be reconfigured to a lower-order, and hence more interference resistant modulation mode, in order to prevent calls from being dropped.

Figure 5.48 shows that the probability of a low quality access was substantially reduced by AQAM for both the FCA scheme and the LOLIA. At lower traffic loads the probability of low quality outage was higher than when using the fixed 4-QAM modulation mode for both of the channel allocation schemes. This was due to the frequent use of the highest order modulation mode, 16-QAM, which was more susceptible to low quality outages. The more frequent usage of the 16-QAM mode by the four element adaptive antenna arrays also explains their greater probability of low quality outage at the lower traffic levels. However, as the traffic levels increased, the lower order modulation modes were invoked more frequently, and hence when combined with the four element arrays, the system guaranteed a lower probability of low quality outage than the two element arrays.

From Figure 5.49 it can be seen that the GOS of the FCA algorithm did not benefit from employing AQAM to the same extent as the LOLIA, except at the lower traffic levels when the new call blocking probability does not dominate the overall GOS performance. The LOLIA, however, benefited substantially, as we have also seen for the probability of low quality outages, since its performance was not constrained by its new call blocking probability observed in Figure 5.46 for both 4-QAM and AQAM.

The employment of AQAM, in Figure 5.50, reduced the mean number of handovers per call of the LOLIA at all traffic loads, and of the FCA for the highest traffic loads, although an increased number of handovers were performed by the FCA at lower traffic loads. At these lower traffic loads, more intra-cell handovers were performed by the FCA algorithm,

Figure 5.48: Probability of low quality access versus mean carried traffic of the LOLIA, with 7 "local" base stations, and of FCA employing a 7-cell reuse cluster, for two and four element antenna arrays, with and without AQAM.

Figure 5.49: GOS performance versus mean carried traffic of the LOLIA, with 7 "local" base stations, and of FCA employing a 7-cell reuse cluster, for two and four element antenna arrays, with and without AQAM.

Figure 5.50: Mean number of handovers per call versus mean carried traffic of the LOLIA, with 7 "local" base stations, and of FCA employing a 7-cell reuse cluster, for two and four element antenna arrays, with and without AQAM.

due to the employment of the 16-QAM modulation mode, which required more frequent intra-cell handovers in order to maintain a sufficiently high SINR. However, as the traffic load increased, the lower-order modulation modes were used more frequently, and hence less intra-cell handovers were required, leading to a reduction in the number of handovers performed.

The mean transmission power results of Figure 5.51 demonstrate how the employment of AQAM can reduce the power transmitted both for the UL and the DL. At low traffic load levels the FCA algorithm performed slightly worse in transmitted power terms, than the LOLIA. However, as the traffic loads increased, the gap became negligible when using two element antenna arrays. By contrast, when using four element antenna arrays, the LOLIA required a higher transmission power at these higher teletraffic loads. When compared to the fixed transmission power of 10 dBm for a network using no power control, the employment of AQAM resulted in a significant reduction of the mean transmission power. Specifically, the minimum reduction of the transmitted power was more than 4 dB and a maximum reduction of more than 7 dB was attained in addition to achieving a superior call quality and an increased mean modem throughput.

The average modem throughput expressed in bits per symbol versus the mean carried teletraffic is shown in Figure 5.52. The figure shows how the mean number of bits per symbol decreased as the network traffic increased. The FCA algorithm offered the lowest throughput with its performance degrading near-linearly upon increasing the network's traffic load. The LOLIA, especially for the lower levels of traffic, offered a greater modem throughput for a given level of teletraffic carried, with the achievable performance gracefully decreasing, as

Figure 5.51: Mean transmit power versus mean carried traffic of the LOLIA, with 7 "local" base stations, and of FCA employing a 7-cell reuse cluster, for two and four element antenna arrays, with and without AQAM.

Figure 5.52: Mean throughput in terms of bits per symbol versus mean carried traffic of the LOLIA, with 7 "local" base stations, and of FCA employing a 7-cell reuse cluster, for two and four element antenna arrays, in conjunction with AQAM.

Table 5.11: Mean modem throughput, when supporting the maximum mean carried traffic, whilst meeting the preset quality constraints of Section 5.3.3.4. The carried traffic is expressed in terms of normalized Erlangs (Erlang/km^2/MHz) for the network described in Table 5.10 in a multipath environment with AQAM.

	Conservative	Lenient
	$P_{FT} = 1\%$, $P_{low} = 1\%$ $GOS = 4\%$, $P_B = 3\%$ Bits per Symbol	$P_{FT} = 1\%$, $P_{low} = 2\%$ $GOS = 6\%$, $P_B = 5\%$ Bits per Symbol
FCA, 2 elements	2.8	2.7
FCA, 4 elements	3.1	2.9
LOLIA ($n = 7$), 2 elements	2.1	≈ 2.0
LOLIA ($n = 7$), 4 elements	2.15	2.05

the carried teletraffic continued to increase. Table 5.11 shows the mean modem throughput in bits per symbol, for the maximum mean carried traffic levels, whilst meeting the predefined quality constraints of Section 5.3.3.4.

From Table 5.12 it can be seen that it is the blocking performance of the network using the FCA algorithm which limits its associated network capacity, thus leading to a relatively high mean modem throughput at its user capacity limits. The increase in the modem throughput for the FCA algorithm varied from 35% to 55%, with corresponding user capacity improvements of 6% and −4%, when comparing the AQAM network to 4-QAM. The table also shows that the number of users supported by the FCA network using two element adaptive antenna arrays increased when using AQAM, which was restricted by the probability of low quality access when using 4-QAM. In contrast, when using four element antenna arrays, the network capacity was limited by the network's new call blocking performance. Hence, using AQAM techniques did not increase the number of users supported. In fact, due to the superior call dropping performance of AQAM, the new call blocking probability increased as a result of the lack of available frequency/timeslot combinations, and hence the number of users supported by the network decreased.

However, the dynamic nature of the LOLIA limited its fixed 4-QAM based network capacity due to its excessive low quality access probability, and thus in all cases, AQAM increased the number of users supported, by 38% to 50%, whilst meeting the required call quality criteria of Section 5.3.3.4. The AQAM-induced improvement in mean modem throughput varied from 0% to 7.5% as a result of the particular AQAM implementation used in the simulations. This can be further verified with the aid of Figure 5.51, which shows that the mean transmission powers were not at their maxima and hence both the modem throughput and the probability of low quality access were sub-optimal. In other words, had the AQAM algorithm been more aggressive in terms of its transmitted power usage, a reduced probability of low quality access and an increased mean modem throughput would have occurred. However, a trade-off existed where both the number of users supported and the mean modem throughput were increased, whilst achieving a significant reduction in the mean transmission powers.

Table 5.12: Maximum mean carried traffic, and maximum number of mobile users that can be supported by each configuration, whilst meeting the preset quality constraints of Section 5.3.3.4. The carried traffic is expressed in terms of normalized Erlangs (Erlang/km^2/MHz) for the network described in Table 5.10 in a multipath environment

Algorithm	Conservative $P_{FT} = 1\%, P_{low} = 1\%$ $GOS = 4\%, P_B = 3\%$			Lenient $P_{FT} = 1\%, P_{low} = 2\%$ $GOS = 6\%, P_B = 5\%$		
	Users	Traffic	Limiting factor	Users	Traffic	Limiting factor
4-QAM with PC						
FCA, 2 elements (el.)	2260	13.30	P_{low}	2455	14.25	P_{FT}
FCA, 4 elements	2510	14.45	P_B	2870	15.95	P_B
LOLIA ($n = 7$), 2 el.	2665	16.30	P_{low}	2935	17.80	P_{low}
LOLIA ($n = 7$), 4 el.	3125	19.08	P_{low}	3295	20.42	P_{FT}
AQAM with PC						
FCA, 2 elements	2400	14.00	P_B	2760	15.75	P_B
FCA, 4 elements	2400	14.10	P_B	2710	15.50	P_B
LOLIA ($n = 7$), 2 el.	3675	23.10	P_{low}	4115	25.4	P_{low}
LOLIA ($n = 7$), 4 el.	4460	27.40	P_{low}	4940	29.6	P_{low}

5.6.2.7 Summary of Non-wraparound Network Performance

The performance results summarized in this section can be gleaned from Tables 5.6–5.12. Specifically, in this section simulation results were obtained for a LOS scenario, for both the FCA algorithm and for the LOLIA, which showed that the FCA algorithm benefited the most from the employment of adaptive antenna arrays, with an increase of 144% in the number of users supported by four element antenna arrays. The corresponding figure was 67% with the aid of two element arrays. The performance of the LOLIA with a 19 base station constraint improved least using adaptive antenna arrays due to the inherently low interference levels present. However, for the LOLIA with a base station constraint of 7, using two element adaptive antenna arrays, an extra 22% additional users were supported with the desired performance metric limits of Section 5.3.3.4 observed. Using four element adaptive antenna arrays at the base stations led to an increase of 58% in the number of users supported.

Identical simulations with the addition of two multipath rays were then performed. These simulations demonstrated that the LOLIA 19 actually performed better in a multipath scenario, than in a LOS situation. This was due to the large reuse distance of the system, resulting in the sum of the powers of the three desired multipath signals versus the sum of the interfering signal powers being higher than the ratio of the LOS desired signal power to the interference power. The FCA algorithm, which offered the lowest network capacity in the LOS simulations, also suffered from the greatest capacity reduction in the multipath scenarios. The corresponding network capacities, expressed in terms of the number of users supported, decreased by between 3% and 17%. The number of users supported by the network using the LOLIA 7, was not significantly affected by the multipath propagation environment, with the highest reduction of almost 7% occurring using a four element antenna array employed in the conservative network scenario of Section 5.3.3.4. The FCA algorithm benefited the most from increasing the number of elements comprising the adaptive antenna arrays, with a minimum increase of 25% in the number of users supported upon doubling the number of antenna elements. The LOLIA employing a reuse cluster size of seven also performed well, with a user capacity increase of at least 15% for each doubling of the number of antenna elements.

Simulations were then performed in the multipath environment, where the network used the power control algorithm to maintain a fairly constant received SINR across the cell area. It was found that the power control algorithm increased the number of users carried in all the scenarios considered. The FCA algorithm exhibited the greatest gains in terms of the number of users supported by the network. When compared to an identical network without power control, the user capacity increased by 28%–72%, with an average increase of 47%. The LOLIA 7 using power control carried more traffic than the equivalent power control assisted FCA networks, and the LOLIA 7 system using no power control. When compared to the LOLIA 7 network using no power control, 9% to 15% more users were carried with a satisfactory performance. With respect to an FCA based network using power control, the increase in the number of supported users varied from 9% to almost 25%.

Further experiments were conducted in order to investigate the potential of AQAM techniques to increase network capacity. The gains achievable by the FCA algorithm were restricted by the number of available frequency/timeslot combinations, for both new calls and handovers, and hence the capacity increases were constrained by the new call blocking probability to 6%. However, this limitation to the number of supported users resulted in an increased mean modem throughput of between 2.7 and 3.1 bits per symbol, a reduced mean transmission power, and an overall improvement in call quality. The LOLIA, however, was not constrained by its new call blocking probability and was able to fully exploit the advantages of adaptive modulation. Thus, the LOLIA achieved a minimum network capacity increase of 38% over an identical scenario not using adaptive modulation.

The next section presents similar results but obtained using the "wraparound" technique in an effort to provide an effectively infinite simulation plane with, on average, constant interference levels present over the entire simulation area.

5.6.3   Wrap-around Network Performance Results

This section presents a range of performance results similar to those obtained in the previous section. However, in this section the "wrap-around" technique of Section 5.6.1 was used to generate results not subjected to the edge effects present at the perimeter of the simulation area. This process was described in Section 5.6.1. Results were obtained for the LOS propagation environment in Section 5.6.3.1 and for the multipath propagation environment of Section 5.6.3.2. Section 5.6.3.3 portrays the results obtained for the multipath propagation environment using power control, and Section 5.6.3.4 presents the network performance using adaptive modulation techniques.

Figure 5.53: New call blocking probability performance versus mean carried traffic, for the LOLIA using 7 and 19 "local" base stations, and for FCA employing a 7-cell reuse cluster, under uniform geographic traffic distribution, for a single antenna element, as well as for two and four element antenna arrays with beamforming in a LOS environment using wrap-around. See Figure 5.26 for the corresponding "desert-island" scenario.

5.6.3.1 Performance Results over a LOS Channel

Firstly we compared the FCA and the LOLIA under uniform geographic traffic distribution conditions using both a single antenna element and adaptive antenna arrays consisting of two and four elements in a LOS propagation environment. The FCA scheme employed a seven-cell reuse cluster, corresponding to one carrier frequency per base station. The LOLIA was used in conjunction with the constraints of seven and nineteen nearest base stations, i.e., $n = 7$ or 19.

As seen in Figure 5.53 the LOLIA using $n = 19$ offered the worst call blocking performance of the three channel allocation schemes, with the AAAs having little beneficial effect. This demonstrated that the limiting factor was not inadequate signal quality for a call to be setup, but the lack of available frequency/timeslot combinations due to the large exclusion zone. The FCA algorithm benefited only to a limited extent from the employment of the AAAs, suggesting that the majority of the blocked calls were as a result of the limited availability of frequency/timeslot combinations. Inadequate signal quality caused the remainder of the blocked calls. The call blocking performance of the LOLIA using $n = 7$ appeared mainly to be interference limited, hence the AAAs guaranteed a significant reduction of the number of blocked calls, particularly for mean carried traffic levels in excess of 9 Erlang/km^2/MHz.

Figure 5.54 shows that—as expected—the FCA algorithm performed the least satisfactorily of the three channel allocation schemes investigated with respect to its call dropping

Figure 5.54: Dropping probability performance versus mean carried traffic, for the LOLIA using 7 and 19 "local" base stations, and for FCA employing a 7-cell reuse cluster, under a uniform geographic traffic distribution, for a single antenna element, as well as for two and four element antenna arrays with beamforming in a LOS environment using wrap-around. See Figure 5.27 for the corresponding "desert-island" scenario.

performance. Even in conjunction with a four-element adaptive antenna array, it exhibited a higher call dropping rate than that of either of the LOLIAs ($n = 19$ and $n = 7$). The large exclusion zone of the LOLIA using $n = 19$ led to a low dropping probability of less than 1×10^{-3} for teletraffic loads below approximately 12 Erlang/km^2/MHz. However, the rapid rise in the call dropping probability upon increasing the teletraffic became unacceptable for teletraffic loads in excess of about 13 Erlang/km^2/MHz. The large exclusion zone of the algorithm prevented from handovers occurring, since there were no free channels available in the vicinity, hence resulting in a high number of dropped calls. Thus, for $n = 19$ the employment of adaptive antenna arrays at the base stations did not improve the performance significantly, unlike for the FCA and LOLIA using $n = 7$, which were predominantly interference limited. The call dropping performance of the LOLIA using $n = 7$ benefited the most from the assistance of adaptive antenna arrays, with the most dramatic gains in call dropping performance at the higher teletraffic levels.

Figures 5.55 and 5.56 show the probability of low quality access and the GOS, which are similar in terms of their trends and are closely related to each other by Equation 5.15. The GOS of the FCA algorithm was dominated by the probability of low quality access, since it had a higher value than the blocking probability. However, the rapid rise of the new call blocking probability of the LOLIA with $n = 19$ caused a steep increase in its GOS, especially when coupled with its rapidly degrading probability of low quality access. All of the algorithms benefited substantially from the employment of adaptive antenna arrays.

Figure 5.55: GOS performance versus mean carried traffic, for the LOLIA using 7 and 19 "local" base
stations, and for FCA employing a 7-cell reuse cluster, under a uniform geographic traffic
distribution, for a single antenna element, as well as for two and four element antenna
arrays with beamforming in a LOS environment using wrap-around. See Figure 5.29 for
the corresponding "desert-island" scenario.

The effect of beamforming on the number of handovers performed can be seen in
Figure 5.57. The LOLIAs required the least frequent handovers, with beamforming barely
altering the results. In contrast, the number of handovers performed when using the FCA
algorithm was reduced significantly due to employing AAAs with a maximum reduction of
72% for two elements, and of 89% for four elements. This translates into a significantly
reduced signaling load for the network, since it has to manage far less handovers, therefore
reducing the complexity of the network infrastructure.

It can be seen from Table 5.13 that in a LOS environment all of the channel allocation
schemes benefit from the use of base station AAAs in terms of an increased level of teletraffic
carried, hence supporting an increased number of users. The FCA algorithm benefited most
from the employment of AAAs, with a 160% increase in terms of the number of users
supported, when using a four-element antenna array. The performance improvements of
the LOLIA in conjunction with $n = 7$ due to using AAAs were more modest than for
the FCA system. Specifically, 44% more users were supported by the four element AAA-
assisted LOLIA using $n = 7$, when compared to the single antenna element based results.
The network capacity of the LOLIA along with a 19-cell exclusion zone was higher than that
of the LOLIA using $n = 7$, until the limited number of channels available in conjunction with
such a large exclusion zone became significant. Up to this point, the AAAs reduced the levels
of interference, thus improving the network capacity. However, when using a four-element
AAA, the new call blocking probability became the dominant network performance limiting
factor.

Figure 5.56: Probability of low quality access performance versus mean carried traffic, for the LOLIA using 7 and 19 "local" base stations, and for FCA employing a 7-cell reuse cluster, under a uniform geographic traffic distribution, for a single antenna element, as well as for two and four element antenna arrays with beamforming in a LOS environment using wrap-around. See Figure 5.28 for the corresponding "desert-island" scenario.

Table 5.13: Maximum mean carried traffic, and maximum number of mobile users that can be supported by each configuration, whilst meeting the preset quality constraints defined in Section 5.3.3.4. The carried traffic is expressed in terms of normalized Erlangs (Erlang/km^2/MHz), for the network described in Table 5.4 in a LOS environment using wrap-around. See Table 5.6 for the corresponding "desert-island" results.

Algorithm	Conservative $P_{FT} = 1\%$, $P_{low} = 1\%$ $P_B = 3\%$, $GOS = 4\%$			Lenient $P_{FT} = 1\%$, $P_{low} = 2\%$ $P_B = 5\%$, $GOS = 6\%$		
	Users	Traffic	Limiting factor	Users	Traffic	Limiting factor
FCA, 1 element (el.)	340	3.6	P_{low}	465	4.9	P_{low}
FCA, 2 elements	575	6.1	P_{low}	755	7.9	P_{low}
FCA, 4 elements	885	9.3	P_{low}	1105	11.2	P_{FT}
LOLIA ($n = 7$), 1 el.	990	10.5	P_{low}	1065	11.45	P_{low}
LOLIA ($n = 7$), 2 el.	1155	12.35	P_{low}	1260	13.5	P_{low}
LOLIA ($n = 7$), 4 el.	1420	14.9	P_{low}	1535	16.5	P_{low}
LOLIA ($n = 19$), 1 el.	1020	10.9	P_{low}	1090	11.6	P_{low}
LOLIA ($n = 19$), 2 el.	1200	12.5	P_{low}	1330	13.35	P_{low}
LOLIA ($n = 19$), 4 el.	1335	13.45	P_B	1400	13.9	P_B

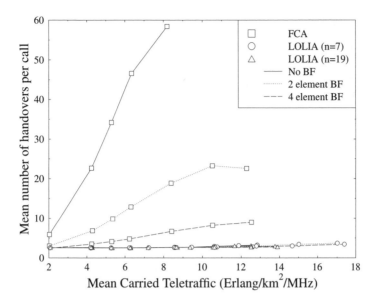

Figure 5.57: Mean number of handovers per call versus mean carried traffic, for comparison of the LOLIA using 7 and 19 "local" base stations, and for FCA employing a 7-cell reuse cluster, under a uniform geographic traffic distribution, for a single antenna element, as well as for two and four element antenna arrays with beamforming in a LOS environment using wrap-around. See Figure 5.30 for the corresponding "desert-island" scenario.

5.6.3.2 Performance Results over a Multipath Channel

Following our previous experiments, where a purely LOS environment existed between the mobiles and their base stations, this section presents results for a multipath environment using two-, four- and eight-element AAAs.

Comparing the call blocking probabilities of the multipath environment, shown in Figure 5.58, with those of the LOS environment, shown in Figure 5.53, reveals that all of the channel allocation algorithms behave similarly for both radio environments. The FCA scheme actually behaved more unfavorably in terms of its new call blocking probability, as the number of AAA elements was increased. However, this is a consequence of the additional antenna elements improving the other performance measures, such as the call dropping rate. This enabled additional calls to be sustained at a given time, leading to a higher call blocking rate. In conjunction with an exclusion zone of 19 cells we found that the LOLIA's blocking performance was barely affected by the adaptive antenna arrays, whilst for $n = 7$ the blocked call rate was improved by a factor of 10 at a traffic load of 14–17 Erlang/km^2/MHz.

Figure 5.59 shows the probability of a dropped call in a multipath propagation environment, which was slightly higher than for the LOS scenario of Figure 5.54, when considered in the context of a given channel allocation algorithm and for a given antenna array size. The call dropping rate was improved with the aid of adaptive antenna arrays for all of the channel allocation algorithms, though the LOLIA using $n = 19$ did not benefit to the same extent as the other algorithms.

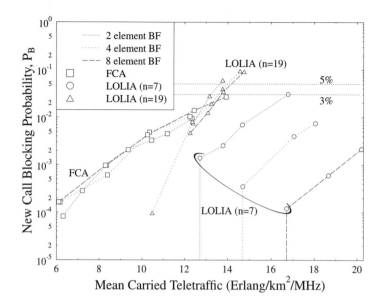

Figure 5.58: New call blocking probability performance versus mean carried traffic, for comparison of the LOLIA using 7 and 19 "local" base stations, and for FCA using a 7-cell reuse cluster, under a uniform geographic traffic distribution, for two, four and eight element antenna arrays with beamforming in a multipath environment using wrap-around. See Figure 5.32 for the corresponding "desert-island" scenario.

Again, as expected, the GOS curves in Figure 5.60 and the probability of low quality access curves of Figure 5.61 are similar in shape, with the differences resulting from the blocked call probability according to Equation 5.15. Hence, the GOS of the LOLIA having an exclusion zone of 19 base stations increases more rapidly than its probability of low quality access. In addition, the gain in its low quality of access performance achieved by using the adaptive antenna arrays is reduced, in terms of the GOS, due to the limited blocking probability improvement offered by the adaptive antenna arrays. All three algorithms benefit significantly in terms of their low quality access performance from the employment of the adaptive antenna arrays. However, the significant blocking performance limitations of the LOLIA using $n = 19$ restricts its GOS performance gains.

Figure 5.62 demonstrates the significant impact that adaptive antenna arrays have on the mean number of handovers per call for the FCA algorithm in a multipath environment. Even in conjunction with adaptive antenna arrays more handovers per call were invoked, when using the FCA system, than for either of the LOLIAs using a single antenna element. Furthermore, a higher number of handovers was required in the multipath environment, than in the LOS scenario, for a given size of adaptive antenna array. The LOLIAs required significantly fewer handovers than the FCA, irrespective of the propagation environment, and did not benefit from the employment of adaptive antenna arrays in terms of the required handovers per call.

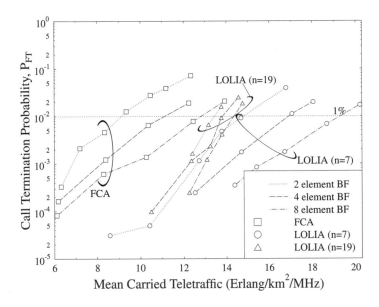

Figure 5.59: Call dropping probability performance versus mean carried traffic, for comparison of the LOLIA using 7 and 19 "local" base stations, and for FCA using a 7-cell reuse cluster, under a uniform geographic traffic distribution, for two, four and eight element antenna arrays with beamforming in a multipath environment using wrap-around. See Figure 5.33 for the corresponding "desert-island" scenario.

Table 5.14: Maximum mean carried traffic, and maximum number of mobile users that can be supported by each configuration, whilst meeting the preset quality constraints defined in Section 5.3.3.4. The carried traffic is expressed in terms of normalized Erlangs (Erlang/km^2/MHz), for the network described in Table 5.4 in a multipath environment using wrap-around. See Table 5.7 for the corresponding "desert-island" results.

Algorithm	Conservative $P_{FT} = 1\%, P_{low} = 1\%$ $P_B = 3\%, GOS = 4\%$			Lenient $P_{FT} = 1\%, P_{low} = 2\%$ $P_B = 5\%, GOS = 6\%$		
	Users	Traffic	Limiting factor	Users	Traffic	Limiting factor
FCA, 2 element (el.)	600	6.0	P_{low}	740	7.65	P_{low}
FCA, 4 elements	790	8.3	P_{low}	995	10.3	P_{low}
FCA, 8 elements	1085	11.2	P_{low}	1250	12.8	P_{FT}
LOLIA ($n = 7$), 2 el.	1195	12.65	P_{low}	1290	13.7	P_{low}
LOLIA ($n = 7$), 4 el.	1370	14.35	P_{low}	1475	15.6	P_{low}
LOLIA ($n = 7$), 8 el.	1555	16.15	P_{low}	1700	17.7	P_{low}
LOLIA ($n = 19$), 2 el.	1235	12.65	P_{low}	1325	13.3	P_{low}
LOLIA ($n = 19$), 4 el.	1360	13.55	P_B	1410	13.8	P_{FT}
LOLIA ($n = 19$), 8 el.	1385	13.7	P_B	1475	14.15	P_B

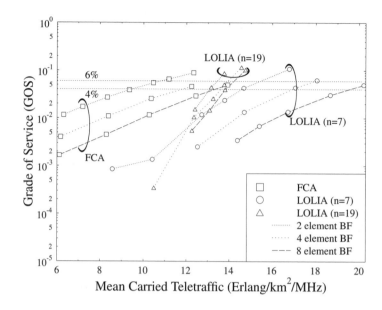

Figure 5.60: GOS performance versus mean carried traffic, for the comparison of the LOLIA with 7 and 19 "local" base stations, and for FCA using a 7-cell reuse cluster, under a uniform geographic traffic distribution, for two, four and eight element antenna arrays with beamforming in a multipath environment using wrap-around. See Figure 5.35 for the corresponding "desert-island" scenario.

Table 5.14 presents results similar to those in Table 5.13, but for a multipath environment, with the bold values highlighting the adaptive antenna array sizes common to both sets of investigations. From this table it can be seen that the LOLIA using $n = 19$ carries approximately the same amount of traffic in the multipath scenario, which translates into a similar network capacity to that of the LOS scenario of Table 5.13. Again, the number of users supported by the network is limited by the probability of a low quality access and by the new call blocking probability. The performance of the LOLIA using $n = 7$ was interference limited, where the smaller reuse distance or exclusion zone led to numerous sources of relatively strong interference, all requiring interference cancellation. Hence, as the number of adaptive antenna array elements increased, so did the number of users supported, with an average improvement of about 15% for each doubling of the number of array elements.

5.6.3.3   Performance over a Multipath Channel using Power Control

This section presents results obtained using the same wrap-around scenario of Section 5.6.1 over a multipath channel using power control. The power control algorithm was the same as that described in Section 5.6.2.3. The power control algorithm implemented attempted to independently adjust the mobile and base station transmit powers, such that the UL and DL SINRs were within a given target SINR window. The use of a target SINR window allowed us to avoid constantly increasing and decreasing the transmission powers, which could lead to potential power control instabilities within the network. Furthermore, the effect of employing

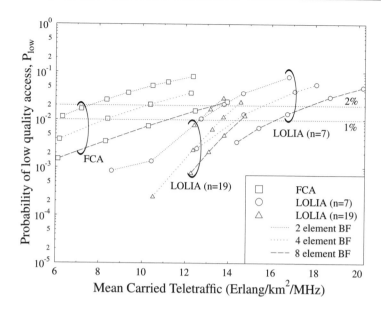

Figure 5.61: Probability of low quality access performance versus mean carried traffic, for the comparison of the LOLIA using 7 and 19 "local" base stations, and for FCA using a 7-cell reuse cluster, under a uniform geographic traffic distribution, for two, four and eight element antenna arrays with beamforming in a multipath environment using wrap-around. See Figure 5.34 for the corresponding "desert-island" scenario.

a range of possible transmission powers is analogous to an inherent power control error plus slow fading phenomenon.

Figure 5.63 portrays the new call blocking probability versus the mean normalized carried traffic, expressed in terms of Erlangs/km^2/MHz. The figure shows that using power control in conjunction with the FCA algorithm resulted in a slight increase in the new call blocking probability as a direct consequence of the improved call dropping probability shown in Figure 5.64. In contrast, the blocking probability of the LOLIA improved significantly due to using power control, achieving a reduction by a factor of 4 to 34.

The new call blocking performance of the LOLIA was superior to that of the FCA algorithm both with and without power control, as seen in Figure 5.63, which is a result of the dynamic nature of the LOLIA. This enabled the LOLIA to allocate any of the available channels not used within the 7-cell exclusion zone (maximum of $7 \times 8 = 56$ channels in this scenario) to a new call request. However, the FCA algorithm only had one carrier frequency per base station, and therefore was less likely to be able to satisfy a new call request. The addition of power control to the LOLIA in conjunction with $n = 7$ led to a reduced new call blocking probability. Specifically, the new call blocking probability with power control was reduced to near that achieved using twice the number of antenna elements without power control. The higher new call blocking probability of the network using no power control can be attributed to the lower average SINR values, which prevent new call initiation, whereas the

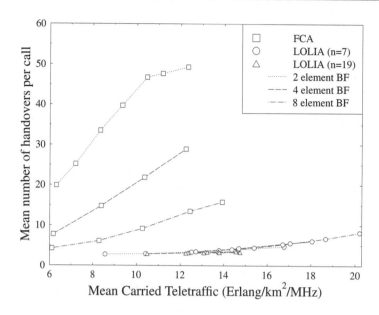

Figure 5.62: Mean number of handovers per call versus mean carried traffic, for comparison of the LOLIA using 7 and 19 "local" base stations, and for FCA using a 7-cell reuse cluster, under a uniform geographic traffic distribution, for two, four and eight element antenna arrays with beamforming in a multipath environment using wrap-around. See Figure 5.36 for the corresponding "desert-island" scenario.

higher average SINR level of the network observed in Figure 5.38 in conjunction with power control enables additional calls to commence.

Figure 5.64 shows that the call dropping probability was significantly reduced for both the FCA algorithm and the LOLIA using $n = 7$, in conjunction with power control. The FCA algorithm in conjunction with power control offered a call dropping probability close to that of a similar network without power control, and using twice the number of adaptive antenna elements. However, at traffic loads of below approximately 7 Erlangs/km^2/MHz the call dropping probability began to level off for the FCA algorithm. This phenomenon was also noticeable in the context of the LOLIA and resulted from the power control algorithm limiting the maximum SINR, leading to a flatter call dropping profile than that of the network without power control. Thus, at lower traffic loads the network without power control had a higher average SINR as was evidenced by Figure 5.38, leading to less dropped calls. However, at higher levels of teletraffic the power control algorithm offered a lower call dropping rate, as a consequence of the lower levels of interference present when using the power control scheme.

The FCA algorithm exhibited the greatest improvement in the probability of a low quality access due to the implementation of power control, as shown in Figure 5.65. Using a two element adaptive antenna array in conjunction with the power control algorithm resulted in a probability of low quality access approximately equal to that obtained using an eight element adaptive array without power control. The LOLIA also benefited from invoking the power

Figure 5.63: New call blocking probability versus mean carried traffic of the LOLIA using 7 "local" base stations, and for FCA employing a 7-cell reuse cluster, for two and four element antenna arrays, with and without power control using wrap-around. See Figure 5.39 for the corresponding "desert-island" scenario.

Figure 5.64: Call dropping probability versus mean carried traffic of the LOLIA using 7 "local" base stations, and for FCA employing a 7-cell reuse cluster, for two and four element antenna arrays, with and without power control using wrap-around. See Figure 5.40 for the corresponding "desert-island" scenario.

Figure 5.65: Probability of low quality outage versus mean carried traffic of the LOLIA using 7 "local" base stations, and for FCA employing a 7-cell reuse cluster, for two and four element antenna arrays, with and without power control using wrap-around. See Figure 5.41 for the corresponding "desert-island" scenario.

control algorithm, but to a lesser extent, offering a performance close to that of an array with twice the number of elements without power control.

The GOS performance gains of the FCA algorithm using power control seen in Figure 5.66, were somewhat reduced compared to those of the probability of a low quality access in Figure 5.65, due to the similar blocking performances of the power-controlled and non-power-controlled scenarios seen in Figure 5.63. Nonetheless, the GOS gains remained quite high in Figure 5.66. The GOS gains of the LOLIA due to power control were also quite substantial, as seen in Figure 5.66.

The effect of power control on the mean number of handovers performed per call becomes explicit in Figure 5.67. From this figure it can be seen for the FCA algorithm that with respect to the number of handovers per call, the performance of the network employing power control significantly exceeded that of the network without power control using an adaptive antenna array of twice the number of antenna elements. The employment of power control in conjunction with the FCA algorithm led to a mean reduction by a factor of 4.4 in the number of handovers. The inherently good performance of the LOLIA was also slightly improved on average.

A further advantage of using power control in a cellular mobile network is portrayed in Figure 5.68, which shows that the mean transmit power was reduced from the fixed transmit power of 10 dBm due to power control. The mean transmit power of the FCA algorithm was reduced the most with reductions varying from 4.5 dB to almost 9 dB at the lowest traffic levels. Doubling the number of antenna elements comprising the base stations' adaptive antenna arrays from two to four, resulted in additional mean transmission power gains of

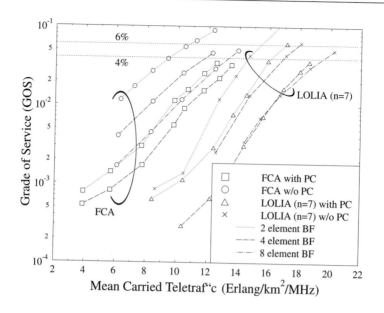

Figure 5.66: GOS performance versus mean carried traffic of the LOLIA using 7 "local" base stations, and for FCA employing a 7-cell reuse cluster, for two and four element antenna arrays, with and without power control using wrap-around. See Figure 5.42 for the corresponding "desert-island" scenario.

Figure 5.67: Mean number of handovers per call versus mean carried traffic of the LOLIA using 7 "local" base stations, and for FCA employing a 7-cell reuse cluster, for two and four element antenna arrays, with and without power control using wrap-around. See Figure 5.43 for the corresponding "desert-island" scenario.

Figure 5.68: Mean transmit power versus mean carried traffic of the LOLIA using 7 "local" base stations, and for FCA employing a 7-cell reuse cluster, for two and four element antenna arrays, with and without power control using wrap-around. See Figure 5.44 for the corresponding "desert-island" scenario.

almost 1 dB at higher traffic loads, which is a consequence of the extra interference rejection capability of the four element array. The mean transmission powers of the LOLIAs were significantly higher due to the higher target SINRs required for maintaining an acceptable call dropping rate. This was a consequence of the dynamic nature of the LOLIA, leading to more rapidly changing interference levels, which required a relatively high target SINR of 31 dB as seen in Table 5.8.

Table 5.15 presents similar results to Table 5.14, but using our power control algorithm, with the bold values highlighting the adaptive antenna array sizes common to both sets of investigations, for the sake of convenient comparison. The table shows the significant performance improvement obtained for both the LOLIA and the FCA algorithm in terms of the number of users supported with the advent of power control, whilst maintaining the desired network quality. In the conservative scenario, for example, the FCA algorithm using a two element adaptive antenna array and power control supported the same number of users as the network using an eight element adaptive antenna array without power control. The LOLIA-based network, however, did not benefit from the employment of the power control algorithm to the same extent, although it still offered similar performance to that of a network without power control and using adaptive antenna arrays having twice the number of antenna elements.

Table 5.15: Maximum mean carried traffic, and maximum number of mobile users that can be supported by each configuration, whilst meeting the preset quality constraints defined in Section 5.3.3.4. The carried traffic is expressed in terms of normalized Erlangs (Erlang/km^2/MHz), for the network described in Table 5.4 in a multipath environment with and without power control using wrap-around.

Algorithm	Conservative $P_{FT} = 1\%, P_{low} = 1\%$ $P_B = 3\%, GOS = 4\%$			Lenient $P_{FT} = 1\%, P_{low} = 2\%$ $P_B = 5\%, GOS = 6\%$		
	Users	Traffic	Limiting factor	Users	Traffic	Limiting factor
4-QAM without PC						
FCA, 2 element (el.)	600	6.0	P_{low}	740	7.65	P_{low}
FCA, 4 elements	790	8.3	P_{low}	995	10.3	P_{low}
FCA, 8 elements	1085	11.2	P_{low}	1250	12.8	P_{FT}
LOLIA ($n = 7$), 2 el.	1195	12.65	P_{low}	1290	13.7	P_{low}
LOLIA ($n = 7$), 4 el.	1370	14.35	P_{low}	1475	15.6	P_{low}
LOLIA ($n = 7$), 8 el.	1555	16.15	P_{low}	1700	17.7	P_{low}
4-QAM with PC						
FCA, 2 elements (el.)	1090	10.6	P_{low}	1120	10.85	P_{FT}
FCA, 4 elements	1370	13.28	P_{FT}	1370	13.28	P_{FT}
LOLIA ($n = 7$), 2 el.	1350	14.05	P_{low}	1445	15.1	P_{low}
LOLIA ($n = 7$), 4 el.	1540	16.15	P_{low}	1640	17.35	P_{low}

5.6.3.4 Performance of an AQAM based Network using Power Control

This section presents our simulation results obtained for a network using burst-by-burst adaptive modulation [13, 209, 398, 399] invoked in order to improve the network's performance. Simulations were conducted for both a standard 7-cell FCA scheme and a 7-cell LOLIA assisted system. The results obtained for a 4-QAM based network using power control were included for comparison purposes.

The new call blocking probability depicted in Figure 5.69 was essentially unchanged for the FCA algorithm using power control in conjunction with 4-QAM or AQAM, suggesting that the new call blocking performance of the FCA algorithm was limited by the lack of available frequency/timeslot combinations, rather than by inadequate signal quality. This hypothesis was confirmed by the improvement in the new call blocking performance of the LOLIA resulting from the superior signal quality of AQAM.

The corresponding call dropping probability is depicted in Figure 5.70. The AQAM LOLIA using $n = 7$ in conjunction with a two element adaptive antenna had, in general, a higher call dropping probability compared to that of power control assisted 4-QAM. However, the power control algorithm, when used in conjunction with AQAM, maintained the call dropping probability below the given threshold for a significantly higher traffic load. Similar performance trends were observed for both the two element and the four element adaptive array, although the higher interference rejection capability offered by the four element array resulted in a substantially reduced call dropping probability. The dropped calls were caused

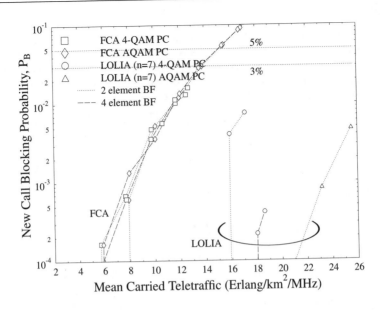

Figure 5.69: New call blocking probability versus mean carried traffic of the LOLIA using 7 "local" base stations, and for FCA employing a 7-cell reuse cluster, for two and four element antenna arrays, with and without AQAM using wrap-around. See Figure 5.46 for the corresponding "desert-island" scenario.

Figure 5.70: Call dropping, or forced termination, performance versus mean carried traffic of the LOLIA using 7 "local" base stations, and for FCA employing a 7-cell reuse cluster, for two and four element antenna arrays, with and without AQAM using wrap-around. See Figure 5.47 for the corresponding "desert-island" scenario.

Figure 5.71: Mean number of handovers per call versus mean carried traffic of the LOLIA using 7 "local" base stations, and for FCA employing a 7-cell reuse cluster, for two and four element antenna arrays, with and without AQAM using wrap-around. See Figure 5.50 for the corresponding "desert-island" scenario.

almost exclusively by insufficient signal quality during the intra-cell handover process, thus increasing the number of adaptive antenna elements from two to four improved the call dropping performance. The high call dropping probability observed for traffic loads between 12 and 20 Erlangs/km²/MHz when using the two element adaptive antenna array was due to the power control and AQAM attempting to trade-off modem throughput and transmit power against each other, whilst attempting to minimize the number of dropped calls. The extra interference suppression capability of the four-element adaptive antenna array led to a reduced call dropping probability. Hence, altering the AQAM mode selection algorithm of Figure 5.45, may improve its performance at these traffic loads, when used in conjunction with a two element antenna array.

The FCA algorithm dropped all of its calls during the inter-cell handover process due to the lack of available slots to handover to. However, since inter-cell handovers could be performed, if necessary, in order to improve the signal quality, the number of dropped calls was reduced when using the four element adaptive array, due to its better interference rejection capability. All the calls were dropped during the inter-cell handover process, which means that no calls were dropped due to insufficient SINR or through the intra-cell handover process. This can be attributed to the AQAM scheme, which enabled users to drop to lower order modulation modes of the AQAM scheme, when the SINR became poor.

Figure 5.71 characterizes the mean number of handovers per call for 4-QAM and AQAM, both using power control. The LOLIA using $n = 7$ performed a lower total number of handovers per call, when using AQAM, due its inherent resilience to poor signal quality conditions.

Figure 5.72: Mean number of inter-cell and intra-cell handovers per call versus mean carried traffic of the LOLIA using 7 "local" base stations, and for FCA employing a 7-cell reuse cluster, for two and four element antenna arrays, in conjunction with AQAM using wrap-around.

The breakdown of the handovers into inter-cell and intra-cell handovers is given in Figure 5.72. Observe that the improved interference rejection capability, and the associated superior SINR of the four-element array results in a lower number of intra-cell handovers for the LOLIA. Since the intra-cell handover process is the primary cause of dropped calls and less intra-cell handovers are performed when using a four-element antenna, more inter-cell handovers are necessitated in the network using four-element adaptive antenna arrays, as the users roam from cell to cell. In other words, since the LOLIA using a four element array drops less calls than when using a two element array, more users are in call at a given time, and hence these users cross more cell boundaries, thus necessitating more inter-cell handovers.

In contrast, the number of intra-cell handovers performed in conjunction with the FCA algorithm decreases, as the teletraffic rises, and as the number of antenna elements is increased from two to four. This is a consequence of the particular implementation of the modulation mode selection/power control algorithm and its interaction with the FCA handover process. The AQAM algorithm attempts to remain in the current modulation mode as long as possible, and hence as the SINR degrades, it will opt for performing an intra-cell handover in an attempt to maintain the SINR, rather than reconfiguring itself in order to use a lower-order modulation mode suitable for the reduced SINR level. Thus, when using a four-element adaptive antenna array, the average (and instantaneous) SINR is typically higher than that of a two-element array, leading to a more frequent employment of the less resilient higher-order modulation modes, which potentially requires additional intra-cell handovers. However, as the mean teletraffic increases, so does the level of interference in the network and a greater proportion of transmission time is spent in the lower-order modulation modes, thus requiring less intra-cell handovers, as illustrated in Figure 5.72.

Figure 5.73: Probability of low quality access versus mean carried traffic of the LOLIA using 7 "local" base stations, and for FCA employing a 7-cell reuse cluster, for two and four element antenna arrays, with and without AQAM using wrap-around. See Figure 5.48 for the corresponding "desert-island" scenario.

The probability of a Low Quality (LQ) access is depicted in Figure 5.73, showing an interesting interaction between the FCA algorithm and the AQAM scheme. The probability of an LQ access occurring is reduced, as the traffic level increases and the number of antenna elements is decreased. This can be attributed to the less frequent usage of the higher-order modulation modes at the higher traffic loads. Hence the lower-order modulation modes are used more frequently and thus the chance of an LQ access taking place is reduced. The four-element adaptive antenna array leads to a higher probability of a low quality access, since its higher associated SINR levels activate a more frequent employment of the less robust, but higher-throughput, higher-order modulation modes. For example, let us consider the FCA AQAM PC scenario supporting 400 users, which corresponded to a traffic load of about 4 Erlang/km^2/MHz. When using two antenna array elements, 85% of the LQ accesses occurred whilst in the 16-QAM mode, however, on increasing the number of antenna array elements to four this rose to 93%.

However, as the network loading rises, an increasing proportion of the LQ outages occur in the BPSK modulation mode. Coupled with the increase in the BPSK modulation mode's employment due to the low SINR constraints, the probability of a low quality outage is expected to increase at a certain traffic load. This can be seen in Figure 5.73, where the LQ outage probability is starting to rise for FCA in conjunction with both two and four elements, though the extra interference suppression capability of the four element array allows extra traffic to be carried, before this phenomenon commences. More specifically, although not explicit in Figure 5.73, we found that for a network supporting 1200 users, corresponding to a traffic load of about 12 Erlang/km^2/MHz, and employing two element adaptive antenna arrays, 43% of the LQ accesses occurred, whilst in the 16-QAM mode, versus 72% with four-

Figure 5.74: GOS performance versus mean carried traffic of the LOLIA using 7 "local" base stations, and for FCA employing a 7-cell reuse cluster, for two and four element antenna arrays, with and without AQAM using wrap-around. See Figure 5.49 for the corresponding "desert-island" scenario.

element antenna arrays. Again, not explicitly shown in the figure, but increasing the number of users to 1400, or a traffic load of just less than 14 Erlang/km^2/MHz, reduced the number of 16-QAM LQ accesses, but increased the BPSK LQ outages to 69% and 31% for the two- and four-element arrays respectively, with reductions to 21% and 53% of the LQ outages in the 16-QAM mode.

From Figure 5.74 it can be seen that the GOS, as defined in Section 5.3.3.4, of the FCA algorithm did not benefit from invoking AQAM to the same extent as the LOLIA. This resulted from the fairly similar probability of low quality access performance of the two and four element antenna array assisted systems in Figure 5.73, and the limiting blocking performance observed in Figure 5.69. However, since the LOLIA did not suffer from these limiting factors, its GOS improved due to the employment of both adaptive antenna arrays and AQAM techniques.

The average modem throughput expressed in bits per symbol versus the mean carried teletraffic is shown in Figure 5.75, demonstrating that the mean number of bits per symbol throughput of the users decreased, as the number of users supported increased. The FCA algorithm offered the lowest throughput and its performance degraded near-linearly upon increasing the number of users supported. At the user capacity limits of 1400 and 1565 users, the mean modem throughput was 2.45 BPS and 2.35 BPS for the conservative and lenient scenarios, respectively, using two element adaptive antenna arrays. Using four element adaptive antenna arrays the corresponding throughputs were 2.7 BPS and 2.6 BPS. The LOLIA, especially for lower levels of traffic, offered a higher modem throughput for a given level of teletraffic carried, with the BPS throughput performance gracefully decreasing, as the

Figure 5.75: Mean throughput of users in terms of bits per symbol versus mean carried traffic of the LOLIA using 7 "local" base stations, and for FCA employing a 7-cell reuse cluster, for two and four element antenna arrays, using AQAM using wrap-around. See Figure 5.52 for the corresponding "desert-island" scenario.

Figure 5.76: Mean transmit power versus mean carried traffic of the LOLIA using 7 "local" base stations, and for FCA employing a 7-cell reuse cluster, for two and four element antenna arrays, using AQAM using wrap-around. See Figure 5.48 for the corresponding "desert-island" scenario.

Table 5.16: Maximum mean carried traffic, and maximum number of mobile users that can be supported by each configuration, whilst meeting the preset quality constraints defined in Section 5.3.3.4. The carried traffic is expressed in terms of normalized Erlangs (Erlang/km^2/MHz), for the network described in Table 5.4 in a multipath environment with and without power control and AQAM using wrap-around.

Algorithm	Conservative $P_{FT} = 1\%, P_{low} = 1\%$ $GOS = 4\%, P_B = 3\%$			Lenient $P_{FT} = 1\%, P_{low} = 2\%$ $GOS = 6\%, P_B = 5\%$		
	Users	Traffic	Limiting factor	Users	Traffic	Limiting factor
4-QAM without PC						
FCA, 2 elements (el.)	600	6.0	P_{low}	740	7.65	P_{low}
FCA, 4 elements	790	8.3	P_{low}	995	10.3	P_{low}
LOLIA ($n = 7$), 2 el.	1195	12.65	P_{low}	1290	13.7	P_{low}
LOLIA ($n = 7$), 4 el.	1370	14.35	P_{low}	1475	15.6	P_{low}
4-QAM with PC						
FCA, 2 elements	1090	10.6	P_{low}	1120	10.85	P_{FT}
FCA, 4 elements	1370	13.275	P_{FT}	1370	13.275	P_{FT}
LOLIA ($n = 7$), 2 el.	1350	14.05	P_{low}	1445	15.1	P_{low}
LOLIA ($n = 7$), 4 el.	1540	16.15	P_{low}	1640	17.35	P_{low}
AQAM with PC						
FCA, 2 elements	1400	13.8	P_B	1565	15.20	P_B
FCA, 4 elements	1415	13.7	P_B	1575	15.15	P_B
LOLIA ($n = 7$), 2 el.	1910	19.75	BPS	1910	19.75	BPS
LOLIA ($n = 7$), 4 el.	2245	23.25	BPS	2245	23.25	BPS

carried teletraffic continued to increase. The capacity limiting factor of the LOLIA was the throughput restriction of 2.0 BPS.

The mean transmission power results of Figure 5.76 demonstrate that the employment of AQAM is capable of reducing the power transmitted, both for the UL and the DL. At low traffic levels the FCA algorithm performed noticeably worse in transmitted power terms, than the LOLIA. However, as the traffic load increased, the difference became negligible. The mean power reduction, when compared to a fixed transmission power of 10 dBm, varied from approximately 1 dB to more than 6 dB. A 1 dB reduction in transmission power is not particularly significant for the mobile user, especially since at this network load a throughput of just 2 bits/symbol is possible. The difference between the network using AQAM and that without, though, is the overall improved call quality that can be achieved in the context of our performance metrics, and the significantly increased number of users that can be supported by the network Again, the constraint of a minimum throughput of 2 bits/symbol was invoked in order to ensure a fair comparison with the fixed 4-QAM based network.

Table 5.16 shows the performance of the various networks using AQAM with power control, as well as 4-QAM with and without power control, in terms of the number of users supported. A mean increase of 61% was achieved in terms of the number of users by the addition of power control to the FCA algorithm based 4-QAM network. Invoking AQAM and power control led to a further average user capacity increase of almost 22%, with any further gains limited by the lack of free frequency/timeslot combinations available for new calls to start. Therefore, since the network capacity of the FCA algorithm when using adaptive modulation was not limited by co-channel interference, it would be possible to reduce the frequency re-use distance to increase the network capacity.

The performance of the LOLIA was not limited in this sense, however, and the addition of power control to the 4-QAM network provided an mean increase of 12% extra users supported. In conjunction with AQAM techniques this user capacity was further extended by an average of 39%, thus supporting an additional 56% more users, when compared to the 4-QAM network using no power control.

5.7 Summary and Conclusions

In this chapter we have examined the network capacity and performance of the FCA algorithm and the LOLIA using an exclusion zone of seven or 19 base stations, in the context of LOS and multipath propagation environments. We have shown that the addition of power control results in a substantially increased number of supported users, additionally benefiting from a superior call quality, and reduced transmission power for a given number of adaptive antenna array elements located at the base stations. The advantages of using AQAM within a mobile cellular network have also been illustrated, resulting in performance improvements in terms of the mean modem throughout, call quality, mean transmission power and the number of supported users. The next chapter involves the investigation of network capacity in the context of a CDMA-based UMTS-type FDD mode network.

Chapter 6

HSDPA-style FDD Networking, Adaptive Arrays and Adaptive Modulation

6.1 Introduction

In January 1998, the European standardization body for third generation mobile radio systems, the European Telecommunications Standards Institute—Special Mobile Group (ETSI SMG), agreed upon a radio access scheme for third generation mobile radio systems, referred to as the Universal Mobile Telecommunication System (UMTS) [11, 59]. Although this chapter was detailed in Chapter 1, here we provide a rudimentary introduction to the system, in order to allow readers to consult this chapter directly, without having to read Chapter 1 first. Specifically, the UMTS Terrestrial Radio Access (UTRA) supports two modes of duplexing, namely Frequency Division Duplexing (FDD) , where the UL and DL are transmitted on different frequencies, and Time Division Duplexing (TDD) , where the UL and the DL are transmitted on the same carrier frequency, but multiplexed in time. The agreement recommends the employment of Wideband Code Division Multiple Access (W-CDMA) for UTRA FDD and Time Division—Code Division Multiple Access (TD-CDMA) for UTRA TDD. TD-CDMA is based on a combination of Time Division Multiple Access (TDMA) and CDMA, whereas W-CDMA is a pure CDMA-based system. The UTRA scheme can be used for operation within a minimum spectrum of 2×5 MHz for UTRA FDD and 5 MHz for UTRA TDD. Both duplex or paired and simplex or unpaired frequency bands have been identified in the region of 2 GHz to be used for the UTRA third generation mobile radio system. Both modes of UTRA have been harmonized with respect to the basic system parameters, such as carrier spacing, chip rate and frame length. Thereby, FDD/TDD dual mode operation is facilitated, which provides a basis for the development of low cost terminals. Furthermore, the interworking of UTRA with GSM [11] is ensured.

3G, HSPA and FDD versus TDD Networking   Second Edition
L. Hanzo, J. S. Blogh and S. Ni   © 2008 John Wiley & Sons, Ltd

In UTRA, the different service needs are supported in a spectrally efficient way by a combination of FDD and TDD. The FDD mode is intended for applications in both macro- and micro-cellular environments, supporting data rates of up to 384 kbps and high mobility. The TDD mode, on the other hand, is more suited to micro and pico-cellular environments, as well as for licensed and unlicensed cordless and wireless local loop applications. It makes efficient use of the unpaired spectrum—for example in wireless Internet applications, where much of the teletraffic is in the DL—and supports data rates of up to 2 Mbps. Therefore, the TDD mode is particularly well suited for environments generating a high traffic density (e.g. in city centres, business areas, airports etc.) and for indoor coverage, where the applications require high data rates and tend to have highly asymmetric traffic again, as in Internet access.

In parallel to the European activities, extensive work has been carried out also in Japan and the USA on third generation mobile radio systems. The Japanese standardization body known as the Association of Radio Industry and Business (ARIB) also opted for using W-CDMA, and the Japanese as well as European proposals for FDD bear strong similarities. Similar concepts have also been developed by the North-American T1 standardization body for the pan-American third generation (3G) system known as cdma2000, which was also described in Chapter 1 [11].

In order to work towards a truly global third generation mobile radio standard, the Third Generation Partnership Project (3GPP) was formed in December 1998. 3GPP consists of members of the standardization bodies in Europe (ETSI), the US (T1), Japan (ARIB), Korea (TTA—Telecommunications Technologies Association), and China (CWTS—China Wireless Telecommunications Standard). 3GPP merged the already well harmonized proposals by the regional standardization bodies and now works towards a single common third generation mobile radio standard under the terminology UTRA, retaining its two modes, and aiming to operate on the basis of the evolved GSM core network. The Third Generation Partnership Project 2 (3GPP2), on the other hand, works towards a third generation mobile radio standard, which is based on an evolved IS-95 type system which was originally referred to as cdma2000 [11]. In June 1999, major international operators in the Operator Harmonization Group (OHG) proposed a harmonized G3G (Global Third Generation) concept, which has been accepted by 3GPP and 3GPP2. The harmonized G3G concept is a single standard with the following three modes of operation:

- CDMA direct spread (CDMA-DS), based on UTRA FDD as specified by 3GPP.

- CDMA multi-carrier (CDMA-MC), based on cdma2000 using FDD as specified by 3GPP2.

- TDD (CDMA TDD) based on UTRA TDD as specified by 3GPP.

6.2   Direct Sequence Code Division Multiple Access

A rudimentary introduction to CDMA was provided in Chapter 1 in the context of single-user receivers, while in Chapter 3 the basic concepts of multi-user detection have been introduced. However, as noted earlier, our aim is to allow reader to consult this chapter directly, without having to refer back to the previous chapters. Hence here a brief overview of the underlying CDMA basics is provided.

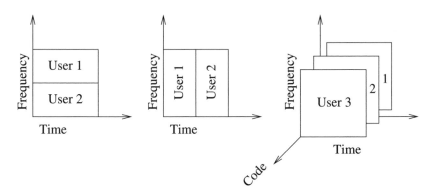

Figure 6.1: Multiple access schemes: FDMA (left), TDMA (middle) and CDMA (right).

Traditional ways of separating signals in time using TDMA and in frequency ensure that the signals are transmitted orthogonal in either time or frequency and hence they are non-interfering. In CDMA different users are separated employing a set of waveforms exhibiting good correlation properties, which are known as spreading codes. Figure 6.1 illustrates the principles of FDMA, TDMA and CDMA. More explicitly, FDMA uses a fraction of the total FDMA frequency band for each communications link for the whole duration of a conversation, while TDMA uses the entire bandwidth of a TDMA channel for a fraction of the TDMA frame, namely for the duration of a timeslot. Finally, CDMA uses the entire available frequency band all the time and separates the users with the aid of unique, orthogonal user signature sequences.

In a CDMA digital communications system, such as that shown in Figure 6.2, the data stream is multiplied by the spreading code, which replaces each data bit with a sequence of code chips. A chip is defined as the basic element of the spreading code, which typically assumes binary values. Hence, the spreading process consists of replacing each bit in the original user's data sequence with the complete spreading code. The chip rate is significantly higher than the data rate, hence causing the bandwidth of the user's data to be spread, as shown in Figure 6.2.

At the receiver, the composite signal containing the spread data of multiple users is multiplied by a synchronized version of the spreading code of the wanted user. The specific auto-correlation properties of the codes allow the receiver to identify and recover each delayed, attenuated and phase-rotated replica of the transmitted signal, provided that the signals are separated by more than one chip period and the receiver has the capability of tracking each significant path. This is achieved using a Rake receiver [5] that can process multiple delayed received signals. Coherent combination of these transmitted signal replicas allows the original signal to be recovered. The unwanted signals of the other simultaneous users remain wideband, having a bandwidth equal to that of the noise, and appear as additional noise with respect to the wanted signal. Since the bandwidth of the despread wanted signal is reduced relative to this noise, the signal-to-noise ratio of the wanted signal is enhanced by the despreading process in proportion to the ratio of the spread and despread bandwidths, since the noise power outside the useful despread signal's bandwidth can be

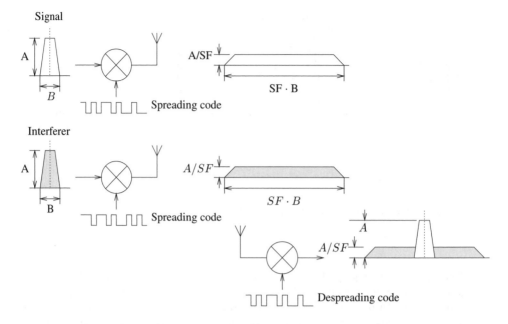

Figure 6.2: CDMA Spreading and Despreading Processes

removed by a low-pass filter. This bandwidth ratio is equal to the ratio of the chip rate to the data rate, which is known as the Processing Gain (PG). For this process to work efficiently, the signals of all of the users should be received at or near the same power at the receiver. This is achieved with the aid of power control, which is one of the critical elements of a CDMA system. The performance of the power control scheme directly affects the capacity of the CDMA network.

6.3 UMTS Terrestrial Radio Access

A bandwidth of 155 MHz has been allocated for UMTS services in Europe in the frequency region of 2.0 GHz. The paired bands of 1920-1980 MHz (UL) and 2110-2170 MHz (DL) have been set aside for FDD W-CDMA systems, and the unpaired frequency bands of 1900-1920 MHz and 2010-2025 MHz for TDD CDMA systems.

A UTRA Network (UTRAN) consists of one or several Radio Network Sub-systems (RNSs), which in turn consist of base stations (referred to as Node Bs) and Radio Network Controllers (RNCs). A Node B may serve one or multiple cells. Mobile stations are known as User Equipment (UE), which are expected to support multi-mode operation in order to enable handovers between the FDD and TDD modes and, prior to complete UTRAN coverage, also to GSM. The key parameters of UTRA have been defined as in Table 6.1.

Table 6.1: Key UTRA Parameters.

Duplex scheme	FDD	TDD
Multiple access scheme	W-CDMA	TD-CDMA
Chip rate	3.84 Mchip/s	3.84 Mchip/s
Spreading factor range	4–512	1–16
Frequency bands	1920–1980 MHz (UL)	1900–1920 MHz
	2110–2170 MHz (DL)	2010–2025 MHz
Modulation mode	4-QAM/QPSK	4-QAM/QPSK
Bandwidth	5 MHz	5 MHz
Nyquist pulse shaping	0.22	0.22
Frame length	10 ms	10 ms
Number of timeslots per frame	15	15

6.3.1 Spreading and Modulation

As usual, the UL is defined as the transmission path from the mobile station to the base station, which receives the unsynchronized channel impaired signals from the network's mobiles. The base station has the task of extracting the wanted signal from the received signal contaminated by both intra- and inter-cell interference. However, as described in Section 6.2, some degree of isolation between interfering users is achieved due to employing unique orthogonal spreading codes, although their orthogonality is destroyed by the hostile mobile channel.

The spreading process consists of two operations. The first one is the channelization operation, which transforms every data symbol into a number of chips, thus increasing the bandwidth of the signal, as seen in Figure 6.2 of Section 6.2. The channelization codes in UTRA are Orthogonal Variable Spreading Factor (OVSF) codes [11] that preserve the orthogonality between a given user's different physical channels, which are also capable of supporting multirate operation. These codes will be further discussed in the context of Figure 6.4. The second operation related to the spreading, namely the "scrambling' process then multiplies the resultant signals separately on the I- and Q-branches by a complex-valued scrambling code, as shown in Figure 6.3. The scrambling codes may be one of either 2^{24} different "long" codes or 2^{24} "short" UL scrambling codes.

The Dedicated Physical Control CHannel (DPCCH) [11, 400] is spread to the chip rate by the channelization code C_c, while the nth Dedicated Physical Data CHannel (DPDCH), namely DPDCH$_n$, is spread to the chip rate by the channelization code $C_{d,n}$. One DPCCH and up to six parallel DPDCHs can be transmitted simultaneously, i.e. $1 \leq n \leq 6$ (as seen in Figure 6.3). However, it is beneficial to transmit with the aid of a single DPDCH, if the required bit-rate can be provided by a single DPDCH for reasons of terminal amplifier efficiency. This is because multi-code transmissions increase the peak-to-average ratio of the transmission, which reduces the efficiency of the terminal's power amplifier [59]. The maximum user data rate achievable with the aid of a single code is derived from the maximum channel bit rate, which is 960 kbps using a spreading factor of four without channel coding in the 1999 version of the UTRA standard. However, at the time of writing a spreading factor of one is being considered by the standardization body. With channel coding the maximum

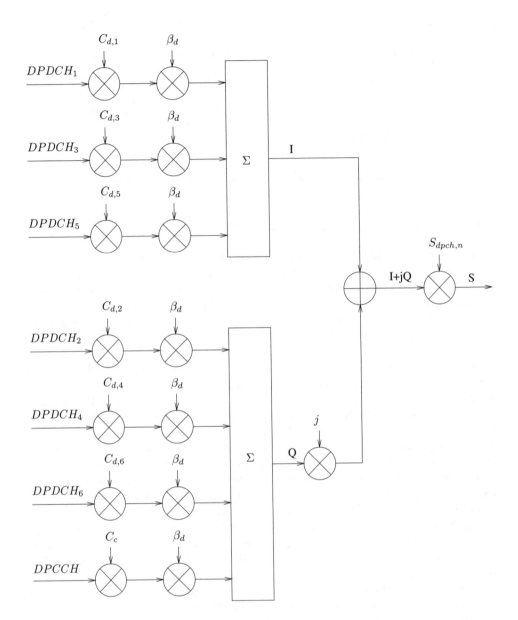

Figure 6.3: Spreading for UL DPCCH and DPDCHs.

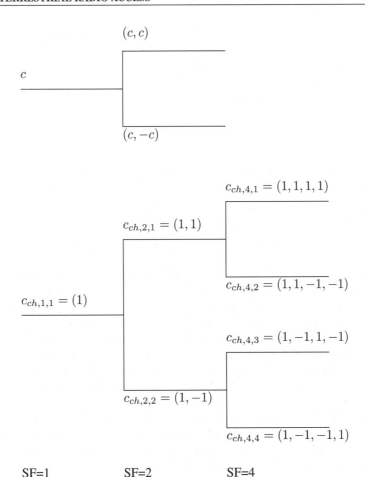

Figure 6.4: Code tree for the generation of Orthogonal Variable Spreading Factor (OVSF) codes.

practical user data rate for single code transmission is of the order of 400–500 kbps. For achieving higher data rates parallel multi-code channels are used. This allows up to six parallel codes to be used, increasing the achievable channel bit rate up to 5740 kbps, which can accommodate a 2 Mbps user data rate or even higher data rates, when the channel coding rate is $1/2$.

The OVSF codes [130] can be defined using the code tree of Figure 6.4. In Figure 6.4, the channelization codes are uniquely described by $C_{ch,SF,k}$, where SF is the spreading factor of the codes, and k is the code index where $0 \leq k \leq SF - 1$. Each level in the code tree defines spreading codes of length SF, corresponding to a particular spreading factor of SF. The number of codes available for a particular spreading factor is equal to the spreading factor itself. All the codes of the same level in the code tree constitute a set and they are orthogonal to each other. Any two codes of different levels are also orthogonal to each other, as long as one of them is not the mother of the other code. For example, the codes $c_{15}(2)$,

$c_7(1)$ and $c_3(1)$ are all the mother codes of $c_{31}(3)$ and hence are not orthogonal to $c_{31}(3)$, where the number in the round bracket indicates the code index. Thus not all the codes within the code tree can be used simultaneously by a mobile station. Specifically, a code can be used by an MS if and only if no other code on the path from the specific code to the root of the tree, or in the sub-tree below the specific node is used by the same MS.

For the DPCCH and DPDCHs the following applies:

- The PDCCH is always spread by code $C_c = C_{ch,256,0}$.

- When only one DPDCH is to be transmitted, DPDCH$_1$ is spread by the code $C_{d,1} = C_{ch,SF,k}$, where SF is the spreading factor of DPDCH$_1$ and $k = SF/4$.

- When more than one DPDCHs have to be transmitted, all DPDCHs have spreading factors equal to four. Furthermore, DPDCH$_n$ is spread by the code $C_{d,n} = C_{ch,4,k}$, where $k = 1$ if $n \subset \{1,2\}$, $k = 3$ if $n \subset \{3,4\}$, and $k = 2$ if $n \subset \{5,6\}$.

A fundamental difference between the UL and the DL is that in the DL synchronization is common to all users and channels of a given cell. This enables us to exploit the cross-correlation properties of the OVSF codes, which were originally proposed in [130]. These codes offer perfect cross-correlation in an ideal channel, but there is only a limited number of these codes available. The employment of OVSF codes allows the spreading factor to be changed and orthogonality between the spreading codes of different lengths to be maintained. The codes are selected from the code tree, which is illustrated in Figure 6.4. As illustrated above, there are certain restrictions as to which of the channelization codes can be used for transmission from a single source. Another physical channel may invoke a certain code from the tree, if no other physical channel to be transmitted employing the same code tree is using a code on an underlying branch, since this would be equivalent to using a higher spreading factor code generated from the spreading code to be used, which are not orthogonal to each other on the same branch of the code tree. Neither can a smaller spreading factor code on the path to the root of the tree be used. Hence, the number of available codes depends on the required transmission rate and spreading factor of each physical channel.

In the UTRA DL a part of the multi-user interference can be orthogonal—apart from the channel effects. The users within the same cell share the same scrambling code, but use different channelization/OVSF codes. In a non-dispersive DL channel, all intra-cell users are synchronized and therefore they are perfectly orthogonal. Unfortunately, in most cases the channel will be dispersive, implying that non-synchronized interference will be suppressed only by a factor corresponding to the processing gain, and thus they will interfere with the desired signal. The interference from other cells which is referred to as inter-cell interference, is non-orthogonal, due to employing different scrambling but possibly the same channelization codes. Therefore inter-cell interference is also suppressed by a factor corresponding to the processing gain.

The channelization code used for the Primary Common PIlot CHannel (CPICH) is fixed to $C_{ch,256,0}$, while the channelization code for the Primary Common Control Physical CHannel (CCPCH) is fixed to $C_{ch,256,1}$ [400]. The channelization codes for all other physical channels are assigned by the UTRAN [400].

A total of $2^{18} - 1 = 262143$ scrambling codes, numbered as $0\ldots262142$ can be generated. However, not all of the scrambling codes are used. The scrambling codes are divided into 512 sets, each consisting of a primary scrambling code and 15 secondary scrambling codes [400].

More specifically, the primary scrambling codes consist of scrambling codes $n = 16 * i$, where $i = 0 \ldots 511$. The ith set of secondary scrambling codes consists of scrambling codes $16*i+k$ where $k = 1 \ldots 15$. There is a one-to-one mapping between each primary scrambling code and the associated 15 secondary scrambling codes in a set, such that the ith primary scrambling code uniquely identifies the ith set of secondary scrambling codes. Hence, according to the above statement, scrambling codes $k = 0 \ldots 8191$ are used. Each of these codes is associated with a left alternative scrambling code and a right alternative scrambling code, that may be used for the so-called compressed frames. Specifically, compressed frames are shortened duration frames transmitted right before a handover, in order to create an inactive period during which no useful data is transmitted. This allows the transceivers to carry out operations necessary for the handover to be successful. The left alternative scrambling code associated with scrambling code k is the scrambling code $k+8192$, while the corresponding right alternative scrambling code is scrambling code $k+16384$. In compressed frames, the left alternative scrambling code is used, if $n < SF/2$ and the right alternative scrambling code is used, if $n \geq SF/2$, where $C_{ch,SF,n}$ is the channelization code used for non-compressed frames.

The set of 512 primary scrambling codes is further divided into 64 scrambling code groups, each consisting of 8 primary scrambling codes. The jth scrambling code group consists of primary scrambling codes $16 * 8 * j + 16 * k$, where $j = 0 \ldots 63$ and $k = 0 \ldots 7$.

Each cell is allocated one and only one primary scrambling code. The primary CCPCH and primary CPICH are always transmitted using this primary scrambling code. The other DL physical channels can be spread and transmitted with the aid of either the primary scrambling code or a secondary scrambling code from the set associated with the primary scrambling code of the cell.

6.3.2 Common Pilot Channel

The Common PIlot CHannel (CPICH) is an unmodulated DL code channel, which is scrambled with the aid of the cell-specific primary scrambling code. The function of the DL CPICH is to aid the Channel Impulse Response (CIR) estimation necessary for the detection of the dedicated channel at the mobile station and to provide the CIR estimation reference for the demodulation of the common channels, which are not associated with the dedicated channels.

UTRA has two types of common pilot channels, namely the primary and secondary CPICHs. Their difference is that the primary CPICH is always spread by the primary scrambling code defined in Section 6.3.1. More explicitly, the primary CPICH is associated with a fixed channelization code allocation and there is only one such channel and channelization code for a cell or sector. The secondary CPICH may use any channelization code of length 256 and may use a secondary scrambling code as well. A typical application of secondary CPICHs usage would be in conjunction with narrow antenna beams intended for service provision at specific teletraffic "hot spots" or places exhibiting a high traffic density [59].

An important application of the primary common pilot channel is during the collection of channel quality measurements for assisting during the handover and cell selection process.

The measured CPICH reception level at the terminal can be used for handover decisions. Furthermore, by adjusting the CPICH power level the cell load can be balanced between different cells, since reducing the CPICH power level encourages some of the terminals to handover to other cells, while increasing it invites more terminals to handover to the cell, as well as to make their initial access to the network in that cell.

6.3.3 Power Control

Agile and accurate power control is perhaps the most important aspect in W-CDMA, in particular on the UL, since a single high-powered rogue mobile can cause serious performance degradation to other users in the cell. The problem is referred to as the "near–far effect" and occurs when, for example, one mobile is near the cell edge, and another is near the cell centre. In this situation, the mobile at the cell edge is exposed to a significantly higher pathloss, say 70 dB higher, than that of the mobile near the cell centre. If there were no power control mechanisms in place, the mobile near the base station could easily "overpower" the mobile at the cell edge, and thus may block a large part of the cell. The optimum strategy in the sense of maximizing the system's capacity is to equalize the received power per bit of all mobile stations at all times.

A so-called open-loop power control mechanism [59] attempts to make a rough estimate of the expected pathloss by means of a DL beacon signal, but this method can be highly inaccurate. The prime reason for this is that the fast fading is essentially uncorrelated between the UL and DL, due to the large frequency separation of the UL and DL band of the W-CDMA FDD mode. Open-loop power control is however, used in W-CDMA, but only to provide a coarse initial power setting of the mobile station at the beginning of a connection.

A better solution is to employ fast closed-loop power control [59]. In closed-loop power control in the UL, the base station performs frequent estimates of the received SIR and compares it to the target SIR. If the measured SIR is higher than the target SIR, the base station commands the mobile station to reduce the power, while if it is too low it will instruct the MS to increase its power. Since each 10 ms UTRA frame consists of 15 timeslots, each corresponding to one power control power adjustment period, this procedure takes place at a rate of 1500 Hz. This is far faster than any significant change of pathloss, including street corner effects, and indeed faster than the speed of Rayleigh fading for low to moderate mobile speeds. The street corner effect occurs when a mobile turns the street corner and hence the received signal power drops markedly. Therefore the mobile responds by rapidly increasing its transmit power, which may inflict sever interference upon other closely located base stations. In response, the mobiles using these base stations increase their transmit powers in an effort to maintain their communications quality. This is undesirable, since it results in a high level of co-channel interference, leading to excessive transmission powers and to a reduction of the battery recharge period.

The same closed-loop power control technique is used on the DL, although the rationale is different. More specifically, there is no near–far problem due to the one-to-many distributive scenario, i.e. all the signals originate from the single base station to all mobiles. It is, however, desirable to provide a marginal amount of additional power to mobile stations near the cell edge, since they suffer from increased inter-cell interference. Hence, the closed loop power control in CDMA systems ensures that each mobile transmits just sufficient power to satisfy the outer-loop power control scheme's SIR target. The SIR target is controlled by an outer-loop power control process that adjusts the required SIR in order to meet the Bit Error Ratio (BER) requirements of a particular service. At higher mobile speeds typically a higher SIR is necessary for attaining a given BER/FER.

6.3.3.1 Uplink Power Control

The UL's inner-loop power control adjusts the mobile's transmit power in order to maintain the received UL SIR at the given SIR target, namely at SIR_{target}. The base stations that are communicating with the mobile generate Transit Power Control (TPC) commands and transmit them, once per slot, to the mobile. The mobile then derives from the TPC commands of the various base stations, a single TPC command, TPC_cmd, for each slot, combining multiple received TPC commands if necessary. In [401] two algorithms were defined for the processing of TPC commands and hence for deriving TPC_cmd.

Algorithm 1: [401]

When not in soft-handover, i.e. when the mobile communicates with a single base station, only one TPC command will be received in each slot. Hence, for each slot, if the TPC command is equal to 0 ($SIR > SIR_{target}$) then $TPC\_cmd = -1$, otherwise, if the TPC command is 1 ($SIR < SIR_{target}$) then $TPC\_cmd = 1$, which implies powering down or up, respectively.

When in soft handover, multiple TPC commands are received in each slot from the different base stations in the active base station set. If all of the base station's TPC commands are identical, then they are combined to form a single TPC command, namely TPC_cmd. However, if the TPC commands of the different base stations differ, then a soft decision W_i is generated for each of the TPC commands, TPC_i, where $i = 1, 2, \ldots, N$, and N is the number of TPC commands. These N soft decisions are then used to form a combined TPC command TPC_cmd according to:

$$TPC\_cmd = \gamma(W_1, W_2, \ldots, W_N) \tag{6.1}$$

where TPC_cmd is either -1 or +1 and $\gamma()$ is the decision function combining the soft values, $W_1, \ldots, W_N$.

If the N TPC commands appear to be uncorrelated, and have a similar probability of being 0 or 1, then function $\gamma()$ should be defined such that the probability that the output of the function $\gamma()$ is equal to 1, is greater than or equal to $1/2^N$, and the probability that the output of $\gamma()$ is equal to -1, shall be greater than or equal to 0.5 [401]. Alternatively, the function $\gamma()$ should be defined such that $P(\gamma() = 1) \geq 1/2^N$ and $P(\gamma() = -1) \geq 0.5$.

Algorithm 2: [401]

When not in soft handover, only one TPC command will be received in each slot, and the mobile will process the maximum 15 TPC commands in a five-slot cycle, where the sets of five slots are aligned with the frame boundaries and the sets do not overlap. Therefore, when not in soft handover, for the first four slots of a five-slot set $TPC\_cmd = 0$ is used for indicating that no power control adjustments are made. For the fifth slot of a set the mobile performs hard decisions on all five of the received TPC commands. If all five hard decisions result in a binary 1, then we set $TPC\_cmd = 1$. In contrast, if all five hard decisions yield a binary 0, then $TPC\_cmd = -1$ is set, else $TPC\_cmd = 0$.

When the mobile is in soft handover, multiple TPC commands will be received in each slot from each of the base stations in the set of active base stations. When the TPC commands of the active base stations are identical, then they can be combined into a single TPC command. However, when the received TPC commands are different, the mobile makes a hard decision concerning the value of each TPC command for three consecutive slots, resulting in N hard decisions for each of the three slots, where N is the number of base stations within the active set. The sets of three slots are aligned to the frame boundaries and do not overlap. Then $TPC\_cmd = 0$ is set for the first two slots of the three-slot set, and then TPC_cmd is determined for the third slot as follows.

The temporary command TPC_temp_i is determined for each of the N sets of three TPC commands of the consecutive slots by setting $TPC\_temp_i = 1$ if all three TPC hard decisions are binary 1. In contrast, if all three TPC hard decisions are binary 0, $TPC\_temp_i = -1$ is set, otherwise we set $TPC\_temp_i = 0$. These temporary TPC commands are then used to determine the combined TPC command for the third slot invoking the decision function $\gamma(TPC\_temp_1, TPC\_temp_2, \ldots, TPC\_temp_N)$ defined as:

$$TPC\_cmd = 1 \quad \text{if } \frac{1}{N} \sum_{i=1}^{N} TPC\_temp_i > 0.5$$

$$TPC\_cmd = -1 \text{ if } \frac{1}{N} \sum_{i=1}^{N} TPC\_temp_i < -0.5 \tag{6.2}$$

$$TPC\_cmd = 0 \quad \text{otherwise.}$$

6.3.3.2 Downlink Power Control

The DL transmit power control procedure simultaneously controls the power of both the DPCCH and its corresponding DPDCHs, both of which are adjusted by the same amount, and hence the relative power difference between the DPCCH and DPDCHs remains constant.

The mobile generates TPC commands for controlling the base station's transmit power and sends them in the TPC field of the UL DPCCH. When the mobile is not in soft handover, the TPC command generated is transmitted in the first available TPC field using the UL DPCCH. In contrast, when the mobile is in soft handover, it checks the DL power control mode (DPC_MODE) before generating the TPC command. If $DPC\_MODE = 0$, the mobile sends a unique TPC command in the first available TPC field in the UL DPCCH. If however, $DPC\_MODE = 1$, the mobile repeats the same TPC command over three consecutive slots of the same frame and the new TPC command is transmitted to the base station in an effort the control its power at the beginning of the next frame. The minimum required transmit power step size is 1 dB, with a smaller step size of 0.5 dB being optional. The power control step size can be increased from 1 dB to 2 dB, thus allowing a 30 dB correction range during the 15 slots of a 10 ms frame. The maximum transmit powers are +21 dBm and +24 dBm, although it is likely that in the first phase of network deployment most terminals will belong to the 21 dBm power class [59].

6.3.4 Soft Handover

Theoretically, the ability of CDMA to despread the interfering signals, and thus adequately operate at low signal-to-noise ratios, allows a CDMA network to have a frequency reuse factor of one [59]. Traditionally, non-CDMA based networks have required adjacent cells to have different carrier frequencies, in order to reduce the co-channel interference to acceptable levels. Therefore, when a mobile hands over from one cell to another, it has to re-tune its synthesizer to the new carrier frequency, i.e. it performs an inter-frequency handover. This process is a "break-before-make" procedure, known as a hard handover, and hence call disruption or interruption is possible. However, in a CDMA based network, having a frequency reuse factor of one, so-called soft handovers may be performed, which is a "make-before-break" process, potentially allowing for a smoother handover between cells. During a soft handover a mobile is connected to two or more base stations simultaneously, thus utilizing more network resources and transmitting more signals, which interfere with other users. Therefore, it is in the network operator's interests to minimize the number of users in soft handover, whilst maintaining a satisfactory QoS. In soft handover, each connected base station receives and demodulates the user's data, and selection diversity is performed between the base stations, i.e. the best version of the UL frame is selected. In the DL, the mobile station performs maximal ratio combining [5] of the signal received from the multiple base stations. This diversity combining improves the coverage in regions of previously low-quality service provision, but at the expense of increased backhaul connections.

The set of base stations engaged in soft handover is known as the *active set*. The mobile station continuously monitors the received power level of the PIlot CHannels (PICHs) transmitted by its neighboring base stations. The received pilot power levels of these base stations are then compared to two thresholds, the acceptance threshold, T_{acc} and the dropping threshold T_{drop}. Therefore, as a mobile moves away from base station 1, and towards base station 2, the pilot signal strength received from base station 2 increases. When the pilot strength exceeds the *acceptance threshold*, T_{acc}, the mobile station enters the soft handover state, as shown in Figure 6.5. As the mobile continues to move away from base station 1, its pilot strength decreases, until it falls below the *drop threshold*. After a given time interval, T_{drop}, during which the signal strength from base station 1 has not exceeded the drop threshold, base station 1 is removed from the active set.

6.3.5 Signal-to-interference plus Noise Ratio Calculations

6.3.5.1 Downlink

The interference received at the mobile can be divided into interference due to the signals transmitted to other mobiles from the same base station, which is known as intra-cell interference, and that received due to the signals transmitted to other mobiles from other base stations, which is termed inter-cell interference. In an ideal case, the intra-cell interference would be zero, since all the signals from the base station are subjected to the same channel conditions, and orthogonal channelization codes are used for separating the users. However, after propagation through a dispersive multipath channel, this orthogonality is eroded. The intra-cell and inter-cell interference values are always non-zero when in a single-user scenario due to the inevitable interference inflicted by the common pilot channels.

The instantaneous SINR is obtained by dividing the received signal powers by the total interference plus thermal noise power, and then by multiplying this ratio by the spreading factor, *SF*, yielding

$$SINR_{DL} = \frac{SF \cdot S}{(1 - \alpha)I_{Intra} + I_{Inter} + N_0},$$

(6.3)

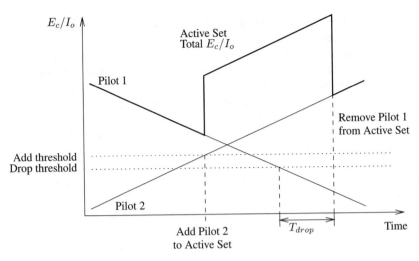

Figure 6.5: The soft handover process showing the process of adding and dropping base stations from the active set.

where $\alpha = 1$ corresponds to the ideal case of perfectly orthogonal intra-cell interference, and $\alpha = 0$ is for completely asynchronous intra-cell interference. Furthermore, N_0 is the thermal noise's power spectral density, S is the received signal power, I_{Intra} is the intra-cell interference and I_{Inter} is the inter-cell interference. Again, the interference plus noise power is scaled by the spreading factor, SF, since after the low-pass filtering the noise bandwidth is reduced by a factor of SF during the despreading process.

When in soft handover, the maximum ratio combining is performed on the N received signals of the N active base stations. Therefore, provided that the active base stations' received signals are independent, the SINR in this situation is:

$$SINR_{DL} = SINR_{DL_1} + SINR_{DL_2} + \cdots + SINR_{DL_N}. \qquad (6.4)$$

6.3.5.2 Uplink

The UL differs from the DL in that the multiple access interference is asynchronous in the UL due to the un-coordinated transmissions of the mobile stations, whereas it may remain quasi-synchronous in the DL. Therefore, the intra-cell UL interference is not orthogonal. A possible solution for mitigating this problem is employing Multi-User Detectors (MUDs) [93] at the base stations.

Thus, we define β as the MUD's efficiency, which effectively gives the percentage of the intra-cell interference that is removed by the MUD. Setting $\beta = 0.0$ implies 0% efficiency, when the intra-cell interference is not reduced by the MUD, whereas $\beta = 1.0$ results in the perfect suppression of all the intra-cell interference. Therefore, the expression for the UL SINR is:

$$SINR_{UL} = \frac{SF \cdot S}{(1 - \beta)I_{Intra} + I_{Inter} + N_0}. \qquad (6.5)$$

When in soft handover, selection diversity is performed on the N received signals at each of the active base stations. Therefore, the SINR in this situation becomes:

$$SINR_{UL} = \max(SINR_{UL_1}, SINR_{UL_2}, \ldots, SINR_{UL_N}). \tag{6.6}$$

6.3.6 Multi-user Detection

Multiple access communications using DS-CDMA is interference limited due to the Multiple Access Interference (MAI) generated by the users transmitting simultaneously within the same bandwidth. The signals received from the users are separated with the aid of the despreader using spreading sequences that are unique to each user. Again, these spreading sequences are usually non-orthogonal. Even if they are orthogonal, the asynchronous UL transmissions of the users or the time-varying nature of the mobile radio channel may partially destroy this orthogonality. The non-orthogonal nature of the codes results in residual MAI, which degrades the performance of the system. The frequency selective mobile radio channel also gives rise to Inter-Symbol Interference (ISI) due to dispersive multipath propagation. This is exacerbated by the fact that the mobile radio channel is time-varying.

Conventional CDMA detectors—such as the matched filter [5, 402] and the Rake combiner [403]—are optimized for detecting the signal of a single desired user. Rake combiners exploit the inherent multi-path diversity in CDMA, since they essentially consist of matched filters combining each resolvable path of the multipath channel. The outputs of these matched filters are then coherently combined according to a diversity combining technique, such as maximal ratio combining [323], equal gain combining or selective diversity combining. These conventional single-user detectors are inefficient, because the interference is treated as noise, and our knowledge concerning the CIR of the mobile channel, or that of the spreading sequences of the interferers is not exploited. The efficiency of these detectors is dependent on the cross-correlation (CCL) between the spreading codes of all the users. The higher the cross-correlation, the higher the MAI. This CCL-induced MAI is exacerbated by the effects of the dispersive multi-path channel and asynchronous transmissions. The utilization of these conventional receivers results in an interference-limited system. Another weakness of the above-mentioned conventional CDMA detectors is the phenomenon known as the "near–far effect" [404, 405]. For conventional detectors to operate efficiently, the signals received from all the users have to arrive at the receiver with approximately the same power. A signal that has a significantly weaker signal strength compared to the other signals will be "swamped" by the relatively higher powers of the other signals and the quality of the weaker signal at the output of the conventional receiver will be severely degraded. Therefore, stringent power control algorithms are needed to ensure that the signals arrive at similar powers at the receiver, in order to achieve a similar QoS for different users [405, 406]. Using conventional detectors to detect a signal corrupted by MAI, while encountering a hostile channel results in an irreducible BER, even if the E_s/N_0 ratio is increased. This is because at high E_s/N_0 values the probability of errors due to thermal noise is insignificant compared to the errors caused by the MAI and the channel. Therefore, detectors that can reduce or remove the effects of MAI and ISI are needed in order to achieve user capacity gains. These detectors also have to be "near–far resistant", in order to avoid the need for stringent power control requirements. In order to mitigate the problem of MAI, Verdú [93] proposed the optimum multi-user detector for asynchronous Gaussian multiple access channels. This optimum detector significantly outperforms the conventional detector and it is near–far resistant, but unfortunately its complexity increases exponentially according to the order of $O(2^{NK})$, where N is the number of overlapping asynchronous bits considered in the detector's window, and K is the number of interfering users. In order to reduce the complexity of the receiver and yet to provide an acceptable BER performance, significant research efforts have been invested in the field of sub-optimal CDMA multiuser receivers [93, 407].

In summary, multi-user detectors reduce the error floor due to MAI and this translates into user capacity gains for the system. These multi-user detectors are also near–far resistant to a certain extent and this results in less stringent power control requirements. However, multi-user detectors are more complex than conventional detectors. Coherent detectors require the explicit knowledge of the channel impulse response estimates, which implies that a channel estimator is needed in the receiver, and hence training sequences have to be included in the transmission frames. Training sequences are specified in the TDD mode of the UTRA standard and enable the channel impulse response of each simultaneously communicating user to be derived, which is necessary for the multi-user detectors to be able to separate the signals received from each user. These multi-user detectors also exhibit an inherent latency, which results in delayed reception. Multi-user detection is more suitable for the UL receiver since the base station has to detect all users' signals anyway and it can tolerate a higher complexity. In contrast, a hand-held mobile receiver is required to be compact and lightweight, imposing restrictions on the available processing power. Recent research into blind MUDs has shown that data detection is possible for the desired user without invoking the knowledge of the spreading sequences and channel estimates of other users. Hence using these detectors for DL receivers is becoming feasible.

6.4   Simulation Results

This section presents simulation results obtained for an FDD mode UMTS type CDMA cellular network, investigating the applicability of various soft handover metrics when subjected to different propagation conditions. This is followed by performance curves obtained using adaptive antenna arrays, when subjected to both non-shadowed as well as shadowed propagation conditions. The performance of adaptive modulation techniques used in conjunction with adaptive antenna arrays in a shadow faded environment is then characterized.

6.4.1   Simulation Parameters

Simulations of an FDD mode UMTS type CDMA based cellular network were conducted for various scenarios and algorithms in order to study the interactions of the processes involved in such a network. As in the standard, the frame length was set to 10 ms, containing 15 power control timeslots. The power control target SINR was chosen to give a Bit Error Ratio (BER) of 1×10^{-3}, with a low quality outage occurring at a BER of 5×10^{-3} and an outage taking place at a BER of 1×10^{-2}. The received SINRs at both the mobile and the base stations were required for each of the power control timeslots, and hence the outage and low quality outage statistics were gathered. If the received SINR was found to be below the outage SINR for 75 consecutive power control timeslots, corresponding to 5 consecutive transmission frames or 50 ms, the call was dropped. The post despreading SINRs necessary for obtaining the target BERs were determined with the aid of physical-layer simulations using a 4-QAM modulation scheme, in conjunction with $1/2$ rate turbo coding and joint detection over a COST 207 seven-path Bad Urban channel [408]. For a spreading factor of 16, the post-despreading SINR required to give a BER of 1×10^{-3} was 8.0 dB, for a BER of 5×10^{-3} it was 7.0 dB, and for a BER of 1×10^{-2} was about 6.6 dB. These values can be seen along with the other system parameters in Table 6.2. The-pre despreading SINR is related to E_b/N_o and to the spreading factor by :

$$SINR = (E_b/N_o)/SF, \tag{6.7}$$

where the spreading factor $SF = W/R$, with W being the chip rate and R the data rate. A receiver noise figure of 7 dB was assumed for both the mobile and the base stations [59]. Thus, in conjunction with a thermal noise density of -174 dBm/Hz and a noise bandwidth of 3.84 MHz, this resulted in a receiver noise power of -100 dBm. The power control algorithm used was relatively simple, and unrelated to the previously introduced schemes of Section 6.3.3. Furthermore, since it allowed a full transmission power change of 15 dB within a 15-slot UTRA data frame, the power control scheme advocated is unlikely to limit the network's capacity.

Specifically, for each of the 15 timeslots per transmitted frame, both the mobile and base station transmit powers were adjusted such that the received SINR was greater than the target SINR, but less than the target SINR plus 1 dB of hysteresis. When in soft handover, a mobile's transmission power was only increased if all of the base stations in the Active Base station Set (ABS) requested a power increase, but was it decreased if any of the base stations in the ABS had an excessive received SINR. In the DL, if the received SINR at the mobile was insufficiently high then all of the active base stations were commanded to increase their transmission powers. Similarly, if the received SINR was unnecessarily high, then the active base stations would reduce their transmit powers. The DL intra-cell interference orthogonality factor, α, as described in Section 6.3.5, was set to 0.5 [409–411]. Due to the frequency reuse factor of one, with its associated low frequency reuse distance, it was necessary for both the mobiles and the base stations, when initiating a new call or entering soft handover, to increase their transmitted power gradually. This was required to prevent sudden increases in the level of interference, particularly on links using the same base station. Hence, by gradually increasing the transmit power to the desired level, the other users of the network were capable of compensating for the increased interference by increasing their transmit powers, without encountering undesirable outages. In an FDMA/TDMA network this effect is less noticeable due to the significantly higher frequency reuse distance.

Since a dropped call is less desirable from a user's viewpoint than a blocked call, two resource allocation queues were invoked, one for new calls and the other—higher priority— queue, for handovers. By forming a queue of the handover requests, which have a higher priority during contention for network resources than new calls, it is possible to reduce the number of dropped calls at the expense of an increased blocked call probability. A further advantage of the Handover Queueing System (HQS) is that during the time a handover is in the queue, previously allocated resources may become available, hence increasing the probability of a successful handover. However, in a CDMA based network the capacity is not hard-limited by the number of frequency/timeslot combinations available, like in an

Table 6.2: Simulation parameters of the UTRA-type CDMA based cellular network.

Parameter	Value	Parameter	Value
Frame length	10 ms	Timeslots per frame	15
Target E_b/N_o	8.0 dB	Outage E_b/N_o	6.6 dB
Low Quality (LQ) Outage E_b/N_o	7.0 dB	BS Pilot Power	−5 dBm
BS/MS Minimum TX Power	−44 dBm	BS Antenna Gain	11 dBi
BS/MS Maximum TX Power	+21 dBm	MS Antenna Gain	0 dBi
Attenuation at 1 m reference point	39 dB	Pathloss exponent	−3.5
Power control SINR hysteresis	1 dB	Cell radius	150 m
Downlink scrambling codes per BS	1	Modulation scheme	4-QAM
Downlink OVSF codes per BS	Variable	Max new-call queue-time	5 s
Uplink scrambling codes per BS	Variable	Average inter-call time	300 s
Uplink OVSF codes per BS	Variable	Average call length	60 s
Spreading factor	Variable	Data/voice bit rate	Variable
Remove BS from ABS threshold	Variable	Add BS to ABS threshold	Variable
User speed	1.34 m/s	Noisefloor	−100 dBm
	(3 mph)	Size of ABS	2

FDMA/TDMA based network, such as GSM. The main limiting factors are the number of available spreading and OVSF codes, where the number of the available OVSF codes is restricted to the spreading factor minus one, since an OVSF code is reserved for the pilot channel. This is because, although the pilot channel has a spreading factor of 256, it removes an entire branch of the OVSF code generation tree. Other limiting factors are the interference levels in conjunction with the restricted maximum transmit power, resulting in excessive call dropping rates. New call allocation requests were queued for up to 5 s, if they could not be immediately satisfied, and were blocked if the request had not been completed successfully within the 5 s.

Similarly to our TDMA-based investigations portrayed in Chapter 5, several network performance metrics were used in order to quantify the QoS provided by the cellular network, namely the:

- New Call Blocking probability, P_B.

- Call Dropping or Forced Termination probability, P_{FT}.

- Probability of low quality connection, P_{low}.

- Probability of Outage, P_{out}.

- Grade Of Service, GOS.

The new call blocking probability, P_B, is defined as the probability that a new call is denied access to the network. In an FDMA/TDMA based network, such as GSM, this may occur because there are no available physical channels at the desired base station or the available channels are subject to excessive interference. However, in a CDMA based network this does not occur, provided that no interference level based admission control is performed and hence the new call blocking probability is typically low.

The call dropping probability, P_{FT}, is the probability that a call is forced to terminate prematurely. In a GSM type network, an insufficiently high SINR, which inevitably leads to dropped calls, may be remedied by an intra- or inter-cell handover. However, in CDMA either the transmit power must be increased, or a soft handover must be performed in order to exploit the available diversity gain.

Again, the probability of a low quality connection is defined as:

$$P_{low} = P\{SINR_{UL} < SINR_{req} \text{ or } SINR_{DL} < SINR_{req}\} \tag{6.8}$$
$$= P\{min(SINR_{UL}, SINR_{DL}) < SINR_{req}\}.$$

The GOS was defined in [331] as:

$$GOS = P\{\text{unsuccessful or low-quality call access}\} \tag{6.9}$$
$$= P\{\text{call is blocked}\} + P\{\text{call is admitted}\}$$
$$\times P\{\text{low signal quality and call is admitted}\}$$
$$= P_B + (1 - P_B)P_{low},$$

and is interpreted as the probability of unsuccessful network access (blocking), or low quality access, when a call is admitted to the system.

In our forthcoming investigations, in order to compare the network capacities of different networks, similarly to our TDMA-based investigations in Chapter 5, it was decided to use two scenarios defined as:

- A *conservative scenario*, where the maximum acceptable value for the new call blocking probability, P_B, is 3%, the maximum call dropping probability, P_{FT}, is 1%, and P_{low} is 1%.

- A *lenient scenario*, where the maximum acceptable value for the new call blocking probability, P_B, is 5%, the maximum call dropping probability, P_{FT}, is 1%, and P_{low} is 2%.

In the next section we consider the network's performance considering both fixed and normalized soft handover thresholds using both received pilot power and received pilot power versus interference threshold metrics. A spreading factor of 16 was used, corresponding to a channel data rate of 3.84 Mbps/16 = 240 kbps with no channel coding, or 120 kbps when using 1/2 rate channel coding. It must be noted at this stage that the results presented in the following sections are network capacities obtained using a spreading factor of 16. The network capacity results presented in the previous chapter, which were obtained for an FDMA/TDMA GSM-like system, were achieved for speech-rate users. Here we assumed that the channel coded speech-rate was 15 kbps, which is the lowest possible Dedicated Physical Data CHannel (DPDCH) rate. Speech users having a channel coded rate of 15 kbps may be supported by invoking a spreading factor of 256. Hence, subjecting the channel data rate of 15 kbps to 1/2 rate channel coding gives a speech-rate of 7.5 kbps, or if protected by a 2/3 rate code the speech-rate becomes 10 kbps, which are sufficiently high for employing the so-called Advanced MultiRate (AMR) speech codec [412–414] capable of operating at rates between 4.7 kbps and 12.2 kbps. Therefore, by multiplying the resultant network capacities according to a factor of 256/16 = 16, it is possible to estimate the number of speech users supported by a speech-rate network. However, with the aid of our exploratory simulations, conducted using a spreading factor of 256, which are not presented here, we achieved network capacities higher than 30 times the network capacity supported in conjunction with a spreading factor of 16. Therefore, it would appear that the system is likely to support more than 16 times the number of 240 kbps data users, when communicating at the approximately 16 times lower speech-rate, employing a high spreading factor of 256. Hence, using the above-mentioned scaling factor of 16 we arrive at the lower bound of network capacity. A mobile speed of 3 mph was used in conjunction with a cell size of 150 m radius, which was necessarily small in order to be able to support the previously assumed 240 kbps high target data rate. The performance advantages of using both adaptive beamforming and adaptive modulation assisted networks are also investigated.

6.4.2   The Effect of Pilot Power on Soft Handover Results

In this section we consider the settings of the soft handover thresholds, for an IS-95 type handover algorithm [58], where the handover decisions are based on DL pilot power measurements. Selecting inappropriate values for the soft handover thresholds, namely for the *acceptance threshold* and the *drop threshold*, may result in an excessive number of blocked and dropped calls in certain parts of the simulation area. For example, if the *acceptance threshold* that has to be exceeded by the signal level for a base station to be added to the active set is too high (Threshold B in Figure 6.6), then a user may be located within a cell, but it would be unable to add any base stations to its active base station set. Hence this user is unable to initiate a call. Figure 6.6 illustrates this phenomenon and shows that the *acceptance thresholds* must be set sufficiently low for ensuring that at least one base station covers every part of the network.

Another consequence of setting the *acceptance threshold* to an excessively high value, is that soft handovers may not be completed. This may occur when a user leaving the coverage area of a cell, since the pilot signal from that cell drops below the *drop threshold*, before the signal from the adjacent cell becomes sufficiently strong for it to be added to the active base station set. However, if the *acceptance threshold*, in conjunction with the *drop threshold*, is set correctly, then new calls and soft handovers should take place as required, so long as the availability of network resources allows it. Care must be taken however, not to set the soft handover threshold too low, otherwise the mobiles occupy additional network resources and create extra interference, due to initiating unnecessary soft-handovers.

6.4.2.1   Fixed Received Pilot Power Thresholds without Shadowing

Figure 6.7 shows the new call blocking probability of a network using a spreading factor of 16, in conjunction with fixed received pilot signal strength based soft handover thresholds without imposing any shadowing effects. The figure illustrates that reducing both the acceptance and the dropping soft handover thresholds results in an improved new call blocking performance. Reducing the threshold at which further base stations may be added to the Active Base station Set (ABS) increases the probability that base stations exist within the ABS, when a new call request is made. Hence, as expected, the new call blocking probability

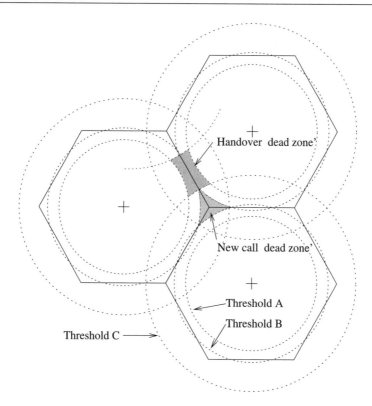

Figure 6.6: This figure indicates that using inappropriate soft handover thresholds may lead to blocked and dropped calls due to insufficient pilot coverage of the simulation area. Threshold A is the drop threshold, which when combined with the acceptance threshold C can fail to cover the simulation area sufficiently well, thus leading to soft handover failure. When combining threshold A with the acceptance threshold B, users located in the "new call dead zone" may become unable to initiate calls.

is reduced, when the acceptance threshold is reduced. Similarly, dropping the threshold at which base stations are removed from the ABS also results in an improved new call blocking probability, since a base station is more likely to be retained in the ABS as a mobile moves away from it. Therefore, should a mobile attempt to initiate a call in this situation, there is a greater chance that the ABS will contain a suitable base station.

The associated call dropping probability is depicted in Figure 6.8, indicating that reducing the soft handover thresholds, and thus increasing the time spent in soft handover, improved the performance up to a certain point. However, above this point the additional interference inflicted by the soft handover process led to a degraded performance. For example, in this figure the performance associated with $T_{acc} = -111$ dBm improved, when T_{drop} was decreased from -112 dBm to -113 dBm. However, at high traffic levels the performance degraded when T_{drop} was decreased further, to -114 dBm. The call dropping probability obtained using $T_{acc} = -113$ dBm and $T_{drop} = -115$ dBm was markedly lower for the lesser levels of traffic carried due to the extra diversity gain provided by the soft handover

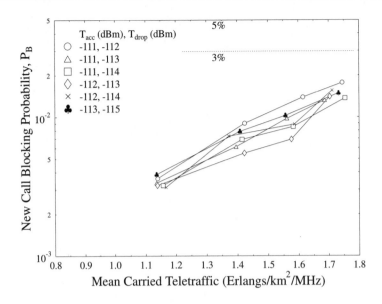

Figure 6.7: New call blocking probability versus mean carried traffic of a CDMA based cellular network using fixed received pilot power based soft handover thresholds without shadowing for $SF = 16$.

process. However, since these soft handover thresholds resulted in a greater proportion of time spent in soft handover, the levels of interference were increased, and thus at the higher traffic levels the performance degraded rapidly, as can be seen in Figure 6.8. Hence, the call dropping performance is based on a trade-off between the diversity gain provided by the soft handover process and the associated additional interference.

The probability of low quality access (not explicitly shown) was similar in terms of its character to the call dropping probability, since reducing T_{drop} improved the performance to a certain point, after which it degraded.

The mean number of base stations in the ABS is shown in Figure 6.9, illustrating that reducing the soft handover thresholds leads, on average, to a higher number of base stations in the ABS. Therefore, a greater proportion of call time is spent in soft handover. The associated diversity gain improves the link quality of the reference user but additional co-channel interference is generated by the diversity links, thus ultimately reducing the call quality, as shown in Figure 6.8. Additionally, this extra co-channel interference required more transmission power for maintaining the target SINR as depicted in Figure 6.10. This figure shows that when lower soft handover thresholds are used, and thus a greater proportion of time is spent in soft handover, greater levels of co-channel interference are present, and thus the required mean transmission powers became higher. It is interesting to note that for the highest soft handover thresholds employed in Figure 6.10, the DL transmission power required for maintaining the target SINR is lower than the UL transmission power, whereas for the lower soft handover thresholds, the required mean UL transmission power is lower than the DL transmission power. The required DL transmission power was, in general, lower than the UL transmission power due to the mobile stations' ability to perform maximal

Figure 6.8: Call dropping probability versus mean carried traffic of a CDMA based cellular network using fixed received pilot power based soft handover thresholds without shadowing for $SF = 16$.

ratio combining when in soft handover. This was observed despite the absence of the pilot interference in the UL, and despite the base stations' ability to perform selective diversity which offers less diversity gain when compared to maximal ratio combining. However, reducing the soft handover thresholds to the lowest levels shown in Figure 6.10, led to increased co-channel interference on the DL, thus requiring higher base station transmission powers, as clearly seen in the figure.

In summary, as seen by comparing Figures 6.7–6.10 the maximum capacity of the network using fixed received pilot power based soft handover thresholds was limited by the call dropping probability. The new call blocking probability remained below the 3% limit, thanks to the appropriate choice of thresholds used, whilst the probability of low quality access was constantly below the 1% mark. Therefore, the maximum normalized teletraffic load was 1.64 Erlangs/km²/MHz, corresponding to a total network capacity of 290 users, while satisfying both QoS constraints, was achieved with the aid of an acceptance threshold of -112 dBm and a dropping threshold of -114 dBm. A mean ABS size of 1.7 base stations was registered at this traffic level, and both the mobile and base stations exhibited a mean transmission power of 5.1 dBm.

6.4.2.2 Fixed Received Pilot Power Thresholds with 0.5 Hz Shadowing

In this section we examine the achievable performance, upon using fixed received pilot power based soft handover thresholds when subjected to log-normal shadow fading having a standard deviation of 3 dB and a maximum frequency of 0.5 Hz.

Figure 6.9: Mean number of base stations in the active base station set versus mean carried traffic of a CDMA based cellular network using fixed received pilot power based soft handover thresholds without shadowing for $SF = 16$.

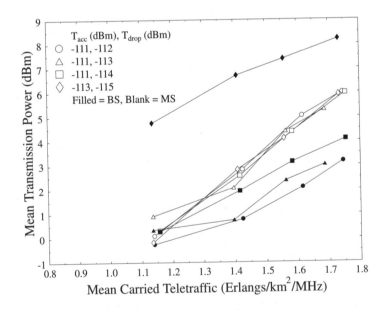

Figure 6.10: Mean transmission power versus mean carried traffic of a CDMA based cellular network using fixed received pilot power based soft handover thresholds without shadowing for $SF = 16$.

Figure 6.11: Call dropping probability versus mean carried traffic of a CDMA based cellular network using fixed received pilot power based soft handover thresholds in conjunction with 0.5 Hz shadowing having a standard deviation of 3 dB for $SF = 16$.

Figure 6.12: Probability of low quality access versus mean carried traffic of a CDMA based cellular network using fixed received pilot power based soft handover thresholds in conjunction with 0.5 Hz shadowing having a standard deviation of 3 dB for $SF = 16$.

The call dropping results of Figure 6.11 suggested that the network's performance was poor when using fixed received pilot power soft handover thresholds in the above mentioned shadow fading environment. The root cause of the problem is that the fixed thresholds must be set such that the received pilot signals, even when subjected to shadow fading, are retained in the active set. Therefore, setting the thresholds too high results in the base stations being removed from the active set, thus leading to an excessive number of dropped calls. However, if the thresholds are set too low, in order to counteract this phenomenon, then the base stations can be in soft handover for too high a proportion of time, and thus an unacceptable level of low quality accesses is generated due to the additional co-channel interference inflicted by the high number of active base stations. Figure 6.11 shows that reducing the soft handover thresholds improved the network's call dropping probability, but Figure 6.12 illustrates that reducing the soft handover thresholds engendered an increase in the probability of a low quality access.

The network cannot satisfy the quality requirements of the conservative scenario, namely that of maintaining a call dropping probability of 1% combined with a maximum probability of low quality access below 1%. However, the entire network supported 127 users, whilst meeting the lenient scenario's set of criteria, which consists of a maximum call dropping probability of 1% and a probability of low quality access of below 2%, using the thresholds of $T_{acc} = -113$ dBm and $T_{drop} = -115$ dBm.

6.4.2.3 Fixed Received Pilot Power Thresholds with 1.0 Hz Shadowing

This section presents results obtained using fixed receiver pilot power based soft handover thresholds in conjunction with log-normal shadow fading having a standard deviation of 3 dB and a maximum fading frequency of 1.0 Hz.

The corresponding call dropping probability is depicted in Figure 6.13, showing that using fixed thresholds in a propagation environment exposed to shadow fading resulted in a very poor performance. This was due to the shadow fading induced fluctuations of the received pilot signal power, which resulted in removing base stations from the ABS mid-call, which ultimately engendered dropped calls. Hence, lowering the fixed thresholds significantly reduced the call dropping probability. However, this led to a deterioration of the low quality access probability, as shown in Figure 6.14. The probability of low quality access was also very poor due to the rapidly fluctuating interference-limited environment. This was shown particularly explicitly in conjunction with $T_{acc} = -113$ dBm and $T_{drop} = -115$ dBm, where reducing the number of users resulted in a degradation of the low quality access performance due to the higher deviation of the reduced number of combined sources of interference. In contrast, adding more users led to a near-constant level of interference that varied less dramatically.

It was found that the network was unable to support any users at the required service quality, since using the thresholds that allowed the maximum 1% call dropping probability restriction to be met, led to a greater than 2% probability of a low quality outage occurring.

6.4.2.4 Summary

In summary of our findings in the context of Figures 6.7–6.14, a disadvantage of using fixed soft handover thresholds is that in some locations all pilot signals may be weak, whereas

Figure 6.13: Call dropping probability versus mean carried traffic of a CDMA based cellular network using fixed received pilot power based soft handover thresholds in conjunction with 1 Hz shadowing having a standard deviation of 3 dB for $SF = 16$.

Figure 6.14: Probability of low quality access versus mean carried traffic of a CDMA based cellular network using fixed received pilot power based soft handover thresholds in conjunction with 1 Hz shadowing having a standard deviation of 3 dB for $SF = 16$.

in other locations, all of the pilot signals may be strong due to the localized propagation environment or terrain. Hence, using relative or normalized soft handover thresholds is expected to be advantageous in terms of overcoming this limitation. An additional benefit of using dynamic thresholds is confirmed within a fading environment, where the received pilot power may drop momentarily below a fixed threshold, thus causing unnecessary removals and additions to/from the ABS. However, these base stations may have been the only base stations in the ABS, thus ultimately resulting in a dropped call. When using dynamically controlled thresholds this scenario would not have occurred. Hence, in the next section we considered the performance of using relative received pilot power based soft handover thresholds under both non-shadowing and shadowing impaired propagation conditions.

To summarize, using fixed received pilot power thresholds in a non-shadowing environment resulted in a total network capacity of 290 users for both QoS scenarios, namely for both the conservative and lenient scenarios considered. However, this performance was severely degraded in a shadow fading impaired propagation environment, where a total network capacity of 127 users was supported in conjunction with a maximum shadow fading frequency of 0.5 Hz. Unfortunately, the network capacity could not be evaluated when using a maximum shadow fading frequency of 1.0 Hz due to the contrasting characteristics of the dropped call and low quality access probability results.

6.4.2.5   Relative Received Pilot Power Thresholds without Shadowing

Employing relative received pilot power thresholds is important in realistic propagation environments exposed to shadow fading. More explicitly, in contrast to the previously used thresholds, which were expressed in terms of dBm, i.e. with respect to 1 mW, in this section the thresholds T_{acc} and T_{drop} are expressed in terms of dB relative to the received pilot strength of the base stations in the ABS. Their employment also caters for situations, where the absolute pilot power may be too low for use in conjunction with fixed thresholds, but nonetheless sufficiently high for reliable communications. Hence, in this section we examine the performance of relative received pilot power based soft handover thresholds in a non-shadow faded environment.

The call dropping performance is depicted in Figure 6.15, which shows that reducing the soft handover thresholds, and thus increasing the time spent in soft handover, improved the call dropping performance. It was also found in the cases considered here, that simultaneously the probability of a low quality access decreased, as illustrated by Figure 6.16. However, it was also evident in both figures, that reducing the soft handover thresholds past a certain point resulted in degraded performance due to the extra interference incurred during the soft handover process.

Since the probability of low quality access was under the 1% threshold, the network capacity for both the lenient and conservative scenarios were the same, namely 1.65 Erlangs/km^2/MHz or a total of 288 users over the entire simulation area of 2.86 km^2. The mean ABS size was 1.7 base stations, with a mean mobile transmission power of 4.1 dBm and an average base station transmit power of 4.7 dBm.

Figure 6.15: Call dropping probability versus mean carried traffic of a CDMA based cellular network using relative received pilot power based soft handover thresholds without shadowing for $SF = 16$.

Figure 6.16: Probability of low quality access versus mean carried traffic of a CDMA based cellular network using relative received pilot power based soft handover thresholds without shadowing for $SF = 16$.

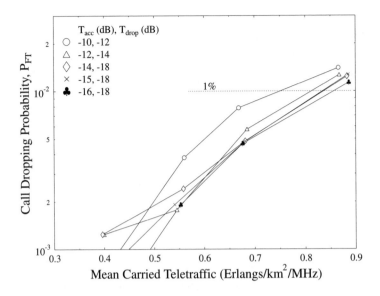

Figure 6.17: Call dropping probability versus mean carried traffic of a CDMA based cellular network using relative received pilot power based soft handover thresholds in conjunction with 0.5 Hz shadowing and a standard deviation of 3 dB for $SF = 16$.

6.4.2.6 Relative Received Pilot Power Thresholds with 0.5 Hz Shadowing

In this section we present results obtained using relative received pilot power based soft handover thresholds in a shadowing-impaired propagation environment. The maximum shadow fading frequency was 0.5 Hz and the standard deviation of the log-normal shadowing was 3 dB.

Figure 6.17 depicts the call dropping probability for several relative thresholds and shows that by reducing both the thresholds, the call dropping performance is improved. This enables the mobile to add base stations to its ABS earlier on during the soft handover process, and to relinquish them at a much later stage than in the case of using higher handover thresholds. Therefore, using lower relative soft handover thresholds results in a longer period of time spent in soft handover, as can be seen in Figure 6.18, which shows the mean number of base stations in the ABS.

The probability of low quality access is shown in Figure 6.19, illustrating that, in general, as the relative soft handover thresholds were reduced, the probability of low quality access increased. This demonstrated that spending more time in soft handover generated more co-channel interference and thus degraded the network's performance. However, the difference between the two thresholds must also be considered. For example, the probability of low quality access is higher in conjunction with $T_{acc} = -16$ dB and $T_{drop} = -18$ dB, than using $T_{acc} = -16$ dB and $T_{drop} = -20$ dB, since the latter scenario has a higher mean number of base stations in its ABS. Therefore, there is a point at which the soft handover gain experienced by the desired user outweighs the detrimental effects of the extra interference generated by base stations' transmissions to users engaged in the soft handover process.

Figure 6.18: Mean number of base stations in the active base station set versus mean carried traffic of a CDMA based cellular network using relative received pilot power based soft handover thresholds in conjunction with 0.5 Hz shadowing and a standard deviation of 3 dB for *SF* = 16.

Figure 6.19: Probability of low quality access versus mean carried traffic of a CDMA based cellular network using relative received pilot power based soft handover thresholds in conjunction with 0.5 Hz shadowing and a standard deviation of 3 dB for *SF* = 16.

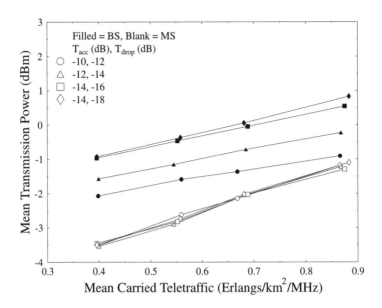

Figure 6.20: Mean transmission power versus mean carried traffic of a CDMA based cellular network
using relative received pilot power based soft handover thresholds in conjunction with
0.5 Hz shadowing and a standard deviation of 3 dB for $SF = 16$.

Figure 6.20 shows the mean transmission powers of both the mobiles and the base
stations. The mobiles are required to transmit at a lower power than the base stations, because
the base stations are not subjected to DL pilot power interference and to soft handover
interference. Furthermore, the mobiles are not affected by the level of the soft handover
thresholds, because only selective diversity is performed in the UL, and hence the mobile
transmits as if not in soft handover. As the soft handover thresholds were reduced, the
time spent in soft handover increased and thus the mean base transmission power had to
be increased in order to overcome the additional DL interference.

The maximum network capacity of 0.835 Erlangs/km²/MHz, or 144 users over the entire
simulation area, was achieved using the soft handover thresholds of $T_{acc} = -14$ dB and
$T_{drop} = -18$ dB for the conservative scenario. The mean ABS size was 1.77 base stations,
while the mean mobile transmit power was -1.5 dBm and 0.6 dBm for the base stations. In
the lenient scenario a maximum teletraffic load of 0.865 Erlangs/km²/MHz, corresponding
to a total network capacity of 146 users was maintained using soft handover thresholds of
$T_{acc} = -16$ dB and $T_{drop} = -18$ dB. The mean number of base stations in the ABS was
1.78, with an average transmit power of -1.5 dBm for the mobile handset, and 1.3 dBm for
the base station.

6.4.2.7   Relative Received Pilot Power Thresholds with 1.0 Hz Shadowing

In this section we present further performance results obtained using relative received
pilot power based soft handover thresholds in a shadowing propagation environment. The

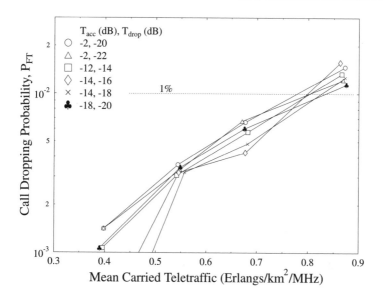

Figure 6.21: Call dropping probability versus mean carried traffic of a CDMA based cellular network using relative received pilot power based soft handover thresholds in conjunction with 1 Hz shadowing and a standard deviation of 3 dB for $SF = 16$.

maximum shadow fading frequency was 1.0 Hz and the standard deviation of the log-normal shadowing was 3 dB.

On comparing the call dropping probability curves seen in Figure 6.21 with the call dropping probability obtained for a maximum shadow fading frequency of 0.5 Hz in Figure 6.17 it was found that the performance of the 1.0 Hz frequency shadowing scenario was slightly worse. However, the greatest performance difference was observed in the probability of low quality access, as can be seen in Figure 6.22.

Using the soft handover thresholds which gave a good performance for a maximum shadow fading frequency of 0.5 Hz resulted in significantly poorer low quality access performance for a maximum shadowing frequency of 1.0 Hz. In order to obtain a probability of low quality access of below 1% it was necessary to use markedly different soft handover thresholds, which reduced the time spent in soft handover and hence also the size of the ABS, as illustrated in Figure 6.23.

For the conservative scenario, where the maximum probability of low quality access, P_{low}, was set to 1%, the maximum network capacity was found to be 0.69 Erlangs/km^2/MHz, equivalent to a total network capacity of 127 users, obtained using $T_{drop} = -2$ dB and $T_{acc} = -16$ dB. In contrast, in the lenient scenario, where the P_{low} limit was 2%, the maximum number of users supported was found to be 144, or 0.825 Erlangs/km^2/MHz, in conjunction with $T_{acc} = -14$ dB and $T_{drop} = -18$ dB.

Figure 6.22: Probability of low quality access versus mean carried traffic of a CDMA based cellular network using relative received pilot power based soft handover thresholds in conjunction with 1 Hz shadowing and a standard deviation of 3 dB for $SF = 16$.

Figure 6.23: Mean number of base stations in the active base station set versus mean carried traffic of a CDMA based cellular network using relative received pilot power based soft handover thresholds in conjunction with 1 Hz shadowing and a standard deviation of 3 dB for $SF = 16$.

6.4.2.8 Summary

In summary, using relative received pilot power as a soft handover metric has resulted in a significantly improved performance in comparison to that of the fixed received pilot power based results in a shadow fading environment. In the non-shadowed environment the network capacity was approximately the same as when using the fixed threshold algorithm, albeit with a slightly improved mean transmission power. Due to the time varying nature of the received signals subjected to shadow fading, using relative thresholds has been found to be more amenable to employment in a realistic propagation environment, than using fixed thresholds. In conclusion, without shadow fading the network supported a total of 288 users, whilst with a maximum shadow fading frequency of 0.5 Hz, approximately 145 users were supported by the entire network, for both the conservative and lenient scenarios. However, different soft handover thresholds were required for each situation, for achieving these capacities. At a maximum shadowing frequency of 1.0 Hz, a total of 127 users were supported in the conservative scenario, and 144 in the lenient scenario. However, again, different soft handover thresholds were required in each scenario in order to maximize the network capacity.

6.4.3 E_c/I_o Power Based Soft Handover Results

An alternative soft handover metric used to determine "cell ownership" is the pilot to DL interference ratio of a cell, which was proposed for employment in the 3rd generation systems [59]. The pilot to DL interference ratio, or E_c/I_o, may be calculated thus as [415]:

$$\frac{E_c}{I_o} = \frac{P_{pilot}}{P_{pilot} + N_0 + \sum_{k=1}^{N_{cells}} P_k T_k},$$ (6.10)

where P_k is the total transmit power of cell k, T_k is the transmission gain, which includes the antenna gain and pathloss as well as shadowing, N_0 is the power spectral density of the thermal noise and N_{cells} is the number of cells in the network. The advantage of using such a scheme is that it is not an absolute measurement that is used, but the ratio of the pilot power to the interference power. Thus, if fixed thresholds were used a form of admission control may be employed for new calls if the interference level became too high. A further advantage is that it takes into account the time-varying nature of the interference level in a shadowed environment.

6.4.3.1 Fixed E_c/I_o Thresholds without Shadowing

The new call blocking probability obtained when using fixed E_c/I_o soft handover thresholds without any form of shadow fading is shown in Figure 6.24, which suggests that in general, lowering the soft handover thresholds reduced the probability of a new call attempt being blocked. However, it was found that in conjunction with $T_{drop} = -40$ dB, dropping the threshold T_{acc} from -20 dB to -24 dB actually increased the new call blocking probability. This was attributed to the fact that the lower threshold precipitated a higher level of co-channel interference, since there was a higher mean number of base stations in the ABS, as evidenced by Figure 6.25. Therefore, since the mean level of interference present in the network is higher, when using a lower threshold, and the threshold determines the value of the pilot to DL interference ratio at which base stations may be added to the ABS, a more

Figure 6.24: New call blocking probability versus mean carried traffic of a CDMA based cellular network using fixed E_c/I_o based soft handover thresholds without shadowing for $SF = 16$.

frequent blocking of calls occurs. Alternatively, a lower threshold resulted in a higher level of DL interference due to the additional interference inflicted by supporting the mobiles in soft handover, which prevented base stations from being included in the ABS due to insufficient pilot to interference "head-room". This then ultimately led to blocked calls due to the lack of base stations in the ABS.

Again, the mean number of base stations in the ABS is given in Figure 6.25, which illustrates that as expected, reducing the soft handover thresholds increased the proportion of time spent in soft handover, and thus reduced the mean number of base stations in the ABS. The average size of the ABS was found to decrease, as the network's traffic load increased. This was a consequence of the increased interference levels associated with the higher traffic loads, which therefore effectively reduced the pilot to interference ratio at a given point, and hence base stations were less likely to be in soft handover and in the ABS.

Figure 6.26 depicts the mean transmission powers for both the UL and the DL, for a range of different soft handover thresholds. These results show similar trends to the results presented in previous sections, with the required average DL transmission power increasing, since a greater proportion of call time is spent in soft handover. Again, the mean UL transmission power varied only slightly, since the selection diversity technique of the base stations only marginally affected the received interference power at the base stations.

Figure 6.27 shows the call dropping performance, indicating that lowering the soft handover thresholds generally improved the call dropping performance. However, reducing the soft handover thresholds too much resulted in a degradation of the call dropping probability due to the increased levels of co-channel interference inherent when a higher proportion of the call time is spent in soft handover. This is explicitly illustrated by

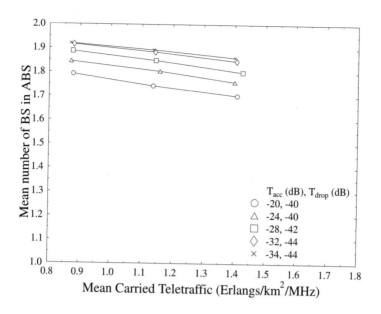

Figure 6.25: Mean number of base stations in the active base station set versus mean carried traffic of a CDMA based cellular network using fixed E_c/I_o based soft handover thresholds without shadowing for $SF = 16$.

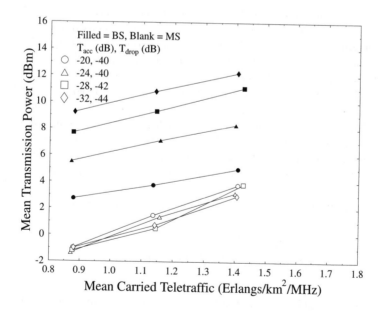

Figure 6.26: Mean transmission power versus mean carried traffic of a CDMA based cellular network using fixed received E_c/I_o based soft handover thresholds without shadowing for $SF = 16$.

Figure 6.27: Call dropping probability versus mean carried traffic of a CDMA based cellular network using fixed received E_c/I_o based soft handover thresholds without shadowing for $SF = 16$.

Figure 6.28, which indicates that reducing the soft handover thresholds caused a significant degradation in the probability of low quality access. This was a consequence of the additional co-channel interference associated with the soft handover process. The figure also shows that there is a point where the diversity gain of the mobiles obtained with the advent of the soft handover procedure outweighs the extra interference that it generates.

On the whole, the capacity of the network when using fixed E_c/I_o soft handover thresholds was lower than when using fixed received pilot power based soft handover thresholds. This can be attributed to the fact that the E_c/I_o thresholds are related to the interference level of the network, which changes with the network load and propagation conditions. Hence using a fixed threshold is sub-optimal. In the conservative scenario, the network capacity was 1.275 Erlangs/km²/MHz, corresponding to a total network capacity of 223 users. In the lenient scenario, this increased to 1.305 Erlangs/km²/MHz, or 231 users. In contrast, when using fixed received pilot power thresholds the entire network supported 290 users.

6.4.3.2   Fixed E_c/I_o Thresholds with 0.5 Hz Shadowing

In this section we consider fixed pilot to DL interference ratio based soft handover thresholds in a propagation environment exhibiting shadow fading in conjunction with a maximum fading frequency of 0.5 Hz and a standard deviation of 3 dB.

Examining Figure 6.29, which shows the call dropping probability, we see, again, that reducing the soft handover thresholds typically resulted in a lower probability of a dropped call. However, since the handover thresholds are dependent upon the interference level,

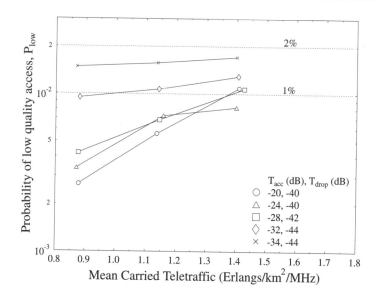

Figure 6.28: Probability of low quality access versus mean carried traffic of a CDMA based cellular network using fixed received E_c/I_o based soft handover thresholds without shadowing for $SF = 16$.

there was some interaction between the handover thresholds and the call dropping rate. For example, it can be seen in the figure that when $T_{drop} = -40$ dB, the call dropping probability fell as T_{acc} was reduced from -20 dB to -24 dB. However, on lowering T_{acc} further, to -26 dB, the call dropping rate at low traffic loads became markedly higher. A similar phenomenon was observed in Figure 6.30, which shows the probability of low quality outage.

It is explicitly seen from Figures 6.29 and 6.30 that the performance of the fixed E_c/I_o soft handover threshold based scheme clearly exceeded that of the fixed received pilot power threshold based system in a shadow fading environment. The network supported a teletraffic load of 0.7 Erlangs/km²/MHz or a total of 129 users in the conservative scenario, which rose to 0.78 Erlangs/km²/MHz, or 140 users, in the lenient scenario. These network capacities were achieved with the aid of a mean number of active base stations in the ABS, which were 1.88 and 1.91, respectively. In order to achieve the total network capacity of 129 users in the conservative scenario, a mean mobile transmit power of -2.4 dBm was required, while the mean base station transmission power was 7 dBm. For the lenient scenario, these figures were -2.4 dBm and 8.7 dBm, respectively.

6.4.3.3 Fixed E_c/I_o Thresholds with 1.0 Hz Shadowing

Increasing the maximum shadow fading frequency from 0.5 Hz to 1.0 Hz resulted in an increased call dropping probability and a greater probability of low quality access, for a given level of carried teletraffic. This is clearly seen by comparing Figures 6.31 and 6.32 with Figures 6.29 and 6.30. Explicitly, Figure 6.31 and 6.32 show that reducing the soft handover

Figure 6.29: Call dropping probability versus mean carried traffic of a CDMA based cellular network using fixed received E_c/I_o based soft handover thresholds in conjunction with 0.5 Hz shadowing and a standard deviation of 3 dB for $SF = 16$.

Figure 6.30: Probability of low quality access versus mean carried traffic of a CDMA based cellular network using fixed received E_c/I_o based soft handover thresholds in conjunction with 0.5 Hz shadowing and a standard deviation of 3 dB for $SF = 16$.

Figure 6.31: Call dropping probability versus mean carried traffic of a CDMA based cellular network using fixed received E_c/I_o based soft handover thresholds in conjunction with 1.0 Hz shadowing and a standard deviation of 3 dB for $SF = 16$.

threshold, T_{acc} from -20 dB to -24 dB led to both an increased call dropping probability and an increased probability of low quality access. This can be attributed to the extra co-channel interference generated by the greater proportion of call time being spent in soft handover. This is also confirmed by the increased probability of low quality access observed in Figure 6.32 for lower soft handover thresholds T_{acc} and T_{drop}.

The network capacity of the conservative scenario was 0.583 Erlangs/km²/MHz, giving an entire network capacity of 107 users. In the lenient scenario the network supported a total of 128 users or a traffic load of 0.675 Erlangs/km²/MHz was carried. The 107 users were serviced in conjunction with a mean ABS size of 1.86, a mean mobile transmit power of -3 dBm and a mean base station transmit power of 4.5 dBm. The 128 users supported in the lenient scenario necessitated an average mobile transmit power of -3 dBm and an average base station transmit power of 9.5 dBm. The mean number of base stations in the ABS was 1.91.

6.4.3.4 Summary

In summary, a maximum network capacity of 290 users was obtained when employing the fixed E_c/I_o soft handover thresholds. This capacity was equal to that when using fixed received pilot power thresholds in the lenient scenario without shadow fading. However, in the conservative scenario the network capacity was reduced from 290 to 231 users. Nevertheless, when a realistic shadowed propagation environment was considered, using the pilot power to interference ratio based soft handover metric improved the network capacity significantly. This was particularly evident in conjunction with the maximum shadow fading frequency of

Figure 6.32: Probability of low quality access versus mean carried traffic of a CDMA based cellular network using fixed received E_c/I_o based soft handover thresholds in conjunction with 1.0 Hz shadowing and a standard deviation of 3 dB for $SF = 16$.

1.0 Hz, when using the fixed received pilot power thresholds no users could be supported whilst maintaining the desired call quality. In contrast, using the fixed E_c/I_o soft handover thresholds led to a total network capacity of between 107 and 128 users, for the conservative and lenient scenarios, respectively. This capacity increase was the benefit of the more efficient soft handover mechanism, which was capable of taking into account the interference level experienced, leading to a more intelligent selection of base stations supporting the call. At a maximum shadow fading frequency of 0.5 Hz the network had a maximum capacity of 129 and 140 users, for the conservative and lenient scenario, respectively, when using the fixed E_c/I_o soft handover thresholds.

6.4.3.5 Relative E_c/I_o Thresholds without Shadowing

In this section we combined the benefits of using the received E_c/I_o ratio and relative soft handover thresholds, thus ensuring that variations in both the received pilot signal strength and interference levels were monitored in the soft handover process.

The call dropping performance is shown in Figure 6.33, illustrating that reducing the soft handover thresholds improved the probability of dropped calls, in particular at higher traffic loads. This phenomenon is also evident in Figure 6.34, which shows the probability of a low quality outage. However, in some cases it was evident that excessive reduction of the thresholds led to increasing the co-channel interference, and hence to a greater probability of outage associated with low quality. Again, this was the consequence of supporting an excessive number of users in soft handover, which provided a beneficial diversity gain for

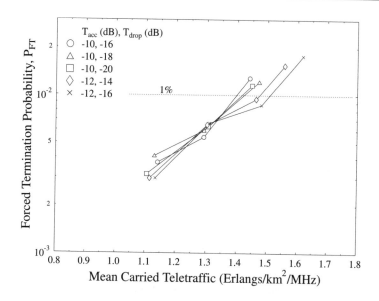

Figure 6.33: Call dropping probability versus mean carried traffic of a CDMA based cellular network using relative received E_c/I_o based soft handover thresholds without shadowing for $SF = 16$.

the mobiles but also increased the amount of DL interference inflicted by the base stations supporting the soft handovers.

The entire network supported a total of 256 users employing soft handover thresholds of $T_{acc} = -12$ dB and $T_{drop} = -16$ dB. The mean number of base stations in the active set was 1.68, and the mean mobile transmit power was 3.1 dBm. The average base station transmit power was 2.7 dBm.

6.4.3.6 Relative E_c/I_o Thresholds with 0.5 Hz Shadowing

Examining the call dropping probability graphs in Figure 6.35 shows that the probability of a dropped call was significantly lower than that of the other soft handover algorithms considered for the same propagation environment. This was because the handover algorithm was capable of taking the current interference levels into account when deciding whether to initiate a handover, additionally, the employment of the relative thresholds minimized the chances of making an inappropriate soft handover decision concerning the most suitable base station to use. The superiority of this soft handover algorithm was further emphasized by the associated low probability of a low quality access, as illustrated in Figure 6.36, which was an order of magnitude lower than that achieved using the alternative soft handover algorithms.

When T_{acc} was set to -10 dB the ultimate capacity of the network was only marginally affected by changing T_{drop}, although some variation could be observed in the call dropping probability. Furthermore, the probability of low quality access increased for the lowest values of T_{drop}. This degradation of the probability of low quality access was due to the higher proportion of time spent in soft handover, as indicated by the correspondingly increased

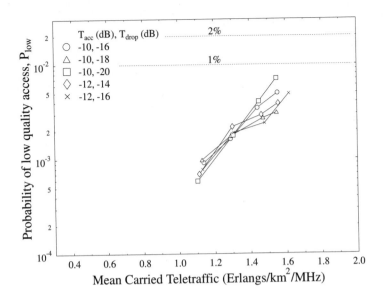

Figure 6.34: Probability of low quality access versus mean carried traffic of a CDMA based cellular network using relative received E_c/I_o based soft handover thresholds without shadowing for $SF = 16$.

Figure 6.35: Call dropping probability versus mean carried traffic of a CDMA based cellular network using relative received E_c/I_o based soft handover thresholds in conjunction with 0.5 Hz shadowing and a standard deviation of 3 dB for $SF = 16$.

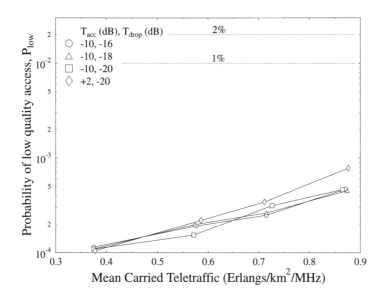

Figure 6.36: Probability of low quality access versus mean carried traffic of a CDMA based cellular
network using relative received E_c/I_o based soft handover thresholds in conjunction with
0.5 Hz shadowing and a standard deviation of 3 dB for $SF = 16$.

ABS size in Figure 6.37, which was a consequence of the associated increased co-channel
interference levels.

The mean transmit power curves of Figure 6.38 exhibited a different characteristic in
comparison to that observed for the other soft handover algorithms. Specifically, at low traffic
loads the mean mobile transmit power was less than that of the base stations, whereas at the
higher traffic loads, the mobile transmit power was greater than that of the base stations.
Although, comparing this graph with Figure 6.20 revealed that the spread and the rate of
change of the mobile transmit power versus the traffic load was similar in both scenarios, the
mean base station transmission power was lower in Figure 6.38. This reduced base station
transmission power, again demonstrated the superiority of this soft handover algorithm, which
manifested itself in its more efficient use of resources.

Since the probability of low quality access fell well below the 1% threshold, both the
conservative and lenient scenarios exhibited the same total network capacity, which was
slightly above 150 users for the entire network. This was achieved on average with the aid of
1.65 base stations, at a mean mobile transmit power of -1.2 dBm and at a mean base station
transmit power of -1.7 dBm.

6.4.3.7   Relative E_c/I_o Thresholds with 1.0 Hz Shadowing

The call dropping probability shown in Figure 6.39 is slightly worse than that obtained in
Figure 6.35 for a maximum shadow fading frequency of 0.5 Hz, with a greater performance
difference achieved by altering T_{drop}. A similar performance degradation was observed for
the probability of low quality access in Figure 6.40, with an associated relatively low impact

Figure 6.37: Mean number of base stations in the active base station set versus mean carried traffic of a CDMA based cellular network using relative received E_c/I_o based soft handover thresholds in conjunction with 0.5 Hz shadowing and a standard deviation of 3 dB for $SF = 16$.

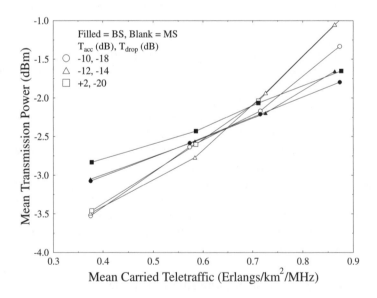

Figure 6.38: Mean transmission power versus mean carried traffic of a CDMA based cellular network using relative received E_c/I_o based soft handover thresholds in conjunction with 0.5 Hz shadowing and a standard deviation of 3 dB for $SF = 16$.

Figure 6.39: Call dropping probability versus mean carried traffic of a CDMA based cellular network using relative received E_c/I_o based soft handover thresholds in conjunction with 1.0 Hz shadowing and a standard deviation of 3 dB for $SF = 16$.

due to varying the soft handover thresholds. Although not explicitly shown, we found that the mean transmission powers were similar to those required for a maximum shadow fading frequency of 0.5 Hz.

6.4.3.8 Summary

In summary, the employment of relative E_c/I_o soft handover thresholds resulted in a superior network performance and capacity under all the propagation conditions investigated. This was achieved whilst invoking the lowest average number of base stations and the minimum mean base station transmit power. A further advantage of this handover scheme is that the same soft handover thresholds excelled in all of the propagation environments studied, unlike the previously considered algorithms, which obtained their best results at different thresholds for different conditions. The entire network capacity was 256 users without shadow fading, with a mean ABS size of 1.68. At a maximum shadowing frequency of 0.5 Hz the network supported just over a total of 150 users, whilst 144 users were served by the entire network, when a maximum shadow fading frequency of 1.0 Hz was encountered.

6.4.4 Overview of Results

Table 6.3 summarizes the results obtained for the various soft handover algorithms over the three different propagation environments considered. The fixed receiver pilot power based algorithm performed the least impressively overall, as expected due to its inherent inability to cope with shadow fading. However, it did offer a high network capacity in a non-shadowed

Figure 6.40: Probability of low quality access versus mean carried traffic of a CDMA based cellular network using relative received E_c/I_o based soft handover thresholds in conjunction with 1.0 Hz shadowing and a standard deviation of 3 dB for $SF = 16$.

Table 6.3: Maximum number of mobile users that can be supported by the network, for different soft handover metrics/algorithms whilst meeting the preset quality constraints. The mean number of base stations in the Active Base station Set (ABS) is also presented, along with the mean mobile and mean base station transmit powers.

Soft handover algorithm	Shadowing	Conservative scenario $P_{FT} = 1\%, P_{low} = 1\%$				Lenient scenario $P_{FT} = 1\%, P_{low} = 2\%$			
		Users	ABS	Power (dBm) MS	BS	Users	ABS	Power (dBm) MS	BS
Fixed pilot pwr.	No	290	1.7	5.1	5.1	290	1.7	5.1	5.1
Fixed pilot pwr.	0.5 Hz, 3 dB	—	—	—	—	127	1.83	−2.0	6.5
Fixed pilot pwr.	1.0 Hz, 3 dB	—	—	—	—	—	—	—	—
Delta pilot pwr.	No	288	1.7	4.1	4.7	288	1.7	4.1	4.1
Delta pilot pwr.	0.5 Hz, 3 dB	144	1.77	−1.5	0.6	146	1.78	−1.5	1.3
Delta pilot pwr.	1.0 Hz, 3 dB	127	1.5	−2.4	−1.9	144	1.72	−1.5	0.8
Fixed E_c/I_o	No	223	1.83	2.0	10.0	231	1.86	2.0	10.3
Fixed E_c/I_o	0.5 Hz, 3 dB	129	1.88	−2.4	7.0	140	1.91	−2.4	8.7
Fixed E_c/I_o	1.0 Hz, 3 dB	107	1.86	−3.0	4.5	128	1.91	−3.0	9.5
Delta E_c/I_o	No	256	1.68	3.1	2.7	256	1.68	3.1	2.7
Delta E_c/I_o	0.5 Hz, 3 dB	≈150	1.65	−1.2	−1.7	≈150	1.65	−1.2	−1.7
Delta E_c/I_o	1.0 Hz, 3 dB	144	1.65	−1.1	−1.6	144	1.65	−1.1	−1.6

environment. Using the relative received pilot power based soft handover algorithm improved the performance under shadow fading, but different fading rates required different thresholds to meet the conservative and lenient quality criteria. The performance of the fixed E_c/I_o based soft handover algorithm also varied significantly, when using the same thresholds for the two different fading rates considered. However, the maximum network capacity achieved under the different shadow fading conditions was significantly higher, than that of the fixed received pilot power based algorithm. This benefit resulted from the inclusion of the interference levels in the handover process, which thus took into account the fading of both the signal and the co-channel interference. Combining the relative threshold based scheme with using E_c/I_o thresholds allowed us to support the highest number of users under the shadow fading conditions investigated. Whilst its performance was not the highest in the non-shadowed environment, this propagation environment is often unrealistic, and hence the relative received E_c/I_o based soft handover algorithm was chosen as the basis for our future investigations, while using the soft handover thresholds of $T_{acc} = -10$ dB and $T_{drop} = -18$ dB. The advantages of this handover algorithm were its reduced fraction of time spent in soft handover, and its ability to perform well under both shadow fading conditions evaluated, whilst utilizing the same soft handover thresholds. Since the constraining factor of these network capacity results was the probability of a dropped call, P_{FT}, which was the same for both scenarios, further network capacity results were only shown for the conservative scenario.

6.4.5 Performance of Adaptive Antenna Arrays in a High Data Rate Pedestrian Environment

In our previous investigations we endeavored to identify the soft handover algorithm, which supports the greatest number of users, at the best call quality, regardless of the propagation conditions. In this section we study the impact of adaptive antenna arrays on the network's performance. The investigations were conducted using the relative E_c/I_o based soft handover algorithm in conjunction with $T_{acc} = -10$ dB and $T_{drop} = -18$ dB, using a spreading factor of 16. Given that the chip rate of UTRA is 3.84 Mchips/sec, this spreading factor corresponds to a channel data rate of $3.84 \times 10^6/16 = 240$ kbps. Applying $1/2$ rate error correction coding would result in an effective data throughput of 120 kbps, whereas utilizing a $2/3$ rate error correction code would provide a useful throughput of 160 kps. As in the previous simulations, a cell radius of 150 m was assumed and a pedestrian walking velocity of 3 mph was used. In our previous results investigations employing adaptive antenna arrays at the base station and using a FDMA/TDMA based network, as in Chapter 5, we observed quite significant performance gains as a direct result of the interference rejection capabilities of the adaptive antenna arrays invoked. Since the CDMA based network considered here has a frequency reuse of 1, the levels of co-channel interference are significantly higher, and hence the adaptive antennas may be able to null the interference more effectively. However, the greater number of interference sources may limit the achievable interference rejection.

Network performance results were obtained using two and four element adaptive antenna arrays, both in the absence of shadow fading, and in the presence of 0.5 Hz and 1.0 Hz frequency shadow fading exhibiting a standard deviation of 3 dB. The adaptive beamforming algorithm used was the Sample Matrix Inversion (SMI) algorithm, as described in Chapter 4 and used in the FDMA/TDMA network simulations of Chapter 5. The specific adaptive

Figure 6.41: Call dropping probability versus mean carried traffic of a CDMA based cellular network using relative received E_c/I_o based soft handover thresholds with and without beamforming and without shadowing for $SF = 16$.

beamforming implementation used in the CDMA based network was identical to that used in the FDMA/TDMA network simulations. Briefly, one of the eight possible 8-bit BPSK reference signals was used to identify the desired user, and the remaining interfering users were assigned the other seven 8-bit reference signals. The received signal's autocorrelation matrix was then calculated, and from the knowledge of the desired user's reference signal, the receiver's optimal antenna array weights were determined with the aid of the SMI algorithm. The reader is referred to Section 5.6.1 for further details. Since this implementation of the algorithm only calculated the receiver's antenna array weights, i.e. the antenna arrays weights used by the base station in the UL, these weights may not be suitable for use in the DL, when independent up/DL shadow fading is experienced. Hence, further investigations were conducted, where the UL and DL channels were identical, in order to determine the potential performance gain that may be achieved by separately calculating the antenna array weights to be used in the DL. The antenna array weights were re-calculated for every power control step, i.e. 15 times per UTRA data frame, due to the potential significant changes in terms of the desired signal and interference powers that may occur during one UTRA frame as a result of the possible 15 dB change in power transmitted by each user.

Figure 6.41 shows the significant reduction in the probability of a dropped call, i.e. the probability of forced termination P_{FT}, achieved by employing adaptive antenna arrays in a non-shadowed propagation environment. The figure has demonstrated that, even with only two antenna elements, the adaptive antenna arrays have considerably reduced the levels of co-channel interference, leading to a reduced call dropping probability. This has been achieved in spite of the numerous sources of co-channel interference resulting from the frequency reuse factor of one, which was remarkable in the light of the limited number of degrees of

Figure 6.42: Probability of low quality access versus mean carried traffic of a CDMA based cellular network using relative received E_c/I_o based soft handover thresholds with and without beamforming and without shadowing for $SF = 16$.

freedom of the two element array. Without employing antenna arrays at the base stations the network capacity was limited to 256 users, or to a teletraffic load of approximately 1.4 Erlangs/km²/MHz. However, with the advent of two element adaptive antenna arrays at the base stations the number of users supported by the network rose by 27% to 325 users, or almost 1.9 Erlangs/km²/MHz. Replacing the two element adaptive antenna arrays with four element arrays led to a further rise of 48%, or 88% with respect to the capacity of the network using no antenna arrays. This is associated with a network capacity of 480 users, or 2.75 Erlangs/km²/MHz. A summary of the network capacities achieved under different conditions is given in Table 6.4.

The probability of low quality outage, presented in Figure 6.42 also exhibited a substantial improvement with the advent of two element adaptive antenna arrays. However, the performance gains obtained when invoking four element adaptive antenna arrays were more involved. It can be seen from the figure that higher traffic loads were carried with at a sufficiently low probability of a low quality occurring, and at higher traffic loads the probability of a low quality access was lower than that achieved using a two element array. However, at lower traffic loads the performance was worse than that obtained when using two element arrays, and the gradient of the performance curve was significantly lower. Further in-depth analysis of the results suggested that the vast majority of the low quality outages were occurring when new calls started. When a user decided to commence communications with the base station, the current interference level was measured, and the target transmission power was determined in order to reach the target SINR necessary for reliable communications. However, in order to avoid disrupting existing calls the transmission power was ramped up slowly, until the target SINR was reached. A network using no adaptive

Figure 6.43: The changes in the antenna array gain, versus time, in the direction of the desired user, the UL and DL transmission powers, and the UL and DL received SINRs, when a new call starts using four element adaptive antenna arrays without shadowing in conjunction with the original power ramping algorithm and $SF = 16$.

antenna arrays, i.e. employing omnidirectional antennas, can be viewed as offering equal gain to all users of the network, which we assumed to be 1.0, or 0 dB. Thus, when a new call is initiated, the level of interference rises gradually, and the power control algorithm ensures that the existing users compensate for the increased level of co-channel interference by increasing their transmission power. In a network using adaptive antenna arrays, the adaptive antenna arrays are used to null the sources of interference, and in doing so the array may reduce the antenna gain in the direction of the desired user, in order to maximize the SINR. Hence a user starting a new call, even if it has low transmission power, can alter the antenna array's response, and thus the antenna gain experienced by the existing users. This phenomenon is more marked when using four element arrays since their directivity, and thus sensitivity to interfering signals, is greater.

Figure 6.43 illustrates this phenomenon, where another user starts a new call at frame 112 suddenly reducing the antenna gain in the direction of the desired user from 0.4 to just above 0.2, a drop of 3 dB. As can be seen from the figure, the DL SINR falls sharply below the low quality outage threshold of 7.0 dB, resulting in several consecutive outages, until the DL transmission power is increased sufficiently. The impact of reducing the initial transmission power, in order to ensure that the power ramping takes place more gently, is depicted in Figure 6.44. In this figure it can be seen that the antenna gain falls much more gently, over a prolonged period of time, thus reducing the number of low quality outages, as the DL

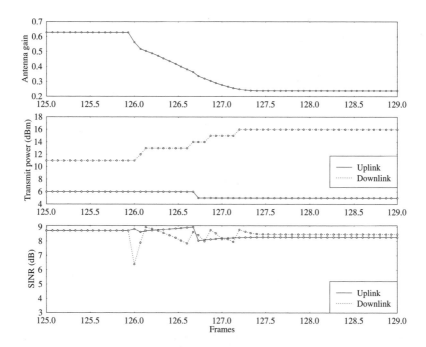

Figure 6.44: The changes in the antenna array gain, versus time, in the direction of the desired user, the UL and DL transmission powers, and the UL and DL received SINRs, when a new call starts using four element adaptive antenna arrays without shadowing in conjunction with a slower power ramping algorithm and $SF = 16$.

transmission power is increased in an effort to compensate for the lower antenna gain. It is of interest to note how the received SINR varies as the antenna gain and the power control algorithm interact, in order to maintain the target SINR.

Even though the employment of adaptive antenna arrays can result in the attenuation of the desired signal, this is performed in order to maximize the received SINR, and thus the levels of interference are attenuated more strongly, ultimately leading to the reduction of the mean transmission power, as emphasized by Figure 6.45. This figure clearly shows the lower levels of transmission power, required in order to maintain an acceptable performance, whilst using adaptive antenna arrays at the base stations. A reduction of 3 dB in the mean mobile transmission power was achieved by invoking two element antenna arrays, and a further reduction of 1.5 dB resulted from using four element arrays. These power budget savings were obtained in conjunction with reduced levels of co-channel interference, leading to superior call quality, as illustrated in Figures 6.41 and 6.42. A greater performance advantage was evident in the UL scenario, suggesting that the selective base station diversity techniques employed in the UL are amenable to amalgamation with adaptive antenna arrays. In contrast, the maximum ratio combining performed at the mobile inherently reduces the impact of co-channel interference, and hence benefits to a lesser extent from the employment of adaptive antenna arrays.

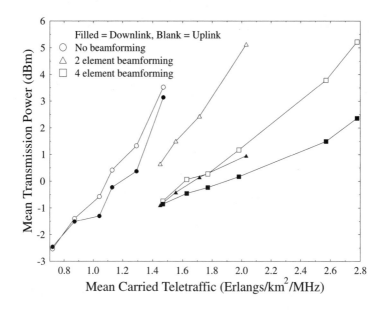

Figure 6.45: Mean transmission power versus mean carried traffic of a CDMA based cellular network using relative received E_c/I_o based soft handover thresholds with and without beamforming and without shadowing for $SF = 16$.

Figure 6.46: Call dropping probability versus mean carried traffic of a CDMA based cellular network using relative received E_c/I_o based soft handover thresholds with and without beamforming and with shadowing having a standard deviation of 3 dB for $SF = 16$.

Figure 6.47: Probability of low quality access versus mean carried traffic of a CDMA based cellular network using relative received E_c/I_o based soft handover thresholds with and without beamforming and with shadowing having a standard deviation of 3 dB for $SF = 16$.

The impact of adaptive antenna arrays in a propagation environment subjected to shadow fading was then investigated. The associated call dropping performance is shown in Figure 6.46. This figure illustrates the substantial network capacity gains achieved with the aid of both two and four element adaptive antenna arrays under shadow fading propagation conditions. Simulations were conducted in conjunction with log-normal shadow fading having a standard deviation of 3 dB, and maximum shadowing frequencies of both 0.5 Hz and 1.0 Hz. As expected the network capacity was reduced at the faster fading frequency. The effect of performing independent UL and DL beamforming, as opposed to using the base station's receive antenna array weights in the DL was also studied, and a small, but not insignificant call dropping probability reduction can be seen in the Figure 6.46. The network supported just over 150 users, and 144 users, when subjected to 0.5 Hz and 1.0 Hz frequency shadow fading, respectively. With the application of two element adaptive antenna arrays, re-using the base station's UL receiver weights on the DL, these capacities increased by 35% and 40%, to 203 users and 201 users. Performing independent UL and DL beamforming resulted in a mean further increase of 13% in the network capacity. The implementation of four element adaptive antenna arrays led to a network capacity of 349 users at a 0.5 Hz shadowing frequency, and 333 users at a 1.0 Hz shadowing frequency. This corresponded to relative gains of 133% and 131% over the capacity provided without beamforming. Invoking independent UL and DL beamforming gave another boost of 7% and 10% to network capacity for 0.5 Hz and 1.0 Hz frequency shadowing environments, respectively, giving final network capacities of just over 375 users and 365 users.

Similar trends were observed regarding the probability of low quality outage to those found in the non-shadowing scenarios. However, the trend was much more prevalent under

Figure 6.48: Mean transmission power versus mean carried traffic of a CDMA based cellular network using relative received E_c/I_o based soft handover thresholds with and without beamforming and shadowing having a standard deviation of 3 dB for $SF = 16$.

shadowing, due to greater variation of the received signal strengths, as a result of the shadow fading, as shown in Figure 6.47. The figure indicates that the trend is also evident, when using two element adaptive antenna arrays in conjunction with shadow fading. As expected, the performance deteriorated as the number of antenna elements increased, and when the maximum shadow fading frequency was increased from 0.5 Hz to 1.0 Hz. It should be noted, however that the probability of low quality access always remained below the 1% constraint of the conservative scenario, and the call dropping probability was considerably reduced by the adaptive antenna arrays.

The mean transmission power performance is depicted in Figure 6.48, suggesting that as for the non-shadowing scenario of Figure 6.45, the number of antenna elements had only a limited impact on the base stations' transmission power, although there was some reduction in the mobile stations' mean transmission power. The mean transmission powers required when using independent UL and DL beamforming are not explicitly shown, but were slightly less than those presented here, with a mean reduction of about 0.4 dB.

A summary of the maximum network capacities of the networks considered in this section both with and without shadowing, employing beamforming using two and four element arrays is given in Table 6.4, along with the teletraffic carried and the mean mobile and base station transmission powers required.

The lower bounds of the maximum network capacities obtained under identical scenarios in conjunction with a spreading factor of 256, are also presented in Table 6.5, leading to a bit rate of 15 kbps, which is suitable for use by speech-rate users. The network capacity calculations were performed by scaling the number of users supported, as presented in

Table 6.4: Maximum mean carried traffic and maximum number of mobile users that can be supported by the network, whilst meeting the conservative quality constraints. The carried traffic is expressed in terms of normalized Erlangs (Erlang/km^2/MHz) for the network described in Table 6.2 both with and without beamforming (as well as with and without independent UL/DL beamforming), and also with and without shadow fading having a standard deviation of 3 dB for $SF = 16$.

				Conservative scenario, P_{FT}=1%, P_{low}=1%		
		Independent		Traffic	Power (dBm)	
Shadowing	Beamforming:	UL/DL	Users	(Erlangs/km^2/MHz)	MS	BS
No	No	—	256	1.42	3.1	2.7
No	2 elements	—	325	1.87	3.75	0.55
No	4 elements	—	480	2.75	4.55	1.85
0.5 Hz, 3 dB	No	—	≈150	0.87	−1.2	−1.7
0.5 Hz, 3 dB	2 elements	No	203	1.16	0.1	−1.1
0.5 Hz, 3 dB	4 elements	No	349	2.0	2.0	0.65
0.5 Hz, 3 dB	2 elements	Yes	233	1.35	0.2	−0.8
0.5 Hz, 3 dB	4 elements	Yes	≈375	2.2	2.15	0.85
1.0 Hz, 3 dB	No	—	144	0.82	−1.1	−1.6
1.0 Hz, 3 dB	2 elements	No	201	1.12	−0.3	−1.1
1.0 Hz, 3 dB	4 elements	No	333	1.88	1.6	0.5
1.0 Hz, 3 dB	2 elements	Yes	225	1.31	0.1	−0.9
1.0 Hz, 3 dB	4 elements	Yes	365	2.05	1.65	0.6

Table 6.4, by the ratio of their spreading factors, i.e. $256/16 = 16$. Further interesting user capacity figures can be inferred for a variety of target bit rates by comparing Tables 6.4, 6.5, 6.7 and 6.8 and applying the appropriate spreading factor related scaling mentioned in the context of estimating the number of 15 kbps speech users supported.

6.4.6 Performance of Adaptive Antenna Arrays and Adaptive Modulation in a High Data Rate Pedestrian Environment

In this section we build upon the results presented in the previous section by applying Adaptive Quadrature Amplitude Modulation (AQAM) techniques. The various scenarios and channel conditions investigated were identical to those of the previous section, except for the application of AQAM. Since in the previous section an increased network capacity was achieved due to using independent UL and DL beamforming, this procedure was invoked in these simulations. AQAM involves the selection of the appropriate modulation mode in order to maximize the achievable data throughput over a channel, whilst minimizing the Bit Error Ratio (BER). More explicitly, the philosophy behind adaptive modulation is the most appropriate selection of a modulation mode according to the instantaneous radio channel quality experienced [12, 13]. Therefore, if the SINR of the channel is high, then a high-order modulation mode may be employed, thus exploiting the temporal fluctuation of the radio channel's quality. Similarly, if the channel is of low quality, exhibiting a low

Table 6.5: A lower bound estimate of the maximum mean traffic and the maximum number of mobile speech-rate users that can be supported by the network, whilst meeting the conservative quality constraints. The carried traffic is expressed in terms of normalized Erlangs (Erlang/km²/MHz) for the network described in Table 6.2 both with and without beamforming (as well as with and without independent UL/DL beamforming), and also with and without shadow fading having a standard deviation of 3 dB for $SF = 256$. The number of users supported in conjunction with a spreading factor of 256 was calculated by multiplying the capacities obtained in Table 6.4 by $256/16 = 16$.

Shadowing	Beamforming:	Independent UL/DL	Users when $SF = 256$	Traffic (Erlangs/km²/MHz)
No	No	—	4096	22.7
No	2 elements	—	5200	29.9
No	4 elements	—	7680	44.0
0.5 Hz, 3 dB	No	—	2400	13.9
0.5 Hz, 3 dB	2 elements	No	3248	18.6
0.5 Hz, 3 dB	4 elements	No	5584	32.0
0.5 Hz, 3 dB	2 elements	Yes	3728	21.6
0.5 Hz, 3 dB	4 elements	Yes	6000	35.2
1.0 Hz, 3 dB	No	—	2304	13.1
1.0 Hz, 3 dB	2 elements	No	3216	17.9
1.0 Hz, 3 dB	4 elements	No	5328	30.1
1.0 Hz, 3 dB	2 elements	Yes	3600	21.0
1.0 Hz, 3 dB	4 elements	Yes	5840	32.8

SINR, a high-order modulation mode would result in an unacceptably high BER or FER, and hence a more robust, but lower throughput modulation mode would be employed. Therefore, adaptive modulation combats the effects of time-variant channel quality, while also attempting to maximize the achieved data throughput, and maintaining a given BER or FER. In the investigations conducted, the modulation modes of the UL and DL were determined independently, thus taking advantage of the lower levels of co-channel interference on the UL, or of the potentially greater transmit power of the base stations.

The particular implementation of AQAM used in these investigations is illustrated in Figure 6.49. This figure describes the algorithm in the context of the DL, but the same implementation was used also in the UL. The first step in the process was to establish the current modulation mode. If the user was invoking 16-QAM and the SINR was found to be below the Low Quality (LQ) outage SINR threshold after the completion of the power control iterations, then the modulation mode for the next data frame was 4-QAM. Alternatively, if the SINR was above the LQ outage SINR threshold, but any of the base stations in the ABS were using a transmit power within 15 dB of the maximum transmit power—which is the maximum possible power change range during a 15-slot UTRA frame—then the 4-QAM modulation mode was selected. This "headroom" was introduced in order to provide a measure of protection, since if the interference conditions degrade, then at least 15 dB of increased transmit power would be available in order to mitigate the consequences of the SINR reduction experienced.

Table 6.6: The target SINR, low quality outage SINR and outage SINR thresholds used for the BPSK, 4-QAM and 16-QAM modulation modes of the adaptive modem.

SINR Threshold	BPSK	4-QAM	16-QAM
Outage SINR	2.6 dB	6.6 dB	12.1 dB
Low Quality Outage SINR	3.0 dB	7.0 dB	12.5 dB
Target SINR	4.0 dB	8.0 dB	13.5 dB

A similar procedure was invoked when switching to other legitimate AQAM modes from the 4-QAM mode. If the SINR was below the 4-QAM target SINR and any one of the base stations in the ABS was within 15 dB (the maximum possible power change during a 15-slot UTRA data frame) of the maximum transmit power, then the BPSK modulation mode was employed for the next data frame. However, if the SINR exceeded the 4-QAM target SINR and there would be 15 dB of headroom in the transmit power budget in excess of the extra transmit power required for switching from 4-QAM to 16-QAM, then the 16-QAM modulation mode was invoked.

And finally, when in the BPSK mode, the 4-QAM modulation mode was selected if the SINR exceeded the BPSK target SINR, and the transmit power of any of the base stations in the ABS was less than the power required to transmit reliably using 4-QAM, while being at least 15 dB below the maximum transmit power. The algorithm was activated at the end of each 15-slot UTRA data frame, after the power control algorithm had performed its 15 iterations per data frame, and thus the AQAM mode selection was performed on a UTRA transmission frame-by-frame basis. When changing from a lower-order modulation to a higher-order modulation mode, the lower-order mode was retained for an extra frame in order to ramp up the transmit power to the required level, as shown in Figure 6.50(a). Conversely, when changing from a higher-order modulation mode to a lower-order modulation mode, the lower-order modulation mode was employed whilst ramping the power down, in order to avoid excessive outages in the higher-order modulation mode due to the reduction of the transmit power, as illustrated in Figure 6.50(b).

Table 6.6 gives the BPSK, 4-QAM and 16-QAM SINR thresholds used in the simulations. The BPSK SINR thresholds were 4 dB lower than those necessary when using 4-QAM, while the 16-QAM SINR thresholds were 5.5 dB higher [408]. In other words, in moving from the BPSK modulation mode to the 4-QAM modulation mode, the target SINR, low quality outage SINR and outage SINR all increased by 4 dB. When switching to the 16-QAM mode from the 4-QAM mode, the SINR thresholds increased by 5.5 dB. However, setting the BPSK to 4-QAM and the 4-QAM to 16-QAM mode switching thresholds to a value 7 dB higher than the SINR required for maintaining the target BER/FER was necessary in order to prevent excessive outages due to sudden dramatic channel-induced variations in the SINR levels.

Performance results were obtained both with and without beamforming in a log-normal shadow fading environment, at maximum fading frequencies of 0.5 Hz and 1.0 Hz, and a standard deviation of 3 dB. A pedestrian velocity of 3 mph, a cell radius of 150 m and a spreading factor of 16 were used, as in our previous investigations.

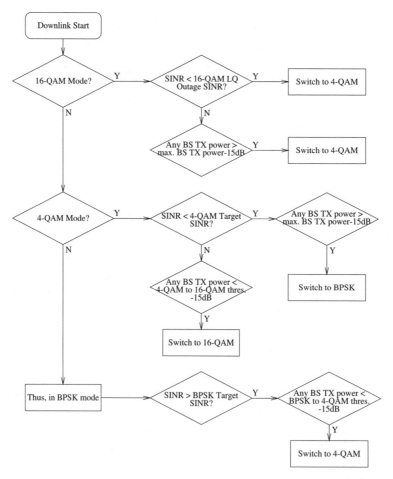

Figure 6.49: The AQAM mode switching algorithm used in the DL of the CDMA based cellular network.

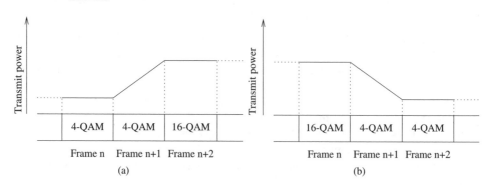

Figure 6.50: Power ramping requirements whilst switching modulation modes: (a) ramping up the transmit power whilst remaining in the lower order modulation mode; (b) ramping down the transmit power whilst switching to the lower order modulation mode.

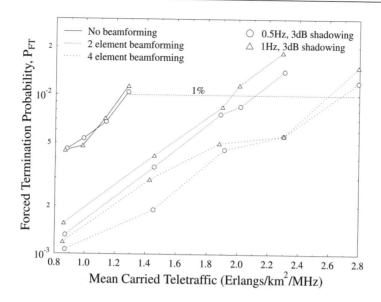

Figure 6.51: Call dropping probability versus mean carried traffic of a CDMA based cellular network using relative received E_c/I_o based soft handover thresholds both with and without beamforming in conjunction with AQAM as well as with shadowing having a standard deviation of 3 dB for $SF = 16$. See Figure 6.46 for corresponding results without adaptive modulation.

Figure 6.51 shows the significant reduction in the probability of a dropped call, achieved by employing adaptive antenna arrays in conjunction with adaptive modulation in a log-normal shadow faded environment. The figure demonstrates that, even with the aid of a two element adaptive antenna array and its limited degrees of freedom, a substantial call dropping probability reduction was achieved. The performance benefit of increasing the array's degrees of freedom, achieved by increasing the number of antenna elements, becomes explicit from the figure, resulting in a further call dropping probability reduction. Simulations were conducted in conjunction with log-normal shadow fading having a standard deviation of 3 dB, and maximum shadowing frequencies of 0.5 Hz and 1.0 Hz. As expected, the call dropping probability was generally higher at the faster fading frequency, as demonstrated by Figure 6.51. The network was found to support 223 users, corresponding to a traffic load of 1.27 Erlang/km²/MHz, when subjected to 0.5 Hz frequency shadow fading. The capacity of the network was reduced to 218 users, or 1.24 Erlang/km²/MHz, upon increasing the maximum shadow fading frequency to 1.0 Hz. On employing two element adaptive antenna arrays, the network capacity increased by 64% to 366 users, or to an equivalent traffic load of 2.11 Erlang/km²/MHz when subjected to 0.5 Hz frequency shadow fading. When the maximum shadow fading frequency was raised to 1.0 Hz, the number of users supported by the network was 341 users, or 1.98 Erlang/km²/MHz, representing an increase of 56% in comparison to the network without adaptive antenna arrays. Increasing the number of antenna elements to four, whilst imposing shadow fading with a maximum frequency of 0.5 Hz, resulted in a network capacity of 2.68 Erlang/km²/MHz or 476 users, corresponding

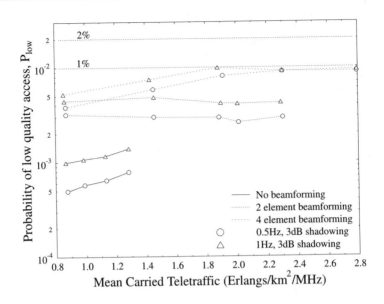

Figure 6.52: Probability of low quality access versus mean carried traffic of a CDMA based cellular network using relative received E_c/I_o based soft handover thresholds both with and without beamforming in conjunction with AQAM as well as with shadowing having a standard deviation of 3 dB for $SF = 16$. See Figure 6.47 for corresponding results without adaptive modulation.

to a gain of an extra 30% with respect to the network employing two element arrays, and of 113% in comparison to the network employing no adaptive antenna arrays. In conjunction with a maximum shadow fading frequency of 1.0 Hz the network capacity was 460 users or 2.59 Erlang/km^2/MHz, which represented an increase of 35% with respect to the network invoking two element antenna arrays, or 111% relative to the identical network without adaptive antenna arrays.

The probability of low quality outage, presented in Figure 6.52, did not benefit from the application of adaptive antenna arrays, or from the employment of adaptive modulation. Figure 6.47 depicts the probability of low quality outage without adaptive modulation, and upon comparing these results to those obtained in conjunction with adaptive modulation shown in Figure 6.52, the performance degradation due to adaptive modulation can be explicitly seen. However, the increase in the probability of low quality access can be attributed to the employment of less robust, but higher throughput, higher-order modulation modes invoked by the adaptive modulation scheme. Hence, under given propagation conditions and using the fixed 4-QAM modulation mode a low quality outage may not occur, yet when using adaptive modulation and a higher order modulation mode, the same propagation conditions may inflict a low quality outage. This phenomenon is further exacerbated by the adaptive antenna arrays, as described in Section 6.4.5, where the addition of a new source of interference, constituted by a user initiating a new call, results in an abrupt change in the gain of the antenna in the direction of the desired user. This in turn leads to low quality outages, which are more likely to occur for prolonged periods of time when using a higher order

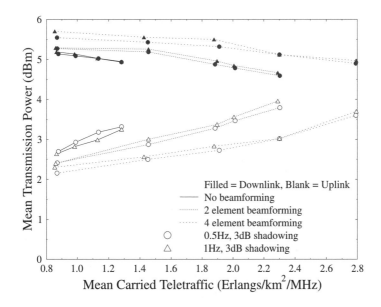

Figure 6.53: Mean transmission power versus mean carried traffic of a CDMA based cellular network using relative received E_c/I_o based soft handover thresholds both with and without beamforming in conjunction with AQAM as well as with shadowing having a standard deviation of 3 dB for $SF = 16$. See Figure 6.48 for corresponding results without adaptive modulation.

modulation mode. Again, increasing the number of antenna elements from two to four results in an increased probability of a low quality outage due to the sharper antenna directivity. This results in a higher sensitivity to changes in the interference incident upon it.

The mean transmission power versus teletraffic performance is depicted in Figure 6.53, suggesting that the mean UL transmission power was always significantly below the mean DL transmission power, which can be attributed to the pilot power interference encountered by the mobiles in the DL. This explanation can be confirmed by examining Figure 6.54, which demonstrates that the mean modem throughput in the DL, without adaptive antenna arrays, was lower than that in the UL even in conjunction with increased DL transmission power. Invoking adaptive antenna arrays at the base stations reduced the mean UL transmission power required in order to meet the service quality targets of the network. The attainable DL power reduction increased as the number of antenna array elements increased, as a result of the superior interference rejection achieved with the aid of a higher number of array elements. A further advantage of employing a larger number of antenna array elements was the associated increase in the mean UL modem throughput, which became more significant at higher traffic loads. In the DL scenario, however, increasing the number of adaptive antenna array elements led to an increased mean DL transmission power, albeit with a substantially improved mean DL modem throughput. This suggests that there was some interaction between the adaptive antenna arrays, the adaptive modulation mode switching algorithm and the maximal ratio combining performed at the mobiles. In contrast, simple switched diversity was performed by the base stations on the UL, thus avoiding such a

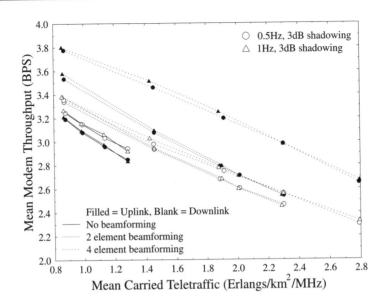

Figure 6.54: Mean modem throughput versus mean carried traffic of a CDMA based cellular network using relative received E_c/I_o based soft handover thresholds both with and without beamforming in conjunction with AQAM as well as with shadowing having a standard deviation of 3 dB for $SF = 16$.

situation. However, the increase in the mean DL transmission power resulted in a much more substantial increase in the mean DL modem throughput, especially with the advent of the four element antenna arrays, which exhibited an approximately 0.5 BPS throughput gain over the two element arrays for identical high traffic loads which can be seen in Figure 6.54.

A summary of the maximum user capacities of the networks considered in this section in conjunction with log-normal shadowing having a standard deviation of 3 dB, with and without employing beamforming using two and four element arrays is given in Table 6.7. The teletraffic carried the mean mobile and base station transmission powers required, and the mean UL and DL modem data throughputs achieved are also shown in Table 6.7. Similarly, the lower bounds of the maximum network capacities obtained under identical scenarios in conjunction with a spreading factor of 256, leading to a bit rate of 15 kbps, suitable for speech-rate users are presented in Table 6.8. The network capacity calculations were performed by scaling the number of users supported, as presented in Table 6.7, by the ratio of their spreading factors, i.e. by $256/16 = 16$.

6.5   Summary and Conclusions

We commenced this chapter with a brief overview of the background behind the 3G UTRA standard. This was followed in Sections 6.2 and 6.3 by an introduction to CDMA and the techniques invoked in the UTRA standard.

Table 6.7: Maximum mean carried traffic and maximum number of mobile users that can be supported by the network, whilst meeting the conservative quality constraints. The carried traffic is expressed in terms of normalized Erlangs (Erlang/km^2/MHz), for the network described in Table 6.2 both with and without beamforming (using independent up/down-link beamforming), in conjunction with shadow fading having a standard deviation of 3 dB, whilst employing adaptive modulation techniques for $SF = 16$.

				Conservative scenario				
				Power (dBm)		Throughput (BPS)	
Shadowing	Beamforming	Users	Traffic (Erlangs /km^2/MHz)	MS	BS	UL	DL
0.5 Hz, 3 dB	No	223	1.27	3.25	4.95	2.86	2.95
0.5 Hz, 3 dB	2 elements	366	2.11	3.55	4.7	2.56	2.66
0.5 Hz, 3 dB	4 elements	476	2.68	3.4	5.0	2.35	2.72
1.0 Hz, 3 dB	No	218	1.24	3.3	4.95	2.87	2.96
1.0 Hz, 3 dB	2 elements	341	1.98	3.5	4.9	2.62	2.73
1.0 Hz, 3 dB	4 elements	460	2.59	3.5	4.95	2.4	2.8

Table 6.8: A lower bound estimate of the maximum mean carried traffic and maximum number of mobile speech-rate users that can be supported by the network, whilst meeting the conservative quality constraints. The carried traffic is expressed in terms of normalized Erlangs (Erlang/km^2/MHz), for the network described in Table 6.2 both with and without beamforming (using independent up/down-link beamforming), in conjunction with shadow fading having a standard deviation of 3 dB, whilst employing adaptive modulation techniques for $SF = 256$. The number of users supported in conjunction with a spreading factor of 256 was calculated by multiplying the capacities obtained in Table 6.7 by $256/16 = 16$.

		Conservative scenario	
Shadowing	Beamforming	Users	Traffic (Erlangs/km^2/MHz)
0.5 Hz, 3 dB	No	3568	20.3
0.5 Hz, 3 dB	2 elements	5856	33.8
0.5 Hz, 3 dB	4 elements	7616	42.9
1.0 Hz, 3 dB	No	3488	19.8
1.0 Hz, 3 dB	2 elements	5456	31.7
1.0 Hz, 3 dB	4 elements	7360	41.4

Network capacity studies were then conducted in Section 6.4, which evaluated the performance of four different soft handover algorithms in the context of both non-shadowed and log-normal shadow faded propagation environments. The algorithm using relative received pilot-to-interference ratio measurements at the mobile, in order to determine the most suitable base stations for soft handover, was found to offer the highest network capacity when subjected to shadow fading propagation conditions. Hence, this algorithm and its associated parameters were selected for use in our further investigations. The impact of adaptive antenna arrays upon the network capacity was then considered in Section 6.4.5 in both non-shadowed and log-normal shadow faded propagation environments. Considerable network capacity gains were achieved, employing both two and four element adaptive antenna arrays. This work was then extended in Section 6.4.6 by the application of adaptive modulation techniques in conjunction with the previously studied adaptive antenna arrays in a log-normal shadow faded propagation environment, which elicited further significant network capacity gains.

Chapter 7

HSDPA-style FDD/CDMA Performance Using Loosely Synchronized Spreading Codes

7.1 Effects of Loosely Synchronized Spreading Codes on the Performance of CDMA Systems

7.1.1 Introduction

In this section we characterize the achievable system performance of a UTRA-like FDD CDMA system employing Loosely Synchronized (LS) spreading codes. Current CDMA systems are interference limited, suffering from Inter-Symbol-Interference (ISI), since the orthogonality of the spreading sequences is destroyed by the channel. They also suffer from Multiple-Access-Interference (MAI) owing to the non-zero cross-correlations of the spreading codes. LS codes exhibit a so-called Interference Free Window (IFW), where both the auto-correlation and cross-correlation values of the codes become zero. Therefore LS codes have the promise of mitigating the effects of both ISI and MAI in time dispersive channels. Hence, LS codes have the potential of increasing the capacity of CDMA networks. This contribution studies the achievable network performance by simulation and compares it to that of a UTRA-like FDD/CDMA system using Orthogonal Variable Spreading Factor (OVSF) codes.

In our previous research [416–418], outlined in the preceding chapters the performance of a UTRA-like FDD CDMA system was quantified, when supported by both adaptive beam-steering and adaptive modulation [419]. In [418], the system employed OVSF spreading codes [420], which offer the benefit of perfect orthogonality in an ideal channel. In a non-dispersive channel, all intra-cell users' signals are perfectly orthogonal. However, upon propagating through a dispersive multipath channel this orthogonality is eroded, hence

Figure 7.1: The LS code structure.

all other users will interfere with the desired signal. Therefore in practice the intra-cell interference is always non-zero.

We will consider the employment of a specific family of spreading codes, which are known as Loosely Synchronized codes [421]. These codes exhibit a so-called Interference Free Window, where the off-peak aperiodic autocorrelation values as well as the aperiodic cross-correlation values become zero, resulting in zero ISI and zero MAI, provided that the delayed asynchronous transmissions arrive within the IFW. More specifically, interference-free CDMA communications become possible, when the total time offset expressed in terms of the number of chip intervals, which is the sum of the time-offset of the mobiles plus the maximum channel-induced delay spread is within the code's IFW [422]. By employing this specific family of codes, we are capable of reducing both the ISI and the MAI, since users roaming in the same cell do not interference with each other, as a benefit of the IFW of the LS codes used, provided that their multipath-induced ISI arrives within the IFW.

The spreading codes of the UTRA CDMA system are based on the Orthogonal Variable Spreading Factor (OVSF) technique, which was originally proposed in [420] and hence in the previous chapters OVSF codes were employed. The UTRA DL employs synchronous transmissions within each cell and hence it is capable of exploiting the orthogonality of OVSF codes [59]. The OVSF codes used in the DL are hence capable of perfectly avoiding intra-cell multiuser interference, provided that no multipath-induced linear distortions are encountered. However, in the presence of wide-band multipath propagation channel-induced linear distortion is encountered and hence the orthogonality of the OVSF codes is destroyed, leading to multiuser interference in the DL [423–426]. More explicitly, the preservation of the OVSF codes' orthogonality is primarily dependent upon the radio channel linking the user population to the base station transmitter. Encountering a high number of multipath components degrades the OVSF codes' orthogonality, unless all the multipath components are resolved, and coherently combined at the receiver.

7.1.2 Loosely Synchronized Codes [427]

There exists a specific family of LS codes [421, 428–430], which exhibits an IFW. Specifically, LS codes exploit the properties of the so-called orthogonal complementary sets [421, 431]. To expound further, let us introduce the notation of $LS(N, P, W_0)$ for denoting the family of LS codes generated by applying a $(P \times P)$-dimensional Walsh-Hadamard matrix to an orthogonal complementary code set of length N, while inserting W_0 number of zeros in the center and at the beginning of the LS code, as shown in Figure 7.1, using the procedure described in [421]. Then, the total length of the $LS(N, P, W_0)$ code is given by $L = 2NP + 2W_0$ and the number of codes available is given by $2P$.

Since the construction method of binary LS codes was described in [421], we will focus our attention on the employment of orthogonal complementary sets [432, 433] for the generation of LS codes. Firstly, we define a sequence set $c_1, \ldots, c_N$, where

$c_n = [c_{n,0}, \ldots, c_{n,L-1}]$ is a spreading sequence having a length of L. These spreading codes exhibit an IFW width of τ_{IFW}, if the cross-correlation of the spreading codes satisfies:

$$R_{jk}(\tau) = \sum_{l=0}^{L-1} c_{j,l} c_{k,(l+\tau) \bmod L} = \begin{cases} L, & \text{for } \tau = 0, j = k \\ 0, & \text{for } \tau = 0, j \neq k \\ 0, & \text{for } 0 < |\tau| \leq \tau_{\text{IFW}}. \end{cases} \qquad (7.1)$$

The aperiodic correlation $R_{j,k}(\tau)$ of two sequences g_j and g_k has to satisfy Equation 7.1 for the sake of maintaining an IFW of τ_{IFW} chip intervals.

For a given complementary pair $\{c_0, s_0\}$ of length N, one of the corresponding mate pairs can be written as $\{c_1, s_1\}$, where we have:

$$c_1 = \tilde{s}_0^*, \qquad (7.2)$$
$$s_1 = -\tilde{c}_0^*, \qquad (7.3)$$

and where $\tilde{s}_0$ denotes the reverse-ordered sequence, while $-s_0$ is the negated version of s_0, respectively. Note that in (7.2) and (7.3) additional complex conjugation of the polyphase complementary sequences is required for deriving the corresponding mate pair in comparison to binary complementary sequences [421]. Having obtained a complementary pair and its corresponding mate pair, we may employ the construction method of [421] for generating a family of LS codes. The LS codes generated exhibit an IFW, where we have $R_{jk}(\tau) = 0$ for $|\tau| \leq min\{N-1, W_0\}$. Hence, we may adopt a choice of $W_0 = N - 1$ in order to minimize the total length of the LS codes generated, while providing as long an IFW as possible.

For example, the LS(4,2,3) codes can be generated based on the complementary pair of [432]:

$$c_0 = + + + - \qquad (7.4)$$
$$s_0 = + + - - . \qquad (7.5)$$

Upon substituting (7.2) and (7.3) into (7.4) and (7.5), the corresponding mate pair can be obtained as:

$$c_1 = \tilde{s}_0^* = + - + + \qquad (7.6)$$
$$s_1 = -\tilde{c}_0^* = + - - - . \qquad (7.7)$$

The generation of this set of the four LS codes can be viewed in Figure 7.2. Upon invoking the 2×2-dimensional Hadamard expansion of [421] in the context of the above orthogonal complementary pairs, we can generate a family of four LS(4,2,3) codes, which are denoted by g_p, $p = 0, \ldots, 3$.

All four different codes of the LS(4,2,3) code family exhibited the same autocorrelation magnitudes, namely that seen in Figure 7.3(a). It can be observed in Figure 7.3(a) that the off-peak autocorrelation $R_p[\tau]$ becomes zero for $|\tau| \leq W_0 = 3$. The crosscorrelation magnitudes $|R_{j,k}(\tau)|$ depicted in Figure 7.3(b) are also zero for $|\tau| \leq W_0 = 3$. Based on the observations made as regards to the aperiodic correlations we may conclude that the LS(4,2,3) codes exhibit an IFW of ± 3 chip durations.

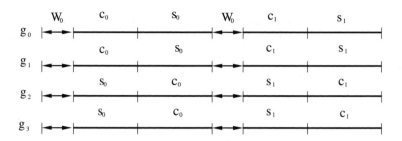

Figure 7.2: Generating four LS codes.

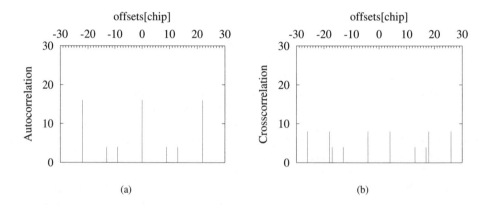

Figure 7.3: Correlation magnitudes of the $LS(4, 2, 3)$ codes. (a) All four codes exhibit the same autocorrelation magnitude. (b) The crosscorrelation magnitudes of g_0 and g_2.

7.1.3   System Parameters

The cell-radius was 78 m, which was the maximum affordable cell radius for the IFW duration of ± 1 chip intervals at a chip rate of 3.84 Mchip/s. The call duration and inter-call periods were Poisson distributed with the mean values shown in Table 7.1. For our initial investigations we have assumed that the basestations and mobiles form a synchronous network.

Furthermore, the post-despreading SINRs required for obtaining the target BERs were determined with the aid of physical-layer simulations using a 4-QAM modulation scheme, in conjunction with 1/2-rate turbo coding for transmission over a COST 207 seven-path Bad Urban channel [434]. Using this turbo-coded transceiver and LS codes having a spreading factor (SF) of 16, the post-despreading SINR required for maintaining the target BER of 1×10^{-3} was 6.2 dB. The BER which was deemed to correspond to low-quality access, was stipulated at 5×10^{-3}. This BER was exceeded for SINRs falling below 5.2 dB. Furthermore, a low-quality outage was declared, when the BER of 1×10^{-2} was exceeded, which was encountered for SINRs below 4.8 dB. These values can be seen along with the other system parameters in Table 7.1. The performance metrics used were defined in Section 5.3.3.4 and

Table 7.1: Simulation parameters.

Parameter	Value	Parameter	Value
Noisefloor	-100 dBm	Pilot power	-9 dBm
Frame length	10 ms	Cell radius	78 m
Multiple access	FDD/CDMA	Number of basestations	49
Modulation scheme	4QAM/QPSK	Spreading factor	16
Min BS transmit power	-48 dBm	Min MS transmit power	-48 dBm
Max BS transmit power	17 dBm	Max MS transmit power	17 dBm
Power control stepsize	1 dB	Power control hysteresis	1 dB
Low quality access SINR	5.2 dB	Outage (1% BER) SINR	4.8 dB
Pathloss exponent	-2.0	Size of Active BS Set (ABS)	2
Average inter-call-time	300 s	Max. new-call queue-time	5 s
Average call length	60 s	Pedestrian speed	3 mph
Max consecutive outages	5	Signal bandwidth	5 MHz
Target SINR	6.2 dB		

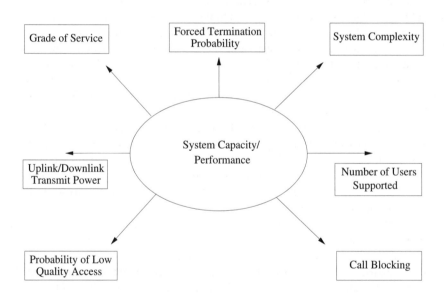

Figure 7.4: System capacity/performance illustration factors.

as before, our network performance studies were conducted with aim of maintaining: $P_B \leq$ 3%, $P_{FT} \leq 1\%$, $P_{low} \leq 1\%$ and GOS $\leq 4\%$.

To elaborate a little further, the design of wireless networks is based on a complex interplay of these four performance metrics as well as on a range of other often contradictory trade-offs, which are summarized in the stylized illustration seen in Figure 7.4. For example, the figure suggests that it is always possible to reduce the call dropping probability by increasing the call blocking probability, since this implies admitting less users to the system. By contrast, we may admit more users to the system for the sake of reducing the call blocking probability, which however results in an increased call dropping probability. Furthermore, the performance of the entire system may also be improved by increasing the system's complexity upon using more intelligent, but more complex signal processing algorithms, such as the beamforming and adaptive modulation aided transceiver techniques advocated in Chapter 8. Similarly, the genetic algorithm based scheduling techniques of Chapter 10 may be used for reducing the co-channel interference experienced by the system and hence for increasing the number of users, and/or for improving the call blocking and call dropping performance. Still continuing our discourse in the spirit of Figure 7.4, the number of users supported may also be increased, provided an increased probability of low quality access value may be tolerated. A whole raft of further similar comments may be made in the context of Figure 7.4, which will emanate from our detailed discourse throughout the forthcoming chapters. Hence we postpone the discussion of these detailed findings to our forthcoming chapters.

7.1.4   Simulation Results

In the investigations of [416], OVSF codes were used as spreading codes. However, the intra-cell interference is only eliminated by employing orthogonal OVSF codes, if the system is perfectly synchronous and provided that the mobile channel does not destroy the OVSF codes' orthogonality. In an effort to prevent intra-cell interference, again, we employ LS codes, which exhibit ideal auto-correlation and cross-correlation functions within the IFW. Thereby, the "near–far effect" may be significantly reduced and hence the user capacity of the system can be substantially enhanced. Figure 7.5 compares the BER performance of OVSF codes and LS codes, which were determined with the aid of physical-layer simulations using a 4QAM modulation scheme, 1/2-rate turbo coding and a Minimum Mean Squared Error Block Decision Feedback Equalizer (MMSE-BDFE) based Multi-User Detector (MUD) [93] joint detection for transmission over a COST 207 seven-path Bad Urban channel [408]. The figure illustrates that the achievable BER performance of LS codes is better than that of OVSF codes. For a spreading factor of 16, the post-despreading SINR required for maintaining a BER of 1×10^{-3} was 6.2 dB in case of LS codes, which is almost 2 dB lower than that necessitated by the OVSF codes.

Figure 7.6 shows the forced termination probability associated with a variety of traffic loads measured in terms of the mean normalized carried traffic expressed in Erlangs/km^2/MHz when subjected to 0.5 Hz frequency shadowing having a standard deviation of 3 dB. The average call duration and inter-call duration are 60 s and 300 s, resulting in maximum 0.2 Erlang/user traffic during the busy hour. The UTRA system's bandwidth is 5 MHz, the SF is 16 and given the cell-radius of 150 m as well as the 49-cell simulation area, the $Erlang$ capacity is computed by recording all the users' call durations and dividing it by the total of duration of the time-interval over which the statistics were collected, while satisfying the network quality constraints of Section 5.3.3.4. The figure illustrates that the network's performance was significantly improved by using LS codes. In conjunction with OVSF codes, the "No beamforming" scenario suffered from the highest forced termination probability of the six traffic scenarios characterized in the figure at a given load. Specifically, the network capacity was limited to 152 users, or to a teletraffic load of approximately 2.65 Erlangs/km^2/MHz. With the advent of employing four-element adaptive antenna arrays at the base stations the number of users supported by the network increased to 428 users, or almost to 7.23 Erlangs/km^2/MHz. However, in

Figure 7.5: BER performance of a UTRA-like system using OVSF codes and LS codes generated with the aid of physical-layer simulations using 4-QAM modulation, 1/2-rate turbo coding and MMSE-BDFE joint detection for transmissions over a COST 207 seven-path Bad Urban channel.

conjunction with LS codes, and even without employing antenna arrays at the base stations, the network capacity was dramatically increased to 581 users, or 10.10 Erlangs/km^2/MHz. When four-element adaptive antenna arrays were employed in the LS-code based scenario, the system was capable of supporting 800 users, which corresponded to a teletraffic load of 13.39 Erlang/km^2/MHz. This is because the LS codes' perfect auto-correlation and cross-correlation properties allowed the system to essentially eliminate the intra-cell interference, as it was discussed in Sections 7.1.1 and 7.1.2.

The probability of low quality access is depicted in Figure 7.7. As expected, a given P_{low} value was associated with a higher traffic load, when the number of antenna elements was increased. In the case of OVSF codes, it can be seen from the figure that without beamforming the system suffered from encountering more multiuser interference as the traffic loads increased. Hence the probability of low quality access became higher. In conjunction with beamforming, both the intra- and inter-cell interference was reduced and hence the probability of low quality access was reduced as well. However, increasing the number of antenna elements from two to four resulted in an increased probability of a low quality outage with the advent of the sharper antenna directivity. As a benefit of employing LS codes, the intra-cell interference was efficiently reduced and therefore the probability of low quality access was found to be lower even without beamforming, than that of the system using OVSF codes and employing 2- or 4-element beamforming. Again, owing to the sharper antenna directivity, the probability of a low quality outage increased, when increasing the number

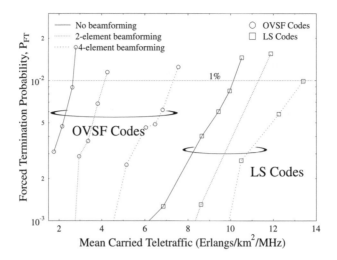

Figure 7.6: Forced termination probability versus mean carried traffic of the UTRA-like FDD cellular network using LS codes and OVSF codes both with as well as without beamforming in conjunction with shadowing having a frequency of 0.5 Hz and a standard deviation of 3 dB for a spreading factor of $SF = 16$.

Table 7.2: Maximum mean carried traffic and maximum number of mobile users that can be supported by the network, whilst meeting the network quality constraints, namely $P_B \leq 3\%$, $P_{FT} \leq 1\%$, $P_{low} \leq 1\%$ and GOS $\leq 4\%$. The carried traffic is expressed in terms of normalized Erlangs (Erlang/km^2/MHz) using OVSF codes and LS codes in conjunction with shadow fading having a standard deviation of 3 dB and a frequency of 0.5 Hz, for a spreading factor of $SF = 16$.

Spreading code	Beamforming	Users	Traffic (Erlangs /km^2/MHz)	Power (dBm) MS	Power (dBm) BS
OVSF codes	No	152	2.65	−9.0	−9.0
OVSF codes	2-elements	242	4.12	−8.28	−7.88
OVSF codes	4-elements	428	7.23	−7.45	−5.40
LS codes	No	581	10.1	−8.19	−5.84
LS codes	2-elements	622	10.6	−9.88	−5.53
LS codes	4-elements	802	13.39	−10.57	−4.49

of antenna elements from two to four. It should be noted that the probability of low quality access always remained below our 1% constraint in the scenarios studied.

Figure 7.8 shows the achievable Grade-Of Service (GOS) for a range of teletraffic loads. Similar trends were observed regarding the probability of low quality access to those shown in Figure 7.7. The grade of service is better (i.e. lower) when the traffic load is low, and vice versa for high traffic loads.

Figure 7.7: Probability of low quality access versus number of users of the UTRA-like FDD cellular network using LS codes and OVSF codes both with as well as without beamforming in conjunction with shadowing having a frequency of 0.5 and a standard deviation of 3 dB for a spreading factor of $SF = 16$.

The mean transmission power versus teletraffic performance is depicted in Figure 7.9. Again, as a benefit of employing LS codes, both the required mean UL and DL transmission power are lower than that necessitated by OVSF codes. It is worth pointing out that the employment of adaptive antenna arrays may in fact result in the attenuation of the desired signal, but this is always performed for the sake of maximizing the received SINR, thus ensuring that the effects of interference are mitigated. As a further benefit, invoking adaptive antenna arrays at the basestation reduced the mean UL transmission power required for meeting the service quality targets of the network. In OVSF code based scenarios, the basestation suffered from more multiuser interference, as the traffic loads increased. This was particularly true for the intra-cell interference, which required an increased UL mean transmission power for the sake of reaching the target SINR. By contrast, the LS codes had potential of eliminating the intra-cell interference, ultimately leading to the reduction of the mean transmission power required.

A summary of the maximum user capacities of the UTRA-like networks using OVSF codes and LS codes in conjunction with log-normal shadowing having a standard deviation of 3 dB and a shadowing frequency of 0.5 Hz as well as both with and without beamforming is given in Table 7.2. The teletraffic carried and the mean mobile and base station transmission powers required are also shown in Table 7.2.

7.1.5   Summary

In this section it was demonstrated that the network performance of a UTRA-like CDMA system employing LS spreading codes was substantially better than that of the system using OVSF codes. Explicitly, a low forced termination probability, low mobile and base

Figure 7.8: Grade-Of-Service (GOS) versus number of users of the UTRA-like FDD cellular network using LS codes and OVSF codes both with as well as without beamforming in conjunction with shadowing having a frequency of 0.5 Hz and a standard deviation of 3 dB for a spreading factor of $SF = 16$.

station transmission power and high call quality has been maintained. In the context of the interference limited 3G CDMA system LS codes [422] might hold the promise of an increased network capacity without dramatic changes of the 3G standards, while dispensing with the employment of high-complexity, power-hungry multiuser detectors. It has to be mentioned, however that the number of available LS codes is limited and hence the system may become code-limited, instead of being interference-limited. Therefore it is necessary to invoke a range of supporting measures for the sake of increasing the number of system users that can be supported. This can be achieved for example by combining time-domain (TD) LS-code based direct-sequence spreading with frequency-domain (FD) OVSF-code based spreading. The employment of FD spreading in the context of multicarrier CDMA has the further advantage of potentially extending the length of the IFW proportionally to the number of FD subcarriers, because the symbols to be transmitted are mapped to a number of parallel subcarriers, where the subcarriers' modulated symbols have an extended duration. Theses physical-layer issues were discussed in [435].

7.2 Effects of Cell Size on the UTRA Performance

7.2.1 Introduction

In this section we embark on exploring the trade-offs between the achievable user capacity and the cell size. In CDMA systems all signals share the entire bandwidth and the users are differentiated by their unique spreading codes. Naturally, the higher the number of users in a

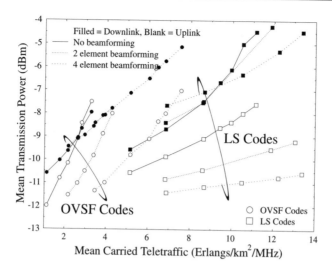

Figure 7.9: Mean transmission power versus number of users of the UTRA-like FDD cellular network using LS codes and OVSF codes both with as well as without beamforming in conjunction with shadowing having a frequency of 0.5 Hz and a standard deviation of 3 dB for a spreading factor of $SF = 16$.

cell, the higher the multiuser interference, which can be modeled by Gaussian noise according to the central limit theorem [436, 437].

Again, in this section we studied the effects of different cell sizes on the user capacity of UTRA-like FDD/CDMA systems, employing cell-radii of 78 m, 150 m, 300 m, 500 m and 800 m. The simulation results were compared for the sake of quantifying how the cell size effects the achievable system performance.

7.2.2 System Model and System Parameters

As in our previous investigations, the mobiles were capable of moving freely, at a speed of 3 mph, in random directions, selected at the start of the simulation from a uniform distribution, within the infinite simulation area of 49 wrapped-around traffic cells [396, 397, 416]. In order to facilitate the employment of an infinite simulation area, a tessellating rhombic simulation area was used [416]. More explicitly, mobile stations about to leave the 49-cell simulation area were reflected back to it at a $180°$-rotated angle, which we refer to as being "wrapped around" from one side of the network to the other [396, 397, 416]. The benefit of employing this technique is that a mobile station in call, which leaves the network at its edge, re-enters the network at the opposite side, whilst continuing to inflict Co-Channel Interference (CCI) to the surrounding users, which may be roaming in its vicinity in the network [416]. Our earlier illustration seen in Figure 5.18 depicts this scenario graphically. By contrast, in the "desert-island"-like scenario of employing no wrap-around, the users near the fringes of the 49-cell simulation area would experience a reduced co-channel interference, resulting in optimistic users capacity estimates.

Table 7.3: Basic simulation parameters.

Parameter	Value	Parameter	Value
Noisefloor	-100 dBm	Frame length	10 ms
Multiple access	FDD/CDMA	Number of basestations	49
Modulation scheme	4QAM/QPSK	Spreading factor	16
Power control stepsize	1 dB	Power control hysteresis	1 dB
Low quality access SINR	7.0 dB	Outage (1% BER) SINR	6.6 dB
Target SINR (at BER = 0.1%)	8.0 dB	Size of Active Basestation Set	2
Average inter-call-time	300 s	Max. new-call queue-time	5 s
Average call length	60 s	Pedestrian speed	3 mph
Maximum consecutive outages	50 ms	Signal bandwidth	5 MHz

Table 7.4: Signal power parameters.

Cell radius	Maximum BS/MS transmit power	Minimum BS/MS transmit power	Pilot power	Pathloss exponent
78 m	17.5 dBm	-47.5 dBm	-8.5 dBm	-2.0
150 m	21.0 dBm	-44.0 dBm	-5.0 dBm	-3.5
300 m	28.0 dBm	-37.0 dBm	2.0 dBm	-3.5
500 m	32.67 dBm	-32.33 dBm	6.67 dBm	-3.5
800 m	39.7 dBm	-25.3 dBm	13.7 dBm	-3.5

As in Section 7.1.3 the basestations are assumed to be equipped with the Minimum Mean Squared Error Block Decision Feedback Equalizer based Multi-User Detector (MUD) [93, 434]. The post-despreading SINRs required by this MUD for obtaining the target BERs were determined with the aid of physical-layer simulations using a 4-QAM/QPSK modulation scheme, in conjunction with $1/2$ rate turbo coding and MUD for transmission over a COST 207 seven-path Bad Urban channel [408].

Using this turbo-coded MUD-assisted transceiver and a spreading factor of 16, the post-despreading SINR required for maintaining the target BER of 1×10^{-3} was 8.0 dB. The BER which was deemed to correspond to low-quality access, was stipulated at 5×10^{-3}. This BER was exceeded for SINRs falling below 7.0dB. Furthermore, a low-quality outage was declared, when the BER of 1×10^{-2} was exceeded, which was encountered for SINRs below 6.6 dB. These values can be seen along with the other system parameters in Table 7.3.

As the cell size changes, the minimum Base Station (BS) and Mobile Station (MS) transmit power as well as the pilot power also has to change for the sake of maintaining an adequate coverage. Table 7.4 summarizes the minimum required BS power associated with the different cell radii. We consider the scenario having a cell radius of 78 m to be associated with Line of Sight (LOS) propagation having a pathloss exponent [436] of 20 dB/decade.

Figure 7.10: Cell radius versus mean carried traffic of the UTRA-like FDD cellular network both with and without beamforming in conjunction with shadowing having a frequency of 0.5 Hz and a standard deviation of 3 dB for a spreading factor of $SF = 16$.

7.2.3 Simulation Results and Comparisons

We investigated various scenarios having different cell radii and compared the QoS by using the performance metrics of Section 5.3.3.4 for estimating how the cell size affected the capacity of the UTRA-like FDD/CDMA system, considered current FDD/CDMA systems are interference limited, suffering from intra-cell interference imposed by the signals transmitted to other mobiles supported by the same basestation, and by the inter-cell interference inflicted by the surrounding base stations. When the cell radius was increased, the maximum/minimum BS/MS transmission powers and the pilot power had to be appropriately adjusted as seen in Table 7.4. Again, observed in Table 7.4 that the cell having a radius of 78 m encountered LOS propagation, which may affect the system's capacity.

7.2.3.1 Network Performance using Adaptive Antenna Arrays

Figure 7.10 shows achievable teletraffic capacity versus the cell radius associated with a variety of traffic loads measured in terms of the mean normalized carried traffic expressed in Erlangs/km²/MHz when subjected to 0.5 Hz frequency shadowing having a standard deviation of 3 dB. The figure illustrates that the network's user capacity was significantly degraded, when the cell radius was increased, which was mitigated by employing adaptive antenna arrays. The scenario having a cell radius of 78 m in Figure 7.11(a), reached a network capacity of 2.65 Erlang/km²/MHz even without employing antenna arrays, which is about 94 times to that of the network having a cell radius of 800 m, which may characterized in Figure 7.11(d) and had a capacity of 0.028 Erlang/km²/MHz. When using "2- or 4-element beamforming", the adaptive antenna arrays have considerably reduced the levels of interference, leading to a higher network capacity. It can be seen in Figure 7.11(a), that scenario having a cell radius of 78 m, the exhibited a network capacity which was increased by 59% to 4.12 Erlang/km^2/MHz with the advent of employing 2-element adaptive antenna arrays at the basestations. Replacing the 2-element adaptive antenna arrays by 4-element arrays led to a further capacity increase of 77%, which is associated with a network capacity of 7.26 Erlangs/km^2/MHz. As it is widely recognized, the high capacity requirements of dense urban environments require a high cell-site density.

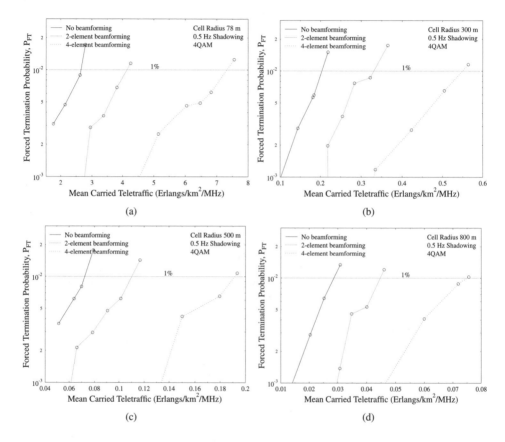

Figure 7.11: Forced termination probability versus mean carried traffic of the UTRA-like FDD cellular network both with and without beamforming in conjunction with shadowing having a frequency of 0.5 Hz and a standard deviation of 3 dB for a spreading factor of $SF = 16$: (a) cell radius is 78 m; (b) cell radius is 300 m; (c) cell radius is 500 m; (d) cell radius is 800 m.

The probability of low quality access experienced in different cell radius scenarios is depicted in Figure 7.12, which also exhibited a substantial improvement with the advent of two-element adaptive antenna arrays. Similar performance trends were observed in all four sub-figures of Figure 7.11. At lower traffic loads the probability of low quality access experienced without employing adaptive antenna arrays is typically better than that of the antenna array-aided scenarios, because of the adaptive antenna arrays' sensitivity to the potentially damaging interfering signals, when a new call commences. This tendency becomes more marked when using 4-element arrays. As the traffic load becomes heavier, the level of multiuser interference increases as well, as seen in Figure 7.12, resulting in a steep rise of the low quality access probability curves, when using no beamforming. By contrast, as expected, using adaptive antenna arrays in case of high traffic loads results in a reduced probability of low quality access. The variation of the cell radius did not dramatically affect the probability of the low quality outage.

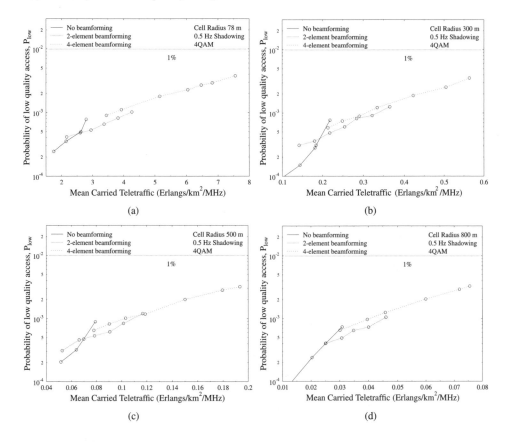

Figure 7.12: Probability of low quality access versus mean carried traffic of the UTRA-like FDD cellular network both with and without beamforming in conjunction with shadowing having a frequency of 0.5 Hz and a standard deviation of 3 dB for a spreading factor of $SF = 16$: (a) cell radius is 78 m; (b) cell radius is 300 m; (c) cell radius is 500 m; (d) cell radius is 800 m.

Figure 7.13 portrays the number of users supported versus the cell radius. In this context the "Number of Users" refers to the simulation area of 49 traffic cells as seen in Figure 5.18, although depending on the cell-radius this corresponds to widely different areas expressed in km^2. Figure 7.13 suggests that varying the cell radius does not dramatically affect the number of users supported within the simulation area, since the number of users supported is approximately proportional to the number of base stations. The more base stations there are in a fixed area, the more mobile users can be supported. As we have mentioned at the beginning

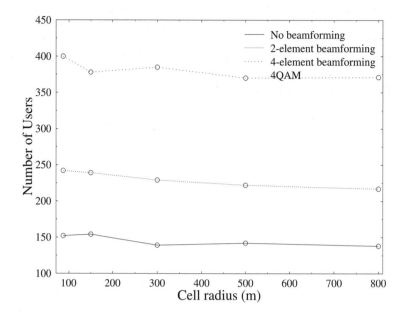

Figure 7.13: Cell radius versus number of users of the UTRA-like FDD cellular network both with and without beamforming in conjunction with shadowing having a frequency of 0.5 Hz and a standard deviation of 3 dB for a spreading factor of $SF = 16$.

of this section, CDMA systems are interference limited. Each additional user admitted to the system constitutes one more source of interference. Again, it can be seen from Figure 7.13 that the scenarios carrying higher traffic loads benefitted from the employment of the 2- or 4-element antenna arrays, which substantially enhanced the achievable network capacity.

The mean UL and DL transmission power versus cell radius is depicted in Figure 7.14, suggesting that a higher average signal power was required for maintaining an acceptable signal to interference plus noise ratio, as the cell radius increased. When the cell radius increased from 78 m to 800 m, the mean transmission power had to be increased by more than 30 dBm. A summary of the maximum achievable user capacities for the UTRA-like network considered, which was subjected to log-normal shadowing having a standard deviation of 3 dB and a frequency of 0.5 Hz, both with and without employing beamforming, is given in Table 7.5. The teletraffic carried and the mean mobile and base station transmission powers required are also shown in Table 7.5.

7.2.3.2   Network Performance using Adaptive Antenna Arrays and Adaptive Modulation

In this section we will quantify the impact of cell radius on the achievable network capacity, while using both adaptive antenna arrays and Adaptive Quadrature Amplitude Modulation (AQAM) [416, 419]. AQAM activates the most appropriate modulation mode

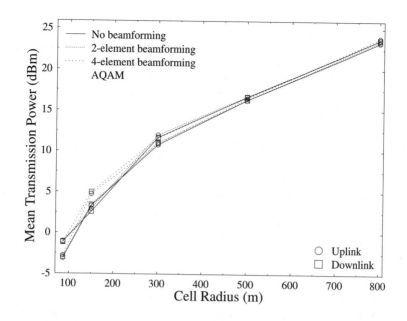

Figure 7.14: Mean UL and DL transmission power versus cell radius of the UTRA-like FDD cellular network both with and without beamforming in conjunction with shadowing having a frequency of 0.5 Hz and a standard deviation of 3 dB for a spreading factor of $SF = 16$.

depending on the near-instantaneous channel quality in order to maximize the achievable data throughput, while maintaining the target bit error ratio.

Figure 7.15 shows the cell radius associated with a variety of traffic loads in terms of the mean normalized carried traffic expressed in Erlang/km^2/MHz. Similar trends were observed in Figure 7.10, namely that the system's capacity degraded quite dramatically, when the cell size was increased. For the sake of comparison with the fixed-modulation based QPSK scenario of Figure 7.10 we note that the highest AQAM-aided mean carried traffic improvement was achieved by the 2-element AAA. Explicitly, in case of the 78 m cell-radius the 2.4 Erlang/km^2/MHz carried traffic was increased to about 5.5 Erlang/km^2/MHz. Upon comparing Figure 7.11 to Figure 7.16, we observe that a further call dropping probability reduction is achieved, when using AQAM.

The probability of low quality outage was shown in Figure 7.17, which in fact indicates a degradation imposed by the employment of adaptive modulation, when compared to the corresponding 4QAM curves seen in Figure 7.12. The increase in the probability of low quality access can be attributed to the employment of less robust, but higher-throughput, higher-order modulation modes invoked by the adaptive modulation scheme, which resulted in an increase of the achievable system capacity. For example, when we use fixed 4QAM modulation mode, as characterized in Figure 7.12, a low quality outage may not occur. By contrast, an outage is more likely to happen for more prolonged periods of time, when a higher-order AQAM mode is invoked, as suggested by Figure 7.17.

Table 7.5: Maximum mean carried traffic and maximum number of mobile users that can be supported by the network, whilst meeting the network quality constraints, namely $P_B \leq 3\%$, $P_{FT} \leq 1\%$, $P_{low} \leq 1\%$ and GOS $\leq 4\%$. The carried traffic is expressed in terms of normalized Erlangs (Erlang/km^2/MHz) in conjunction with shadow fading having a standard deviation of 3 dB and a frequency of 0.5 Hz for a spreading factor of $SF = 16$. The average transmission power of the MSs and BSs are also summarized.

Cell radius (m)	Beamforming	Users	Traffic (Erlangs/km^2/MHz)	Power (dBm) MS	BS
78	No	152	2.65	−8.30	−8.57
78	2-elements	242	4.12	−8.28	−7.88
78	4-elements	430	7.26	−7.41	−5.38
150	No	150	0.87	−1.05	−1.51
150	2-elements	239	1.39	−0.43	−0.58
150	4-elements	348	1.99	1.94	0.67
300	No	139	0.19	8.76	7.69
300	2-elements	229	0.32	9.49	8.85
300	4-elements	385	0.54	10.12	10.91
500	No	142	0.07	16.30	15.15
500	2-elements	222	0.10	16.89	16.27
500	4-elements	370	0.19	17.37	18.11
800	No	138	0.02	23.53	22.39
800	2-elements	217	0.04	23.84	23.42
800	4-elements	371	0.07	24.53	25.56

Figure 7.18 shows the number of users supported versus the cell radius. From Figure 7.18 we can see that the achievable user capacity of FDD/CDMA employing adaptive modulation was improved in comparison to the system using fixed 4QAM. Figure 7.19 portrays the mean transmission power versus the cell radius. Naturally, the required signal power increased, as the cell radius increased.

7.2.4 Summary and Conclusion

In Section 7.2 we quantified the impact of cell radius on the achievable network capacity of UTRA-like FDD/CDMA systems, while using both adaptive antenna arrays and adaptive modulation. The simulation results demonstrate that the high capacity requirement of dense urban environments necessitates a high cell-site density. The variation of the cell size did not dramatically affect the probability of low call quality access. However, a higher average signal power was required for maintaining an acceptable signal to interference plus noise ratio, as the cell radius increased. Considerable network capacity gains were achieved, employing both 2- and 4-element adaptive antenna arrays in conjunction with adaptive modulation techniques.

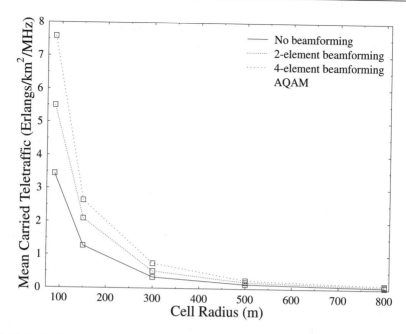

Figure 7.15: Cell radius versus mean carried traffic of the UTRA-like FDD cellular network both with and without beamforming in conjunction with AQAM as well as shadowing having a frequency of 0.5 Hz and a standard deviation of 3 dB for a spreading factor of $SF = 16$.

7.3 Effects of SINR Threshold on the Performance of CDMA Systems

7.3.1 Introduction

The third generation wireless networks are capable of adjusting the transmission integrity for the sake of providing multimedia services. The term multimedia encompasses a number of diverse media to be combined in novel ways for the sake of communicating using text, voice, video, graphics, images, audio etc [438]. In [439] Acampora and Naghshineh refer to three types of wireless communications services, namely: (a) real-time connections using voice and low-rate video, (b) non-real-time delay-sensitive connection-oriented services with limited delay bounds, such as using remote login and the File Transfer Protocol (FTP), (c) message-oriented, delay-insensitive traffic such as paging, electronic mail, voice mail and fax. These different types of services have diverse the target SINR requirements. As the target SINR requirement are changing, so does the user capacity. In this section we study the effects of the SINR threshold on the user capacity of CDMA systems, since this allows us to directly quantify the impact of more or less error-resilient transceivers on the network's performance.

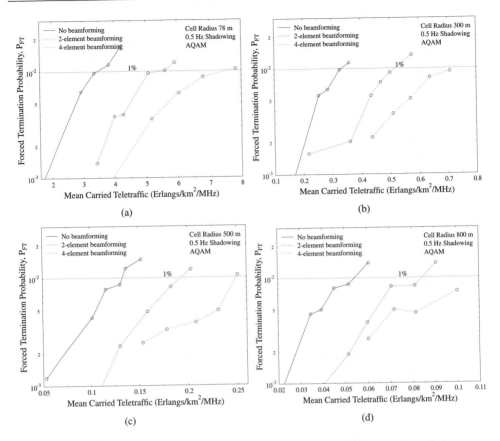

Figure 7.16: Forced termination probability versus mean carried traffic of the UTRA-like FDD cellular network both with and without beamforming in conjunction with AQAM as well as shadowing having a frequency of 0.5 Hz and a standard deviation of 3 dB for a spreading factor of $SF = 16$: (a) cell radius is 78 m; (b) cell radius is 300 m; (c) cell radius is 500 m; (d) cell radius is 800 m.

7.3.2   Simulation Results

This simulation conditions were the same as in Section 7.2.2. Again, mobiles were capable of roaming freely, at a speed of 3 mph, in random directions, selected at the start of the simulation from a uniform distribution, within the simulation area of 49 traffic cells. The cell radius was 150 m. The propagation environment was modeled using the a pathloss model having a pathloss exponent of -3.5. The mobile and base station transmit powers were restricted to the range of -44 dBm to $+21$ dBm for the power control assisted and adaptive modulation based simulations. If a channel allocation request for a new call could not be satisfied immediately, this request was queued for a duration of up 5 s, after which time, if not satisfied, it was classed as blocked.

In Section 7.2 we observed significant performance gains with the advent of employing adaptive antenna arrays at the base station. The CDMA based network considered here has

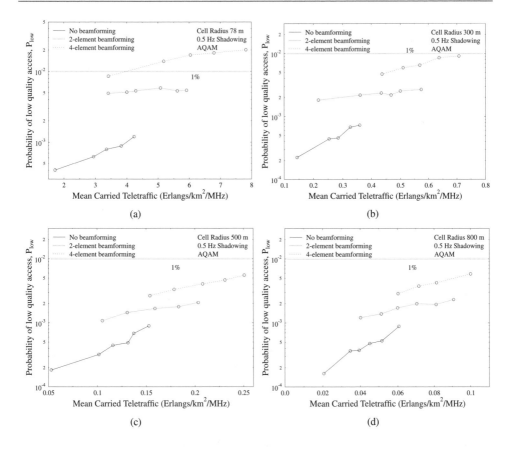

Figure 7.17: Probability of low quality access versus mean carried traffic of the UTRA-like FDD cellular network both with and without beamforming in conjunction with AQAM as well as shadowing having a frequency of 0.5 Hz and a standard deviation of 3 dB for a spreading factor of $SF = 16$. (a) Cell radius is 78 m; (b) cell radius is 300 m; (c) cell radius is 500 m; (d) cell radius is 800 m.

a frequency reuse factor of unity, therefore the level of co-channel interference is high, and hence the adaptive antennas are expected to provide substantial performance benefits. The SINR threshold used by the network control algorithms is determined by the error resilience of the wireless transceiver used, namely, by the SINR value required for maintaining the target BER of the service concerned. For example, if a more error resilient transceiver is used, the SINR requirements may be reduced and hence more users can be supported. The same is true, when the service can tolerate a higher BER.

Figure 7.20 shows the target SINR threshold associated with a variety of traffic loads measured in terms of the mean normalized carried traffic expressed in Erlangs/km^2/MHz when subjected to 0.5 Hz frequency shadowing having a standard deviation of 3 dB as a function of the target SINR threshold. As expected, the figure illustrates that the network's user capacity degrades, when the target SINR requirement is increased. When the target SINR threshold was 6 dB, it can be seen in Figure 7.21(a) that the network capacity reached 1.87

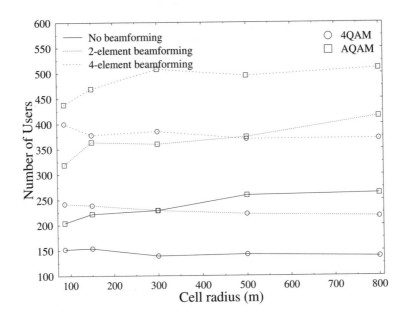

Figure 7.18: Cell radius versus number of users of the UTRA-like FDD cellular network both with and without beamforming in conjunction with AQAM as well as shadowing having a frequency of 0.5 Hz and a standard deviation of 3 dB for a spreading factor of $SF = 16$.

Erlang/km^2/MHz, which is about 27 times the capacity, when the SINR value was set to 12 dB, as evidenced by Figure 7.21(d), where the corresponding carried traffic was 0.069 Erlang/km^2/MHz without employing antenna arrays at the base station.

When using 2- or 4-element beamforming, the adaptive antenna arrays have considerably reduced the levels of interference, leading to a higher carried traffic. As it can be seen in Figure 7.21(a), when the SINR threshold was 6 dB, the carried traffic increased by 33% to 2.80 Erlang/km^2/MHz with the advent of employing 2-element adaptive antenna arrays at the basestations. Replacing the 2-element adaptive antenna arrays with 4-element arrays led to a further capacity increase of 35%, which is associated with a network capacity of 4.34 Erlangs/km^2/MHz. When the target SINR threshold was increased to 12 dB, it can be observed in Figure 7.21(d) that the carried traffic became extremely poor without the employment of adaptive antenna arrays. This is because the target SINR is high, hence the required transmitted power is increased, inevitably increasing the interference level imposed. Hence an error-sensitive transceiver, which requires a high SINR for maintaining the target integrity may lead to an unstable, low-capacity system. The benefits of using adaptive antenna arrays are clearly demonstrated in this scenario. With the advent of using 2- or 4-element beamforming the carried traffic becomes a factor four or eight higher than that of the "No beamforming" scenario, supporting 43 and 78 users, respectively.

Four different probability of low quality access scenarios associated with various target SINR threshold were presented in Figure 7.22. Specifically, in Figure 7.22(a) and

Figure 7.19: Mean transmission power versus cell radius of the UTRA-like FDD cellular network both with and without beamforming in conjunction with AQAM as well as shadowing having a frequency of 0.5 Hz and a standard deviation of 3 dB for a spreading factor of $SF = 16$.

Figure 7.22(b) similar trends were observed, namely that at lower traffic loads the achievable low quality outage performance was worse than that of no beamforming when using adaptive antenna arrays. This likely to be due to the addition of a new source of interference, constituted by a user initiating a new call, which results in an abrupt change in the gain of the antenna in the direction of the desired user, when invoking adaptive antenna arrays. Increasing the number of antenna elements from two to four results in an increased probability of low quality outage due to the sharper antenna directivity. In contrast to Figure 7.22(a) and Figure 7.22(b), Figure 7.22(c) and Figure 7.22(d) portray a better performance, when employing adaptive antenna arrays than that of no beamforming. However, the price to be paid for this is that a network using no adaptive antenna arrays requires a higher transmission power for maintaining the target SINR level, as it will be discussed in more detail in the context of Figure 7.23 and Figure 7.24. This results in a higher overall interference level. In conclusion, in a network having a high target SINR threshold, employment of adaptive beamforming holds the promise of a reduced probability of low quality outage.

The mean transmission power performance versus carried traffic is depicted in Figure 7.23 and Figure 7.24. Figure 7.23 clearly shows the lower levels of transmission power required for maintaining an acceptable SINR, while using adaptive antenna arrays at the base stations. These power budget savings were obtained in conjunction with reduced levels of co-channel interference, leading to superior call quality, as illustrated in Figure 7.21 and Figure 7.22. This phenomenon can be seen more clearly in Figure 7.24, where we recorded the system's

Figure 7.20: Mean carried traffic of the UTRA-like FDD cellular network versus the target SINR threshold both with and without beamforming in conjunction with shadowing having a frequency of 0.5 Hz and a standard deviation of 3 dB for a spreading factor of $SF = 16$.

mean transmission power when the teletraffic load was 1 Erlang/km^2/MHz. The mean transmission power was increasing rapidly, as the required target SINR was increased. An additional average transmission power of 5 dBm was required, when the target SINR threshold was increased from 6 dB to 8 dB. A further transmission power increment of 5 dBm was necessitated, when the required SINR threshold was increased from 8 dB to 10 dB. Figure 7.25 characterizes the mean transmission power versus SINR threshold performance, when the cellular network achieved the maximum user-capacity values shown in Figure 7.20. A summary of the maximum user capacities of the UTRA-like FDD networks in conjunction with log-normal shadowing having a standard deviation of 3 dB and a frequency of 0.5 Hz, both with and without employing beamforming is given in Table 7.6. The teletraffic carried and the mean mobile as well as base station transmission powers required for attaining these user capacities are also shown in Table 7.6.

7.3.3 Summary and Conclusion

In this section we studied the effects of the SINR threshold on the achievable user capacity of the UTRA-like FDD/CDMA systems studied, in order to quantify the impact of the error-resilience of the transceivers employed on the network's performance. From the simulation results we observed that increasing the required SINR for the sake of invoking higher throughput, but less error resilient modems may in fact lead to an unstable, easily overloaded, low-capacity system, which is associated with a high power consumption.

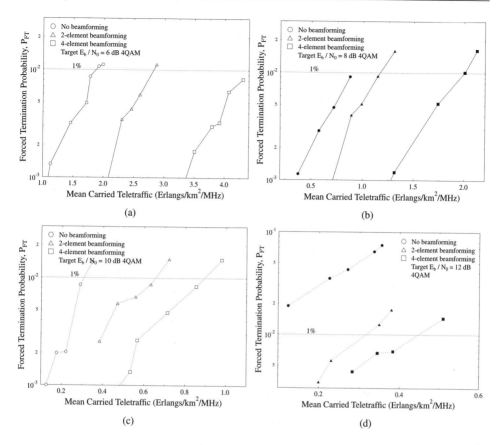

Figure 7.21: Forced termination probability versus mean carried traffic of the UTRA-like FDD cellular network both with and without beamforming in conjunction with shadowing having a frequency of 0.5 Hz and a standard deviation of 3 dB for a spreading factor of $SF = 16$. This figure demonstrates the effects of the different target SINRs of 6, 8, 10 and 12 dB and may be compared to Figure 7.11, where the effects of different cell radii were studied. (a) Target SINR of 6 dB; (b) target SINR of 8 dB; (c) target SINR of 10 dB; (d) target SINR of 12 dB.

However, the advantages of using adaptive antenna arrays within a mobile cellular network result in substantial performance improvements in terms of the achievable call quality, mean transmission power and the number of supported users.

7.4 Network-layer Performance of Multi-carrier CDMA

7.4.1 Introduction [440]

A range of novel techniques combining DS-CDMA and OFDM have been presented in the literature [440–445].

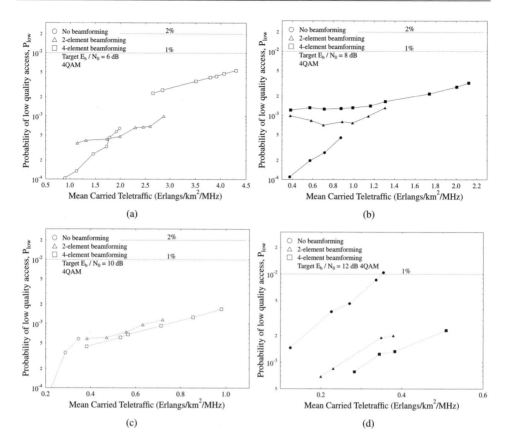

Figure 7.22: Probability of low quality access versus mean carried traffic of the UTRA-like FDD
cellular network both with and without beamforming in conjunction with shadowing
having a frequency of 0.5 Hz and a standard deviation of 3 dB for a spreading factor
of $SF = 16$. This figure demonstrates the effects of the different target SINRs of 6, 8, 10
and 12 dB and may be compared to Figure 7.12, where the effects of different cell radii
were studied. (a) Target SINR of 6 dB; (b) target SINR of 8 dB; (c) target SINR of 10 dB;
(d) target SINR of 12 dB.

A DS-CDMA system applies spreading sequences in the time domain and uses Rake
receivers for optimally combining the time-dispersed energy in order to combat the effects
of multi-path fading. However, in indoor wireless environments the time dispersion is
low, on the order of nano seconds, and hence a high chip rate, on the order of tens of
MHz, is required for resolving the multi-path components. This implies a high clock-rate,
improving a high power consumption as well as a range of implementation difficulties.
In order to overcome these difficulties, several techniques have been proposed, which
combine DS-CDMA and multi-carrier modulation, such as MC-CDMA [441–443], MC-DS-
CDMA [444] and Multi-Tone CDMA (MT-CDMA) [445]. This overview is mainly based on
references [446, 447] by Prasad and Hara, as well as on [448] by Scott, Grant, McLaughlin,
Povey and Cruickshank.

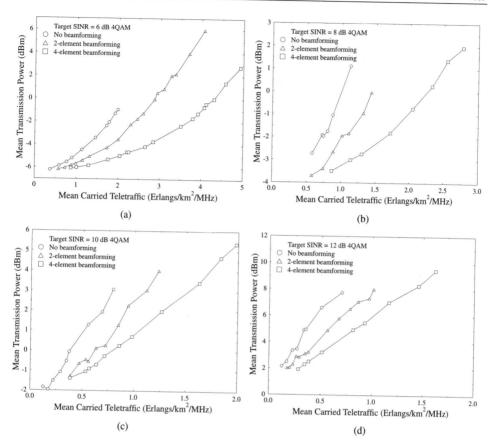

Figure 7.23: Mean transmission power versus mean carried traffic of the UTRA-like FDD cellular network both with and without beamforming in conjunction with shadowing having a frequency of 0.5 Hz and a standard deviation of 3 dB for a spreading factor of $SF = 16$. This figure demonstrates the effects of the different target SINRs of 6, 8, 10 and 12 dB. (a) Target SINR of 6 dB; (b) target SINR of 8 dB; (c) target SINR of 10 dB; (d) target SINR of 12 dB.

In MC-CDMA, instead of applying spreading sequences in the time domain, we apply them in the frequency domain, mapping a different chip of a spreading sequence to an individual OFDM subcarrier. Hence each OFDM subcarriers [440] has a data rate identical to the original input data rate and the multicarrier system "absorbs" the increased chip-rate due to spreading to a wider frequency band. The transmitted signal of the ith data symbol of the jth user $s_i^j(t)$ is written as [441, 449] :

$$s_i^j(t) = \sum_{k=0}^{K-1} b_i^j \, c_k^j \cos\{2\pi(f_0 + kf_d)t\} \, p(t - iT), \qquad (7.8)$$

where

Figure 7.24: Mean transmission power versus target SINR threshold of the UTRA-like FDD cellular network while carried traffic is 1 Erlangs/km^2/MHz, both with and without beamforming in conjunction with shadowing having a frequency of 0.5 Hz and a standard deviation of 3 dB for a spreading factor of $SF = 16$.

- K is the number of subcarriers,

- b_i^j is the ith message symbol of the jth user,

- c_k^j represents the kth chip, $k = 0, \ldots, K - 1$, of the spreading sequence of the jth user,

- f_0 is the lowest subcarrier frequency,

- f_d is the subcarrier separation and

- $p(t)$ is a rectangular signaling pulse shifted in time given by:

$$p(t) \triangleq \begin{cases} 1 & \text{for } 0 \leq t \leq T \\ 0 & \text{otherwise.} \end{cases} \qquad (7.9)$$

If $1/T$ is used for f_d, the transmitted signal can be generated using the IFFT, as in the case of an OFDM system [440]. The overall transmitter structure can be implemented by concatenating a DS-CDMA spreader [434] and an OFDM transmitter, as shown in Figure 7.26. At the spreader, the information bit, b_i^j, is spread in the time domain by the jth user's spreading sequence, c_k^j, $k = 0, \ldots, K - 1$. In this implementation, high-speed operations are required at the output of the spreader in order to carry out the chip-related

Figure 7.25: Mean transmission power versus target SINR threshold of the UTRA-like FDD cellular network while the maximum user-capacity achieved, both with as well as without beamforming in conjunction with shadowing having a frequency of 0.5 Hz and a standard deviation of 3 dB for a spreading factor of $SF = 16$.

Figure 7.26: Transmitter schematic of MC-CDMA.

operations. The spread chips are fed into the serial-to-parallel (S/P) block and IFFT is applied to these K parallel chips. The output values of the IFFT in Figure 7.26 are time domain samples in parallel form. After parallel to serial (P/S) conversion these time domain samples are low-pass-filtered, in order to obtain the continuous time domain signal. The signal modulates the carrier and is transmitted to the receiver.

Table 7.6: Maximum mean carried traffic and maximum number of mobile users that can be supported by the network, whilst meeting the network quality constraints, namely $P_B \leq 3\%$, $P_{FT} \leq 1\%$, $P_{low} \leq 1\%$ and GOS $\leq 4\%$. The carried traffic is expressed in terms of normalized Erlangs (Erlang/km²/MHz) in conjunction with shadow fading having a standard deviation of 3 dB and a frequency of 0.5 Hz for a spreading factor of $SF = 16$.

Target SINR (dB)	Beamforming	Users	Traffic (Erlangs/km²/MHz)	Power (dBm) MS	BS
6	No	320	1.87	−1.55	−3.03
6	2-elements	489	2.81	−0.31	−1.58
6	4-elements	758	4.34	0.23	0.65
8	No	155	0.90	−1.19	−1.63
8	2-elements	203	1.16	−0.40	−0.56
8	4-elements	350	2.00	0.10	1.46
10	No	53	0.30	−0.91	−0.81
10	2-elements	113	0.65	0.15	1.15
10	4-elements	156	0.89	0.36	1.26
12	No	9	0.07	1.64	1.48
12	2-elements	43	0.25	2.01	2.53
12	4-elements	78	0.44	5.61	2.79

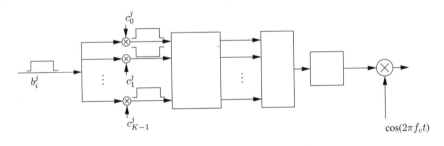

Figure 7.27: Alternative transmitter schematic of MC-CDMA.

Figure 7.27 shows another implementation, which removes the time domain spreader. In this implementation, the spreading sequence is applied directly to the identical parallel input bits. Hence, the high speed spreading operation is not required.

The unique, user-specific spreading sequences in MC-CDMA separate other users' signals from the desired signal, provided that their spreading sequences are orthogonal to each other. Orthogonal codes have zero cross correlation and hence they are particularly suitable for MC-CDMA.

At the MC-CDMA receiver shown in Figure 7.28 each carrier's symbol, i.e. the corresponding chip c_k^j of user j, is recovered using FFT after sampling at a rate of K/T samples/sec and the recovered chip sequence is correlated with the desired user's spreading code in order to recover the original information bit, b_i^j. Let us define the ith received symbol

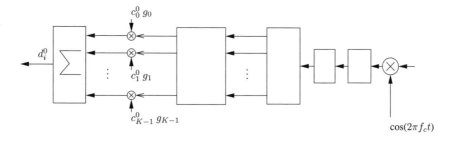

Figure 7.28: Receiver schematic of MC-CDMA.

at the kth carrier in the DL as:

$$r_{k,i} = \sum_{j=0}^{J-1} H_k \, b_i^j \, c_k^j + n_{k,i}, \tag{7.10}$$

where J is the number of users, H_k is the frequency response of the kth subcarrier and $n_{k,i}$ is the corresponding noise sample. The MC-CDMA receiver of the 0-th user multiplies $r_{k,i}$ of (7.10) by its spreading sequence chip, c_k^0, as well as by the gain, g_k, which is given by the reciprocal of the estimated channel transfer factor of subcarrier k, for each received subcarrier symbol for $k = 0, \ldots, K - 1$. It sums all these products, in order to arrive at the decision variable, d_i^0, which is given by:

$$d_i^0 = \sum_{k=0}^{N-1} c_k^0 \, g_k \, r_{k,i}. \tag{7.11}$$

Without the frequency domain equalization of the received subcarrier symbols discussed in great detail in [13,434], the orthogonality between the different users cannot be maintained. Several methods have been proposed for advantageously choosing the value of g_k [441,446, 449]. The associated BER analysis was performed using various equalization methods over both Rayleigh channels and Rician channels by Yee and Linnartz [441]. The comparative summary of numerical results for various equalization strategies was given, for example, by Prasad and Hara [446,447] and by Hanzo [440] *et al.*

7.4.2 Simulation Results

In this section simulations using an MC-CDMA [440] based cellular network were conducted in various scenarios employing adaptive antenna arrays [416] as well as adaptive modulation techniques [419]. Network performance results were obtained using 2- and 4-element adaptive antenna arrays in conjunction with adaptive modulation, in the presence of 0.5 Hz frequency shadow fading exhibiting a standard deviation of 3 dB. As in the context of our previous investigations presented in Section 7.2, the expected carried traffic gains were quantified. For the sake of comparing to the achievable network performance to a UTRA-like wideband CDMA system, the similar parameters were adopted to those summarized in

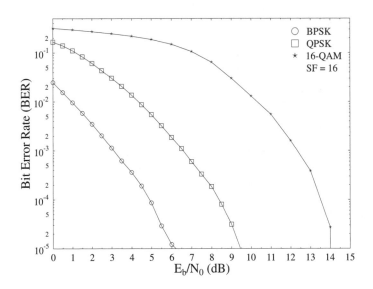

Figure 7.29: BER performance of a MC-CDMA system generated with the aid of physical-layer simulations using BPSK, 4QAM, 16-QAM modulation, 1/2-rate turbo coding and MMSE-BDFE joint detection for transmissions over a COST207 Bad-Urban Reduced-mode A (BU-RA) channel.

Table 7.1. A cell radius of 150 m was assumed, 49 wrapped-around traffic cells constituted the simulation area, as it was shown in Figure 5.18. Figure 7.29 portrays the BER performance of the MC-CDMA system using various modulation schemes for a spreading factor of 16, where the number of subcarriers was also 16. These results were determined with the aid of physical-layer simulations using BPSK, 4QAM and 16-QAM modulation schemes [13], $1/2$-rate turbo coding [419] and a Minimum Mean Squared Error Block Decision Feedback Equalizer (MMSE-BDFE) based Multi-User Detector [93, 434] joint detection for transmission over a COST207 Bad-Urban Reduced-mode A (BU-RA) channel [450]. The system was configured to operate at a target BER of 0.1%, a low-quality outage was recorded for BERs in excess of 0.5%, while an outage was declared for BER $\geq$ 1%. Table 7.7 summarizes the corresponding BPSK, 4QAM and 16-QAM SINR thresholds used in our simulations, when employing AQAM [419].

Figure 7.30 shows the forced termination probability associated with a variety of traffic loads, measured in terms of the mean normalized carried traffic expressed in Erlangs/km^2/MHz. The figure shows that the carried traffic was significantly improved by using adaptive antenna arrays [416] and adaptive modulation [419]. In Figure 7.30(a), the curve labelled as "No beamforming" presents the achievable carried traffic of the MC-CDMA network without the aid of AAAs and AQAM techniques, which was limited to 323 users, or to a teletraffic load of approximately 1.83 Erlangs/km^2/MHz. However, with the advent of 2- and 4-element adaptive antenna arrays at the base stations, the number of users supported by the network increased by 44% to 466 users and by 127%

Table 7.7: The target SINR, low quality outage SINR and outage SINR thresholds used for the BPSK, 4-QAM and 16-QAM modulation modes of the adaptive modem in MC-CDMA based cellular networks.

SINR Threshold	BPSK	4-QAM	16-QAM
Outage SINR (1% BER)	0.87 dB	4.20 dB	10.3 dB
Low Quality Outage SINR (0.5% BER)	1.60 dB	4.85 dB	11.0 dB
Target SINR (0.1% BER)	2.75 dB	6.15 dB	12.2 dB

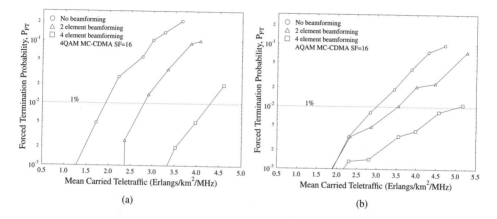

Figure 7.30: Forced termination probability versus mean carried traffic of the MC-CDMA cellular network both with and without beamforming in conjunction with 4QAM and AQAM for a spreading factor of $SF = 16$. (a) Using adaptive antenna arrays; (b) using AAAs and AQAM.

to 733 users, which corresponded to carried traffic values of 2.72 Erlangs/km²/MHz and 4.18 Erlangs/km²/MHz, respectively. When the network employs AQAM techniques without the assistance of adaptive antenna arrays, the attainable performance is characterized by the curve labelled as "No beamforming" in Figure 7.30(b). A carried traffic gain of 60% corresponding to supporting a total of 517 users was achieved compared to the "No beamforming" scenario of Figure 7.30(a). This carried traffic was higher than the carried traffic supported upon 2-element beamforming in the 4QAM scenario of Figure 7.30(a). When both AAAs and AQAM techniques were invoked in the MC-CDMA system, the maximum user capacity reached 869 subscribers, which corresponded to a teletraffic load of 4.98 Erlangs/km²/MHz, which was attained upon using 4-element beamforming in conjunction with AQAM.

The probability of low quality access versus the mean carried teletraffic load was presented in Figure 7.31, where we observe that the system did not benefit from the application of adaptive antenna arrays, in fact on the contrary. This is likely to be a consequence of the potentially more rapidly fluctuating SINR levels imposed the AAAs, which may experience abrupt SINR level variations as a detriment of their spatial selectivity, which subscribers move in and out of the high-gain beams. This low quality access degradation would be less pronounced in conjunction with omni-directional antennas.

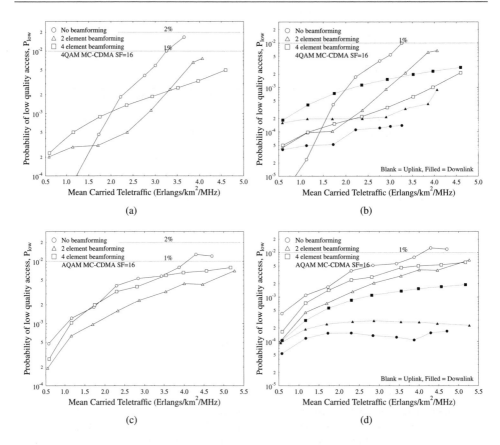

Figure 7.31: Probability of low quality access versus mean carried traffic of the MC-CDMA cellular network both with and without beamforming in conjunction with 4QAM and AQAM for a spreading factor of $SF = 16$. (a) Averaged UL/DL P_{low} for 4QAM and AAAs; (b) separate UL and DL P_{low} for 4QAM and AAAs; (c) averaged UL/DL P_{low} for AQMA and AAAs; (d) separate UL and DL P_{low} AQAM and AAAs.

Furthermore, Figure 7.31(a) depicted the probability of low quality access without employing adaptive modulation, i.e. when using fixed 4QAM. Upon comparing these results to those obtained in conjunction with adaptive modulation in Figure 7.31(c), the probability of low quality access degradation imposed by the employment of adaptive modulation can be explicitly seen. This increased probability of low quality access can be attributed to the employment of less robust, but higher-throughput, higher-order modulation modes invoked by the adaptive modulation scheme, which are more vulnerable to sudden SINR changes, than 4QAM. Hence, under given propagation and SINR conditions encountered in the interference-resilient fixed 4QAM modulation mode characterized Figure 7.31(a), a low quality outage event may be avoided. By contrast, when using adaptive modulation invoking a less resilient, but higher-throughput and higher-order modulation mode, the same propagation and SINR conditions may inflict a low quality outage. Upon comparing Figure 7.31(b) to Figure 7.31(d), it can be seen that this scenario was more often encountered in the

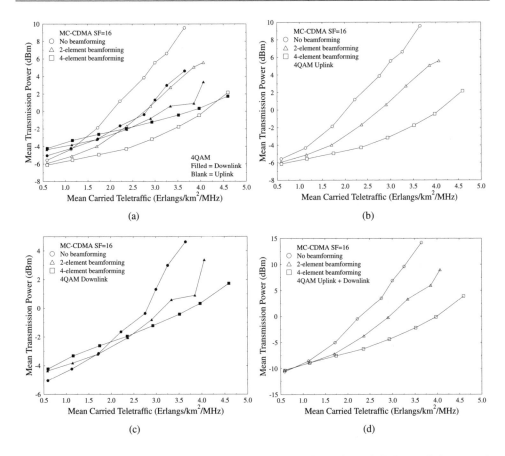

Figure 7.32: Mean transmission power versus mean carried traffic of the MC-CDMA cellular network both with as well as without beamforming for a spreading factor of $SF = 16$. (a) Mean UL and DL transmission power; (b) mean UL transmission power; (c) mean DL transmission power; (d) mean UL+DL transmission power.

UL transmission. This phenomenon will be discussed in more detail in the context of Figure 7.34.

The mean transmission power versus teletraffic performance using both fixed and adaptive modulation in conjunction with AAAs is depicted in Figure 7.32 and Figure 7.33, respectively. The employment of AAAs may result in the attenuation of the desired signal, while maximizing the received SINR, hence the levels of interference are efficiently reduced, ultimately leading to the reduction of the mean transmission power, as seen in Figure 7.32(b) and Figure 7.32(c). Figure 7.33(a) suggests that the mean UL transmission power was below the mean DL transmission power when the traffic loads were low, which may be attributed to encountering interfered pilot signals by the mobiles in the DL. At higher traffic loads the mean required UL transmission power had to be increased for the sake of maintaining an acceptable SINR, as evidenced by Figure 7.33(b). However, it is seen in Figure 7.33(c) that the mean DL transmission power requirement was reduced, as the traffic load became

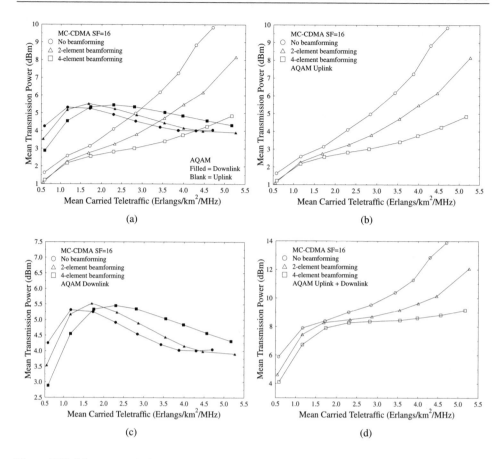

Figure 7.33: Mean transmission power versus mean carried traffic of the MC-CDMA cellular network both with and without beamforming in conjunction with AQAM for a spreading factor of $SF = 16$. (a) Mean UL and DL transmission power; (b) mean UL transmission power; (c) mean DL transmission power; (d) mean UL+DL transmission power.

higher. That is because when the traffic load increased, the level of interference rose, resulting in a low SINR. In this scenario the AQAM control regime is expected to switch from 16-QAM to 4-QAM or from 4-QAM to BPSK, hence requiring a reduced average power level for the network. This hypothesis may be confirmed by examining Figure 7.34, which portrays the mean SINR versus the mean carried teletraffic as well as the discrete histogram modeling the probability density function (PDF) of the mean SINR. From Figure 7.34(a) we observe that the mean SINR of the network reduced as the traffic load became higher. The PDFs suggests that at lower traffic loads typically higher target SINRs were maintained, which facilitated the employment of higher-throughput but more vulnerable modulation modes, such as 16-QAM. By contrast, at higher traffic loads typically low target SINRs were maintained, requiring lower-throughput but more robust modulation modes, such as BPSK.

Figure 7.34: Mean SINR versus mean carried traffic and the SINR histogram modeling the probability density function of the MC-CDMA cellular network's SINR both with and without beamforming in conjunction with AQAM for a spreading factor of $SF = 16$.

A summary of the maximum achievable user capacity of the MC-CDMA networks considered in this section under the detrimental effect of a log-normal shadowing having a standard deviation of 3 dB, both with and without employing beamforming in conjunction with AQAM is given in Table 7.8. The teletraffic carried and the mean mobile as well as base station transmission powers required for attaining these user capacities are also shown in Table 7.8.

7.4.3 Summary and Conclusions

In this section we have examined the achievable network capacity and the overall performance of the MC-CDMA based cellular network benefiting from both adaptive antenna arrays and adaptive modulation techniques. We have shown that a substantially increased number of users may be supported, who benefit from a superior call quality, and reduced transmission

Table 7.8: Maximum carried traffic and maximum number of mobile users that can be supported by the MC-CDMA network, whilst meeting the target network quality constraints, namely $P_{FT} \leq$ 1%, $P_{low} \leq 1\%$ and GOS $\leq 4\%$. The carried traffic is expressed in terms of normalized Erlangs (Erlang/km^2/MHz) in conjunction with shadow fading having a standard deviation of 3 dB and a frequency of 0.5 Hz, whilst employing adaptive modulation techniques [408] for a spreading factor of $SF = 16$. The average transmission power of the MSs and BSs are also summarized.

Modulation mode	Beamforming	Users	Traffic (Erlangs/km^2/MHz)	Power (dBm) MS	BS
4-QAM	No	323	1.83	-1.15	-2.79
4-QAM	2-elements	466	2.72	-0.18	-1.22
4-QAM	4-elements	733	4.18	0.46	0.82
AQAM	No	517	2.95	5.20	4.48
AQAM	2-elements	594	3.50	4.66	4.47
AQAM	4-elements	869	4.98	4.65	4.39

power requirements for a given number of AAA array elements located at the base stations.

Chapter 8

HSDPA-style TDD/CDMA Network Performance

8.1 Introduction

In January 1998, the European standardization body created for the definition of the third-generation (3G) mobile radio system, namely the European Telecommunications Institute's Special Mobile Group (ETSI SMG), ratified a radio access scheme referred to as the Universal Mobile Telecommunications System (UMTS) [416]. The UMTS Terrestrial Radio Access (UTRA) supports two duplexing modes, namely the Frequency Division Duplexing (FDD) mode, where the UL and DL are transmitted on different frequencies, and the Time Division Duplexing (TDD) mode, where the UL and the DL are transmitted on the same carrier frequency, but multiplexed in time [416]. UMTS networks will introduce into wide area using a completely new high bit rate radio technology: wideband CDMA (WCDMA).

In UTRA, the different services are expected to be supported in a spectrally efficient manner, either by FDD or TDD. The FDD mode is intended for applications in both macro- and micro-cellular environments, supporting data rates of up to 384 Kbps both at relatively high velocity. The TDD mode, on the other hand, is more suited to micro and pico-cellular environments, as well as for licensed and unlicensed cordless and wireless local loop applications. It makes efficient use of the unpaired spectrum, for example in wireless Internet applications, where much of the teletraffic is expected to be on the DL and supports data rates of up to 2 Mbps. Therefore, the TDD mode is particularly well suited for environments generating a high traffic density, e.g. in city centres, business areas, airports, etc., and for indoor coverage, where the applications require high data rates and tend to have highly asymmetric traffic, again, as in wireless Internet access.

3G, HSPA and FDD versus TDD Networking   Second Edition
L. Hanzo, J. S. Blogh and S. Ni   © 2008 John Wiley & Sons, Ltd

8.2 UMTS FDD versus TDD Terrestrial Radio Access

A bandwidth of 215 MHz in the region of 2.0 GHz has been allocated for UMTS services in Europe. The paired bands of 1920–1980 MHz (UL) and 2110–2170 MHz (DL) have been set aside for FDD W-CDMA systems, and the unpaired frequency bands of 1900–1920 MHz and 2010–2025 MHz for TDD CDMA systems.

A UTRA Network (UTRAN) consists of one or several Radio Network Sub-systems (RNSs), which in turn consist of Base Stations (referred to as Node Bs) and Radio Network Controllers (RNCs). A Node B may serve a single or multiple cells. Mobile stations (MSs) are known as User Equipment (UE), which are expected to support multi-mode operation in order to enable handovers between the FDD and TDD modes and, prior to complete UTRAN coverage, also to GSM. The two modes differ in a number of ways in the physical layer, but for compatibility and implementation reasons they are harmonized as far as possible, especially in higher layers. More details on the differences and distinctions can be found in [451]. The key parameters of UTRA have been defined in Table 6.1. The harmonization enables the same services to be offered over both modes, while the differences lead to one mode being best utilized in certain system scenarios while the other mode may perform better in other scenarios.

8.2.1 FDD versus TDD Spectrum Allocation of UTRA

The FDD versus TDD spectrum allocation of UTRA is shown in Figure 8.1. As can be seen, UTRA is unable to utilize the full frequency spectrum allocated for the 3G mobile radio systems during the WARC'92 conference [436], since those frequency bands have also been partially allocated to the Digital Enhanced Cordless Telecommunications (DECT) system [452]. The frequency spectrum was originally allocated based on the assumption that speech and low data rate transmission would become the dominant services offered by IMT-2000 [416, 453]. However, in recent years a paradigm has been experienced towards services that require high-speed data transmission, such as wireless Internet access and multimedia services. A study conducted by the UMTS Forum [454] forecast that the current frequency bands allocated for IMT-2000 are only sufficient for the initial deployment until the year 2005 although this was not the case even at the time of writing. According to the current demand estimates, it was foreseen that an additional frequency spectrum of 187 MHz might be required for IMT-2000 in high-traffic areas by the year 2010. Among of numerous candidate extension bands, the band 2520–2670 MHz has been deemed to be the most likely. Unlike other bands, which have already been allocated for use in other applications, this band was allocated to mobile services in all regions. Furthermore, the 150 MHz bandwidth available is sufficiently wide for satisfying most of the forecast spectrum requirements.

Again, the UMTS radio access supports both FDD and TDD operations [416]. The operating principles of these two schemes are augmented here in the context of Figure 8.2.

Specifically, the UL and DL signals are transmitted using different carrier frequencies, namely f_{UL} and f_{DL}, respectively, separated by a frequency guard band in the FDD mode. On the other hand, the UL and DL messages in the TDD mode are transmitted using the same carrier frequency f_{TDD}, but in different timeslots, separated by a guard period. As seen from the spectrum allocation of Figure 8.1, the paired bands of 1920–1980 MHz and 2110–2170 MHz are allocated for FDD operation in the UL and DL, respectively, whereas the

GSM 1800 Uplink		GSM 1800 Dowlink	DECT RX/TX	UTRA TDD RX/TX	UTRA FDD Uplink	MS	UTRA TDD RX/TX		UTRA FDD Dowlink	MS

1710   1785   1805   1885 1900   1920   1980 2010   2025 2110   2170 2200

Frequency (MHz)

Figure 8.1: The proposed spectrum allocation in UTRA.

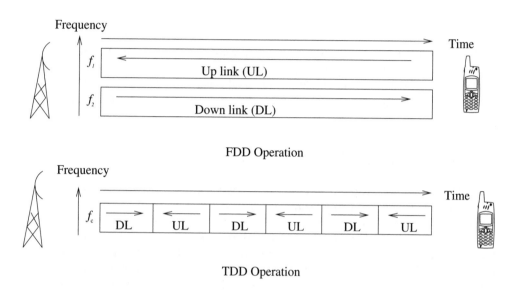

FDD Operation

TDD Operation

Figure 8.2: Principle of FDD and TDD operation.

TDD mode is operated in the remaining unpaired bands. The parameters designed for FDD and TDD operations are mutually compatible so as to ease the implementation of a dual-mode terminal capable of accessing the services offered by both FDD and TDD operators.

8.2.2 Physical Channels

The transport channels are transmitted using the UTRA physical channels [416,455,456]. The physical channels are typically organized in terms of radio frames and timeslots, as shown in Figure 8.3. While in GSM [457] each TDMA user had an exclusive slot allocation, in WCDMA the number of simultaneous users supported is dependent on the users' required bit rate and their associated spreading factors. The MSs can transmit continuously in all slots or discontinuously, for example, when invoking a Voice Activity Detector (VAD) [457].

As seen in Figure 8.3, there are 15 timeslots within each radio frame. The duration of each timeslot is 2/3 ms, which yields a total duration of 10 ms for the radio frame. As we show later in this section, the configuration of the information in the timeslots of the physical channels differs from one another in the UL and DL, as well as in the FDD and TDD modes. In the FDD

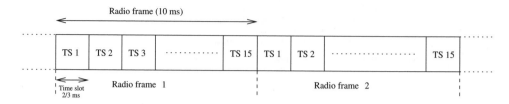

Figure 8.3: UTRA physical channel structure.

mode, a DL physical channel is defined by its spreading code and frequency. Furthermore, in the UL, the modem's orthogonal in-phase (I) and quadrature-phase (Q) branches are used for delivering the data and control information simultaneously in parallel [416]. On the other hand, in the TDD mode, a physical channel is defined by its spreading code, frequency and timeslot.

8.3   UTRA TDD/CDMA System

The UTRA TDD mode is partly a result of the original UMTS spectrum allocation, which consists of one paired and two unpaired bands. This led to an ETSI decision in 1998 that not just one but two of the proposed access technologies should be adopted for the UMTS standard. Hence, the FDD mode should be used in the paired band and the TDD mode in the unpaired band. The TDD UTRA scheme will be deployed in the unpaired IMT-2000 frequency bands. The so-called band A is the 3G unpaired frequency allocation in Europe: 1900–1920 MHz and 2010–2025 MHz. In the United States it is the so-called band B, namely the PCS spectrum allocation encompasses the range of 1850–1910 MHz and 1930–1990 MHz. Furthermore, the United States also allocated band C, an unlicensed band from 1910 to 1930 MHz. The nominal channel spacing in UTRA is 5 MHz, with a channel raster of 200 kHz, which means that the carrier frequency is a multiple of 200 kHz.

There are a few characteristics that are typical of TDD systems and different from the characteristics of FDD systems. These characteristics are listed below.

- *Utilization of unpaired bands.* The TDD system can be invoked in unpaired bands, while the FDD system always requires a pair of bands. It is more likely that in the future unpaired spectrum resources will be made available for UMTS.

- *Possible interference between UL and DL.* Since both the UL and DL share the same carrier frequency in TDD, any timeslot can be used in any direction and, hence, the signals of the two transmission directions may interfere with each other.

- *Flexible capacity allocation between the UL and DL.* In the TDD mode, the UL and DL are divided in the time domain. It is possible to control the switching point [458] between the UL and DL, as seen in Figure 8.2, and move capacity from the UL to DL, or vice versa, if the capacity requirement is asymmetric between the UL and DL.

- *Discontinuous transmission.* The mobile and the base station transmissions are discontinuous in TDD. Discontinuous transmissions impose specific requirements on the implementation. Switching between the transmission directions requires a reflecting time, since the effects of switching effects of transients must be avoided. Hence, in order to avoid overlapping of the UL and DL transmissions, a guard period is used at the end of each slot.

- *UL/DL channel properties.* In case of frequency selective fading the channel's function depends on the frequency and, therefore, in the FDD mode the fast fading is typically uncorrelated between the UL and DL. Since the same frequency is used both for the UL and DL in the TDD mode, the fast fading properties are more similar in the UL and DL. The similarity of the fast fading between the UL and DL can be exploited in both power control and adaptive antenna arrays used in TDD.

It is unlikely that any of the service providers would operate standalone wide-area TDD networks, but rather they would invoke the FDD UTRA mode and possibly GSM to provide continuous wide-area coverage, while using TDD to serve as a separate capacity-enhancing layer in the network [459]. Furthermore, as a benefit of being able to arbitrarily adjust the UL/DL asymmetry, the TDD mode is also capable of supporting high bit rates, ranging from 144 kbps to 2 Mbps in wireless Internet-type applications.

8.3.1 The TDD Physical Layer

The UTRA TDD mode has a similar frame structure to that of the UTRA FDD mode. As seen in Figure 8.3, there are 15 slots in a frame, which has a period of 10 ms. Each slot has 2560 chips and lasts for 0.667 ms. A superframe consists of 72 frames and lasts for 720 ms. A physical channel consists of bursts that are transmitted in the same slot of each frame. For specifying a physical channel explicitly, we also have to define its so-called repetition period, repetition length and superframe offset, which are exemplified below. The number of frames between slots belonging to the same physical channel is the repetition period of a given physical channel, which must be sub-multiple of 72, i.e. $1, 2, 3, 4, 6, 8, 9, 12, 18, 24, 36$ and 72. An example is given by the physical channel occupying slot 0 in every 12th frame. The superframe offset defines the repetition period offset within a superframe, with respect to the beginning of the frame. Returning to our example, if the superframe offset is 3, the physical channel will occupy slot 0 in frames $3, 15, 27, 39, \ldots$, since it was offset by 3 frames, where the corresponding slots are 12 frames apart. The repetition length defines the number of slots associated with each repetition, and may have values of $1, 2, 3, 4$. For the example where the physical channel occupies slot 0, the repetition period is 12, the superframe offset is 3, and say, the repetition length is 4, the physical channel will occupy slot 0 in frames $(3, 4, 5, 6)$, $(15, 16, 17, 18), (27, 28, 29, 30)$, etc.

8.3.2 Common Physical Channels of the TDD Mode

The UTRA TDD mode employs TDD for creating bidirectional transmission links. Each slot in a frame can be used for carrying either UL or DL information. The switching point or points between UL and DL slots may be variable, as is the number of slots allocated to each link. At least one slot must be allocated in each direction.

In TDD operation, the burst structure of Figure 8.4 is used for all of the physical channels, where each timeslot's transmitted information can be arbitrarily allocated to the UL or DL,

Burst type	Data	Midamble	Guard Period
Burst Type 1	976	512	96
Burst Type 2	1104	256	96

Figure 8.4: Timeslot of the physical channels.

as shown in the four possible TDD allocations of Figure 8.5. A symmetric UL/DL allocation refers to a scenario in which an approximately equal number of DL and UL bursts is allocated within a TDD frame, while in case of asymmetric UL/DL allocation there is an unequal number of UL and DL bursts, such as services, etc., for example, in "near-simplex" file download from the wireless Internet or in the case of video-on-demand.

In UTRA, two different TDD burst structures, known as *Burst Type 1* and *Burst Type 2*, are defined, which are shown in Figure 8.4. The Type 1 burst has a longer midamble of 512 chips than the Type 2 burst of length 256 chips. However, both types of bursts have an identical *Guard Period* (GP) of 96 chips. The midamble sequences that are allocated to the different TDD bursts in each timeslot belong to a so-called *midamble code set*. The codes in each midamble code set are derived from a unique *Basic Midamble Code*. Adjacent cells are allocated different midamble code sets. This can be exploited to assist in cell identification.

8.3.3   Power Control

Power control of the UTRA TDD mode is performed on a per-frame basis, namely using a power control update per 10 ms frame, which is carried out differently for the UL and DL.

Specifically, the UL power control uses an open loop technique, which exploits the similarity of the UL and DL channel in a TDD system, in particular as regards to the pathloss. In each cell there is at least one beacon, i.e. a physical channel having a known transmit power. Furthermore, during unallocated UL timeslots the base station is capable of estimating the UL interference by exploiting the knowledge of the required target SIR, the MS can set its transmission power in order to fulfill the transmission integrity requirements at the BS. A first-order predictor corresponding to a weighting factor can be used for taking into account the expected delay between the DL pathloss estimate and the actual UL pathloss. At the BS, an outer power control loop is used for estimating the SIR of the received signal, which is compared with the target SIR requirements. Then the necessary

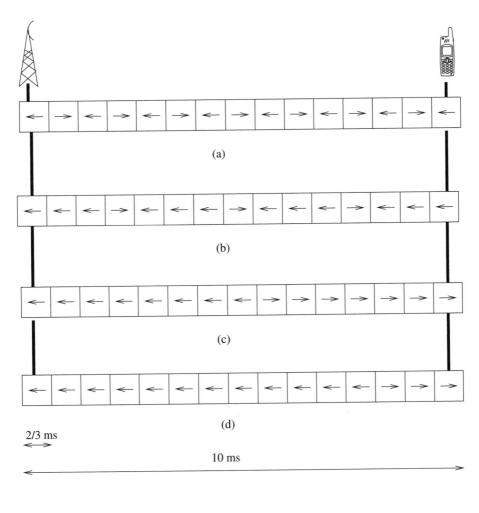

(a)

(b)

(c)

(d)

2/3 ms

10 ms

Uplink: ←

Downlink: →

Figure 8.5: Multiple switching points per frame for different slot per frame allocations. (a) Symmetric UL/DL allocation with multiple switching points; (b) asymmetric UL/DL allocation with multiple switching points; (c) symmetric UL/DL allocation with single switching point; (d) asymmetric UL/DL allocation with single switching point.

MS transmit power is calculated, which is signaled to the MS. This requirement allows the SIR-based outer loop to compensate for the long-term fluctuation of the associated pathlosses.

8.3.4   Time Advance

Timing advance is the mechanism used in UTRA for controlling the transmit time instant of signals from different MSs for mitigating leakage between timeslots. During the initial access, the base station estimates the instant of reception for the MSs and advances their instant of transmission by the estimated propagation delay, so that all signals arrive approximately within the expected time window at the BS. The UTRA TDD system can be used in wide area cells, where the employment of this timing advance mechanism is necessary for preventing the UL burst collisions at the BS receiver. The timing advance operates to a resolution of four chips or 1.04 μs, since the chip rate is 3.84 Mchip/s. The BS estimates the time offset associated with the PRACH transmissions [416] and calculates the required initial timing advance. The timing advance parameter is transmitted as an 8-bit number, catering for a maximum timing advance of 256×1.04 μs corresponding to the UL transmissions from the MS. This maximum propagation delay of approximately 256 μs potentially allows for a cell size of 80 km.

There are proposals to have an enhanced timing advance mechanism with a resolution of one-eighth of a chip period. This potentially holds the promise of quasi-synchronous UL transmission, which would dramatically decrease the multiple access interference, since all the transmitted codes of the MSs would remain quasi-orthogonal.

When performing a handover to another TDD cell, which is generally synchronized to a reference cell, the MS is capable of autonomously applying the right timing advance in the new cell. In any case, the MS has to signal the timing advance it applies to the BS in the new cell.

8.4   Interference Scenario in TDD CDMA

One of the major attractions for the UTRA TDD mode system is that it allows the UL and DL capacities to be allocated asymmetrically. The UL and DL are transmitted on the same carrier frequency, which creates additional interference scenarios compared with UTRA FDD, and as seen in Figure 8.6, the UL/DL transmission directions of adjacent co-channel BSs may severely interfere with each other. This kind of interference may become particularly detrimental if the base stations are not synchronized or if a different ratio of UL and DL timeslots is used in adjacent cells, even if the base stations are frame synchronized. Frame synchronization requires an accuracy of a few symbols, rather than an accuracy of a few chips.

The interference between UL and DL can also occur between adjacent frequencies. Therefore, the interference between UL and DL can take place within one operator's band, and also between two operators.

The interference between UL and DL can occur between two MSs and between two base stations. In FDD operation the duplex separation prevents the interference between UL and DL. In a TDD system there are four types of inter-cell/inter-operator interference. These are:

- MS → MS;

- BS → BS;

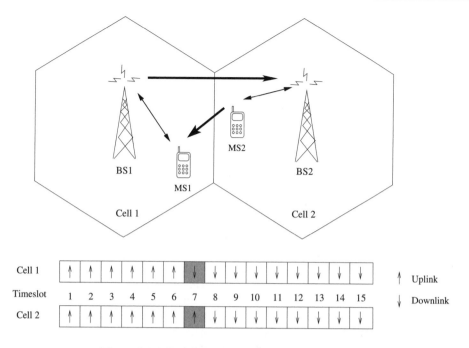

Figure 8.6: MS-to-MS BS-to-BS inter-cell interference.

- MS → BS;

- BS → MS.

The interference between a MS and a base station is the same both in TDD and in FDD operation. The extent of the interference is dependent on many parameters such as the cell locations and user distributions; however, there are two parameters that can have a major effect on the system performance and can potentially be managed by the network. There are synchronization between cells, and the asymmetry across the network.

8.4.1 Mobile-to-Mobile Interference

Mobile-to-mobile interference occurs in Figure 8.6, at the timeslot 7 the mobile MS_2 is transmitting and the mobile MS_1 is receiving in the same frequency in adjacent cells. Mobile to mobile interference is statistical because the locations of the mobiles cannot be controlled. Therefore, mobile to mobile interference cannot be avoided completely by the network planning.

8.4.2 Base Station-to-Base Station Interference

In Figure 8.6, base station-to-base station interference occurs, at the timeslot 7: the base station BS_1 is transmitting and the base station BS_2 is receiving in the same frequency in adjacent cells. Base station to base station interference depends heavily on the pathloss between the two base stations and, therefore, the network planning has a major effect on this interference scenario [460].

8.5   Simulation Results

A number of studies have been conducted in order to characterize the network capacity of WCDMA-assisted 3G networks [409–411]. *The Timeslot (TS) opposing technique* proposed by Haas, McLaughlin and Povey [32, 461] enables asynchronous cells to overlap without a significant capacity degradation in comparison with the more idealistic scenario, when the base stations of all cells transmit and receive slot-synchronously in the UTRA-TDD system. Furthermore, the Dynamic Channel Allocation (DCA) [462] aided TS-opposing algorithm [32] enables neighboring cells to adopt different grades of UL/DL asymmetry without inflicting a significant capacity loss. The co-existence of the UTRA-TDD and FDD modes was studied in [463–465], since they are expected to co-exist in the same geographical area. Owing to the presence of increased levels of interference, capacity degradations are expected. It is crucial to estimate this potential capacity degradation and to identify appropriate countermeasures. Power control is a standard technique of improving the performance of wireless systems. Different power control techniques and their application within the UMTS were presented in [466–469]. More specifically, in [466], received signal level-based and interference level-based power control algorithms were introduced and the achievable system performance was compared by means of simulations. In [467], the UTRA TDD mode was studied in conjunction with an open loop power control algorithm combined with outer loop power control functions, which resulted in an improved rate of successful call establishment in the network. An Optimum Power Control (OPC) method was proposed in [468], which achieved the same performance as Wu's approach [470] at the cost of a lower complexity. Kurjenniemi *et al.* [469] studied UL power control in the context of the UTRA TDD system by means of system-level simulations, demonstrating that the UTRA TDD UL power control substantially benefited from exploiting accurate interference measurements and, hence, achieved a high capacity, even in the presence of implementation errors. A pre-Rake smart antenna system designed for TDD CDMA was studied in [471]. The study demonstrated that incorporating an antenna array at the base station significantly improves the achievable capacity by reducing the interference between the UL and DL of adjacent cells, which is a consequence of potentially using all timeslots in an arbitrary uncoordinated fashion both in the UL and DL. Conventional single-user detectors, such as the Rake receiver are expected to result in a low network capacity owing to the excessive TDD-induced Multiple Access Interference. In contrast, Multi-User Detectors (MUDs) have the potential of increasing the network capacity at the cost of a higher complexity [93, 419, 472].

This section presents our simulation results obtained for a TDD mode UTRA-like CDMA cellular network, investigating the achievable user capacity of the TDD mode in both non-shadowed and shadowed propagation environments. This is described in Section 8.5.2 followed by our performance investigations using adaptive antenna arrays, when subjected to both non-shadowed as well as shadowed propagation conditions. Finally, the performance of adaptive modulation techniques used in conjunction with adaptive antenna arrays in shadow faded environments is then characterized in Section 8.5.3.

8.5.1 Simulation Parameters [416]

In this section simulations were conducted for various scenarios and algorithms in the context of a TDD mode UTRA-like CDMA-based cellular network in order to study the interactions of the processes involved in such a network. The simulation parameters are as follows [416]. As in the UTRA standard, the frame length was set to 10 ms, containing 15 power control timeslots. The power control target SINR was chosen to give a Bit Error Ratio (BER) of 1×10^{-3}, with a low-quality outage occurring at a BER of 5×10^{-3} and an outage taking place at a BER of 1×10^{-2}. The received SINRs at both the mobile and the base stations were required for each of the power control timeslots, and hence the outage and low-quality outage statistics were gathered. If the received SINR was found to be below the outage SINR for 75 consecutive power control timeslots, corresponding to 5 consecutive transmission frames or 50 ms, the call was dropped. The post-despreading SINRs necessary for obtaining the target BERs were determined with the aid of physical-layer simulations using a 4QAM modulation scheme, in conjunction with $1/2$ rate turbo coding and joint detection over a COST 207 seven-path Bad Urban channel [408]. For a spreading factor of 16, the post-despreading SINR required for maintaining BER of 1×10^{-3} was 8.0 dB, while for a BER of 5×10^{-3} it was 7.0 dB, and for a BER of 1×10^{-2} was about 6.6 dB. These values can be seen along with the other system parameters specified earlier in Table 6.2. The pre-despreading SINR is related to E_b/N_o and to the spreading factor by

$$SINR = (E_b/N_o)/SF, \tag{8.1}$$

where the spreading factor is given by $SF = W/R$, with W being the chip rate and R the data rate. A receiver noise figure of 7 dB was assumed for both the mobile and the base stations [59]. Thus, in conjunction with a thermal noise density of -174 dBm/Hz and a noise bandwidth of 5 MHz, this resulted in a receiver noise power of -100 dBm. The power control algorithm used was relatively simple, and unrelated to the previously introduced schemes of Section 8.3.3. Furthermore, since it allowed a full transmission power change of 15 dB within a 15-slot UTRA data frame, the power control scheme advocated is unlikely to limit the network's capacity.

Specifically, for each of the 15 timeslots per transmitted frame, both the mobile and base station transmit powers were adjusted such that the received SINR was higher than the target SINR, but less than the target SINR plus a 1 dB hysteresis. When in handover, a mobile's transmission power was only increased if all of the base stations in the Active Base station Set (ABS) requested a power increase, but was it decreased if any of the base stations in the ABS had an excessive received SINR. In the DL, if the received SINR at the mobile was insufficiently high, then all of the active base stations were commanded to increase their transmission powers. Similarly, if the received SINR was unnecessarily high, then the active base stations would reduce their transmit powers. The DL intra-cell interference orthogonality factor α, was set to 0.5 [409–411]. Owing to the use of a frequency reuse factor of one, with its associated low-frequency reuse distance, it was necessary for both the mobiles and the base stations to increase their transmitted power gradually when initiating a new call or entering handover. This was required to prevent sudden increases in the level of interference, particularly on links using the same base station. Hence, by gradually increasing the transmit power to the desired level, the other users of the network were capable of compensating for the increased interference by increasing their transmit powers, without encountering undesirable outages. In an FDMA/TDMA network this effect is less noticeable owing to the significantly higher frequency reuse distance.

Since a dropped call is less desirable from a user's viewpoint than a blocked call, two resource allocation queues were invoked, one for new calls and the other (higher-priority) queue for handovers. By forming a queue of the handover requests, which have a higher priority during contention for network resources than new calls, it is possible to reduce the number of dropped calls at the expense of an increased blocked call probability. A further advantage of the Handover Queueing System (HQS) is that during the time a handover is in the queue, previously allocated resources may become available, hence increasing the probability of a successful handover. However, in a CDMA-based network the capacity is not hard-limited by the number of frequency/timeslot combinations available, like in a FDMA/TDMA-based network such as GSM. The main limiting factors are the number of available spreading or OVSF codes, or the interference levels in conjunction with the restricted maximum transmit power, resulting in excessive forced termination rates. New call allocation requests were queued for up to 5 s, if they could not be immediately satisfied, and were blocked if the request had not been completed successfully within the 5 s.

There are several performance metrics that can be used for quantifying the QoS provided by a cellular network. The following performance metrics have been widely used in the literature and were also advocated by Chuang and Sollenberger [397]:

- New call blocking probability, P_B,

- Call dropping or forced termination probability, P_{FT},

- Probability of low-quality connection, P_{low},

- Probability of outage, P_{out},

- Grade of service, GOS.

The new call blocking probability, P_B, is defined as the probability that a new call is denied access to the network. In an FDMA/TDMA-based network, such as GSM, this may occur because there are no available physical channels at the desired base station or the available channels are subject to excessive interference. However, in a CDMA-based network this does not occur, and hence the new call blocking performance is typically very high.

The forced termination probability, P_{FT}, is the probability that a call is forced to terminate prematurely. In a GSM-type network, an insufficiently high SINR, which inevitably leads to dropped calls, may be remedied by an intra- or inter-cell handover. However, in CDMA either the transmit power must be increased, or a soft handover must be performed in order to exploit the available diversity gain.

Again, the probability of a low quality connection is defined as

$$P_{low} = P\{SINR_{UL} < SINR_{req} \text{ or } SINR_{DL} < SINR_{req}\} \qquad (8.2)$$
$$= P\{min(SINR_{UL}, SINR_{DL}) < SINR_{req}\}.$$

The *GOS* was defined in [397] as

$$GOS = P\{\text{unsuccessful or low-quality call access}\} \qquad (8.3)$$
$$= P\{\text{call is blocked}\} + P\{\text{call is admitted}\}$$
$$\times P\{\text{low signal quality and call is admitted}\}$$
$$= P_B + (1 - P_B)P_{low},$$

and is interpreted as the probability of unsuccessful network access (blocking), or low-quality access, when a call is admitted to the system. However, since the new call blocking probability of CDMA-based networks is negligible, this metric has been omitted.

In our forthcoming investigations, in order to compare the network capacities of different networks, it was decided to use two scenarios defined as follows.

(i) A *conservative scenario*, where the maximum acceptable value for the new call blocking probability, P_B, is 3%, the maximum forced termination probability, P_{FT}, is 1% and P_{low} is 1%.

(ii) A *lenient scenario*, where the maximum acceptable value for the new call blocking probability, P_B, is 5%, the maximum forced termination probability, P_{FT}, is 1% and P_{low} is 2%.

In the next section we characterize the capacity of an adaptive modulation [13] assisted, beam-steering aided TDD/CDMA system. In TDD/CDMA the mobiles suffer from interference inflicted by the other MSs both in the reference cell the MS is roaming in (intra-cell interference) as well as those in the neighboring cells (inter-cell interference). Furthermore, in contrast to FDD/CDMA, where the BSs transmit in an orthogonal frequency band, in TDD/CDMA there is additional interference imposed by other BSs of the adjacent cells, since all times-slots can be used in both the UL and DL. In return for this disadvantage TDD/CDMA guarantees the flexible utilization of all of the available bandwidth, which meets the demand for the support of asymmetric UL and DL services, such as high data rate file download in mobile Internet services, etc. In wireless systems the link quality fluctuates owing to either fading- and dispersion-induced channel impairments or as a consequence of the time-variant co-channel interference imposed by the teletraffic fluctuations owing to the varying number of users supported. Owing to these impairments conventional wireless systems often drop the call. In contrast, a particular advantage of employing adaptive modulation is that the transceiver is capable of automatically reconfiguring itself in a more error-resilient transmission mode, instead of dropping the call. Here we study the achievable network performance by simulation and compare it to that of the FDD/UTRA system.

8.5.2 Performance of Adaptive Antenna Array Aided TDD CDMA Systems

In this section we study the impact of adaptive antenna arrays on the network's performance. The investigations were conducted using a spreading factor of 16. Given that the chip rate of UTRA is 3.84 Mchips/s, this spreading factor corresponds to a channel data rate of $3.84 \times 10^6/16 = 240$ kbps. Applying 1/2 rate error correction coding would result in an effective data throughput of 120 kbps, whereas utilizing a 2/3 rate error correction code would provide a useful throughput of 160 kbps. A cell radius of 150 m was assumed, and a pedestrian walking velocity of 3 mph was used.

The advanced UTRA FDD system-level simulator [416] employing adaptive antenna arrays at the BS as well as adaptive modulation [13] was extended to the UTRA TDD mode for evaluating the system's achievable performance. We observed quite significant performance gains as a direct result of the interference rejection capabilities of the adaptive antenna arrays and adaptive modulation invoked. Network performance results were obtained using two- and four-element adaptive antenna arrays, both in the absence of shadow fading, and in the presence of 0.5 Hz and 1.0 Hz frequency shadow fading exhibiting a standard deviation of 3 dB. The adaptive beamforming algorithm used was the Sample Matrix Inversion (SMI) algorithm [416]. The specific adaptive beamforming implementation used in our TDD/CDMA-based network was identical to that used in the network simulations of [416]. Briefly [416], one of the eight possible 8-bit BPSK reference signals was used for uniquely and unambiguously identifying the desired user, while the remaining interfering users (up to seven of them) were assigned the other seven 8-bit reference signals. The received signal's autocorrelation matrix was then calculated, and from the knowledge of the desired user's reference signal the receiver's optimal antenna array weights were determined with the aid of the SMI algorithm. This implementation of the algorithm only calculated the receiver's antenna array weights, namely the antenna array weights used by the BS for receiving the mobiles' UL transmissions. However, it was demonstrated in [416] that further performance gains are attainable, if the BS's UL and DL array patterns, namely the transmit and receive beamforms, are optimized individually. The antenna array weights were re-calculated for every power control step, i.e. 15 times per UTRA data frame, owing to the potential significant changes in terms of the desired signal and interference powers that may occur during one UTRA frame as a result of the maximum possible 15 dB change in the power transmitted by each user.

Figure 8.7 shows the forced termination probability associated with a variety of traffic loads without shadowing, measured in terms of the mean normalized carried traffic expressed in Erlangs/km^2/MHz. The figure suggests that the TDD network's performance was poor in comparison with the FDD mode both with and without employing antenna arrays at the base stations. As expected, the "No beamforming" scenario suffered from the highest forced termination probability of the three beamforming scenarios at a given traffic load, which was valid for both the TDD and FDD modes. Our discussions are focused here on the TDD mode, using FDD as the benchmark. When using "two-element beamforming", the adaptive antenna arrays have considerably reduced the levels of interference, leading to a reduced forced termination probability. Without employing antenna arrays at the BSs the network capacity was limited to 142 users, or to a teletraffic load of approximately 0.81 Erlangs/km^2/MHz. However, with the advent of employing two-element adaptive antenna arrays at the BSs the number of users supported by the network increased by 45% to 206 users, or almost to 1.18 Erlangs/km^2/MHz. Replacing the two-element adaptive antenna arrays with four-element arrays led to a further capacity increase of 56%, or 127% with respect to the capacity of the network using no antenna arrays. This is associated with a network capacity of 322 users, or 1.85 Erlangs/km^2/MHz. We can also see in Figure 8.7 that the capacity of the UTRA-like TDD/CDMA cellular system is significantly worse than that of the UTRA-like FDD/CDMA system under the same propagation conditions. The "TDD four-element beamforming" scenario has a similar performance to the "FDD two-element

Figure 8.7: Forced termination probability versus mean carried traffic of the UTRA-like FDD and TDD/CDMA-based cellular network of Table 6.2 both with and without beamforming and without shadowing for $SF = 16$.

beamforming" scenario. This is because the TDD system suffers from the effects of the extra inter-cell interference, which we alluded to in Section 8.4.

Figure 8.8 portrays the probability of low-quality access versus various traffic loads. It can be seen from the figure that higher traffic loads were carried with the aid of the four-element array at a sufficiently low probability of a low quality, than that achieved using a two-element array. Again, the user-capacity of the TDD mode is often a factor two lower than that of the FDD mode close to the 1% P_{low} limit and TDD system is more prone to rapid performance degradation. However, at lower traffic loads the FDD mode performance with four elements was worse than that using two elements. This is because in a network using adaptive antenna arrays, when new calls started, the adaptive antenna arrays are used to null the sources of interference, and the array may reduce the antenna gain in the direction of the desired user, in order to maximize the SINR. This phenomenon was more marked when using four-element arrays since the directivity, and thus sensitivity to interfering signals is greater.

Figure 8.9 shows the achievable *GOS* for a range of teletraffic loads. Similar trends were observed regarding the probability of call blocking to those shown in Figure 8.7. The *GOS* is better (i.e. lower) when the traffic load is low, and vice versa for high traffic loads. This is mainly attributable to the higher call blocking probability of the "No beamforming" scenario, particularly in the region of the highest traffic loads. As before, the TDD mode is more prone to rapid interference-level fluctuations as well as to avalanche-like teletraffic overload and its teletraffic capacity is up to a factor two lower than that of the FDD mode. Our expectation is that this performance trend may be partially mitigated with the aid of the adaptive modulation

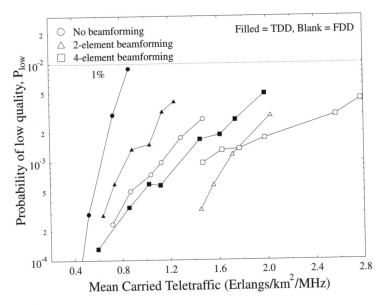

Figure 8.8: Probability of low-quality access versus mean carried traffic of the UTRA-like FDD and TDD/CDMA-based cellular network both with and without beamforming and without shadowing for $SF = 16$.

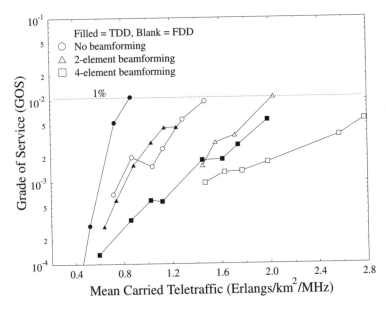

Figure 8.9: *GOS* versus mean carried traffic of the UTRA-like FDD and TDD/CDMA-based cellular network both with and without beamforming and without shadowing for $SF = 16$.

Figure 8.10: Forced termination probability versus mean carried traffic of the UTRA-like FDD and TDD/CDMA-based cellular network both with and without beamforming and with shadowing for $SF = 16$.

techniques of Section 8.5.3 [13], because when the instantaneous SINR is low, we activate a robust, but low-throughput modulation mode, and vice versa.

The impact of adaptive antenna arrays recorded in a propagation environment subjected to shadow fading was then investigated. The associated forced termination performance is shown in Figure 8.10. This figure illustrates the substantial network capacity gains achieved with the aid of both two- and four-element adaptive antenna arrays under shadow fading propagation conditions. Simulations were conducted in conjunction with log-normal shadow fading having a standard deviation of 3 dB, experiencing maximum shadowing frequencies of both 0.5 Hz and 1.0 Hz. As expected, the network capacity was reduced at the higher shadow fading frequency in both the FDD and TDD modes. Without employing adaptive antenna arrays, the TDD network supported just over 71 users and 62 users, when subjected to 0.5 Hz and 1.0 Hz frequency shadow fading, respectively. With the application of two-element adaptive antenna arrays, these capacities increased by 111% and 113%, to 151 users and 131 users, respectively. The employment of four-element adaptive antenna arrays led to a TDD network capacity of 245 users at a 0.5 Hz shadowing frequency, and 234 users at a 1.0 Hz shadowing frequency. This corresponded to relative gains of 62% and 78% over the capacity provided in the TDD mode with the aid of two-element adaptive antenna arrays. In comparison with the FDD benchmark we have again recorded a factor of up to two lower teletraffic capacity.

The probability of low-quality access performance is depicted in Figure 8.11. As expected, a given P_{low} value was associated with a higher traffic load, as the number of antenna elements increased. When the maximum shadow fading frequency was increased from 0.5 to 1.0 Hz, P_{low} also increased. The probability of low-quality access seen in Figure 8.11 is similar in the scenarios employing adaptive antenna arrays in the UTRA TDD and FDD CDMA systems. It should be noted, however, that the probability of low-

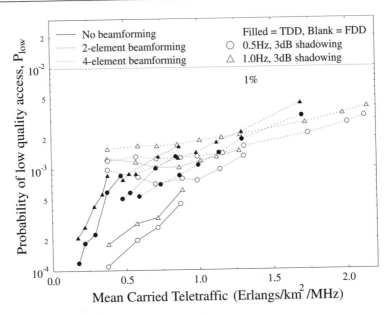

Figure 8.11: Probability of low-quality access versus mean carried traffic of the UTRA-like FDD and TDD/CDMA-based cellular network both with and without beamforming and with shadowing for $SF = 16$.

quality access always remained below the 1% constraint of the conservative scenario under the scenarios studied, and the forced termination probability was considerably reduced by the adaptive antenna arrays, as is demonstrated in our discussion in the context of Figure 8.13 below. When using beamforming, the inferiority of the TDD mode was less pronounced than in the context of the previously studied performance metrics.

Figure 8.12 presents the *GOS* for a range of teletraffic loads with and without beamforming as well as in conjunction with shadowing. A summary of the maximum network capacities of the various scenarios considered in this section both with and without shadowing having a standard deviation of 3dB, as well as with and without employing beamforming using two and four element arrays is given in Table 8.1. Throughout this section we have observed that the capacity of the TDD mode was consistently lower than that of the FDD mode owing to the fact that any timeslot may be used both in the UL and in the DL. In the next section we invoke adaptive modulation as a further countermeasure for mitigating this deficiency.

8.5.3   Performance of Adaptive Antenna Array and Adaptive Modulation Aided TDD HSDPA-style Systems

In this section we build upon the results presented in the previous section by applying Adaptive Quadrature Amplitude Modulation (AQAM) techniques [13]. The various scenarios and channel conditions to be investigated here are identical to those of the previous section, except for the application of AQAM. Since in the previous section an increased network capacity was achieved by using independent UL and DL beamforming, this procedure was

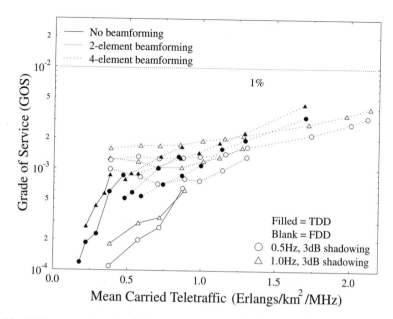

Figure 8.12: *GOS* versus mean carried traffic of the UTRA-like FDD and TDD/CDMA-based cellular network both with and without beamforming and without shadowing for *SF* = 16.

Table 8.1: Maximum mean carried traffic and maximum number of mobile users that can be supported by the FDD/TDD network, whilst meeting the conservative quality constraints. The carried traffic is expressed in terms of normalized Erlangs (Erlang/km²/MHz), for the network described in Table 6.2 both with and without beamforming, and also with and without shadow fading having a standard deviation of 3 dB for *SF* = 16. The FDD benchmark results were adopted from [416].

		Conservative scenario			
		Number of users		Traffic (Erlangs/km²/MHz)	
Shadowing	Beamforming	FDD	TDD	FDD	TDD
No	No	256	142	1.42	0.81
No	Two elements	325	206	1.87	1.18
No	Four elements	480	322	2.75	1.85
0.5 Hz, 3 dB	No	150	72	0.87	0.41
0.5 Hz, 3 dB	Two elements	203	151	1.16	0.87
0.5 Hz, 3 dB	Four elements	349	245	2.0	1.39
1.0 Hz, 3 dB	No	144	62	0.82	0.35
1.0 Hz, 3 dB	Two elements	201	131	1.12	0.75
1.0 Hz, 3 dB	Four elements	333	234	1.88	1.33

Figure 8.13: Forced termination probability versus mean carried traffic of the UTRA-like FDD and
TDD/CDMA-based cellular network both with and without beamforming in conjunction
with AQAM as well as with shadowing having a standard deviation of 3 dB for $SF = 16$.

invoked in these simulations. AQAM involves the selection of the appropriate modulation
mode in order to maximize the achievable data throughput over a given channel, whilst
maintaining a given target the BER. More explicitly, the philosophy behind adaptive
modulation is the most appropriate selection of a modulation mode according to the
instantaneous radio channel quality experienced [13]. Therefore, if the SINR of the channel
is high, then a high-throughput high-order modulation mode may be employed for exploiting
the high instantaneous quality of the radio channel. Similarly, if the channel is instantaneously
of low quality, exhibiting a low SINR, a high-order modulation mode would result in an
unacceptably high BER or FER, and hence a more robust, but lower throughput modulation
mode would be employed. Therefore, adaptive modulation combats the effects of time-
variant channel quality, while also attempting to maximize the achieved data throughput,
and maintaining a given BER or FER. In the investigations conducted, the modulation modes
of the UL and DL were determined independently, thus taking advantage of the lower levels
of co-channel interference on the UL, or of the potentially higher transmit power of the BSs.
The particular implementation of AQAM used in these investigations is illustrated in [416].

A comparison of Figure 8.13 with Figure 8.10 shows the significant reduction in the
probability of a dropped TDD call, achieved by employing adaptive antenna arrays in
conjunction with adaptive modulation [416, 419] in a log-normal shadow faded environment.
Figure 8.13 demonstrates that even with the aid of a two-element adaptive antenna array,
a substantial forced termination probability reduction was achieved. The single-antenna
based TDD network was found to support 153 users, corresponding to a traffic load of
0.875 Erlang/km^2/MHz, when subjected to 0.5 Hz frequency shadow fading. The capacity

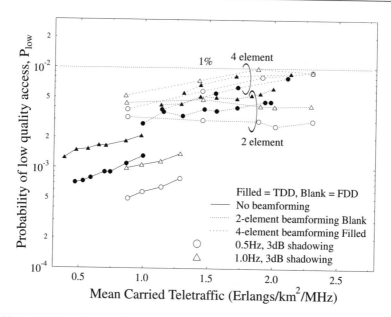

Figure 8.14: Probability of low-quality access versus mean carried traffic of the UTRA-like FDD and TDD/CDMA-based cellular network both with and without beamforming in conjunction with AQAM as well as with shadowing having a standard deviation of 3 dB for *SF* = 16.

of the single-antenna aided TDD network was slightly reduced to 152 users, corresponding to 0.874 Erlang/km^2/MHz, when increasing the maximum shadow fading frequency to 1.0 Hz. Upon employing two-element adaptive antenna arrays, the TDD network capacity increased by 109% to 320 users, or to an equivalent traffic load of 1.834 Erlang/km^2/MHz, when subjected to 0.5 Hz frequency shadow fading. When the maximum shadow fading frequency was increased to 1.0 Hz, the number of users supported by the TDD network was 307, or 1.82 Erlang/km^2/MHz, representing an increase of 102% in comparison to the network refraining from using adaptive antenna arrays. It is seen in Figure 8.13 that the forced termination probability of the UTRA-like TDD/CDMA scenarios is close to that of the FDD/CDMA scenarios, when employing adaptive antenna arrays in conjunction with adaptive modulation.

The probability of low-quality outage, presented in Figure 8.14, did not benefit from the application of adaptive antenna arrays, in fact the opposite occurred. Furthermore, recall that Figure 8.11 depicted the probability of low-quality outage without adaptive modulation, i.e. using fixed modulation, and upon comparing these results to those obtained in conjunction with adaptive modulation shown in Figure 8.14, the performance degradation owing to the employment of adaptive modulation can be seen explicitly. Similar trends may be observed also in Figure 8.15 in the context of the GOS. This is because the increase in the probability of low-quality access can be attributed to the employment of less robust, but higher throughput, higher-order modulation modes invoked by the adaptive modulation scheme. Hence, under given propagation conditions and using the interference-resilient fixed 4QAM modulation mode, as in Figure 8.11, a low-quality outage may not occur. In contrast, when using adaptive

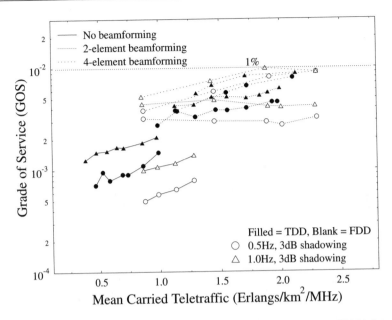

Figure 8.15: *GOS* versus mean carried traffic of the UTRA-like FDD and TDD/CDMA-based cellular network both with and without beamforming in conjunction with AQAM as well as with shadowing having a standard deviation of 3 dB for *SF* = 16.

modulation invoking a less resilient, but higher-throughput and higher-order modulation mode, the same propagation conditions may inflict a low-quality outage.

A summary of the maximum user capacities of the FDD and TDD networks considered in this section both with and without shadowing having a standard deviation of 3 dB as well as with and without employing beamforming using two- and four-element arrays, whilst employing adaptive modulation is given in Table 8.2.

8.6 Loosely Synchronized Spreading Code Aided Network Performance of UTRA-like TDD/CDMA Systems

8.6.1 Introduction

In this section we investigate the achievable capacity of a UTRA-like TDD/CDMA system employing Loosely Synchronized (LS) spreading codes. The family of operational CDMA systems is interference limited, suffering from Inter-Symbol Interference (ISI), since the orthogonality of the spreading sequences is destroyed by the frequency selective channel. They also suffer from Multiple-Access Interference (MAI) owing to the non-zero cross-correlations of the spreading codes. In contrast, the family of LS codes exhibits a so-called Interference-Free Window (IFW), where both the auto-correlation and cross-correlation values of the codes become zero. Therefore, LS codes have the promise of mitigating the effects of both ISI and MAI in time dispersive channels. Hence, LS codes have the potential

Table 8.2: Maximum mean carried traffic and maximum number of mobile users that can be supported by the FDD and TDD network, whilst meeting the conservative quality constraints. The carried traffic is expressed in terms of normalized Erlangs (Erlang/km^2/MHz), for the network described in Table 6.2 both with and without beamforming, in conjunction with shadow fading having a standard deviation of 3 dB, whilst employing adaptive modulation techniques for $SF = 16$. The FDD benchmark results were adopted from [416].

		\multicolumn{4}{c}{Conservative scenario}			
		\multicolumn{2}{c}{Number of users}	\multicolumn{2}{c}{Traffic (Erlangs/km^2/MHz)}
Shadowing	Beamforming	FDD	TDD	FDD	TDD
0.5 Hz, 3 dB	No	223	153	1.27	0.875
0.5 Hz, 3 dB	Two elements	366	320	2.11	1.834
0.5 Hz, 3 dB	Four elements	476	420	2.68	2.41
1.0 Hz, 3 dB	No	218	152	1.24	0.874
1.0 Hz, 3 dB	Two elements	341	307	1.98	1.758
1.0 Hz, 3 dB	Four elements	460	393	2.59	2.234

of increasing the capacity of CDMA networks. This section studies the achievable network performance in comparison to that of a UTRA-like TDD/CDMA system using Orthogonal Variable Spreading Factor (OVSF) codes.

The air interface of UMTS supports both FDD and TDD mode [416], in order to facilitate an efficient exploitation of the paired and unpaired band of the allocated spectrum. The FDD mode is intended for applications in both macro- and micro-cellular environments, when supporting both medium data rates and high mobility. In contrast to the FDD mode, the TDD mode was contrived for environments associated with a high traffic density and asymmetric UL as well as DL indoor coverage. Although the UTRA/TDD mode was contrived for the sake of improving the achievable network performance by assigning all of the timeslots on a demand basis to the UL and DL [436], this measure may result in an excessive BS $\rightarrow$ BS interference and, hence, in a potentially reduced number of system users [473, 474]. As seen in Figure 8.6, if BS$_1$ is transmitting and BS$_2$ is receiving at the same time in a given timeslot, BS $\rightarrow$ BS interference takes place, provided that these BSs are in adjacent cells. In [473] we demonstrated that the employment of adaptive arrays in conjunction with AQAM limited the detrimental effects of co-channel interference on the UTRA-like TDD/CDMA system and resulted in performance improvements both in terms of the achievable call quality and the number of users supported. However, in comparison with a UTRA-like FDD/CDMA system, the capacity of the UTRA-like TDD/CDMA cellular system was shown to remain somewhat poorer than that of the UTRA-like FDD/CDMA system under the same propagation conditions.

The network performance of the UTRA-like FDD/CDMA systems was quantified in our previous research [427], when supported by adaptive beam-steering [419] and LS [421] spreading codes. It was demonstrated that the network performance of a UTRA-like FDD/CDMA system employing LS spreading codes was substantially better than that of the system using OVSF codes [420]. We consider the employment of this specific family of LS spreading codes in the UTRA-like TDD/CDMA system. The LS spreading codes exhibit a so-called IFW, where the off-peak aperiodic autocorrelation values as well as the aperiodic cross-correlation values become zero. With the advent of the IFW we may encounter both zero ISI and zero MAI, provided that all of the delayed asynchronous transmissions arrive within the IFW. More specifically, interference-free CDMA communications become possible, when the total time offset expressed in terms of the number of chip intervals, which is the sum of the time-offset of the mobiles plus the maximum channel-induced delay spread is within the code's IFW [422]. By employing this specific family of codes, we are capable of reducing the ISI and MAI, since users in the same cell do not interference with each other, as a benefit of the IFW provided by the LS codes used.

8.6.2   LS Codes in UTRA TDD/CDMA

There exists a specific family of LS codes [421], which exhibit an IFW, where both the auto-correlation and cross-correlation values of the codes become zero. Specifically, LS codes exploit the properties of the so-called orthogonal complementary sets [421, 431]. An example of the design of LS spreading codes can be found in [427]. In the UTRA TDD mode, the UL and DL timeslots are transmitted on the same carrier frequency, which creates additional undesirable and grave interference infested scenarios compared to UTRA FDD. More explicitly, as argued in the context of Figure 8.6, both transmission directions may interfere with each other, resulting in MS → MS and BS → BS interference, respectively. The interference experienced at the mobile may be divided into two categories. First, interference is imposed by the signals transmitted to other mobiles from the same base station, which is known as intra-cell interference. Secondly, interference is encountered owing to the signals transmitted to other mobiles from other basestations, as well as to other basestations from other mobiles, which is termed inter-cell interference.

The instantaneous SINR is obtained upon dividing the received signal powers by the total interference plus thermal noise power, and then by multiplying this ratio by the spreading factor, *SF*, yielding [416]

$$SINR_{DL} = \frac{SF \cdot P_{BS}}{(1 - \alpha)I_{Intra} + I_{Inter} + N_0}, \tag{8.4}$$

where $\alpha = 1$ corresponds to the ideal case of perfectly orthogonal intra-cell interference and $\alpha = 0$ to completely asynchronous intra-cell interference. Furthermore, P_{BS} is the signal power received by the mobile user from the base station, N_0 is the thermal noise, I_{Intra} is the intra-cell interference and I_{Inter} is the inter-cell interference. Again, the interference plus noise power is scaled by the spreading factor, *SF*, since during the despreading process the associated low-pass filtering reduces the noise bandwidth by a factor of *SF*. The inter-cell interference is not only due to the MSs, but also due to the BSs illuminating the adjacent cells by co-channel signals. Owing to invoking LS spreading codes in our UTRA-like TDD/CDMA system, the intra-cell interference may be completely eliminated, hence we have $\alpha = 1$. Our current research is building on our previous findings recorded in the context of an UTRA-like TDD system [473], where we found that invoking adaptive modulation as well as beam-steering proved to be a powerful means of enhancing the capacity of TDD/CDMA. In the investigations of [473], OVSF codes were used as spreading codes. However, the intra-cell interference is only eliminated by employing orthogonal OVSF codes,

Table 8.3: Simulation parameters [427].

Parameter	Value	Parameter	Value
Noisefloor	−100 dBm	Pilot power	−9 dBm
Frame length	10 ms	Cell radius	78 m
Multiple access	TDD/CDMA	Number of basestations	49
Modulation scheme	4QAM/QPSK	Spreading factor	16
Min BS transmit power	−48 dBm	Min MS transmit power	−48 dBm
Max BS transmit power	17 dBm	Max MS transmit power	17 dBm
Low quality access SINR	5.2 dB	Outage (1% BER) SINR	4.8 dB
Pathloss exponent	−2.0	Target SINR	6.2 dB
Average inter-call time	300 s	Max. new-call queue-time	5 s
Average call length	60 s	Pedestrian speed	3 mph
Max consecutive outages	5	Signal bandwidth	5 MHz

if the system is perfectly synchronous and provided that the mobile channel does not destroy the OVSF codes' orthogonality. In an effort to prevent intra-cell interference, again, in this section we employ LS codes, which exhibit ideal auto-correlation and cross-correlation functions within the IFW. Thereby, the "near–far effect" may be significantly reduced and hence the user capacity of the system can be substantially enhanced. As a benefit of the LS codes' interference resilience, it was shown in [427] that the achievable BER performance of LS codes is better than that of OVSF codes. For a spreading factor of 16, the post-despreading SINR required for maintaining a BER of 1×10^{-3} was 6.2 dB in case of LS codes, which is almost 2 dB lower than that necessitated by the OVSF codes.

8.6.3   System Parameters

The cell radius was 78 m, which was the maximum affordable cell radius for the IFW duration of ±1 chip intervals at a chip rate of 3.84 Mchip/s. The mobiles were capable of moving freely, at a speed of 3 mph, in random directions, selected at the start of the simulation from a uniform distribution, within the infinite simulation area of 49 wrapped-around traffic cells [416]. Furthermore, the post-despreading SINRs required for obtaining the target BERs were determined with the aid of physical-layer simulations using a 4QAM modulation scheme, in conjunction with 1/2-rate turbo coding for transmission over a COST 207 seven-path Bad Urban channel [434]. Using this turbo-coded transceiver and LS codes having a *SF* of 16, the post-despreading SINR required for maintaining the target BER of 1×10^{-3} was 6.2 dB. The BER, which was deemed to correspond to low-quality access, was stipulated at 5×10^{-3}. This BER was exceeded for SINRs falling below 5.2 dB. Furthermore, a low-quality outage was declared when the BER of 1×10^{-2} was exceeded, which was encountered for SINRs below 4.8 dB. These values can be seen along with the other system parameters in Table 8.3. All other experimental conditions were identical to those in [416].

Figure 8.16: Forced termination probability versus mean carried traffic of the UTRA-like TDD cellular network using LS codes and OVSF codes both with and without beamforming in conjunction with shadowing having a frequency of 0.5 Hz and a standard deviation of 3 dB for a spreading factor of $SF = 16$.

8.6.4   Simulation Results

Figure 8.16 shows the forced termination probability associated with a variety of traffic loads quantified in terms of the mean normalized carried traffic expressed in Erlangs/km²/MHz, when subjected to 0.5 Hz frequency shadowing having a standard deviation of 3 dB. As observed in the figure, nearly an order of magnitude reduction of the forced termination probability has been achieved by employing LS spreading codes compared with those of using OVSF spreading codes. In conjunction with OVSF codes, the "No beamforming" scenario suffered from the highest forced termination probability of the four traffic scenarios characterized in the figure at a given load. Specifically, the network capacity was limited to 50 users, or to a teletraffic density of approximately 0.55 Erlangs/km²/MHz. With the advent of employing four-element adaptive antenna arrays at the BSs the number of users supported by the TDD system increased to 178 users, or a teletraffic density of 2.03 Erlangs/km²/MHz. However, in conjunction with LS codes, and even without employing antenna arrays at the BSs, the TDD system was capable of supporting 306 users, or an equivalent traffic density of 3.45 Erlangs/km²/MHz.

Figure 8.17 portrays the probability of low-quality access versus various traffic loads. In conjunction with OVSF codes, it can be seen from the figure that without beamforming the system suffered from encountering more multiuser interference, as the traffic loads increased. Hence, the probability of low-quality access became higher. When invoking beamforming, both the intra- and inter-cell interference was reduced and hence the probability of low-quality access was reduced as well. As a benefit of employing LS codes, the intra-cell interference

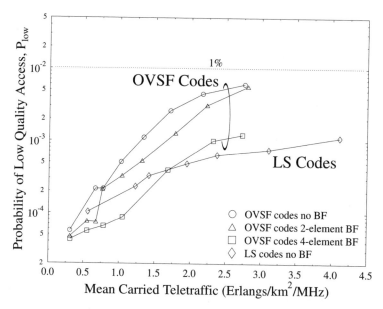

Figure 8.17: Probability of low-quality access versus number of users of the UTRA-like TDD cellular network using LS codes and OVSF codes both with and without beamforming in conjunction with shadowing having a frequency of 0.5 and a standard deviation of 3 dB for a spreading factor of $SF = 16$.

was efficiently reduced and therefore the probability of low-quality access was found to be lower even without beamforming than that of the system using OVSF codes and employing two-element beamforming. We also observed that at lower traffic loads the probability of low-quality access for the "LS codes no BF" scheme is higher than that of "OVSF codes 4-element BF" scheme. This is a consequence of the associated high probability of forced termination for the "LS codes no BF" scheme, as shown in Figure 8.16, because the higher the probability of forced termination, the lower the number of users supported by the TDD system and hence the effects of co-channel interference imposed by the existing connections remain more benign when a new call starts.

For the sake of also characterizing the achievable system performance from a different perspective, the mean transmission power versus teletraffic performance is depicted in Figure 8.18. Again, as a benefit of employing LS codes, both the required mean UL and DL transmission power are lower than that necessitated by OVSF codes. The TDD system using OVSF codes required an average 10–20 dBm more signal power compared with the TDD system using LS codes. In [474] it was shown that the major source of interference is constituted by the BS-to-BS interference as a consequence of the BS's high signal power and the near-LOS propagation conditions prevailing between BSs. Even though the employment of LS codes can only reduce the intra-cell interference, it results in a substantial reduction of the BSs' power consumption, as shown in Figure 8.18. Hence the source of BS $\rightarrow$ BS inter-cell interference was also reduced. In other words, the employment of LS codes indirectly

Figure 8.18: Mean transmission power versus number of users of the UTRA-like TDD cellular network
using LS codes and OVSF codes both with and without beamforming in conjunction with
shadowing having a frequency of 0.5 Hz and a standard deviation of 3 dB for a spreading
factor of $SF = 16$.

Table 8.4: Maximum mean carried traffic and maximum number of mobile users that can be supported
by the network, whilst meeting the network quality constraints of Section 8.6.3, namely
$P_B \leq 3\%$, $P_{FT} \leq 1\%$, $P_{low} \leq 1\%$ and GOS $\leq 4\%$. The carried traffic is expressed in terms
of normalized Erlangs (Erlang/km²/MHz) using OVSF codes and LS codes in conjunction
with shadow fading having a standard deviation of 3 dB and a frequency of 0.5 Hz for a
spreading factor of $SF = 16$.

| | | | Traffic | Power (dBm) | |
Spreading code	Beamforming	Users	(Erlangs/km²/MHz)	MS	BS
OVSF codes	No	50	0.55	0.54	−0.28
OVSF codes	Two-elements	113	1.18	1.33	0.90
OVSF codes	Four-elements	178	2.03	2.07	1.81
LS codes	No	306	3.45	−9.11	−9.21

reduced the severe BS → BS inter-cell interference by keeping the BSs' transmission power
at a low level.

Figure 8.19 shows the achievable *GOS* for a range of teletraffic loads. We observe
similar trends regarding the probability of low-quality access, as shown in Figure 8.17.
In Equation 5.15, the GOS performance is jointly determined by P_B and P_{low}, which is
interpreted as the probability of unsuccessful network access (blocking), or the probability
of encountering a low-quality access, provided that a call is admitted to the system.

Figure 8.19: *GOS* versus number of users of the UTRA-like TDD cellular network using LS codes and OVSF codes both with and without beamforming in conjunction with shadowing having a frequency of 0.5 Hz and a standard deviation of 3 dB for a spreading factor of $SF = 16$.

The employment of the LS codes may cause the shortage of spreading codes and, hence, may lead to the blocking of a new call, since there are only eight LS codes that can be used, when the IFW duration is ± 1 chip length. The call duration and inter-call periods were Poisson distributed having the mean values shown in Table 8.3. When encountering this call arrival distribution, we observe that the new call blocking probability is negligible, as shown in Figures 8.17 and 8.19.

A summary of the maximum user capacities of the UTRA-like TDD/CDMA system using OVSF codes and LS codes in conjunction with log-normal shadowing having a standard deviation of 3 dB and a shadowing frequency of 0.5 Hz as well as both with and without beamforming is given in Table 8.4. The teletraffic carried and the MS and BS transmission powers required are also shown in Table 8.4.

8.6.5 Summary and Conclusions

In this section we studied the network performance of a UTRA-like TDD/CDMA system employing LS spreading codes. The computer simulation results provided showed that the TDD system invoking LS codes had a better performance compared with the system using OVSF codes. We designed a 49-cell "wrapped around" simulation area, constituted by sufficiently small 78 m radius cells, which guaranteed that the delayed asynchronous transmissions arrive within the IFW, where the auto-correlation and cross-correlation values of the LS codes became zero and hence eliminated the effects of intra-cell interference. The SINR required by the LS codes for the sake of maintaining a BER of 1×10^{-3} was almost 2 dB lower than that necessitated by the OVSF codes. Furthermore, a low MS and BS transmission power has been maintained. Hence, the average intra- and inter-cell interference level has become low, the severe $BS \rightarrow BS$ interference has been reduced and this resulted in TDD system performance improvements both in terms of the achievable call quality and the number of users supported. Our future research will focus on further improving the performance of TDD systems using GA-based timeslot scheduling.

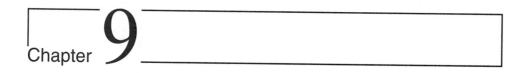

Chapter **9**

The Effects of Power Control and Hard Handovers on the UTRA TDD/CDMA System

9.1 A Historical Perspective on Handovers

The terminology of handover [475] is synonymous in mobile communications to the process of transferring a MS from one BS or channel to another. It is typically initiated by experiencing a degraded signal quality in the current cell. Handovers (HOs) may be divided into two broad categories, namely hard HOs and soft HOs, which are also often characterized informally by the terms "break before make" and "make before break", respectively. Therefore, in the context of hard HOs, the current transmission resources are released before the new resources are reserved, while in soft HOs, both the existing and new resources are engaged in the HO process. Poorly designed HO schemes may generate very heavy signaling traffic, hence potentially imposing a dramatic QoS degradation. The increasing interest and mass-market for mobile communications as well as the limited available spectrum has motivated the employment of cellular architectures based on small cells. As a result, the number of mobile users crossing the cell boundaries is increased and, hence, the resultant HO rate is also increased. Therefore, the efficiency of HO algorithms is expected to play a crucial role in the overall system performance [476–479]. This suggests that efficient HO algorithms constitute a cost-effective way of enhancing both the capacity and the QoS in cellular systems.

The radio propagation environment and the related HO algorithms are different in different cellular systems [480, 481]. Hence, a particular HO algorithm endowed with a specific set of parameters cannot perform equally well in different communication scenarios [482]. First- and second-generation cellular systems provide wide-area coverage even in cities using macro-cells [483–489]. The brief history of HO algorithms designed for cellular systems is summarized in Table 9.1.

3G, HSPA and FDD versus TDD Networking   Second Edition
L. Hanzo, J. S. Blogh and S. Ni   © 2008 John Wiley & Sons, Ltd

Table 9.1: Contributions on HO algorithms designed for cellular systems.

Year	Author	Contribution
1992	Tekinay and Jabbari [488]	Studied the performance of non-preemptive priority queueing for HO calls.
1993	Vijayan and Holtzman [480]	A HO algorithm based on signal strength measurements made by the MSs in a lognormal fading environment was proposed.
1994	Viterbi, Gilhousen and Zehavi [484]	The effect of HO techniques on cell coverage and UL capacity was investigated in a CDMA system.
1995	Nakano, Umeda and Ohno [485]	A BS-diversity aided HO algorithm was proposed for high-capacity DS-CDMA cellular systems.
1996	Kim, Lee and Chin [486]	An adaptive HO algorithm taking into account the velocity of MSs was proposed.
1997	Calin and Zeghlache [489]	An analytical model characterizing a non-preemptive priority queueing system incorporating both voice and data users was proposed.
1998	Wong and Cox [487]	A handoff algorithm using pattern recognition was proposed.
1999	Benvenuto and Santucci [490]	A least squares pathloss estimation approach to HO algorithms was proposed.
2000	Santucci, Pratesi, Ruggieri and Graziosi [491]	A range of statistical parameters used in the performance analysis of a relative signal strength based HO algorithm.
2001	Yang, Ghaheri-Niri and Tafazolli [492]	The performance of power-triggered and E_c/N_0-triggered soft HO algorithms designed for UTRA was investigated.
2002	Wang, Sridhar and Green [493]	An adaptive soft HO algorithm using the location information of mobile stations was proposed.
2003	Wang, Liu and Cen [481]	HO algorithms designed for a dynamic spreading aided WCDMA multimedia system were proposed.
2004	Lugara, Tartiere and Girard [482]	Inter-system HO algorithms supporting both UMTS and GSM were proposed.

9.2   Hard HO in UTRA-like TDD/CDMA Systems

The UTRA network supports different types of HO, where the HO control procedure may be divided in the following types [459]:

- *Intra-system HO* [494, 495] occurring within a UTRA system, which can be further subdivided into:

  - *Intra-frequency* HO [496] between cells using to the same UTRA carriers;
  - *Inter-frequency* HO [23] between cells employing different UTRA carriers.

- *Inter-system HO* [497,498] taking place between cells belonging to two different Radio Access Technologies (RATs) or different Radio Access Modes (RAMs). The most typical case of inter-system HO is expected to take place between UTRA [416] and GSM/EDGE systems [457]. A typical example of inter-RAM HO is likely to take place between the UTRA FDD [416] and UTRA TDD [416] modes.

Furthermore, the following HO procedures can be identified [459].

- *Hard HOs* [499, 500] represent a family of HO procedures, where the old radio link of a MS is released before the new radio link is established. For real-time interactive voice-type bearers this hard HO procedure implies encountering a brief disconnection of the bearer. In contrast, for non-interactive data-type bearers hard HOs appear to be seamless. Invoking soft HOs is also a design option for the TDD mode, as suggested in [501]. Soft HOs improve the QoS, since a diversity gain is provided by combining the signals received from both links [484]. However, being engaged in communications with two BSs introduces more interference [502], since in the TDD mode portrayed in Figure 8.6 a number of gravely detrimental interference scenarios may exist. The interference imposed is typically higher if the network is asynchronous [503] or if the neighboring cells carry different asymmetric traffic loads [32, 451, 460, 474]. Furthermore, in the UTRA system the legitimate spreading factor range of the FDD mode is 4 to 256 in the UL and 4 to 512 in the DL, while in the TDD mode the corresponding range is 1 to 16 in both the UL and DL [59, 504], as seen in Table 6.1. Hence, soft HOs carried out in the TDD mode would need two of the 16 spreading codes, because the communications between the MS and the serving as well as target BSs would have to take place concurrently via two air interface channels to distinguish between the signals [59] arriving from the serving and target BSs. This may be expected to severely limit the number of users supported. Hence hard HOs constitute a more appropriate solution for a TDD CDMA system, despite having no diversity gain.

- *Soft HOs* [492, 505–507] and softer HOs [508, 509] constitute a category of HO procedures in which a MS maintains at least one radio link all of the time, typically establishing a new link to the target BS, before relinquishing the previous link. More explicitly, during soft HO the MS is simultaneously controlled by two or more BSs of the same or different Radio Network Controllers (RNCs). Softer HOs constitute a special case of soft HOs, where the radio links that are added and removed belong to co-located BSs managed by the same BS controller. Soft and softer HOs are only possible, when using the same carrier frequency.

9.2.1 Relative Pilot Power-based Hard HO

The UTRA TDD/CDMA system supports both inter-system HOs and intra-system HOs. All of these HOs are mobile-assisted hard HOs and hence their philosophy is clearly different from that of the UTRA FDD mode, since in the latter protocol structure has been designed to support soft HOs. The generic HO procedure is typically described in four phases [510]: the related signal-quality measurements, HO initiation, HO decision and HO execution. A range of signal-quality related parameters, such as the received power, BER and the MS's distance can be evaluated and processed as the related HO criteria. It is anticipated that the UTRA TDD system's hard HO is likely to be predominantly MS assisted HO, where the MS performs signal-quality measurements that are signaled to the RNC that makes the actual HO decisions. The conception of efficient HOs is one of the main challenges in UTRA TDD/CDMA networks, since it has a substantial impact on the system's performance and capacity. The philosophy of soft HO and macro diversity cannot be utilized in the TDD mode, since the active target BS set size is always equal to one during the HO procedure [511]. This is because HOs in the UTRA TDD mode are hard HOs, where the MS is connected to one BS only [451].

In our simulations, we rely on a single RAT or a single RAM, hence we only consider intra-system hard HOs. Again, hard HOs cause a temporary disconnection of real-time interactive access bearers, but they appear to be seamless for non-interactive bearers. The hard HO algorithm used is based on the relative[1] pilot power of E_c/I_o in the serving cell and in the neighboring cells, where a minimum E_c/I_o HO margin is used as a threshold for preventing repetitive hard HOs between cells [59,512]. A hard HO can be processed when the following condition is satisfied [459]:

$$E_c/I_{o(serving\ cell)} + E_c/I_{o(margin)} < E_c/I_{o(new\ cell)}, \tag{9.1}$$

where $E_c/I_{o(serving\ cell)}$ represents the average relative pilot power of the serving cell, while $E_c/I_{o(new\ cell)}$ is the average pilot power of the best potential target cell. The parameter $E_c/I_{o(margin)}$ is the margin by which the E_c/I_o value of the best HO target cell has to exceed the E_c/I_o value of the serving cell before the hard HO is activated. The so-called HO *acceptance threshold* T_{acc} and call *drop threshold* T_{drop} are the corresponding hard HO thresholds, where a HO is enabled and the call is dropped, respectively. Again, a relative rather than absolute received pilot power scheme is used in our investigations, which exhibits performance benefits in realistic propagation environments exposed to shadow fading [416]. The relative hard HO thresholds T_{acc} and T_{drop} are expressed in terms of dB, which are normalized to the received pilot strength E_c/I_o of the best potential neighboring HO target cell. Again, when the E_c/I_o value of the best neighboring HO target cell exceeds the E_c/I_o value of the serving cell, the hard HO may be enabled, provided that it is necessary due to the serving cell's signal quality degradation. Hence the relative *acceptance threshold* T_{acc} is set to 0 dB, which implies that the best neighboring cell can be accepted as a candidate HO target cell for hard HO, provided that it has at least as high a pilot power as the serving cell. The call *drop threshold* T_{drop} is the $E_c/I_{o(margin)}$ value shown in Equation 9.1. When the received signal quality of the serving cell degrades and, hence, the relative pilot power E_c/I_o of the serving cell becomes by at least T_{drop} dB lower than the E_c/I_o value of the HO candidate cell, the hard HO will be activated.

9.2.2 Simulation Results

In this section the effect of having a hysteresis in the TDD system's hard HO candidate/active BS set update procedure is evaluated. The appropriate choice of the HO hysteresis threshold is critical for the sake of achieving an attractive tradeoff between the transmit power required

---

[1]The reason for using a relative rather than absolute pilot power-based hard HO procedure is because it was found in [416] that in some cell areas all pilot signals may be weak, while in other locations they all may be strong, and this phenomenon may generate either too many or too few potential target HO cells. This potential deficiency may be overcome by normalizing the pilot powers of the potential HO target cells to that of the serving cell or to those of the other BSs in the active BS set, which allows the inclusion of at least one target HO BS.

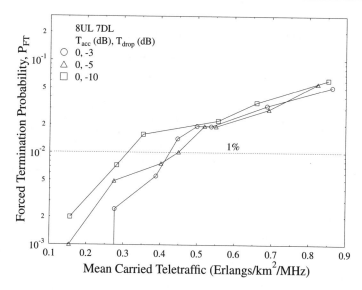

Figure 9.1: Forced termination probability versus mean carried traffic of the UTRA-like TDD/CDMA-based cellular network with shadowing and using relative received pilot power, E_c/I_o, based hard-HO thresholds in conjunction with a near-symmetric traffic load of 8:7 (UL:DL) timeslots. The system parameters are summarized in Table 9.2.

for supporting the HO process and seamless call continuity [513]. Having a low HO hysteresis threshold would enable a mobile user to capture and register a larger number of pilots, hence creating a high number of potential target BSs, but if they tend to provide a relatively low signal quality, the undesirable "ping-pong" effect of switching back and forth between calls may not be avoided. On the other hand, an excessive HO hysteresis threshold would practically eliminate the "ping-pong" effect, hence reducing the HO-related signaling, but this is achieved at the risk of not finding any better-quality BSs. This would consequently generate an excessive call dropping rate. Three different traffic scenarios are studied, namely having a near-symmetric UL:DL traffic load of 8:7 timeslots, UL-dominated asymmetric traffic loads and DL-dominated asymmetric traffic loads.

9.2.2.1 Near-symmetric UL/DL Traffic Loads

The forced termination performance at the near-symmetric traffic load ratio of 8:7 (UL:DL) timeslots is shown in Figure 9.1, illustrating that reducing the absolute value of the HO threshold T_{drop} to -3 dB and -5 dB improved the forced termination performance compared with $T_{drop} = -10$ dB, in particular at low traffic loads. The reduced force termination probability is a benefit of handing over to potential target BSs earlier, without jeopardizing terminating the call. This phenomenon is also evident in Figure 9.2, which shows the probability of a low-quality outage versus mean the carried teletraffic of a UTRA-like TDD/CDMA-based cellular network carrying a near-symmetric traffic load of 8 : 7 (UL:DL). However, the reduction of the absolute value of T_{drop} from -10 to -3 dB led to an increased number of HO events, as shown in Figure 9.3. This is predictable, since having

Table 9.2: Parameters used for the hard HO investigations.

Parameter	Value	Parameter	Value
Cell radius	150 m	Noisefloor	−100 dBm
Chip rate	3.84 Mcps	Spreading factor	16
BS/MS minimum TX power	−44 dBm	BS/MS maximum TX power	+21 dBm
Modulation scheme	4-QAM	Pathloss exponent	−3.5
Target E_b/N_o	8.0 dB	Low-quality outage E_b/N_o	7.0 dB
Outage E_b/N_o	6.6 dB	HO margin	3,5,10 dB

Figure 9.2: Probability of low-quality access versus mean carried traffic of the UTRA-like TDD/CDMA-based cellular network with shadowing and using relative received pilot power, E_c/I_o, based hard-HO thresholds in conjunction with a near-symmetric traffic load of 8:7 (UL:DL) timeslots. The system parameters are summarized in Table 9.2.

higher hysteresis requires a higher received pilot power from the neighboring HO target cells to be selected. Consequently, the rate of active BS set update becomes lower, leading to an increased average BS and MS transmission power, as shown in Figure 9.4. The increased BS and MS transmission power results in an increased co-channel interference level and, hence, in a higher probability of outage associated with a low call quality. It was observed in Figure 9.3 that the number of HO events started to reduce, when the traffic loads exceeded 0.6 Erlang/km²/MHz in Figure 9.3. This is a consequence of the associated high forced termination probability, as seen in Figure 9.1. This phenomenon ultimately led to the reduction of the number of supported users, while most users also suffered from experiencing a high level of interference.

Figure 9.3: Number of HO events per 49-cell simulation area versus mean carried traffic of the UTRA-like TDD/CDMA-based cellular network with shadowing and using relative received pilot power, E_c/I_o, based hard-HO thresholds in conjunction with a near-symmetric traffic load of 8:7 (UL:DL) timeslots. The system parameters are summarized in Table 9.2.

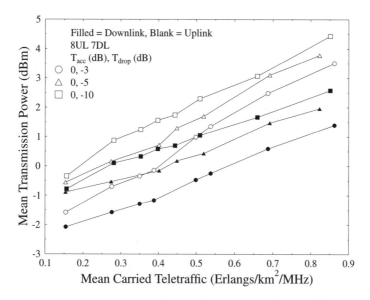

Figure 9.4: Mean BS and MS transmission power versus mean carried traffic of the UTRA-like TDD/CDMA-based cellular network with shadowing and using relative received pilot power, E_c/I_o, based hard-HO thresholds in conjunction with a near-symmetric traffic load of 8:7 (UL:DL) timeslots. The system parameters are summarized in Table 9.2.

Figure 9.5: Forced termination probability versus mean carried traffic of the UTRA-like TDD/CDMA-based cellular network with shadowing and using relative received pilot power, E_c/I_o, based hard-HO thresholds in conjunction with an asymmetric traffic load of 13:2 (UL:DL) timeslots. The system parameters are summarized in Table 9.2.

9.2.2.2   Asymmetric Traffic loads

In this section the effects of both UL-dominated as well as DL-dominated asymmetric traffic loads on the achievable teletraffic performance are examined, while using different hard HO hysteresis thresholds.

Figures 9.5 and 9.6 demonstrate that the system's forced termination probability was improved, when T_{drop} was adjusted from -10 to -3 dB. It is observed in Figure 9.6 that when carrying predominantly DL traffic loads associated with a traffic load ratio of 1:14 (UL:DL) the system benefited more from reducing the absolute value of the hard HO hysteresis from -10 to -3 dB compared with the predominantly UL traffic load scenario of 13:2 (UL:DL) timeslots, as seen in Figure 9.5. A carried traffic improvement of 0.2 Erlangs/km²/MHz was achieved in the 1:14 (UL:DL) timeslot scenario, which is twice as high as the performance gain observed in the 13:2 (UL:DL) scenario, when the HO hysteresis was reduced from -10 to -5 dB.

This is because the higher the HO hysteresis, the slower the HO process and hence a forced termination event may occur, before the hard HO can be completed owing to the insufficiently high signal power received from the serving BS. More explicitly, during the hard HO process, the MS tends to recede from the serving BS and approaching the HO target BS. The received pilot signal level of the serving BS may gradually reduce, while the received pilot signal level of the HO target BS may be increased, as shown in Figure 9.7. The SINR of the mobile station was gradually reduced as it receded from the serving BS. When the received pilot power approaches $T_{drop} = -10$ dB, the SINR is often already below the outage SINR, before the mobile can be handed over to the HO target BS, hence

Figure 9.6: Forced termination probability versus mean carried traffic of the UTRA-like TDD/CDMA-based cellular network with shadowing and using relative received pilot power, E_c/I_o, based hard-HO thresholds in conjunction with an asymmetric traffic load of 1:14 (UL:DL) timeslots. The system parameters are summarized in Table 9.2.

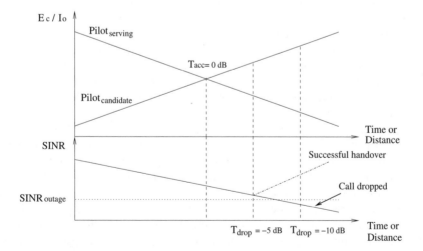

Figure 9.7: The hard-HO process in the UTRA/TDD CDMA system.

Figure 9.8: Probability of low-quality access versus mean carried traffic of the UTRA-like
TDD/CDMA-based cellular network with shadowing and using relative received pilot
power, E_c/I_o, based hard-HO thresholds in conjunction with an asymmetric traffic load
of 13:2 (UL:DL) timeslots. The system parameters are summarized in Table 9.2.

the call is likely to be terminated, as shown in Figure 9.7. However, still considering the
same special case, if we have $T_{drop} = -5$ dB, the mobile station may have been handed
over to the HO target BS, before the SINR degraded further and the call was terminated. For
a predominantly DL traffic load of 1:14 (UL:DL) timeslot scenario, most of the terminated
calls encountered during the hard HO process occurred owing to encountering a poor UL
connection quality, which is typically caused by the routinely incurred severe BS-to-BS
interference of the UTRA TDD mode. Only when the UL/DL TS allocation of the interfering
cells is the as same as that of the serving cell can the BS-to-BS interference be avoided,
since in this scenario the serving BS is not receiving during the interfering BS's transmit
TS. However, for the predominantly DL traffic load of 1:14 (UL:DL) a grave BS-to-BS
interference is encountered with a probability of $93.33\% = 14/15$, when a mobile user
invoking a hard HO is transmitting in a serving cell's UL timeslot, since in the interfering
cell 14 DL timeslots out of the total 15 timeslots may be inflicting interference. For the
predominantly UL traffic load of 13:2 (UL:DL), the probability of BS-to-BS interference
occurring is $13.33\% = 2/15$, when a mobile user invoking a hard HO is transmitting in a
serving cell's UL timeslot, and only two DL timeslots may be contaminated by the BS-to-
BS interference. Hence, carrying a predominantly DL traffic load is more beneficial, since it
allows us to reduce the HO hysteresis, which in turn reduces the call dropping probability, as
seen in Figure 9.6.

Figures 9.8 and 9.9 portray the probability of low-quality access versus various traffic
loads. It is observed that the specific choice of the HO hysteresis T_{drop} does not significantly
affect the probability of low-quality access. Similar trends were found for various traffic

Figure 9.9: Probability of low-quality access versus mean carried traffic of the UTRA-like TDD/CDMA-based cellular network with shadowing and using relative received pilot power, E_c/I_o, based hard-HO thresholds in conjunction with an asymmetric traffic load of 1:14 (UL:DL) timeslots. The system parameters are summarized in Table 9.2.

loads. The low-quality access performance associated with predominantly UL traffic loads is poorer than that recorded for predominantly DL traffic loads. This is because the specific UL and DL timeslot allocation is random in every cell and leads to a randomly fluctuating interference load for each timeslot. In the case of closed-loop power control [514], it is difficult to accurately adjust the power level if the received interfering signal is uncorrelated between timeslots. We provide more detailed justifications for this issue in the next section.

The number of HO events recorded across the entire 49-cell simulation area is shown in Figures 9.10 and 9.11, illustrating that reducing the absolute value of the hard-HO hysteresis threshold T_{drop} leads to a higher number of HO events, which is beneficial in terms of reducing the call dropping probability. However, as a penalty, the associated HO signaling traffic is increased and a higher proportion of the call duration is spent in the process of hard HO. From the perspective of the radio resource management, a high number of HO events will decrease the overall available resources, since the control channels of both the serving BSs and the HO target BSs are more likely to be engaged by the mobile stations in HO [515,516]. Having a high HO hysteresis also results in high BS and MS power consumption. The serving BS and the MS have to increase their transmission powers in an effort to maintain the SINR value required for sustaining the current connection quality during the process of hard HO, as depicted in Figures 9.12 and 9.13. For example, it may be inferred from Figures 9.12 and 9.13 that a 7 dB increase of the hysteresis threshold T_{drop} results in an average transmission power increase of 1.5 dBm for both the BS and mobile station.

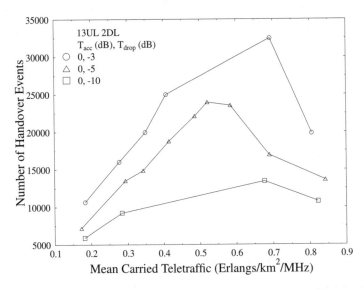

Figure 9.10: Number of HO events per 49-cell simulation area versus mean carried traffic of the UTRA-like TDD/CDMA-based cellular network with shadowing and using relative received pilot power, E_c/I_o, based hard-HO thresholds in conjunction with an asymmetric traffic load of 13:2 (UL:DL) timeslots. The system parameters are summarized in Table 9.2.

Figure 9.11: Number of HO events per 49-cell simulation area versus mean carried traffic of the UTRA-like TDD/CDMA-based cellular network with shadowing and using relative received pilot power, E_c/I_o, based hard-HO thresholds in conjunction with an asymmetric traffic load of 1:14 (UL:DL) timeslots. The system parameters are summarized in Table 9.2.

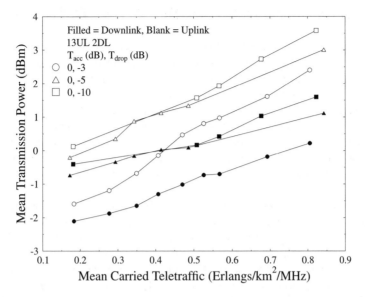

Figure 9.12: Mean transmission power versus mean carried traffic of the UTRA-like TDD/CDMA-based cellular network with shadowing and using relative received pilot power, E_c/I_o, based hard-HO thresholds in conjunction with an asymmetric traffic load of 13:2 (UL:DL) timeslots. The system parameters are summarized in Table 9.2.

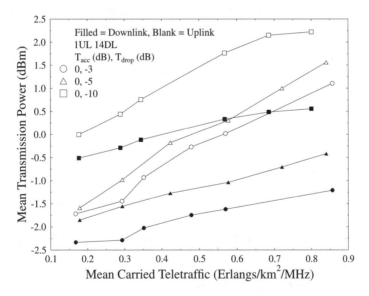

Figure 9.13: Mean transmission power versus mean carried traffic of the UTRA-like TDD/CDMA-based cellular network with shadowing and using relative received pilot power, E_c/I_o, based hard-HO thresholds in conjunction with an asymmetric traffic load of 1:14 (UL:DL) timeslots. The system parameters are summarized in Table 9.2.

9.3   Power Control in UTRA-like TDD/CDMA Systems

Agile and accurate power control is one of the key factors predetermining the attainable performance of the UTRA TDD/CDMA system, where all users share the same frequency. The power control regime has a vital influence on controlling the interference. Without power control a MS roaming in the vicinity of the BS and transmitting at an excessive power may overwhelm mobiles that are at the cell edge, a phenomenon which is often referred to as the near–far problem [517–520]. It is essential to keep the transmission power at the minimum level ensuring adequate signal quality at the receiver end. Power control may be classified as open-loop power control, inner-loop power control sometimes also referred to as closed-loop power control and outer-loop power control, all of which may be used in both the UL and DL [59].

The preferred solution to power control in the UTRA FDD/CDMA system is based on the philosophy of inner-loop power control or fast closed-loop power control in both the UL and DL. In the UL the BS generates frequent estimates of the received SINR and compares it with the target SINR required for maintaining the quality of a specific service. If the measured SINR is higher than the target SINR, the BS will instruct the MS to decrease its power. This measure–instruct–react cycle is executed at a rate of 1500 times per second (1.5 kHz) for each mobile station and thus operates faster than any significant change of pathloss could possibly happen. In fact, typically it is even faster than the typical Doppler frequency of fast Rayleigh fading for low to moderate mobile speeds. Hence, closed-loop power control will prevent any power imbalance among all of the UL signals received at the BS.

In the DL the same closed-loop power control is used as in the UL. However, there is no near–far problem owing to the one-to-many broadcast-type transmission scenario in the DL. All of the signals received by the MS within a specific traffic cell originate from the same BS transmitting to all MSs. The motivation of using closed-loop power control in the DL is firstly to provide the minimum amount of additional power to MSs roaming at the cell edge, since they suffer from increased inter-cell interference. Secondly, closed-loop DL power control is capable of enhancing the signals attenuated by Rayleigh fading with the aid of transmitting an additional power with the aim of augmenting the action of error-correcting methods [59].

9.3.1   UTRA TDD Downlink Closed-loop Power Control

As argued in the previous section, the aim of the UTRA TDD DL power control scheme is to limit the effects of interference. The transmitter typically uses a signal-quality-based power control on the DL [514]. Closed-loop power control facilitates for the BS transmitter to adjust the power in response to the MS's specific request. Downlink closed-loop power control is based on SIR measurements at the MS receiver and the corresponding Transmit Power Control (TPC) commands are generated by the MS. The power control step size determines the change in the DL power in response to a TPC message received from the MS, where the legitimate DL power steps are 1, 2 and 3 dB [514].

The inter-cell interference encountered is not only due to the MSs, but also due to the BSs contaminating the adjacent cells by co-channel signals. The DL closed-loop power control adjusts the MSs' transmit power in order to maintain the DL SINR near the SINR target, namely near $SINR_{target}$. As discussed in Section 9.2, the HOs in the UTRA TDD mode are based on hard HOs [59]. The mobile communicates with a single BS, and only one TPC

Table 9.3: TDD DL power control stepsize tolerance [514].

Stepsize	Tolerance	Range of average rate of change in code domain power per 10 steps	
		Minimum	Maximum
1 dB	± 0.5 dB	± 8 dB	± 12 dB
2 dB	± 0.75 dB	± 16 dB	± 24 dB
3 dB	± 1 dB	± 24 dB	± 36 dB

Table 9.4: FDD DL power control tolerance for the 1 dB stepsize mode [521].

DL TPC commands	Tolerance		Range of average rate of change in code domain power per 10 steps	
	Lower	Upper	Lower	Upper
For powering up	+0.5 dB	+1.5 dB	+8 dB	+12 dB
For powering down	−0.5 dB	−1.5 dB	−8 dB	−12 dB

command will be received in each DL timeslot. When we have $SINR_{DL} > SINR_{target}$, the TPC command is set to 0, otherwise, if $SINR_{DL} < SINR_{target}$, then the TPC command is set to 1. When the MS receives a TPC command, the MS is instructed to power down or up according to the "stepsize" typically expressed in decibels. The tolerance of the transmit power and the highest average rate of change in code domain power according to the power control stepsize should be within the range shown in Table 9.3 [514].

For the sake of comparison, Table 9.4 shows the tolerance of the code domain power and the highest average rate of change in the UTRA FDD mode in conjunction with a power stepsize of 1 dB [521]. Upon comparing Tables 9.3 and 9.4, it can be seen that the tolerance and range of power control is the same in the FDD and TDD modes. However, the power control agility of the TDD and FDD modes is different. In the FDD mode, there may be 15 dB power change across a 15-timeslot, 10 ms FDD frame. In contrast, in the TDD mode the achievable power control agility in the DL depends on the ratio of the number of UL/DL timeslots. To maintain the maximum possible flexibility, while facilitating closed-loop power control whenever deemed useful, the Synchronization Channel (SCH) has two TSs per 15-timeslot, 10 ms TDD frame for DL transmission in cellular usage, which corresponds to the most extreme UL asymmetry of TS allocation having a ratio of 2:13 DL:UL allocation. On the other hand, at least one TS has to be allocated for the UL transmission of the Random Access Channel (RACH), which corresponds to a maximum DL asymmetry of 14:1 (DL:UL). The procedure of carrying out one power-control step requires a pair of UL and DL TSs. Then a TPC command is transmitted in an UL TS, when the received signal power evaluated during the previous DL TS has to be adjusted. Hence, the TDD power-control rate in the DL ranges from 100 to 700 Hz, corresponding to having access to 1–7 DL TSs. The ability to support asymmetric UL/DL capacity allocations is the most attractive feature of the TDD mode.

However, the TDD mode imposes a problem in terms of the associated low power-control rate. Hence the MS may require a relatively high BS transmitted power, when the mobile suffers from experiencing a high level of interference or when it enters a building, which substantially attenuates the signal power received from the BS. This phenomenon may also lead to a high call dropping probability, when the BS is unable to satisfy the MS's received power requirement owing to the relatively low power-control agility of the TDD mode. For example, in the FDD mode a 15 dB power change can be achieved within a 15-timeslot 10 ms frame, if the power control stepsize is 1 dB. In contrast, in the TDD mode a 15 times longer duration corresponding to 15 frames and 150 ms may be needed for the extreme DL asymmetry of 14:1 (DL:UL) TS allocation. A dropped call would be encountered, when there are five consecutive 10 ms frames having an SINR value below the target SINR value. For the above-mentioned extreme asymmetric TDD traffic situation, after requesting a power increase in five consecutive frames only 5 dB power change has been achieved, potentially requiring a further $10 \, \text{dB} = 15 - 5 \, \text{dB}$ powering up, which eventually leads to this connection being terminated owing to having an insufficiently high received power level. One possible option of compensating for this relatively slow feedback loop is using a higher power-control stepsize. In the 3GPP initiative [522], using 2 bits for the TPC command was proposed for the DL, allowing for a more flexible power-control stepsize adjustment, ranging from 1 to 3 dB. Hence, we could adjust the power-control stepsize commensurately with the difference between the measured and target SINR, namely according to $\Delta SINR$:

$$\Delta SINR = |SINR_{target} - SINR_{DL}|. \tag{9.2}$$

The relationship between the stepsize and $\Delta SINR$ is shown in Table 9.5 based on Table 9.3 and the 3GPP initiative [523]. Using a flexible power-control stepsize adjustment is a desirable feature in the UTRA system's TDD mode due to the associated reduced power-control feedback rate. However, a higher power-control stepsize may impose a possible increase of the interference level inflicted upon other MSs in both the same as well as in the adjacent cells. In Section 9.3.3 we provide a comparative study of using a 1 dB fixed power control stepsize and a flexible power-control stepsize in order to investigate the achievable system performance, when invoking a higher PC stepsize.

9.3.2   UTRA TDD Uplink Closed-loop Power Control

Closed-loop power-control may also be used in the UTRA TDD mode's 1.28 Mchip/s option [523]. The UL closed-loop power control is used to set the power of both the UL Dedicated Physical Control Channel (DPCH) and Physical Common Packet Channel (PCPCH). Both the SINR measurement and power adjustment phase of the UTRA TDD UL closed-loop PC scheme is similar to that described in Section 9.3.1.

9.3.3   Closed-loop Power Control Simulation Results

In this section the effects of the closed-loop power-control stepsize on the UTRA TDD/CDMA system's performance was studied. As we have discussed in Section 9.3.1, the asymmetric teletraffic load of the UTRA TDD mode results in a potentially lower power-control rate compared to the FDD mode. The slow feedback loop of the power control command may lead to calls being terminated owing to the insufficiently high transmitted

Table 9.5: Closed-loop power control stepsizes.

Stepsize (dB)	$\Delta SINR$ (dB)	The number of TPC bits needed
+1	$0 < \Delta SINR \leq 1.5$	1
+2	$1.5 < \Delta SINR \leq 2.75$	2
+3	$2.75 < \Delta SINR \leq 4$	2
+4	$4 < \Delta SINR \leq 5.25$	3
+5	$5.25 < \Delta SINR \leq 6.5$	3
+6	$6.5 < \Delta SINR \leq 7.75$	3
	$\vdots$	
−1	$1 < \Delta SINR \leq 2.5$	1
−2	$2.5 < \Delta SINR \leq 3.75$	2
−3	$3.75 < \Delta SINR \leq 5$	2
−4	$5 < \Delta SINR \leq 6.25$	3
−5	$6.25 < \Delta SINR \leq 7.5$	3
−6	$7.5 < \Delta SINR \leq 8.75$	3
	$\vdots$	

power [524]. The limited TPC command feedback rate can be compensated for by an appropriate stepsize selection, since it is possible to use three different power-control stepsizes in the UTRA TDD mode [522], as we have seen in Table 9.3. In our simulations both the symmetric and asymmetric traffic loads of the UTRA-like TDD/CDMA system are studied. Both the UL and DL use closed-loop power control. Three different power-control stepsize algorithms are used in our simulations, namely:

- fixed power-control stepsize of 1 dB, one TPC bit is needed;

- flexible power-control stepsize of 1–3 dB, two TPC bits are needed;

- flexible power-control stepsize of 1–6 dB, three TPC bits are needed.

The stepsizes of the UL/DL commands and the required number of TPC bits are presented in Table 9.5 for each legitimate scenario.

9.3.3.1 UL/DL Symmetric Traffic Loads

For a near-symmetric traffic load, we used an 8:7 (UL:DL) TS allocation ratio in each 15-slot TDD frame, where the eight UL TSs and seven DL TSs are allocated randomly in each TDD frame. Hence, the achievable power control rate is 700 Hz, allowing a 7 dB power correction range during the 15 slots of a 10 ms frame. Figure 9.14 portrays the forced termination probability versus mean carried traffic of the UTRA-like TDD/CDMA-based cellular network in conjunction with the above-mentioned symmetric traffic load of 8:7 (UL:DL). It may be observed that the system's achievable traffic load did not benefit from invoking an adaptive PC stepsize in the scenario considered, in fact, it performed slightly worse compared with using a fixed 1 dB PC stepsize. The reason for this observation is outlined below. For the

Figure 9.14: Forced termination probability versus mean carried traffic of the UTRA-like TDD/CDMA-based cellular network with shadowing and using various closed-loop PC schemes in conjunction with a near-symmetric traffic load of 8:7 (UL:DL) timeslots. The system parameters are summarized in Table 9.6.

Table 9.6: Parameters used for the simulation of the power control.

Parameter	Value	Parameter	Value
Cell radius	150 m	Noisefloor	−100 dBm
Chip rate	3.84 Mcps	Spreading factor	16
BS/MS minimum TX power	−44 dBm	BS/MS maximum TX power	+21 dBm
Modulation scheme	4-QAM	Pathloss exponent	−3.5
Target E_b/N_o	8.0 dB	Low-quality (LQ) Outage E_b/N_o	7.0 dB
Outage E_b/N_o	6.6 dB	HO margin	5 dB
Power control SINR hysteresis	1 to 6 dB		

"PC Stepsize 1 dB" scenario of Figure 9.14, the TDD system was capable of supporting 78 users at $P_{FT} = 1\%$, corresponding to a teletraffic density of 0.46 Erlang/km^2/MHz. The "Adaptive PC Stepsize 1 dB to 3 dB" and "Adaptive PC Stepsize 1 dB to 6 dB" scenario of the TDD network was found to support 73 users and 70 users, corresponding to a normalized traffic load of 0.42 and 0.40 Erlang/km^2/MHz, respectively. The percentage of forced call terminations entirely deemed to be due to encountering an insufficiently high signal power (rather than due to violating any of the other performance requirements) within the total number of forced termination scenarios was found to be zero, although this is not explicitly demonstrated here. This is because for the near-symmetric traffic scenario considered, the TDD system was capable of maintaining a relatively high power-control feedback rate of

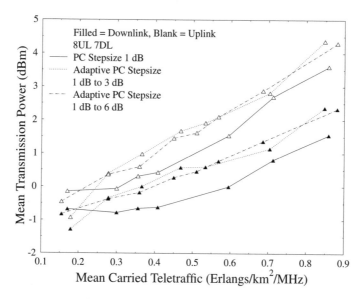

Figure 9.15: Mean MS and BS transmission power versus mean carried traffic of the UTRA-like TDD/CDMA-based cellular network with shadowing and using various closed-loop PC schemes in conjunction with a near-symmetric traffic load of 8:7 (UL:DL) timeslots. The system parameters are summarized in Table 9.6.

700 Hz, which prevented the calls from being dropped due to experiencing an insufficiently high signal power. This is why the adaptive power control stepsize adjustment algorithm failed to improve the achievable system performance.

Figure 9.15 characterizes the mean transmission power versus mean carried traffic of the UTRA-like TDD/CDMA-based cellular network. It may be observed in the figure that using the adaptive PC stepsize-based algorithm required on average 0.1–0.5 dBm more signal power as the traffic load became higher, which is the reason that invoking the adaptive PC stepsize-based algorithm slightly degraded the overall carried traffic load of a near-symmetrically loaded TDD system, as shown in Figure 9.14. The higher PC stepsize-based algorithm resulted in an increased average transmission power, which increased the system's interference level and led to the degradation of the TDD system's carried traffic.

Figure 9.16 portrays the probability of low-quality access versus various traffic loads, where most of the connections appear to have a poor call quality. Even though the system is capable of potentially achieving 0.46 Erlang/km²/MHz normalized traffic density when judged purely on the basis of the 1% forced termination probability shown in Figure 9.14, the overall system's carried traffic is reduced to 0.27 Erlang/km²/MHz, when considering the probability of low-quality access, as shown in Figure 9.16. This is because the closed-loop power-control algorithm is unable to sufficiently accurately compensate for the SINR variations imposed by the dynamically fluctuating timeslot allocations of the BSs. It was observed that the probability of low-quality access is the limiting factor of the overall system throughput, rather than the forced termination probability values shown in Figure 9.14.

Figure 9.16: Probability of low-quality access versus mean carried traffic of the UTRA-like TDD/CDMA-based cellular network with shadowing and using various closed-loop PC schemes in conjunction with a near-symmetric traffic load of 8:7 (UL:DL) timeslots. The system parameters are summarized in Table 9.6.

9.3.3.2   UL Dominated Asymmetric Traffic Loads

The adaptive PC stepsize scenario of "1 dB to 3 dB" and "1 dB to 6 dB" summarized in Table 9.5 is employed to compensate for the slowly acting power-control feedback loop associated with the asymmetric TDD traffic loads, as highlighted in Section 9.3.1. In this section we present the achievable carried traffic improvement of the TDD system, when invoking the adaptive PC stepsize algorithm of Table 9.5. The UL:DL TS allocation ratio of 13:2 (UL:DL) was studied in our simulations.

The associated forced termination probability versus mean carried traffic of the UTRA-like TDD CDMA system conveying an asymmetric traffic load of 13:2 (UL:DL) is portrayed in Figure 9.17. Observe in the figure that a significant forced termination probability improvement was achieved by employing the adaptive PC stepsize algorithm of Table 9.5. When using the fixed 1 dB power control stepsize, the achievable performance of the TDD system was gravely degraded, because the MSs and BSs are unable to sufficiently increase the transmission power at the 200 Hz power control rate facilitated by having only two DL TSs, hence only allowing a 2 dB power correction range during the 15-slot 10 ms frame. More specifically, although not explicitly shown here for reasons of space economy, we observed that about 90% of the forced call terminations of the total number of dropped calls were due to the slowly acting power-control feedback when the traffic load was low, which was reduced to about 21%, when the traffic load was high. When using the fixed 1 dB power-control stepsize, the TDD system is capable of supporting only 50 users at $P_{FT} = 1\%$, corresponding to a normalized teletraffic density of 0.30 Erlang/km^2/MHz. In conjunction with the "1 dB to 3 dB adaptive PC stepsize" algorithm, the number of users supported by the network increased by

Figure 9.17: Forced termination probability versus mean carried traffic of the UTRA-like TDD/CDMA-based cellular network with shadowing and using various closed-loop PC schemes in conjunction with an asymmetric traffic load of 13:2 (UL:DL) timeslots. The system parameters are summarized in Table 9.6.

46% to 73 users, corresponding to a traffic load of 0.44 Erlang/km^2/MHz. When invoking the "1 dB to 6 dB adaptive PC stepsize" algorithm, the achievable forced termination probability becomes similar to that of the "Adaptive PC Stepsize 1 dB to 3 dB" scenario. The number of users supported by the network was 72, corresponding to a normalized traffic load of 0.43 Erlang/km^2/MHz. Using the "Adaptive PC Stepsize 1 dB to 3 dB" scheme, a maximum of 6 dB power correction range per TDD frame is possible, which is close to the power correction range of the near-symmetric traffic load of 8:7 (UL:DL) and statistically speaking avoids the forced termination events potentially inflicted by the provision of insufficient transmit power. Hence no further performance improvement is achieved by employing the higher power control stepsize of 6 dB, as seen in Figure 9.17.

Figure 9.18 portrays the mean transmission power versus mean carried traffic performance of the UTRA-like TDD network. Similar trends are observed to those seen in Figure 9.15. Using a high-power-control stepsize may promptly compensate for the associated signal power variations based on the estimated channel quality, but the increased transmit power inflicts an increased interference upon the other mobile users at the same time. Hence, the other users may also have to increase their transmit power owing to this sudden interference change. An additional 0.3 dBm signal power is required, when invoking the adaptive PC stepsize algorithm. Figure 9.19 shows the associated low-quality access performance, both with and without adaptive stepsize control. Regardless of the presence or absence of adaptive stepsize control, the various traffic loads result in a similar low-quality access performance. Again, as we have discussed in Section 9.3.3.1, most of the low-quality access events are imposed, because the closed-loop power control is incapable of accurately

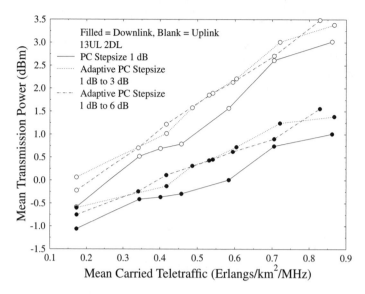

Figure 9.18: Mean MS and BS transmission power versus mean carried traffic of the UTRA-like TDD/CDMA-based cellular network with shadowing and using various closed-loop PC schemes in conjunction with an asymmetric traffic load of 13:2 (UL:DL) timeslots. The system parameters are summarized in Table 9.6.

Figure 9.19: Probability of low-quality access versus mean carried traffic of the UTRA-like TDD/CDMA-based cellular network with shadowing and using various closed-loop PC schemes in conjunction with an asymmetric traffic load of 13:2 (UL:DL) timeslots. The system parameters are summarized in Table 9.6.

compensating for the SINR variations imposed by the erratically varying timeslot allocations of the different BSs. However, the SINR fluctuation imposed by an asymmetric traffic load of 13:2 (UL:DL) for example is typically significantly smaller than that of the near-symmetric traffic load of 8:7 (UL:DL). Hence, the overall probability of low-quality access seen in Figure 9.19 is better than the corresponding performance associated with a near-symmetric traffic load, as seen in Figure 9.16.

9.3.3.3 DL Dominated Asymmetric Traffic Loads

For a TS allocation ratio of 1:14 (UL:DL), Figure 9.20 presents the forced termination probability versus the mean carried traffic performance of the UTRA-like TDD/CDMA system. In [474] it was shown that the major source of interference is constituted by the BS-to-BS interference as a consequence of the near-LOS propagation conditions prevailing between the high-elevation BSs. The co-channel interference is typically more severe in urban areas, owing to the typically high number of interfering BSs and MSs. Hence, more frequent power adjustments may be needed for maintaining the target SINR in the UL. In Figure 9.20 we can see that there is a high probability of forced termination when using a low power-control stepsize of 1 dB. Although not shown graphically, a nearly 98% outage probability was recorded owing to the insufficiently high signal power received from the MSs when the traffic load is as low as 0.25 Erlang/km^2/MHz, and 70% when the traffic load is as high as 0.5 Erlang/km^2/MHz. However, it is observed in Figure 9.20 that with the advent of the "1 dB to 3 dB" PC stepsize control regime, the TDD network can support a teletraffic density of 0.40 Erlang/km^2/MHz, corresponding to 72 users. As also seen in Figure 9.20, the employment of the "1 dB to 6 dB" regime led to a TDD network that supported a traffic load of 0.52 Erlang/km^2/MHz and handled 93 users. This corresponded to a relative gain of 26% over the performance improvement provided in the TDD mode by the "1 dB to 3 dB" PC stepsize control regime. This suggests again that a 6 dB power-control correction range per TDD frame is needed for both symmetric and asymmetric TDD traffic loads for the sake of avoiding a high forced termination probability imposed by an insufficiently responsive power ramping. A TDD system using a TS allocation of 1:14 (UL:DL) has a rather limited 100 Hz power-control rate imposed by the 1 dB transmit power adjustment per TDD frame. In conjunction with a power-control stepsize of 1 dB, this system can hardly handle any sudden power variations in excess of 5 dB, since a call is terminated within 50 ms or five TDD frames if the target SINR cannot be maintained. Hence, using a sufficiently high power-control stepsize is the key factor in maintaining an adequate system performance in the case of carrying DL-dominated traffic loads.

The mean transmission power versus teletraffic performance achieved in conjunction with the asymmetric traffic load of 1:14 (UL:DL) is depicted in Figure 9.21. Again, a higher than necessary power may increase the interference imposed upon other MSs supported by both the serving BS and by the BSs in the adjacent cells. Observe in Figure 9.21 that an additional 0.3 dBm signal power is needed for both UL and DL transmission, when invoking the adaptive PC stepsize scheme. Upon comparing Figure 9.22 with both Figure 9.19 and Figure 9.16 we observed that the probability of low-quality access in Figure 9.22 has been reduced as a benefit of the reduced channel quality fluctuations imposed by the various TS allocations of the different cells, since only a single timeslot can be allocated to either the UL or the DL in conjunction with a TS-allocation ratio of 1:14 (UL:DL). This TS-allocation ratio

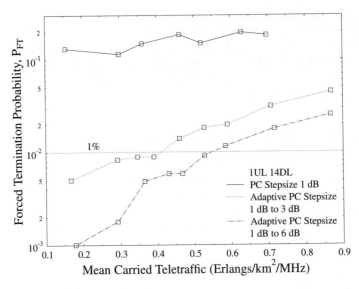

Figure 9.20: Forced termination probability versus mean carried traffic of the UTRA-like TDD/CDMA-based cellular network with shadowing and using various closed-loop PC schemes in conjunction with an asymmetric traffic load of 1:14 (UL:DL) timeslots. The system parameters are summarized in Table 9.6.

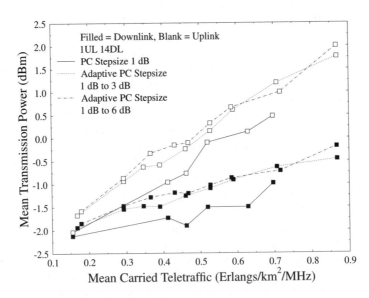

Figure 9.21: Mean transmission power versus mean carried traffic of the UTRA-like TDD/CDMA-based cellular network with shadowing and using various closed-loop PC schemes in conjunction with an asymmetric traffic load of 1:14 (UL:DL) timeslots. The system parameters are summarized in Table 9.6.

Figure 9.22: Probability of low-quality access versus mean carried traffic of the UTRA-like TDD/CDMA-based cellular network with shadowing and using various closed-loop PC scheme in conjunction with an asymmetric traffic load of 1:14 (UL:DL) timeslots. The system parameters are summarized in Table 9.6.

indirectly limits the rate of interference variation engendered by the TS-allocation variations in the interfering cells.

9.3.4 UTRA TDD UL Open-loop Power Control

One of the inherent benefits of open-loop power control is that it makes a rough estimate of the pathloss encountered by means of a DL beacon signal. However, in the UTRA FDD mode this pathloss estimation technique is far too inaccurate, because the fast fading is essentially uncorrelated between the UL and DL, owing to the large frequency separation of the UL and DL bands of the UTRA FDD mode. Hence in the UTRA FDD mode, open-loop power control is only used for providing a coarse initial power setting of the mobile station at the beginning of a call. In contrast, in the 3.84 Mcps UTRA TDD mode, the reciprocity of the UL/DL channel may be exploited for assisting the operation of the open-loop power control in the UL. Based on the estimated interference level at the BS as well as on the pathloss estimate of the DL, the mobile weights the pathloss measurements by taking into account its interference estimate and sets the UL transmission power accordingly. The estimated interference level and the BS transmitter power used are signaled to the MS for the sake of calculating the required transmit power [514]. The transmitter power of the mobile is calculated according to [523]:

$$P_{UE} = \alpha L_{PCCPCH} + (1 - \alpha)L_0 + I_{BS} + SINR_{TARGET} + C, \qquad (9.3)$$

where P_{UE} is the transmitter power level expressed (in dBm), L_{PCCPCH} is the measured pathloss (in dB), L_0 is the long-term average pathloss (in dB), I_{BS} is the estimated interference power level at the BS's receiver (quantified in terms of dBm), and, finally, α is a weighting parameter which represents the BS's confidence in the pathloss measurements, which may be impaired by both interference and fading. Furthermore, $SINR_{TARGET}$ is the target SINR (expressed in dB), while C is a constant to be set by the higher Open Systems Interconnection (OSI) layers. To elaborate a little further, the weighting factor α is a function of the time delay between the UL TS of the MS, for which the power is being calculated and the most recent DL PCCPCH TS. The specific value of the parameter α should also reflect the fading channel's Doppler frequency, which depends on the speed of the MS. More explicitly, the weighting factor α is defined as a function of the time delay, d, which is expressed in terms of the number of the TSs between the UL TS and the most recent DL TS [469], obeying

$$\alpha = 1 - \frac{(d-1)}{6}. \tag{9.4}$$

In our UTRA-like TDD system [473] a lognormally distributed slow fading obeying an average frequency of 0.5 Hz using the sum-of-sinusoid-like shadowing model of [416], and a pedestrian walking velocity of 3 mph were used. The MSs' positions and the fading parameters are updated on a frame-by-frame basis. The measured pathloss L_{PCCPCH} is assumed to be constant during the 15 timeslots of a 10 ms frame. Hence, we have $L_{PCCPCH} = L_0$, in Equation 9.3, yielding

$$P_{UE} = L_{PCCPCH} + I_{BS} + SINR_{TARGET} + C. \tag{9.5}$$

9.3.5 Frame-delay-based Power Adjustment Model

In the FDD mode the UL and DL traffic is transmitted on different frequencies, which prevents encountering interference between the UL and DL. Hence, only two different interference scenarios exist, namely the BS-to-MS interference encountered during DL transmissions and the MS-to-BS interference engendered during UL transmissions. The interference received from other cells is near-constant during an FDD 10 ms frame. Figure 9.23 illustrates this phenomenon, where the DL SINR is below the target SINR of 8 dB at TS_0. Hence the closed-loop power control scheme starts to increase the DL transmit power, seen in the middle trace of Figure 9.23, while the interference plotted at the top does not change between TS_1 and TS_2, as seen in the top trace of Figure 9.23. Therefore, the SINR reaches the target SINR value of 8 dB at TS_2 and hence improves the call quality, as shown in the bottom trace of Figure 9.23.

In contrast, in TDD mode two additional interference scenarios exist, since the UL and DL TSs are transmitted on the same carrier frequency. The received interference is imposed either by a BS or a mobile station in the interfering cell. Hence the interference level may change, for example, due to the movement of the dominant interfering source, as shown in Figure 9.24(a). The table seen at the bottom of Figure 9.24(d) presents the UL and DL timeslots' allocation in both the interfered cell and the interfering cell. The BS and mobile station of the interfered cell are denoted as BS_A and MS_A, while BS_B and MS_B represent the interfering cell's BS and mobile station, respectively. Figures 9.24(a), (b) and (c) portray the interference level, the received power and the instantaneous SINR value at the receiver of both BS_A (solid line) and MS_A (dotted line), respectively. At TS_0 and TS_2, the transmit direction is the UL in the interfered cell. In contrast, in the interfering cell the transmit direction is the DL at TS_0 and

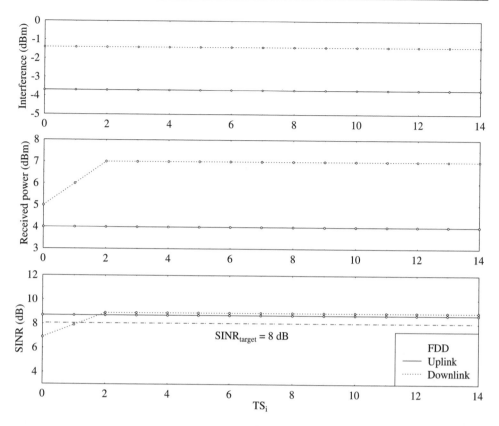

Figure 9.23: A "snap-shot" of the interference, UL and DL received powers, extracted from simulations, also showing the UL and DL received SINRs, versus TS index in a FDD frame using closed-loop power control in conjunction with a near-symmetric traffic load of 8:7 (UL:DL) timeslots.

the UL at TS_2. Hence, the UL and DL interference scenarios are different from each other, as seen in Figure 9.24(a), which results in having an UL SINR that is below the target SINR of 8 dB at TS_2. Even though the closed-loop power control regime is capable of compensating for the interference degradation by increasing the transmit power at TS_4, the same scenario is encountered again in Figure 9.24(c) at TS_{10}, which results in an inadequate SINR value of 5 dB. The closed-loop power control scheme is incapable of predicting the interference level variations imposed by the various TS allocations of the different cells, potentially leading to a poor call quality, as shown in Figure 9.16.

As seen in Figure 9.24(d), for example, in TS_3, MS_A roaming in the interfered cell receives in the DL and its signal is contaminated by the dominant interfering BS_B of the interfering cell, which is also transmitting in TS_3 in the DL. Owing to the dominant interferer BS_B, MS_A would request BS_A to increase its power transmitted to MS_A. However, as depicted in Figure 9.24(d), during TS_5 the interference scenario has changed, since now BS_B is no longer interfering with MS_A, because it is also receiving. Therefore, the previously

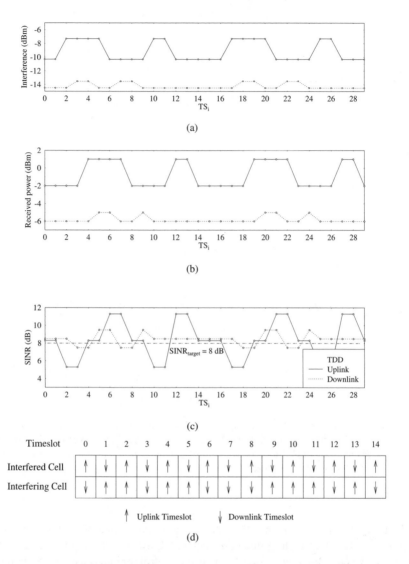

Figure 9.24: A "snap-shot" of the interference, the UL and the DL received powers, extracted from our simulations, also showing the UL and DL received SINRs, versus the TS index in a TDD frame using closed-loop power control in conjunction with a near-symmetric traffic load of 8:7 (UL:DL). The system parameters are summarized in Table 9.6.

requested BS_A transmit power is likely to become excessive, since in reality now a BS_A transmit power reduction would be required. In order to circumvent this problem it may be beneficial to postpone implementing the increased power request of MS_A until the same TS (namely TS_3 (TS_{18})) of the next TDD frame at which point the interference level may be expected to be similar to that experienced at TS_3 of MS_A during the previous frame, as seen in Figure 9.24(a), unless MS_A or MS_B become inactive or, alternatively, another dominant interferer initiates a call.

To elaborate on the related power control actions in a little more detail, let us now consider Figures 9.24(a)–(c), commencing from TS_0, where MS_A is transmitting in the UL and MS_B is receiving in the DL. During TS_0 the target SINR of 8 dB is met at both the serving BS's UL receiver, namely at BS_A and at the DL receiver of MS_A. More explicitly, the UL interference level at the receiver of BS_A in Figure 9.24(a) is -10.3 dBm and the received power (P_{RX}) of MS_A is -2 dBm in Figure 9.24(b), yielding an SINR of 8.3 dB at the input of BS_A, as seen in Figure 9.24(c). When the interference pattern changes during TS_1 in Figure 9.24(d), in the example considered in Figure 9.24(c) the target $SINR_A$ remained unchanged. To elaborate on the associated scenario, MS_A is receiving in TS_1 of Figure 9.24(d), and the interference level at its input is -14.5 dBm in Figure 9.24(a). The corresponding received power of MS_A is -6 dBm in Figure 9.24(b), yielding an SINR of 8.5 dB at the input of MS_A.

Let us now proceed to TS_2, when MS_A is transmitting in the UL and so is MS_B. Observe in Figure 9.24(a) that as shown by the continuous line, the interference at the input of BS_A increases from $-10.3--7.3$ dBm, which is an indication that MS_B is likely to be closer to BS_A than to MS_A, because in TS_1, MS_A was receiving and yet its received interference level was lower, namely -14.5 dBm as shown by the dotted line. Therefore the SINR at the input of BS_A is reduced to 5.3 dB, as shown using the continuous line in Figure 9.24(c), which is below the target SINR of 8 dB, hence necessitating a transmit power increase by MS_A in time for its next UL transmission during TS_4. Hence, the adaptive PC stepsize regime arranges for a 3 dB power increase in time for MS_A's transmission during TS_4 as seen in Figure 9.24(b), which meets the 8 dB target SINR requirement depicted in Figure 9.24(c), because the interference level plotted in Figure 9.24(a) remained unchanged.

In contrast, the interference scenario encountered during TS_3 and TS_5 will highlight a deficiency of this PC regime. Explicitly, in TS_3 both MS_A and MS_B are receiving in the DL, while in TS_5 MS_B switches to UL transmission. In TS_3 the interference level experienced by MS_A is seen to increase in Figure 9.24(a) owing to the interference imposed by BS_B transmitting to the DL receiver of MS_B. This degrades the SINR to 7.5 dB during TS_3 at the input of MS_A, which hence requests a higher transmit power from BS_A, as indicated by the ramp up of the dotted curve of Figure 9.24(b) to 5 dBm, showing an increased received power of -5 dBm during TS_5. The resultant DL SINR of TS_5 plotted using the dotted lines is increased to 9.5 dB. When MS_B switches to UL transmit mode in TS_5, BS_B has to switch to its receiver mode, inevitably ceasing its transmission to MS_B and, hence, the interference level plotted by the dotted line is seen to decease to -14.5 dBm in Figure 9.24(a). At the same time the received power of MS_A printed using dotted lines in Figure 9.24(b) is seen to be increased to -5 dBm, which results in an unnecessarily high SINR of 9.5 dB, as plotted by the dotted line in Figure 9.24(c). This unnecessarily high SINR is a consequence of a change in the interference scenario between TS_3 and TS_5, which resulted in MS_A requesting an excessive transmit power from BS_A. An even more undesirable deficiency of this PC regime is encountered when the power requested by MS_A becomes insufficient owing to an unexpected rise in the interference level. This deficiency may be mitigated by a less agile power control regime, which does not react prematurely on the basis of the erratically fluctuating UL/DL interference pattern, it rather acts during the same TS of the next TDD frame, which is likely to have the same SINR as the specific TS, when the SINR estimate was generated. More explicitly, the benefit of this deferred power-control philosophy is that the UL/DL TS configuration as well as the received signal level are likely to be similar to those experienced when the SINR estimate was generated. Whilst the advantages of this frame-delay based PC regime might appear less tangible at this stage, the results of Section 9.3.5 demonstrate its virtues in more quantitative terms. We demonstrate namely that its benefits manifest themselves in terms of reducing the probability of low-quality access, which became excessive owing to the often deficient premature power adjustments imposed by the unpredictable UL/DL TS assignments.

As we have discussed above, the erratic interference-level fluctuation imposed by the time-variant TS allocations governed by the different TDD cells results in a relatively high power-control inaccuracy, as highlighted for example in Figure 9.24. However, the TSs having the same index in consecutive TDD frames typically have the same UL/DL TS allocation pattern and their associated interference load may also be expected to be similar. Hence, if the power adjustment takes place in the TS having the same index in the next frame, the accuracy of the power control may be improved, especially in the absence of shadowing. It can be observed in Figure 9.25 that the interference-level fluctuation imposed by the interfering cells results in a low UL SINR between TS_2 and TS_5. When we invoke the above-mentioned "frame-delay"-based power control scheme, the transmitted power seen in Figure 9.25(b) is not adjusted in the current frame, namely in frame N, it is rather postponed until frame $(N + 1)$. More specifically, TS_2 and TS_{17} are the third timeslots in frame N and frame $(N + 1)$, respectively, and the associated TS allocation pattern as well as interference pattern are identical, as seen in the context of TS_2^{FrameN} and $TS_2^{FrameN+1}$ in Figure 9.25(d) and in terms of the interference-level seen in Figure 9.25(a). Hence, a power adjustment carried out at TS_{17} based on the interference-related measurements conducted during TS_2 has the potential of compensating for the interference load increase experienced, as shown in Figure 9.25(c).

In the frame-delay-based power-control simulations open-loop power control was used in the UL, as discussed in Section 9.3.4. The estimated interference level and the BS transmitter power used are signaled to the mobile station for the sake of calculating the required UL transmit power. The required transmit power is calculated during each UL TS based on the information generated during the TS having the same index in the previous frame. The DL power control also operates in a closed-loop fashion, but the power adjustment is frame-delayed. In other words, each TPC command is processed in the same TS of the next frame. In the next section we embark on studying the achievable performance of the frame-delay based open-loop UL power control regime in case of different near-symmetric as well as asymmetric traffic loads and compare the associated results to the best system performance obtained in Section 9.3.3, where closed-loop power control was used for both the UL and the DL without frame-delay power-based adjustment.

9.3.5.1 UL/DL Symmetric Traffic Loads

Recall from the simulation results of Section 9.3.3.1 that the system's performance was seriously limited by the poor probability of low-quality access in the case of supporting near-symmetric traffic loads, as evidenced by Figure 9.16. Again, the reason for this phenomenon was that the power control was unable to compensate for the erratic interference-level variations imposed by the rapidly fluctuating TS allocations of the interfering cells. For a traffic load ratio of 8:7 (UL:DL), there are $C_{15}^7 = 6435$ possible TS allocations in a TDD

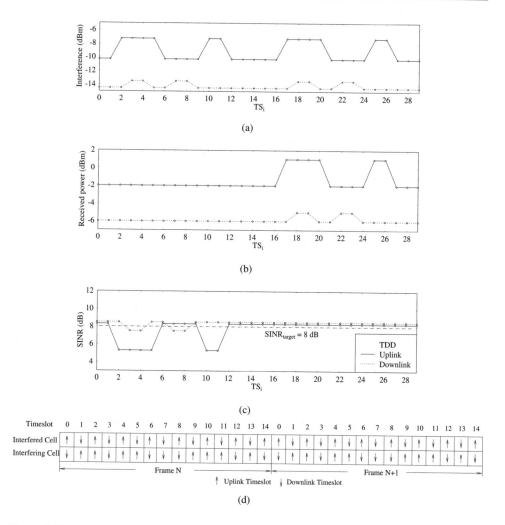

Figure 9.25: A "snap-shot" of the interference, the UL and the DL received powers, extracted from our simulations, also showing the UL and DL received SINRs, versus the TS index in a TDD frame using frame-delay-based power control in conjunction with a near-symmetric traffic load of 8:7 (UL:DL) timeslots. The system parameters are summarized in Table 9.6.

frame. In Figure 9.26 the forced termination probability of the system using closed-loop UL power control was found to be close to the one using open-loop UL power control. In contrast, using open-loop power control instead of closed-loop power control in the UL did not reduce the number of users supported. Explicitly, at $P_{FT} = 1\%$, a teletraffic density of 0.46 Erlang/km^2/MHz was achieved, corresponding to 78 users.

It is observed in Figure 9.27 that the probability of low-quality access reduced dramatically when invoking the frame-delay-based power-control scheme in comparison with the system without frame-delay-based power-control scheme. The system's performance is no longer dominated by the excessive number of low-quality outage events, which is a valuable

Figure 9.26: Forced termination probability versus mean carried traffic of the UTRA-like
TDD/CDMA-based cellular network with shadowing as well as with and without frame-
delay-based power control in conjunction with a near-symmetric traffic load of 8:7
(UL:DL) timeslots. The system parameters are summarized in Table 9.6.

Figure 9.27: Probability of low-quality access versus mean carried traffic of the UTRA-like
TDD/CDMA-based cellular network with shadowing as well as with and without using
frame-delay-based power control in conjunction with a near-symmetric traffic load of 8:7
(UL:DL) timeslots. The system parameters are summarized in Table 9.6.

Figure 9.28: The SINR histogram modeling the probability density function of the UTRA-like TDD/CDMA cellular network's SINR both with and without using frame-delay-based power control in conjunction with a near-symmetric traffic load of 8:7 (UL:DL) timeslots.

benefit of the frame-delay-based power adjustment. This phenomenon was confirmed by examining Figure 9.28, which portrays the discrete histogram modeling the probability density function of the instantaneous SINR. The majority of the users reaches the target SINR of 8 dB by employing the frame-delay-based power-control scheme. The enhanced call quality associated with a 10^{-4} low-quality access probability has the benefit of a low probability of retransmission requests and enhances the system's carried traffic.

9.3.5.2 Asymmetric Traffic Loads

Figure 9.29 characterizes the forced termination probability versus asymmetric traffic loads of 13:2 and 1:14 (UL:DL) timeslots using different power control schemes. The UL versus DL traffic load ratio of 13:2 is characteristic of uploading data files from mobile users to the BS, which typically requires a higher channel quality than the classic speech service, if an excessive retransmission probability is to be avoided. In Figure 9.29 both the closed-loop power control refraining from frame-delay based power adjustment and the UL open-loop power control employing frame-delay-based power control have similar forced termination performances, which suggests that invoking the frame-delay-based power adjustment does not degrade the system's overall performance, quite the opposite. There are three propagation-related phenomena, which may affect the accuracy of the power control in our system, namely the shadow fading, pathloss and the interference variations. Since power update is carried out only once per 10 ms frame duration, the effects of channel quality fluctuations due to both shadowing and pathloss are similar [525] in a TDD frame. The interference variations may be compensated by the frame-delay based power adjustment, hence both the closed-loop and open-loop power control have a similar forced termination probability in Figure 9.29.

Figure 9.29: Forced termination probability versus mean carried traffic of the UTRA-like TDD/CDMA-based cellular network with shadowing as well as with and without frame-delay-based power control in conjunction with an asymmetric traffic load of 13:2 and 1:14 (UL:DL) timeslots. The system parameters are summarized in Table 9.6.

Similar trends are observed regarding the probability of forced termination call at an asymmetric traffic load ratio of 1:14 (UL:DL) in Figure 9.29 to those found for the traffic load of 13:2 (UL:DL) scenario. Again, the specific choice of employing the UL open-loop power control and the frame-delay-based power adjustment does not dramatically influence the system's performance. However, the mean DL dominated carried teletraffic corresponds to a better total throughput than the UL dominated traffic load. The TDD network carrying UL dominated traffic is found to support a traffic load of 0.43 Erlang/km²/MHz at $P_{FT} = 1\%$, corresponding to 72 users. In contrast, the TDD network conveying DL dominated traffic supports an equivalent traffic load of 0.52 Erlang/km²/MHz, corresponding to 93 users. The difference is mainly caused by the co-channel interference imposed by the UL. In Figure 9.30 the mean UL transmission power associated with a traffic load of 13:2 (UL:DL) requires on average 1.2 dBm higher power than the traffic loads of 1:14 (UL:DL). Since 86.7% of the total traffic load is generated for UL transmission, the interference engendered by the mobile users degrades the achievable system performance.

The probability of low-quality access recorded for the asymmetric traffic loads of 13:2 and 1:14 (UL:DL) versus the mean carried teletraffic load is portrayed in Figure 9.31. Observe that a substantial performance improvement has been achieved, when invoking the frame-delay-based power adjustment scheme, which is a benefit of the significantly higher call quality shown in Figure 9.31. The low-quality outage probability was seen in Figure 9.31 to be below 10^{-5}–10^{-6} for a traffic load ratio of 1:14, when supporting 72 users or a traffic load of 0.43 Erlang/km²/MHz. This phenomenon was observed because the mean UL transmission power is higher than that of the DL, as seen in Figure 9.31, and the more symmetric the

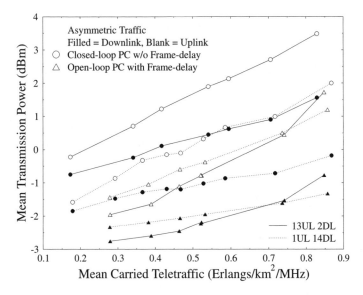

Figure 9.30: Mean transmission power versus mean carried traffic of the UTRA-like TDD/CDMA-based cellular network with shadowing as well as with and without frame-delay-based power control in conjunction with an asymmetric traffic load of 13:2 and 1:14 (UL:DL) timeslots. The system parameters are summarized in Table 9.6.

Figure 9.31: Probability of low-quality access versus mean carried traffic of the UTRA-like TDD/CDMA-based cellular network with shadowing as well as with and without frame-delay-based power control in conjunction with an asymmetric traffic load of 13:2 and 1:14 (UL:DL) timeslots. The system parameters are summarized in Table 9.6.

Table 9.7: Maximum mean carried traffic and maximum number of mobile users that can be supported
by the network, whilst meeting the network quality constraints of Section 8.6.3, namely
$P_B \leq 3\%$, $P_{FT} \leq 1\%$, $P_{low} \leq 1\%$ and $GOS \leq 4\%$. The carried traffic is expressed in
terms of Erlang/km^2/MHz using both closed-loop and open-loop power control with as well
as without frame-delay power adjustment. Shadow fading with a standard deviation of 3 dB
and a frequency of 0.5 Hz was encountered and a spreading factor of $SF = 16$ was used.

PC mode	Traffic ratio (UL:DL)	Stepsize	Users	Traffic (Erlangs/km^2/MHz)	Limiting factor	Power (dBm) MS	BS
Closed-loop	8:7	1 dB	45	0.27	P_{low}	−0.09	−0.66
Closed-loop	13:2	1–3 dB	73	0.44	P_{FT}	1.19	0.01
Closed-loop	1:14	1–6 dB	93	0.52	P_{FT}	0.43	−0.98
Open-loop	8:7	1–6 dB	78	0.46	P_{FT}	−1.05	−2.33
Open-loop	13:2	1–6 dB	72	0.43	P_{FT}	−1.54	−2.60
Open-loop	1:14	1–6 dB	93	0.52	P_{FT}	−0.43	−1.98

traffic load, the lower the overall co-channel interference level. Hence, a better link quality is
typically achieved.

A summary of the maximum number of users supported by the UTRA-like TDD/CDMA
system at various traffic load ratios using both closed-loop power control and UL open-loop
power control in conjunction with log-normal shadowing having a standard deviation of 3 dB
and a shadowing frequency of 0.5 Hz was summarized in Table 9.7 both with and without
frame-delay-based power adjustment. The teletraffic carried and the mean mobile as well as
base station transmission powers required are also shown in Table 9.7.

9.4   Summary and Conclusion

In this chapter, we have studied the effects of both the hard HO margin and different
power control schemes on the UTRA TDD/CDMA system's performance. In Sections 9.3.1–
9.3.4 both closed-loop power control as well as open-loop power control schemes were
developed. In Section 9.3.5 a frame-delay based power adjustment algorithm was proposed
to overcome the channel quality variations imposed by the erratically fluctuating timeslot
allocations in the different interfering radio cells. To elaborate a little further, we commenced
our discourse in Sections 9.1 and 9.2 with a brief introduction to hard HOs in the context
of the UTRA TDD/CDMA system. In Section 9.2.1 a relative pilot power-based hard
HO algorithm [59, 512] was employed. The related simulation results were provided in
Section 9.2.2. A HO margin range of 3–10 dB was considered in three different near-
symmetric and asymmetric traffic load scenarios. The best hard HO margin was found to
be 5 dB in conjunction with $T_{acc} = 0$ dB and $T_{drop} = −5$ dB, whilst meeting the network
quality constraints of Section 8.6.3, as evidenced by Figures 9.1, 9.5 and 9.6 of Section 9.2.2.

We then continued our discourse with a power control study of the UTRA-like TDD/
CDMA system in Section 9.3. We described a closed-loop power scheme designed for the
DL and UL in Sections 9.3.1 and 9.3.2, respectively. Although it is a beneficial feature of the

UTRA TDD mode that it is capable of supporting both asymmetric traffic and a flexible timeslot allocation, the associated low power-control rate often results in a high forced termination probability owing to the associated insufficiently high transmit power. However, this deficiency may be compensated by employing a flexible power-control stepsize, as evidenced by the simulation results of Section 9.3.3. Furthermore, an open-loop UL power-control scheme was also developed based on the 3GPP standard [526] in Section 9.3.4. Again, the main advantage of the TDD mode is its flexible timeslot allocation regime capable of adopting to the prevalent traffic requirements. However, this may impose erratic channel quality fluctuations and result in inaccurate power control. As a countermeasure, in Section 9.3.5, we proposed a frame-delay-based power adjustment algorithm, which substantially improved the system's performance, as evidenced by Figures 9.27 and 9.31 of Section 9.3.5.

Chapter 10

Genetically Enhanced UTRA/TDD Network Performance

10.1 Introduction

In Chapter 8 we demonstrated that although the UTRA/TDD mode was contrived for the sake of improving the achievable network performance by assigning all of the timeslots on a demand basis to the UL and DL, this measure may result in excessive BS-to-BS interference and, hence, in a potentially reduced number of system users. In Section 8.5 we therefore invoked both adaptive modulation and adaptive beamforming for the sake of mitigating this TDD-specific problem and demonstrated that with their advent the number of users supported may become similar but still somewhat inferior in comparison to that of an FDD system. In this chapter our research evolves further and as a design alternative we apply a Genetic Algorithm (GA) to improve the achievable performance of the UTRA-TDD mode. More specifically, in Figure 8.13 we demonstrated that the employment of adaptive arrays in conjunction with AQAM limited the detrimental effects of co-channel interference on the UTRA-like TDD/CDMA system and resulted in performance improvements both in terms of the achievable call quality and the number of users supported. However, in comparison with an UTRA-like FDD/CDMA system, the capacity of the UTRA-like TDD/CDMA cellular system was shown to remain somewhat lower than that of the UTRA-like FDD/CDMA system under the same propagation conditions. It was shown, for example, in Figure 8.10 of Chapter 8 that the TDD mode is more prone to avalanche-like teletraffic overload and its carried teletraffic is up to a factor two lower than that of the FDD mode. Again, this is because in the TDD mode MSs can interfere both with BSs as well as with each other. The same holds for BSs, which can interfere with both MSs and other BSs [474] owing to using all timeslots in both the UL and DL. The resultant additional interference has a significant detrimental impact on the system's capacity owing to the employment of the interference-limited CDMA technique. These conclusions were also corroborated by Wu [474], who also pointed out that the inter-cell BS-to-BS interference substantially decreases the system's user capacity. Hence, we can increase the total system capacity by reducing the BS-to-BS interference. One way of achieving reduced BS-to-BS interference is by invoking beamforming, since the BS can focus its transmitted signal energy on the MSs, while creating a radiation null in the direction of the adjacent BSs, as we have investigated in Chapter 8. However, this can only be achieved if there are no MSs roaming in the immediate vicinity of the line between the serving BS and the neighboring BS. Hence, the achievable capacity of the beamforming-aided UTRA-like TDD/CDMA system remains limited.

In order to mitigate these performance limitations, in this chapter we design a GA-assisted UTRA-like TDD/CDMA system. A DCA algorithm is developed, which minimizes the amount of MUI experienced at the BSs by employing GAs [434, 527–533].

The structure of this chapter is as follows. We will first study the effect of timeslot allocation on the system performance. Then the GA-aided UTRA-like TDD/CDMA system model used in this chapter is described in Section 10.2. The numerical results characterizing the various interference scenarios and the number of users supported by the GA-assisted TDD/CDMA system is quantified and compared with that of the TDD/CDMA system dispensing with GAs in Section 10.3. Let us now commence our discourse by briefly highlighting how GAs may be used for enhancing the UTRA/TDD system's performance.

10.2 The Genetically Enhanced UTRA-like TDD/CDMA System

Recently substantial advances have been made in the context of diverse wireless receivers, such as in CDMA multiuser detectors [434], beamforming [534] and SDMA-aided OFDM [535]. GAs have been used as robust guided stochastic search algorithms for solving various optimization problems, such as multiprocessor scheduling [536], topology design and bandwidth allocation in ATM networks [537], for improving the performance of channel allocation in cellular networks [538], for code design [539] and code set selection in optical CDMA networks [540]. Despite establishing themselves as useful optimization tools in numerous applications, the employment of GAs in the network layer of mobile communications has been extremely rare. In order to probe further in this promising field, in this chapter GAs have been utilized by a UTRA-like TDD/CDMA system, where the GA-assisted timeslot allocator assigns either UL or DL timeslots to MSs or BSs, while maintaining certain QoS guarantees.

The aim of this design is to maximize the achievable UTRA-like TDD/CDMA network's capacity, measured in terms of the mean normalized carried traffic expressed in units of Erlang/km^2/MHz. The performance metrics used to quantify the QoS have been described in Section 5.3.3.4. Recall that the call dropping probability, P_{FT}, quantifies the probability that a call is forced to be prematurely terminated. This may be the consequence of an insufficiently high SINR encountered during the call, which is not remedied by an intra-cell HO, either due to the lack of available channels or due to an insufficient improvement of the SINR, which leads to successive outages and eventually to a dropped call. Calls may also suffer from forced termination, when a mobile enters a heavily loaded cell, which either suffers from a poor average SINR or has no available channels for the mobile to HO to. The main limiting factors are the number of available spreading or OVSF codes, or high interference levels and low maximum affordable transmit power, resulting in excessive call dropping rates. Since a

f_{ij} Timeslot Index

	1	2	3	4	5	$\cdots$						m
1	0	1	0	0	1	$\cdots$	0	0	0	0	0	1
2	0	0	0	0	0	$\cdots$	0	1	0	0	0	0
3	0	0	1	0	0	$\cdots$	0	0	0	1	0	0
Cell Index $\cdots$						$\vdots$						
	1	0	1	0	0	$\cdots$	0	0	1	0	1	0
	0	0	0	1	0	$\cdots$	1	0	0	0	0	0
n	1	0	0	0	0	$\cdots$	0	0	0	1	0	0

Figure 10.1: Timeslot allocation matrix used by the GA.

dropped call constitutes an annoyance from a user's viewpoint, the users' SINR value has to be maintained safely above the target SINR value.

The interference experienced at the mobile can be divided into interference due to the signals transmitted to other mobiles from the same base station, which is known as intra-cell interference, and that encountered due to the signals transmitted to other mobiles from other BSs as well as to other BSs from other mobiles, which is termed inter-cell interference.

The instantaneous SINR is obtained by dividing the received signal powers by the total interference plus thermal noise power, and then by multiplying this ratio by the spreading factor, *SF*, yielding [416]

$$SINR_{DL} = \frac{SF \cdot P_{BS}}{(1 - \alpha)I_{Intra} + I_{Inter} + N_0}, \tag{10.1}$$

where $\alpha = 1$ corresponds to the ideal case of perfectly orthogonal intra-cell interference and $\alpha = 0$ to completely asynchronous intra-cell interference. Furthermore, P_{BS} is the signal power received by the mobile user from the base station, N_0 is the thermal noise, I_{Intra} is the intra-cell interference and I_{Inter} is the inter-cell interference. Again, the interference plus noise power is scaled by the spreading factor, *SF*, since during the despreading process low-pass filtering reduces the noise bandwidth by a factor of *SF*. The inter-cell interference is not only due to the MSs, but also due to the BSs illuminating the adjacent cells by co-channel signals.

Following the above introductory considerations, let us represent the GA's solution space F as $n \times m$-dimensional binary matrix, where n is the number of radio cells and m is the total number of timeslots. Explicitly, the total number of timeslots is the product of the number of traffic cells, the number of RF carriers per cell and the number of timeslots per carrier. Each element f_{ij} in the matrix is either one or zero, as shown in Figure 10.1.

The UL differs from the DL in that the multiple access interference is asynchronous in the UL due to the un-coordinated transmissions of the mobile stations, whereas it may remain quasi-synchronous in the DL. To elaborate a little further, all of the synchronous DL signals of the users sharing a given timeslot are assumed to arrive at the MS via the same propagation channel. The pathloss and shadow fading are updated on a 10 ms frame-by-frame basis every 15 timeslots. If this channel is dispersive, it does affect the orthogonality of each of the OVSF codes, but the amount of interference inflicted remains lower than in case of the asynchronous UL, where all multipath components of the asynchronous users arrive at different times at the BS, as discussed in [541]. A possible solution for mitigating the problem of OVSF code orthogonality degradation imposed by channel induced linear distortion is employing Multi-User Detectors (MUDs) [93, 434] at the base stations.

Thus, we define β as the MUD's efficiency, which quantifies the percentage of the intra-cell interference that is removed by the MUD. Setting $\beta = 0.0$ implies 0% efficiency, implying that the intra-cell interference is not reduced by the MUD, whereas $\beta = 1.0$ results in perfect suppression of all of the intra-cell interference. Therefore, based on Equation 10.1 UL SINR expression becomes

$$SINR_{UL} = \frac{SF \cdot P_{MS}}{(1 - \beta)I_{Intra} + I_{Inter} + N_0},$$ (10.2)

in conjunction with a MUD, where P_{MS} is the signal power received by BS from the mobile user. Again, the inter-cell interference is imposed by the MSs and the BSs in the adjacent cells.

In our previous investigations [473] we quantified the achievable performance of the UTRA TDD/CDMA system, demonstrating that significant performance improvements can be achieved as a direct result of the interference rejection capabilities of the adaptive antenna arrays and adaptive modulation invoked. Hence, the reduction of the interference improved the system's performance. The amount of inter-cell interference imposed depends on the angle of arrival of the interference imposed by the adjacent radio cell. If the timeslot in the interfering cell is used as an UL timeslot, then we have $f_{ij} = 1$ in Table 10.1, and vice versa.

A simple example of the possible timeslot allocation scenarios is given in Figure 10.2, portraying four possible timeslot allocation scenarios for two BSs and two MSs. More specifically in the scenario of Figures 10.2(a) and (c), BS_1 experiences two types of inter-cell interference, namely $MS_2 \rightarrow BS_1$ and $BS_2 \rightarrow BS_1$, respectively. Similarly, in the scenario seen in Figures 10.2(b) and (d), MS_1 also experiences two different types of inter-cell interferences imposed by the neighboring cell, which is experienced as $MS_2 \rightarrow MS_1$ and $BS_2 \rightarrow MS_1$, respectively. In [474] it was shown that the major source of interference is constituted by the BS-to-BS interference as a consequence of the BS's high signal power and the near-LOS propagation conditions prevailing between BSs. Hence, we can avoid BS's encountering a high BS $\rightarrow$ BS inter-cell interference by appropriately scheduling the allocation of timeslots. Interference is inherent in cellular systems, and it is challenging to control it in practice owing to the presence of random propagation effects. Interference is more severe in urban areas, owing to the typically large number of interfering BSs and MSs. If there are n BSs in an area, for each timeslot, there are 2^n ways of allocating it to a specific BS either in the UL or DL. An optimal timeslot allocation algorithm would have to tentatively invoke all possible 2^n TS allocations, in order to find the best, when a new TS has to be allocated to a user who is initiating a new call. However, since this new TS allocation affects the entire system's interference patterns, the complexity of the optimum full-search algorithm would become excessive. In order to reduce the complexity of the associated decision, we

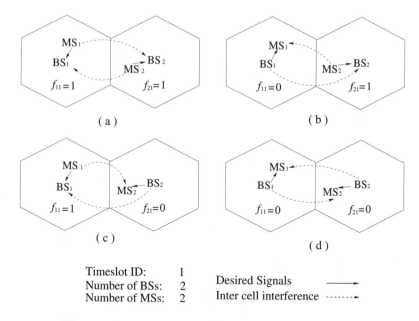

Figure 10.2: An example of UL/DL timeslot allocation options.

invoked a GA for determining the advantageous scheduling of UL and DL timeslots. The GA uses an objective function to determine how "fit" each UL/DL TS allocation is for survival in the consecutive generations of the GA. For instance, the aim of the GA is to determine, in our example provided in Section 10.2, which UL/DL TS allocation of the total of four different options has the best overall connection quality, lower UL/DL average power consumption and lower interference level. The GA's objective function will be evaluated for a small fraction of the entire set of possible TS allocations, while aiming for a near-optimum solution.

In the following, several definitions are introduced for the sake of describing the GA's objective function. There are n radio cells and each radio cell is illuminated by a BS belonging to the set

$$BS \overset{\text{def}}{=} \{BS_1, BS_2, \ldots, BS_n\}. \tag{10.3}$$

Several actively communicating mobile users belong to a radio cell and there are n sets of MSs, where each set is constituted by the MSs roaming in a specific cell and the entire set of MSs is defined as

$$M \overset{\text{def}}{=} \{M_1, M_2, \ldots, M_n\}. \tag{10.4}$$

The GA-assisted timeslot allocation scheme decides on the UL/DL transmit direction of each timeslot of a carrier, as shown in each column of Figure 10.1. Then an individual of the GA, which is also often referred to as a genome, can be defined as:

$$f \overset{\text{def}}{=} \{f_1, f_2, \ldots, f_n\}, \tag{10.5}$$

where f_j $(j = 1, 2, \ldots, n)$ denotes the UL/DL transmit direction of each timeslot of a carrier in a radio cell. As mentioned earlier in Section 10.2, each gene f_j of an individual is either one or zero, where

$$f_j = \begin{cases} 1 & \text{if } cell_j \text{ is dedicated to UL transmission,} \\ 0 & \text{if } cell_j \text{ is assigned to DL transmission.} \end{cases}$$

Hence, in Equations 10.1 and 10.2, I_{Inter} can now be written as

$$I_{Inter}^k = \sum_{j=1:j \neq k}^{n} (f_j \cdot I_{M_j} + (1 - f_j) \cdot I_{BS_j}), \tag{10.6}$$

where I_{M_j} and I_{BS_j} are the inter-cell interference received from the MSs and BS of cell j, respectively. When experiencing an instantaneous $SINR_{UL}^k$ or $SINR_{DL}^k$ for the kth active connection, the received signal power, the amount of intra-cell interference and thermal noise power cannot be readily altered. However, the inter-cell interference can be minimized by advantageously scheduling the UL/DL transmit direction in other radio cells, which maintains the value of $SINR^k$ above the target $SINR$. The performance of the network may be characterized on the basis of the probability of having a sufficiently high SINR for a timeslot. This is defined as

$$P_{Satisfied} = \frac{N_{adequate}}{N_{outage} + N_{low\text{-}quality} + N_{adequate}}, \tag{10.7}$$

where N_{outage}, $N_{low\text{-}quality}$ and $N_{adequate}$ are the number of timeslots experiencing an outage, a low-quality and adequate SINRs. More explicitly, the probability of $P_{Satisfied}$ quantifies, how "fit" a specific GA-assisted timeslot allocation is. The values of N_{outage}, $N_{low\text{-}quality}$ and $N_{adequate}$ will be determined by comparing each slot's SINR to the thresholds of 6.6, 7.0 and 8.0 dBs. The flowchart of the GA invoked in this chapter is depicted in Figure 10.3. First, an initial population consisting P number of so-called individuals is created in the "Initialization" block, where P is known as the population size. Each individual is defined according to Equation 10.5, which represents a legitimate timeslot allocation. The size of each individual of the GA is n ($0 < n \leqslant 49$), which is the number of active BSs in the simulation area containing binary flags corresponding the specific UL/DL TS allocation. There are 49 cells in the simulation area, hence the size of the full search space is 2^{49}. However, since not all of the BSs are in active status, the GA-assisted TS allocation mechanism will detect the number of active BSs and decide upon the specific size of the search space given by 2^n, which reduces the complexity, when we have $n < 49$. Each binary bit of an individual represents the transmission direction in a cell, and it is a logical one for UL transmission and vice versa. This initial population of individuals is generated randomly. The fitness value is evaluated by substituting the candidate solution into the objective function, as indicated by the "Evaluation" block of Figure 10.3. The evaluation process is invoked according to Equation 10.7. The SINR value of each active connection is calculated according to Equations 10.1 and 10.2. Then the SINR value is classified by comparing it with the SINR thresholds of outage, low-quality access and adequate SINRs. The probability of $P_{Satisfied}$ in Equation 10.7 is the individual's fitness value.

10.3 Simulation Results

In our initial investigations we do not impose any user requirements concerning the number of UL and DL TSs requested, we simply aim for determining the best possible UL/DL

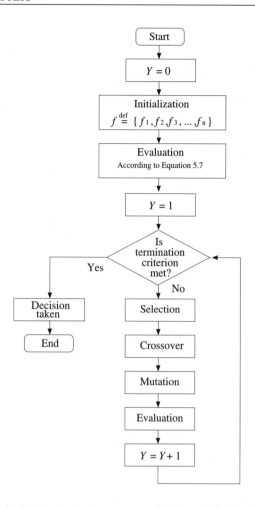

Figure 10.3: A flowchart depicting the structure of a GA used for function optimization.

system configuration, which would allow us to estimate the capacity of the system. The associated UTRA/TDD system parameters are described in Table 6.2 of Section 8.5.1. These investigations were conducted using a spreading factor of 16. Given that the chip rate of UTRA is 3.84 Mchips/s, this spreading factor corresponds to a channel data rate of $3.84 \times 10^6/16 = 240$ kbps. Applying 1/2-rate error correction coding would result in an effective data throughput of 120 kbps. A cell radius of 150 m was assumed and a pedestrian walking velocity of 3 mph was used. The simulation area was constituted by 49 traffic cells using the wrapped-around structure of Section 7.2.2.

It was shown for example in Chapter 9 of [434] that the GA's performance is dependent on numerous factors, such as the population size P, the number of generations Y, the choice of the parents' selection method, as well as on a number of other genetic operations employed. In this section, we quantify the system's achievable performance with the advent of GAs,

Table 10.1: Configuration of the GA used to obtain the results of Figure 10.4.

Set-up/parameter	Method/value
Individual initialization method	Uniform random
Selection method	Fitness—proportionate
Cross-over operation	Single point
Mutation operation	Uniform random bit flip
Population size	Variable $P = 4, 10, 20$
Generation size	Variable $Y = 25, 10, 5$
Probability of mutation	0.1
Probability of crossover	0.9
Computational complexity	100

Figure 10.4: Forced termination probability versus mean carried traffic of the UTRA-like TDD/CDMA-based cellular network both with and without GA-assisted timeslots allocation as well as with shadowing having a standard deviation of 3 dB for $SF = 16$.

and attempt to find an appropriate GA set-up and parameter configuration that are best suited for our optimization problem. The GA's parameters are summarized in Table 10.1. Our performance metrics are, as before, the call dropping or forced termination probability P_{FT}, the probability of low-quality access P_{low} and the mean transmission power, which were defined in Section 5.3.3.4. The complexity of the GA is governed by the number of generations Y required in order to achieve a reliable decision. For the sake of simplicity, the computational complexity of the GA is quantified here in the context of the total number of objective function evaluations, given by $P \times Y$.

Figure 10.4 shows the forced termination probability associated with a variety of traffic loads quantified in terms of the mean normalized carried traffic expressed in Erlangs/km^2/MHz, when subjected to 0.5 Hz frequency shadowing having a standard deviation of 3 dB. As observed in the figure nearly an order of magnitude reduction of the forced termination probability has been achieved by employing GA-assisted timeslots scheduling compared with the "No GA" scheme refraining from using UL/DL TS optimization. In the context of the "No GA" scheme, the allocation of the UL and DL timeslots for each BS was fixed to a ratio 7:8 (UL:DL). This fixed timeslot allocation may inflict a high BS→BS interference, when the serving cell is using UL timeslots and the interfering cell is using DL timeslots, as portrayed in Figures 10.2(b) and (c). The associated high inter-cell interference may result in a poor SINR, which fails to satisfy the system's target SINR required for maintaining a high-quality connection and, hence, increases the probability of forced termination. In contrast, in the GA-assisted UTRA TDD/CDMA system each timeslot in a frame can be allocated to either the UL or DL, depending on the associated slot-SINR, potentially allowing us to allocate the timeslot by minimizing the inter-cell interference inflicted. As we mentioned in the previous section, for a UTRA/TDD system having n BSs, there are 2^n possible UL/DL TS allocation schemes for each timeslot. In our simulated scenario there are 49 wrapped-around traffic cells, as was shown in Figure 5.18, creating a search space of size 2^{49}. As argued before, the size of this search space is excessive, preventing a full search. As a more attractive design option, a GA is utilized for finding a suboptimum, but highly beneficial UL or DL TS allocation. The computational complexity of GA-aided search was set to $P \cdot Y = 100$, while using different P and Y values. The "No GA" based TDD network was found to support 58 users, at $P_{FT} = 1\%$, corresponding to a traffic load of 0.3 Erlang/km^2/MHz. Upon employing GA-assisted UL/DL timeslot allocation, the number of users supported by the TDD network increased to 185 users, or to an equivalent traffic load of 1.05 Erlang/km^2/MHz, when invoking a population size of 10 and 10 generations. When the population size was reduced to 4 in conjunction with 25 generations, the TDD system was capable of supporting 174 users, corresponding to a teletraffic density of 1.01 Erlang/km^2/MHz.

Figure 10.5 portrays the probability of low-quality access versus various traffic loads. It can be seen from the figure that the probability of low-quality access for the "No GA" scheme becomes better than that of systems using GA-assisted UL/DL timeslot scheduling. This is a consequence of the associated high probability of the forced termination "No GA" scheme, as shown in Figure 10.4, because the higher the probability of forced termination, the lower the number of users supported by the TDD system and, hence, the effects of co-channel interference imposed by the existing connections remain more benign when a new call starts. Hence, a better connection quality is maintained compared with that of the "GA-assisted" scheme. From the figure we observe that the GA-aided TDD system's teletraffic density was limited to 0.87 Erlangs/km^2/MHz, corresponding to 151 users, which was limited by the performance metric P_{low}, as mentioned in Section 5.3.3.4.

For the sake of characterizing the achievable system performance also for a different perspective, the mean transmission power versus teletraffic performance is depicted in Figure 10.6. We observe in the figure that both the "GA-assisted" and "No GA" scenarios obey a similar trend in terms of their DL power consumption. However, in terms of UL power consumption, the "No GA" scheme requires an average of 2–5 dB more signal power than the "GA-assisted" scheme as the traffic load becomes higher. Again, this is because the

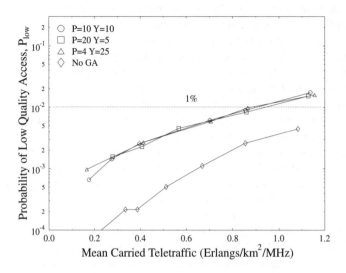

Figure 10.5: Probability of low-quality access versus mean carried traffic of the UTRA-like TDD/CDMA-based cellular network both with and without GA-assisted UL/DL TS-allocation as well as with shadowing having a standard deviation of 3 dB for $SF = 16$.

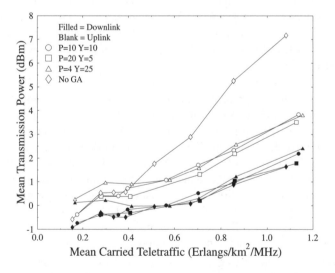

Figure 10.6: Mean transmission power versus mean carried traffic of the UTRA-like TDD/CDMA-based cellular network both with and without GA-assisted UL/DL TS-allocation as well as with shadowing having a standard deviation of 3 dB for a spreading factor of $SF = 16$.

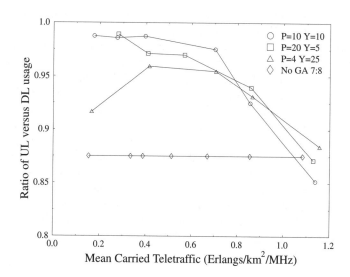

Figure 10.7: Ratio of UL timeslots to DL timeslots versus the mean carried traffic of the UTRA-like TDD/CDMA-based cellular network both with and without GA-assisted UL/DL TS allocation and with shadowing having a standard deviation of 3 dB for a spreading factor of $SF = 16$.

severe BS $\rightarrow$ BS inter-cell interference degrades the quality of the call. Hence, for the sake of achieving the target SINR and maintain the existing connections, the MSs have to increase their transmission power, which results in an increased interference level imposed on other connections, hence inflicting a performance degradation upon the whole system. The "GA-assisted" system is capable of avoiding the presence of severe interference by advantageously scheduling the UL/DL timeslots, and keep the system's average power as low as possible for the sake of supporting more MSs.

Figure 10.7 shows the ratio of UL to DL timslots versus various traffic loads. In the context of the "No GA" scheme we fixed the UL to DL timeslot utilization ratio to 0.875, since there are seven UL timeslots and eight DL timeslots in each frame. In contrast, in the "GA-assisted" scheme we did not specify the UL to DL timeslots ratio. The GA-assisted timeslot scheduling scheme determined whether a timeslot was used in the UL or DL of the system. From the resultant statistical results we observe that the UL/DL ratio of the "GA-assisted" schemes was between 0.9 and 1.0, which is close to the symmetric traffic load allocation.

10.4 Summary and Conclusion

In this chapter, we introduced a GA-assisted UL/DL timeslot scheduling scheme for the sake of avoiding the severe inter-cell interference caused by using the UTRA TDD/CDMA air interface. The system model and simulation parameters used in this chapter were highlighted in Section 10.2. The GA-aided UTRA TDD/CDMA system's performance was then examined using computer simulations in Section 10.3. Summaries of the various parameters and the GA configuration that were used in our simulations are listed in Table 10.1. Significant system performance gains have been achieved by employing the GA-aided UL/DL TS scheduling scheme, as seen in Figure 10.4. The "No GA"-based TDD network was found to support 58 users at $P_{FT} = 1\%$, corresponding to a traffic load of 0.6 Erlang/km^2/MHz. Upon employing GA-assisted UL/DL timeslot allocation in conjunction with the computational complexity of $P \cdot Y = 100$ objective function evaluations, while using a population size of 10 and 10 generations, the number of users supported by the TDD network increased to 185, or to an equivalent traffic load of 2.11 Erlang/km^2/MHz. In Figure 10.5 we observed that it was the probability of low-quality access, not the probability of forced termination, which imposed the more severe constraint on the system's capacity. In Figure 10.6, we compared the power consumption between the "No GA" and the "GA-assisted" TDD systems. We observed a similar trend in terms of their DL power consumption. However, in terms of UL power consumption the "No GA" scheme requires on average 2–5 dB more signal power than the "GA-assisted" scheme as the traffic load is increased.

Chapter **11**

Conclusions and Further Research

11.1 Summary of FDD Networking

In this book we have discussed the performance implications of adaptive antenna arrays and adaptive modulation techniques in both FDMA/TDMA and CDMA cellular mobile communications networks.

Following Chapters 1 and 2 dedicated to the 3G and HSDPA/HSUPA standards, in Chapter 4 we investigated antenna arrays and adaptive beamforming algorithms. We commenced, in Section 4.2.2, by considering the possible applications of antenna arrays and their related benefits. The signal model used was then described in Section 4.2.3 and a rudimentary example of how beamforming operates was presented. Section 4.3 highlighted the process of adaptive beamforming using several different temporal reference techniques, along with the approaches used in spatial reference techniques. The challenges that must be overcome before beamforming for the DL becomes feasible were also discussed in Section 4.3.5. Results were presented showing how the SMI, ULMS and NLMS beamforming algorithms behaved for a two-element adaptive antenna in conjunction with varying eigenvalue spread and reference signal length. The SMI algorithm was shown to converge rapidly, irrespective of the eigenvalue spread. The performance of the ULMS beamformer was shown to be highly dependent upon the input signal power presented to the antenna, rendering it impractical. However, the NLMS algorithm was found to be far superior in this respect and it was later shown to approach the performance of the SMI beamformer for a three-element adaptive array. A low SNR gives a poor estimate of the received signal's cross-correlation matrix, resulting in similar performance for all three algorithms. However, as the SNR improves, the SMI technique guarantees a stronger interference rejection. The SMI algorithm is more complex for a large number of antenna elements, but for a realistic number of elements, such as four, its complexity is below that of the LMS routines.

In Chapter 5 the performance gains achieved using adaptive antenna arrays at the BSs in a cellular network were investigated for both LOS and multipath environments. An exposure to modeling an adaptive array was provided in Section 5.2, before an overview of fixed and dynamic channel allocation schemes was conducted in Section 5.3. Section 5.5 then reviewed

3G, HSPA and FDD versus TDD Networking   Second Edition
L. Hanzo, J. S. Blogh and S. Ni   © 2008 John Wiley & Sons, Ltd

some of the different models available for simulating multipath environments, followed by a more detailed portrayal of the Geometrically Based Single-Bounce Elliptical Model (GBSBEM). The metrics used for characterizing the performance of mobile cellular networks were presented under both LOS and multipath propagation conditions, with and without adaptive antenna arrays. The network capacity was found to increase when using adaptive antenna arrays, with further increases achieved owing to the adoption of power control. An adaptive modulation mode switching algorithm with combined power control was developed and network capacity investigations were conducted. Employing adaptive modulation using adaptive antenna arrays was found to increase the network's capacity significantly, whilst providing a superior call quality and a higher mean modem throughput.

Our investigations in Chapter 5 initially focused on the non-wraparound or "desert island" type networks, where the outer cells of the simulation area are subjected to lower levels of co-channel interference, a scenario that may be encountered in the suburbs of large conurbations. Simulations were carried out for the FCA algorithm, and the LOLIA using nearest BS constraints of 7 and 19, when exposed to LOS propagation conditions. The FCA algorithm offered the lowest network capacity, but benefited the most from employing adaptive antenna arrays. Specifically, the network capacity of FCA increased by 67%, when employing two-element antenna arrays at the BSs, and 144%, when using four element arrays. The LOLIA using a nearest BS constraint of 7 cells supported a higher number of users, but the adaptive antenna arrays did not result in such dramatic improvements in network capacity. Explicitly, a 22% increase was observed for the two-element case, and a 58% when using four elements. However, the network capacity supported by the LOLIA in conjunction with $n = 7$ always exceeded that of the FCA algorithm. When using a 19 BS constraint, the LOLIA resulted in the highest network capacity without employing adaptive antenna arrays, although the large frequency reuse distance of this algorithm resulted in a modest increase of the network capacity.

We then conducted further simulations in Section 5.6.2.2 using a more realistic three-ray multipath propagation environment. Again, the FCA algorithm supported the lowest number of users, and gained the most from invoking adaptive antenna arrays. Using a four-element array instead of a two-element array led to a network capacity increase of 35%, and replacing the four-element array with one employing eight elements resulted in a 24–34% increase in the number of users supported. The LOLIA employing $n = 7$ supported the greatest number of users, but did not benefit from the same capacity increases as the FCA algorithm with the advent of adaptive antenna arrays. The number of users supported increased by 18% upon upgrading the system from two- to four-element adaptive antenna arrays, and by between 5% and 15% upon using eight-element arrays in place of the four-element arrays. Using a frequency reuse constraint of 19 in conjunction with the LOLIA resulted in a network whose capacity was restricted by the high new call blocking probability associated with its large frequency reuse distance. This large frequency reuse distance led to low levels of co-channel interference, which could not be nulled effectively by the adaptive antenna arrays and, hence, the network capacity did not increase by more than 5% upon doubling the number of antenna elements comprising the array. Hence, our future studies only considered the FCA algorithm and the LOLIA in conjunction with $n = 7$.

The network capacity gains accruing from the implementation of power control over the same three-ray multipath channel, as in the previous section, were then investigated for the FCA algorithm and the LOLIA using $n = 7$. Significant network capacity increases were observed for all of the scenarios considered. Specifically, the network capacity without power control and using a given number of antenna elements, was frequently exceeded by that of an identical scenario using power control and half the number of antenna elements. On comparing otherwise identical scenarios, an increase in the network capacity of between 28% and 72% was attributed to the implementation of power control, whilst using the FCA algorithm. When employing the LOLIA and power control, the number of users supported increased by between 8.5% and 15%. The network capacity gains resulting from increasing the number of elements in the adaptive antenna arrays were reduced, however, to 11% and 17% for the FCA algorithm. In contrast, the adaptive nature of the LOLIA enabled it to maintain the network capacity increases of 12–17%, achieved due to increasing the number of elements comprising the adaptive arrays.

The implementation of adaptive modulation techniques was then investigated in Section 5.6.2.4, since they allow the exploitation of good near-instantaneous channel conditions, whilst providing resilience when subjected to poor quality channels. The network capacity of the FCA algorithm was found to increase by 6–12%, when invoking adaptive modulation in conjunction with two-element adaptive antenna arrays. However, when using four-element adaptive antenna arrays the network capacity was reduced upon invoking adaptive modulation. This was due to the improved call dropping probability accruing from employing adaptive modulation, leading in turn to a lower number of frequency/timeslot combinations available for new calls. Since the new call blocking probability was the factor limiting the network's capacity, the capacity was reduced. This phenomenon was not observed when employing the LOLIA, which supported 43% more users on average upon invoking adaptive modulation techniques. Doubling the number of antenna elements led to an extra 20% supported users.

In summary, the network using the FCA algorithm supported 2400 users, or 14 Erlangs/km²/MHz, in the conservative scenario, and approximately 2735 users, or 15.6 Erlangs/km²/MHz, in the lenient scenario. When using the LOLIA 7 channel allocation algorithm and two-element adaptive antenna arrays, 3675 users (23.1 Erlangs/km²/MHz) were carried under the conservative conditions, and 4115 users (25.4 Erlangs/km²/MHz) under the lenient specifications. When invoking four-element adaptive antenna arrays, 4460 users (27.4 Erlangs/km²/MHz) and 4940 users (29.6 Erlangs/km²/MHz) were supported under the conservative and lenient scenarios, respectively.

In Section 5.6.3 our investigations then led us to consider results obtained for an infinite network using the so-called "wraparound" technique, which allows a cellular network to be simulated as if part of a much larger network, thus inflicting similar levels of co-channel interference upon all cells within the network. The FCA algorithm again supported the lowest number of users, but benefited the most from the employment of adaptive antenna arrays, resulting in network capacity increases of between 46% and 70%, when employing adaptive antenna arrays, or when using four rather than two elements. The LOLIA using a nearest base station constraint of 7, supported an extra 17–23% of users due to the application of adaptive antenna arrays at the base stations. As in the "desert island" scenarios, the LOLIA in conjunction with a frequency reuse constraint of 19 base stations, offered the greatest network capacity without adaptive antenna arrays. However, when using two-element arrays, the network capacity grew by almost 20%, since the limiting factor was the co-channel interference, not the new call blocking probability. The extra interference rejection potential offered by the four-element arrays was also exploited, but was also somewhat limited, since the new call blocking probability became the capacity limiting constraint once again.

Under three-ray multipath propagation conditions the network capacities of both the FCA algorithm and the LOLIAs were limited by the probability of low-quality access and, hence, invoking adaptive beamforming techniques increased the number of users supported. For an adaptive antenna array consisting of a given number of elements, the FCA algorithm supported the least number of users, and although exhibiting the greatest capacity gains due to the adaptive antenna arrays, the LOLIA 7 employing two-element arrays exceeded the capacity of the FCA algorithm using eight-element arrays. The LOLIA in conjunction with a frequency reuse of 19 base stations benefited from doubling the number of antenna elements from two to four and from four to eight, but the network capacity was then limited by the new call blocking probability and, hence, further increases in the number of antenna array elements would have had no impact on the network's capacity.

The addition of power control in the "infinite" network was then considered under the above three-ray multipath conditions. The capacity gains were significant for both the FCA algorithm and the LOLIA 7, when compared with our identical investigations conducted without power control. Again, the network capacity when using the FCA algorithm benefited the most, with the number of users supported increasing by between 38% and 82%, exhibiting a mean increase of 61%. However, the LOLIA 7 based network still supported the greatest number of users, although the capacity gains of the power control were limited to around 12%.

The employment of adaptive modulation techniques led to the saturation of network resources for the FCA algorithm, with the network capacity limited by the number of frequency/timeslot combinations available for new calls. Hence, increasing the number of antenna elements from two to four resulted in an increase in the mean modem throughput from 2.4 BPS to 2.7 BPS, and a small reduction in the mean transmission power. The adaptive nature of the LOLIA allowed it to fully exploit the potential of adaptive modulation and supported more than 32% extra users. The limiting factor of the LOLIA's network capacity was the requirement of a minimum mean modem throughput of 2.0 BPS.

Therefore, the FCA algorithm supported 1400 users and carried a teletraffic load of 13.8 Erlangs/km^2/MHz in the conservative scenario and 1570 users, or 15.2 Erlangs/km^2/MHz of traffic under the lenient conditions. The LOLIA however supported an extra 35% of users, giving a network capacity of 1910 users, or 19.75 Erlangs/km^2/MHz, when using two-element adaptive antenna arrays for both the conservative and lenient scenarios. Utilizing four-element antenna arrays at the base stations allowed 2245 users, or 23.25 Erlangs/km^2/MHz of network traffic to be supported at the required quality levels of the conservative and lenient scenarios.

Thus, the network capacity was found to substantially increase, when using adaptive antenna arrays, with further increases achieved through the adoption of power control. An adaptive modulation mode switching algorithm combined with power control was developed and network simulations were conducted. Employing adaptive modulation in conjunction with adaptive antenna arrays was found to increase the network capacity significantly, whilst providing superior call quality and a greater mean modem throughput.

Chapter 6 examined the performance of a CDMA-based cellular mobile network, very similar in its nature to the FDD-mode of the proposed UTRA standard. A comparison of various soft handover algorithms was conducted in both non-shadowed and shadowed propagation environments. The algorithm that was found to offer the highest network capacity, i.e. the highest number of users supported at a given QoS, used the relative received E_c/I_o for determining cell ownership. The impact of using adaptive antenna arrays at the base stations was then investigated, in both non-shadowed and shadowed environments for high data rate users. This work was then extended by the application of adaptive modulation techniques, in conjunction with adaptive antenna arrays.

The network capacity in terms of the number of users supported was 256 when experiencing no log-normal shadow fading and using no adaptive antenna arrays. However, with the application of two-element adaptive antenna arrays the network capacity increased by 27% to 325 users, and when upgrading the system to four-element arrays, the capacity of the network increased by a further 47% to 480 users. When subjected to log-normal shadow fading having a standard deviation of 3 dB in conjunction with a maximum fading frequency of 0.5 Hz, the network capacity without adaptive antennas was reduced to about 150 users. Again, invoking adaptive antenna arrays at the base stations increased the network capacity to 203 users and 349 users when employing two and four array elements, respectively.

We then applied independent UL and DL beamforming. This implied determining separately the optimum weights for both the UL and the DL, rather than re-using the antenna array weights calculated for the UL scenario in the DL. This measure led to further network capacity gains. Specifically, employing independent UL and DL beamforming resulted in 15% and 7% network capacity increases, for the two- and four-element arrays, respectively, giving total network capacities of 349 and 375 users. Increasing the maximum shadow fading frequency from 0.5 to 1.0 Hz slightly reduced the maximum number of users supported by the network, resulting in a network capacity of 144 users without beamforming, and capacities of 201 and 333 users, when invoking two- and four-element arrays, respectively. These absolute network capacity increases corresponded to relative network capacity gains of 40% and 131%, respectively. Again, performing independent UL and DL beamforming increased the network capacities, with 225 and 365 users supported by the two- and four-element adaptive antenna arrays, respectively. Hence, these results show that applying both two and four element adaptive antenna arrays have led to significant network capacity increases both with and without log-normal shadow fading. Furthermore, the capacity of the network was found to be reduced by approximately 40%, when subjected to log-normal shadow fading having a standard deviation of 3 dB. However, increasing the maximum log-normal fading frequency from 0.5 to 1.0 Hz had little impact on the total network capacity.

These results were then extended by applying adaptive modulation techniques, both with and without adaptive antenna arrays, which were performing independent UL and DL beamforming in conjunction with log-normal shadow fading having a standard deviation of 3 dB as well as maximum fading frequencies of 0.5 and 1.0 Hz. Without adaptive antenna arrays the network supported 223 users, at a mean UL modem throughput of 2.86 BPS. The mean throughput of the DL was 2.95 BPS. Upon increasing the maximum shadowing frequency from 0.5 to 1.0 Hz the network capacity fell slightly to 218 users, whilst the mean modem throughputs remained essentially unchanged. However, invoking two-element adaptive antenna arrays enhanced the network capacities by 64% upon encountering 0.5 Hz shadow fading, and by 56% when subjected to 1.0 Hz shadowing. In both cases the mean modem throughput dropped by approximately 0.3 BPS. A further 0.2 BPS reduction of the mean modem throughput occurred when applying four-element adaptive antenna arrays. However, this allowed an extra 30% of users to be supported when subjected to shadow fading fluctuating at a maximum frequency of 0.5 Hz and 35% in conjunction with 1.0 Hz frequency shadowing. Therefore, these results have shown the significant network capacity increases achieved by invoking adaptive modulation techniques. These network capacity improvements have been achieved in conjunction with a higher mean modem throughput, albeit at a slightly higher mean transmission power.

The performance results obtained for the UTRA-type network of Chapter 6 were obtained for high data rate users communicating at a raw data rate of 240 kbps, using a spreading factor of 16. However, as described in Section 6.4, some exploratory investigations not presented in this book demonstrated that the increase in the number of users supported by the network, was up to a factor of two higher than expected on the basis of simple spreading factor proportionate scaling. Specifically, the expected increase in switching from a spreading factor of 16 to 256 was a factor of $256/16 = 16$, and hence Tables 6.5 and 6.8 were presented showing the potential worst-case network capacities achieved by multiplying the high data rate results by 16. Even when considering these user capacities, the teletraffic carried by the network normalized with respect to both the occupied bandwidth and the network's area, was found to be higher than that achieved by the FDMA/TDMA-based networks considered in Chapter 5.

11.2   Summary of FDD versus TDD Networking

In the second half of this book we have investigated the performance of a CDMA-based cellular mobile network, similar in its nature to the FDD and TDD mode of the UTRA standard. We also characterized the performance benefits of HSDPA-style adaptive modulation and beamforming.

Chapter 7 examined the performance of a FDD/CDMA-based cellular mobile network. In Section 7.1 we characterized the achievable capacity of a UTRA-like FDD CDMA system employing LS spreading codes [416] in comparison to OVSF codes. We noted that the intra-cell interference may only be eliminated by employing orthogonal OVSF codes, if the system is perfectly synchronous and, hence, the mobile channel does not destroy the OVSF codes' orthogonality. The currently operational CDMA systems are interference limited, suffering from ISI, since the orthogonality of the spreading sequences is destroyed by the dispersive channel. They also suffer from MAI owing to the non-zero cross-correlations of the spreading codes. LS codes exhibit a so-called IFW, where both the auto-correlation and cross-correlation values of the codes become zero. Therefore, LS codes have the promise of mitigating the effects of both ISI and MAI in time dispersive channels. Hence, LS codes have the potential of increasing the attainable capacity of CDMA networks.

A quantitative comparison of the OVSF and LS codes was provided in Section 7.1.4. In conjunction with OVSF codes, the number of users supported by the "No beam-forming" scenario was limited to 152 users, or to a teletraffic load of approximately 2.65 Erlangs/km^2/MHz. With the advent of employing four-element adaptive antenna arrays at the base stations the number of users supported by the network increased to 428 users, or almost to 7.23 Erlangs/km^2/MHz. However, in conjunction with LS codes, and even without employing antenna arrays at the base stations, the network capacity was dramatically increased to 581 users, or 10.10 Erlangs/km^2/MHz, provided that the cell-size was sufficiently small for ensuring that all multipath components of the interfering users arrived within the IFW of the code. When four-element adaptive antenna arrays were employed in the above-mentioned LS-code-based scenario, the system was capable of supporting 800 users, which is equivalent to a teletraffic load of 13.39 Erlang/km^2/MHz. It was demonstrated that the network performance of the UTRA-like system employing LS spreading codes was substantially better than that of the system using OVSF codes.

Explicitly, as evidenced by Figures 7.6, 7.7, 7.9, respectively, a low call dropping probability, low MS and BS transmission power and high call quality has been maintained. In Section 7.2 we studied the network performance of different FDD/CDMA systems having various cell sizes, i.e. a cell radius of 78, 150, 300, 500 and 800 m. The simulation results were compared for the sake of quantifying how the cell size affects the achievable system performance. From the results of Figure 7.11 we observed that, as expected, the network's performance became worse, when the cell radius increased and a further improvement of the system's performance was achieved by using adaptive antenna arrays and adaptive modulation, as evidenced by Figure 7.16. The teletraffic density of the scenario having a cell radius of 78 m and employing no antenna arrays at the BS reached 2.65 Erlang/km^2/MHz, which is about 94 times higher than that of the system having a cell radius of 800 m, which supported a traffic density of 0.028 Erlang/km^2/MHz. When using two- or four-element beamforming, the adaptive antenna arrays have considerably reduced the levels of interference, leading to a higher network capacity, as seen in Figure 7.16. In practice, the coverage and capacity requirements within suburban and dense urban environments lead directly to high BS site densities. Hence, microcells constitute attractive practical solutions in terms of their relative ease of site acquisition and increased air interface capacity. In Section 7.3 the performance of a UTRA-like FDD/CDMA cellular network was investigated as a function of various target SINR thresholds. As expected, the comparisons seen in Figure 7.21 illustrate that the network's traffic-density performance became worse, when the target SINR was increased, resulting in supporting less links at a better quality. When the target SINR threshold was set to 6 dB, without employing antenna arrays the achievable traffic density reached 1.87 Erlang/km^2/MHz, which is about 27 times higher than that of the scenario, when the SINR value was set to 12 dB, which yielded 0.069 Erlang/km^2/MHz. When using two- or four-element beamforming, the adaptive antenna arrays have considerably reduced the levels of interference, leading to a higher network capacity, as evidenced by Figure 7.21. When the SINR threshold was set to 6 dB, with the advent of employing two-element adaptive antenna arrays at the BSs the achievable traffic density increased by 33% to 2.80 Erlang/km^2/MHz. Replacing the two-element adaptive antenna arrays with four-element arrays led to a further traffic density increase of 35%, which is associated with a density of 4.34 Erlangs/km^2/MHz, as seen in Figure 7.21. When the target SINR threshold was increased to 12 dB, we observed in Figure 7.21 that the user-capacity became extremely poor without the employment of adaptive antenna arrays, and only a total of nine users can be supported in the whole area of 49 base stations. This is because the target SINR was excessive and, hence, the required transmitted power increased rapidly, which then increased the interference level imposed on other users, until the system became unstable. Hence, the receivers' SINR cannot reach the target SINR and the vast majority of the calls have to be dropped. The great advantage of using adaptive antenna arrays was clearly demonstrated in this scenario. In conjunction with two- or four-element beamforming the number of users supported by the system became a factor four or eight higher than that of "no beamforming", supporting 43 and 78 users, respectively. Hence, a low value of the target SINR results in a substantially increased number of supported users, additionally benefiting from a superior call quality and from a reduced transmission power at a given number of adaptive antenna array elements installed at the base stations. When the target SINR is excessive, the overall required transmitted power rapidly increases, imposing an increased interference level and, hence, resulting in a degradation of the network capacity.

We continued our discourse in Section 6.3 by studying the characteristics of the UTRA FDD and TDD modes. Recall that the UTRA system supports two modes, the FDD mode, where the UL and DL signals are transmitted on different frequencies, and the TDD mode, where the UL and the DL signals are transmitted on the same carrier frequency, but multiplexed in time [416]. The operating principles of these two schemes were described in Figure 8.2. The UTRA TDD system was then further detailed in Section 8.3 and a comparison between the UTRA TDD system and FDD system was carried out. The UTRA TDD physical layer and physical channels were highlighted in Sections 8.3.1 and 8.3.2, while the power control regime of the TDD system was discussed in Section 8.3.3. One of the major attractions of the UTRA TDD mode is that it allows the UL and DL capacities to be adjusted asymmetrically. Recall from Figure 8.6 that the UL and DL are supported by the same carrier frequency, which creates additional interference compared with the UTRA FDD mode of the system. Two additional interference scenarios were described in Section 8.4.

We then conducted simulations in Section 8.5.1 for the sake of investigating the achievable performance of the TDD mode in both non-shadowed and shadowed propagation environments, in conjunction with both adaptive antenna arrays and HSDPA-style adaptive modulation techniques. As seen in Figure 8.7, the TDD network supported 256 users when experiencing no log-normal shadow fading and using no adaptive antenna arrays. However, with the advent of two-element adaptive antenna arrays the number of users supported was increased by 27% to 325 users, and when upgrading the system to four-element arrays, the TDD network supported a further 47% more users, increasing their number to 480 users. As seen in Figure 8.10, when subjected to log-normal shadow fading having a standard deviation of 3 dB in conjunction with a maximum fading frequency of 0.5 Hz, the TDD mode supported about 150 users without adaptive antennas. Again, invoking adaptive antenna arrays at the base stations increased the number of users supported to 203 and 349 when employing two and four array elements, respectively. These results were then improved by applying adaptive modulation techniques, both with and without adaptive antenna arrays. The beamforming-based investigations were performed in conjunction with log-normal shadow fading having a standard deviation of 3 dB as well as maximum fading frequencies of both 0.5 and 1.0 Hz. As seen in Figure 8.13, without adaptive antenna arrays the AQAM-aided TDD network supported 223 users at a mean UL modem throughput of 2.86 BPS. The mean throughput of the DL was 2.95 BPS. Upon increasing the maximum shadowing frequency from 0.5 to 1.0 Hz the number of users supported by the TDD network reduced slightly to 218 users, whilst the mean modem throughput remained essentially unchanged. However, invoking two-element adaptive antenna arrays enhanced the TDD network's user population by 64% upon encountering 0.5 Hz shadow fading, and by 56% when subjected to 1.0 Hz shadowing, as evidenced by Figure 8.13. In both cases the mean TDD throughput dropped by approximately 0.3 BPS. A further 0.2 BPS reduction of the mean TDD throughput occurred, when applying four-element adaptive antenna arrays. However, this allowed an extra 30% of TDD users to be supported, when subjected to shadow fading fluctuating at a maximum frequency of 0.5 Hz and 35% in conjunction with 1.0 Hz frequency shadowing, as supported by Figure 8.13.

Therefore, the results of Table 6.7 have shown the significant TDD user-population increases achieved by invoking adaptive modulation techniques, which allowed us to achieve a FDD-like network performance. In Section 7.4 our discussions evolved further by examining the achievable network performance of a MC-CDMA-based cellular network benefiting from both adaptive antenna arrays and adaptive modulation techniques. A brief introduction of MC-CDMA was given in Section 7.4.1. The adaptive beamforming and adaptive modulation assisted MC-CDMA network's performance was quantified in Section 7.4.2.

In Chapter 9, we studied the effects of both the hard HO margin and of different power control schemes on the UTRA TDD/CDMA system's performance. In Sections 9.3.1–9.3.4 both closed-loop power control as well as open-loop power control schemes were developed, respectively. In Section 9.3.5 a frame-delay-based power adjustment algorithm was proposed to overcome the channel quality variations imposed by the erratically fluctuating timeslot allocations in the different interfering radio cells. To elaborate a little further, we commenced our discourse in Sections 9.1 and 9.2 with a brief introduction to hard HOs in the context of the UTRA TDD/CDMA system. In Section 9.2.1 a relative pilot power based hard HO algorithm [59, 512] was employed. The related simulation results were provided in Section 9.2.2. A handover margin range of 3–10 dB was considered in three different near-symmetric and asymmetric traffic load scenarios. The best hard handover margin was found to be 5 dB in conjunction with $T_{acc} = 0$ dB and $T_{drop} = -5$ dB, whilst meeting the network quality constraints of Section 8.6.3, as evidenced by Figures 9.1, 9.5 and 9.6 of Section 9.2.2.

We then continued our discourse with a power control study of UTRA-like and HSDPA-style TDD/CDMA systems in Section 9.3. We described a closed-loop power scheme designed for the DL and UL in Sections 9.3.1 and 9.3.2, respectively. Although it is a beneficial feature of the UTRA TDD mode that it is capable of supporting both asymmetric traffic and a flexible timeslot allocation, the associated low power-control rate often results in a high forced termination probability owing to the associated insufficiently high transmit power. However, this deficiency may be compensated by employing a flexible power-control stepsize, as evidenced by the simulation results of Section 9.3.3. Furthermore, An open-loop UL power control scheme was also developed based on the 3GPP standard [526] in Section 9.3.4. Again, the main advantage of the TDD mode is its flexible timeslot allocation regime capable of adopting to the prevalent traffic requirements. However, this may impose erratic channel quality fluctuations and result in inaccurate power control. As a countermeasure, in Section 9.3.5, we proposed a frame-delay-based power adjustment algorithm, which substantially improved the system's performance, as evidenced by Figures 9.27 and 9.31 of Section 9.3.5.

In Chapter 10, we introduced a GA-assisted UL/DL timeslot scheduling scheme for the sake of avoiding the severe inter-cell interference caused by using the UTRA TDD/CDMA air interface. The system model and simulation parameters used in this chapter were highlighted in Section 10.2. The GA-aided UTRA TDD/CDMA system's performance was then examined using computer simulations in Section 10.3. Summaries of the various parameters and the GA configuration that were used in our simulations were listed in Table 10.1. Significant system performance gains have been achieved by employing the GA-aided UL/DL TS scheduling scheme, as seen in Figure 10.4. The "No GA"-based TDD network was found to support 58 users at $P_{FT} = 1\%$, corresponding to a traffic load of 0.6 Erlang/km^2/MHz. Upon employing GA-assisted UL/DL timeslot allocation in conjunction with the computational complexity of $P \cdot Y = 100$ objective function evaluations, while

Table 11.1: Summary of network performance results using the system parameters of Tables 7.1 and 7.3.

Duplex method	Spreading codes	Cell radius (m)	Target SINR (dB)	Number of AAA elements	Extracted from Figure	Modulation mode	Users	Erlang Traffic/ km²/MHz
FDD	OVSF	78	8	1	7.6	4QAM	152	2.65
FDD	OVSF	78	8	2	7.6	4QAM	242	4.12
FDD	OVSF	78	8	4	7.6	4QAM	428	7.23
FDD	LS	78	6	1	7.6	4QAM	581	10.1
FDD	LS	78	6	2	7.6	4QAM	622	10.6
FDD	LS	78	6	4	7.6	4QAM	802	13.39
FDD	OVSF	150	8	1	7.10	4QAM	150	0.87
FDD	OVSF	150	8	2	7.10	4QAM	239	1.39
FDD	OVSF	150	8	4	7.10	4QAM	348	1.99
FDD	OVSF	300	8	1	7.10	4QAM	139	0.19
FDD	OVSF	300	8	2	7.10	4QAM	229	0.32
FDD	OVSF	300	8	4	7.10	4QAM	385	0.54
FDD	OVSF	500	8	1	7.10	4QAM	142	0.07
FDD	OVSF	500	8	2	7.10	4QAM	222	0.10
FDD	OVSF	500	8	4	7.10	4QAM	370	0.19
FDD	OVSF	800	8	1	7.10	4QAM	138	0.02
FDD	OVSF	800	8	2	7.10	4QAM	217	0.04
FDD	OVSF	800	8	4	7.10	4QAM	371	0.07
FDD	OVSF	150	6	1	7.20	4QAM	320	1.87
FDD	OVSF	150	6	2	7.20	4QAM	489	2.81
FDD	OVSF	150	6	4	7.20	4QAM	758	4.34
FDD	OVSF	150	10	1	7.20	4QAM	53	0.30
FDD	OVSF	150	10	2	7.20	4QAM	113	0.65
FDD	OVSF	150	10	4	7.20	4QAM	156	0.89
FDD	OVSF	150	12	1	7.20	4QAM	9	0.07
FDD	OVSF	150	12	2	7.20	4QAM	43	0.25
FDD	OVSF	150	12	4	7.20	4QAM	78	0.44

using a population size of 10 and 10 generations, the number of users supported by the TDD network increased to 185, or to an equivalent traffic load of 2.11 Erlang/km²/MHz. In Figure 10.5 we observed that it was the probability of low-quality access, not the probability of forced termination, which imposed the more severe constraint on the system's capacity. In Figure 10.6, we compared the power consumption between the "No GA" and the "GA-assisted" TDD system. We observed a similar trend in terms of their DL power consumption. However, in terms of UL power consumption the "No GA" scheme requires on average 2–5 dB more signal power than the "GA-assisted" scheme, as the traffic load is increased.

The number of users and the corresponding Erlang capacities of the various cellular systems and various system environments considered are given in Tables 11.1 and 11.2.

Table 11.2: Summary of network performance results using the system parameters of Tables 7.7, 6.2 and 8.3.

Duplex method	Spreading codes	Cell radius (m)	Target SINR (dB)	Number of AAA elements	Extracted from Figure	Modulation mode	Users	Erlang Traffic/ km²/MHz
TDD	OVSF	150	8	1	8.10	4QAM	72	0.41
TDD	OVSF	150	8	2	8.10	4QAM	151	0.87
TDD	OVSF	150	8	4	8.10	4QAM	245	1.39
GA-TDD	OVSF	150	8	1	10.4	4QAM	185	1.05
TDD	OVSF	78	8	1	8.16	4QAM	50	0.55
TDD	OVSF	78	8	2	8.16	4QAM	113	1.18
TDD	OVSF	78	8	4	8.16	4QAM	178	2.03
TDD	LS	78	6	1	8.16	4QAM	306	3.45
FDD	OVSF	150	variable	1	7.15	AQAM	223	1.27
FDD	OVSF	150	variable	2	7.15	AQAM	366	2.11
FDD	OVSF	150	variable	4	7.15	AQAM	476	2.68
TDD	OVSF	150	variable	1	8.13	AQAM	153	0.88
TDD	OVSF	150	variable	2	8.13	AQAM	320	1.83
TDD	OVSF	150	variable	4	8.13	AQAM	420	2.41
MC-CDMA	OVSF	150	6	1	7.30(a)	4QAM	323	1.83
MC-CDMA	OVSF	150	6	2	7.30(a)	4QAM	466	2.72
MC-CDMA	OVSF	150	6	4	7.30(a)	4QAM	733	4.18
MC-CDMA	OVSF	150	variable	1	7.30(b)	AQAM	517	2.95
MC-CDMA	OVSF	150	variable	2	7.30(b)	AQAM	594	3.50
MC-CDMA	OVSF	150	variable	4	7.30(b)	AQAM	869	4.98

11.3 Further Research

Future research that builds upon the investigations considered here includes applying beamforming techniques to the pilot signals, or developing a method by which the pilot signals received at the mobile may be cancelled. In future systems the carrier frequency may be sufficiently high so that two antenna elements may be incorporated into the mobile handset, thus enabling beamforming to be performed at both ends of the data link. Further research is required for optimizing the AQAM mode switching criteria, which could amalgamate the power control and beamforming algorithms. This could be further developed to a joint optimization of the adaptive modulation mode switching, power control and beamforming, and potentially could also be incorporated into multi-user detection algorithms. In addition, the performance of multi-rate networks is worthy of investigation, especially when combined with adaptive modulation and adaptive beam-forming techniques, which are particularly suitable for mitigating the significant levels of interference inflicted by the high data rate users. Since the high-rate users impose the majority of interference on the numerous low-rate users, the employment of interference reduction techniques is of vital importance. This book has only considered the employment of uniform linear antenna arrays having an antenna element spacing of $\lambda/2$. However, other antenna geometries, exhibiting no symmetry and possibly relying on antenna elements, which are not omni-directional may result is higher network capacity gains. More sophisticated propagation models tailored for different environments, such as macro- and pico-cells also have to be considered. The TDD mode of UTRA offers a further rich ground for system optimization in conjunction with various timeslot allocation techniques, whilst endeavoring to maintain the advantages of the asymmetric UL/DL data rate nature of the TDD mode.

As a further topic, the network performance of High Altitude Platform Stations (HAPS) [542] remains to be investigated, especially in the context of adaptive modulation and, finally, future networks may be *ad-hoc* [542] in nature, which currently is a promising unexplored region of research.

Further research topics include increasing the achievable total system capacity by invoking space-time coding aided sophisticated MC-CDMA networks [543–547]. In addition, the performance evaluation of AD-HOC networks [542, 548–553] is a promising unexplored area of research.

In the context of the interference limited 3G CDMA system LS codes might hold the promise of an increased network capacity without dramatic changes to the 3G standards. However, LS codes exhibit two impediments. First, the number of spreading codes exhibiting a certain IFW is limited and, hence, under high user-loads the system may become code-limited, rather than interference-limited. The number of LS codes may be increased using the procedure proposed in [428], but further research is required for increasing the number of codes. A particularly attractive solution is to invoke both DS-CDMA TD and FD spreading [554] to multiple carriers in MC-CDMA. This can be achieved, for example, using LS and OVSF codes in the TD and FD, respectively. Then no MUD is required in the TD and the MUD employed in the FD has a low complexity owing to using an OVSF code having a low SF. The total number of users supported becomes the product of the number of LS and OVSF codes.

The second deficiency of LS codes is that they tend to exhibit a short IFW duration. However, this deficiency is also eliminated with the aid of the above-mentioned joint TD and FD spreading regime, because upon spreading to information to multiple carriers the TD chip-duration may be commensurately extended by a factor corresponding to the number of carriers.

We have concentrated our efforts on studying the performance of UTRA-like and HSDPA-style TDD/CDMA systems in both symmetric and asymmetric traffic scenarios by employing various GA-assisted timeslot scheduling schemes. The most influential factor in determining the achievable system performance is the specific choice of the GA's objective function invoked for capable of determining the near-optimum UL/DL TS allocation-based user scheduling. With the aid of a properly designed objective function it is possible to incorporate additional information about the UTRA TDD/CDMA system in the context of different system constraints, such as handover algorithms, user mobility, power-control algorithms, average power consumption, call connection quality, symmetric or asymmetric traffic load requirements, etc. Since the system's performance depends on the dimensionality of the GA, it is a meritorious future research item to document the achievable network performance as a function of the GA's affordable complexity. It is also informative to determine the histogram of various network performance metrics for different GA configurations. The further GA-assisted UTRA TDD/CDMA system's achievable performance under asymmetric traffic load constitutes the subject of our further interest.

11.3.1 Advanced Objective Functions

The employment of appropriate objective functions and fitness scaling aided GAs is often more attractive in network optimization than using the family of classic gradient search-based methods, because they do not require the solution of differential equations or a smooth search surface. The GA requires only a single measure of how meritorious a single individual is compared with the other individuals [434, 527, 555]. The objective function provides a goodness measure, given a single solution to a problem. In Section 10.3 our objective function evaluates how meritorious a genome is based on two aspects, the average power consumption and the call's connection quality. The mean UL and DL transmission power are calculated and compared with that of the previous timeslot, respectively. A mean transmission power value, which requires the lowest power increment compared with that of the previous timeslot is deemed to have a better fitness. To estimate the call's connection quality, the SINR of each UL/DL timeslot is compared with the target SINR and the number of low-quality outages is monitored. The lower the number of low-quality outages, the better the fitness of an individual. From Figures 10.4 and 10.6 we observed that significant system performance gains have been achieved and in terms of UL power consumption, since the "No GA" scheme requires an average of 2–5 dB more signal power than the "GA-assisted" scheme, as the traffic load is increased by employing the GA-aided UL/DL TS scheduling scheme. Hence, we may conclude that our scheduling scheme is capable of maintaining a low average power consumption in the context of a UTRA-like TDD/CDMA system. Furthermore, we speculate that a higher system capacity gain can be achieved by invoking more advanced objective functions, since only two aspects of the TDD system, namely its average power consumption and call connection quality were taken into account in optimizing the attainable TDD system performance. In our future research the effects of HOs, power control, the users' movement and other factors on the TDD system's achievable performance will be studied. The effects of the GA's population size, the probability of mutation, the choice of crossover operation, incest prevention and elitism will also be studied with the aid of computer simulations.

11.3.2 Other Types of GAs

In Section 10.3 we used a "simple" GA, employing so-called non-overlapping populations [527]. Some other types of GA, namely Steady-State GAs and so-called Deme GAs [555] may be worth investigating.

To elaborate a little further, "Steady-State" GAs using overlapping populations are similar to the algorithms described by DeJong [556], where the amount of population overlap may be controlled for the sake of adjusting the GA's properties. The algorithm creates a new temporary population of individuals and adds these to the previous population, then removes the lowest-fitness individuals in order to reduce the population to its original size. Again, the amount of overlap between generations may be controlled. Newly generated offspring are added to the population, then the lowest-fitness individuals are removed, hence the new offspring may or may not survive until the new generation, depending on whether they are more meritorious than the least promising individuals in the current population.

The so-called "Deme" GA [555] has multiple independent populations. Each population evolves using a steady-state GA, but in each generation some individuals migrate from one population to another. More specifically, each population migrates a fixed number of its best individuals to the neighboring population. The master population is updated in each generation with the best individuals from each population. Our future work may investigate a range of migration methods or migration operators.

Our computer simulations will comparatively study the above-mentioned types of GAs at different complexities of $P \cdot Y = C$.

Glossary

16QAM   16 Quadrature Amplitude Modulation. A modulation scheme that conveys 4 bits of data by modulating the signal amplitude and phase.

3G   Third Generation. The third-generation standard of wireless communications.

3GPP   Third Generation Partnership Project. A collaboration between groups of telecommunications associations, to make a globally applicable 3G mobile phone system specification.

ACK   Acknowledgement. Indicates the received data has passed the CRC check

AG   Absolute Grant. A grant access to the UE which determines the maximum transmit power of the scheduled E-DCH transmission.

AMC   Adaptive Modulation Coding. Adaptively select the modulation scheme and coding rate to match the varying channel conditions.

APMD   Average Path Metric Difference. A HS-SCCH tie-breaking algorithm that works by comparing the average path metric difference of the detected HS-SCCHs.

ARIB   Association of Radio Industries and Businesses. A standardization organization in Japan.

ATIS   Alliances for Telecommunications Industry Solutions. It is a standardization organization in USA.

AWGN   Additive White Gaussian Noise

BO   Buffer Occupancy. The number of bits in the buffer.

BPSK   Binary Phase Shift Keying. A modulation scheme that conveys 1 bit of data by modulating the signal phase.

BS	A common abbreviation for Base Station
CCSA	China Communications Standards Association. A standardization organization in China.
CDMA	Code Division Multiple Access
CMA	Constant Modulus Algorithm
CPICH	Common Pilot Channel. Reference channel used by UE to measure the DL channel quality.
CQI	Channel Quality Indicator. A value range of 0 to 30 to indicate the channel quality.
CRC	Cyclic Redundancy Check. A code which adds redundancy to the data for detecting errors.
DCS1800	A digital mobile radio system standard, based on GSM, but operates at 1.8GHz at a lower power.
DL	Downlink. Transmission from Node B to UE.
DOA	Direction Of Arrival
DPDCH	Dedicated Physical Data Channel. R99 DL channel which carries the data.
DTX	Discontinuous Transmission. No transmission of data.
E-AGCH	E-DCH Absolute Grant Channel. HSUPA DL channel which carries absolute grant value.
E-DCH	Enhanced Dedicated Channel. HSUPA transport channel which carries the data.
E-DPCCH	E-DCH Dedicated Physical Control Channel. HSUPA UL channel which carries control signal for decoding E-DPDCH successfully.
E-DPDCH	E-DCH Dedicated Physical Data Channel. HSUPA UL channel which carries the data.
E-HICH	E-DCH Hybrid ARQ Indicator Channel. HSUPA DL channel which carries HARQ result ACK or NACK.
E-RGCH	E-DCH Relative Grant Channel. HSUPA DL channel which carries relative grant command up, down or hold.
E-TFC	E-DCH Transport Format Combination. HSUPA transport format combination of transport block size, modulation scheme, number of physical codes.
E-TFCI	E-DCH Transport Format Combination Indicator. An index to the HSUPA transport block size table.

ETSI   European Telecommunications Standard Institute. It is a standardization organization in Europe.

FDD   Frequency Division Duplex

FPGA   Field Programmable Gate Array. A type of logic chip that can be programmed to perform specific signal processing operations.

FPMD   Frequency Path Metric Difference. A HS-SCCH tie-breaking algorithm by comparing the number path metric difference which exceeds certain threshold.

GSM   Global System of Mobile communications.

HARQ   Hybrid Automatic Repeat Request. Combination of error control codes and automatic repeat request to improve performance.

HIPERLAN   High Performance Radio Local Area Network

HSDPA   High Speed Downlink Packet Access. 3G evolution to increase the DL speed to 13.976 Mbps.

HS-DPCCH   High Speed Dedicated Physical Control Channel. HSDPA UL channel which carries control signals HARQ result and CQI value.

HS-DSCH   High Speed Downlink Shared Channel. HSDPA transport channel which carries the data.

HS-PDSCH   High Speed Physical Downlink Shared Channel HSDPA DL channel which carries the data and it is shared by all UEs.

HS-SCCH   High Speed Shared Control Channel. HSDPA DL channel which carries control signal for decoding HS-PDSCH successfully.

HSUPA   High Speed Uplink Packet Access. 3G evolution to increase the UL speed to 5.742 Mbps.

IF   Intermediate Frequency

ISI   Inter Symbol Interference. Additional interference caused by multipath environment.

LMS   Least Mean Square, a stochastic gradient algorithm used in adapting coefficients of a system

LPMD   Last Path Metric Difference. An HS-SCCH tie-breaking algorithm by comparing the last path metric difference of the detected HS-SCCHs.

MAC   Medium Access Control. Layer 2 of the data communication protocol sub-layer in the seven-layer OSI model.

MAC-es/e   Medium Access Control-es/e. A MAC entity which handles E-DCH transport channel.

MAC-hs   Medium Access Control-hs. A MAC entity which handles HS-DSCH transport channel.

MMSE   Minimum Mean Square Error. An equalizer algorithm which gives minimum mean square error.

MPMD   Minimum Path Metric Difference. An HS-SCCH tie-breaking algorithm by comparing the minimum path metric difference of the detected HS-SCCHs.

MS   A common abbreviation for Mobile Station

MSE   Mean Square Error, a criterion used to optimized the coefficients of a system such that the noise contained in the received signal is minimized.

NACK   Negative Acknowledgement. Indicates the received data has failed the CRC check.

NLMS   Normalized Least Mean Square. An equalizer algorithm which gives least mean square error.

P-CCPCH   Primary Common Control Physical Channel. Reference channel used by UE to determine Node B's system frame number.

PDF   Probability Density Function

PDU   Protocol Data Unit. A data block constructed within a MAC entity which consists of its header and payload from higher layer.

QPSK   Quadrature Phase Shift Keying. A modulation scheme that conveys 2 bits of data by modulating the signal phase.

RF   Radio Frequency

RLS   Recursive Least Square

RSN   Retransmission Sequence Number. A counter to count the number of HARQ retransmission.

RV   Redundancy Version. Parameters which control transmission modulation scheme and the way systematic and parity bits are punctured.

SAW   Stop And Wait. A HARQ method where the sender transmits one frame at a time and it does not transmit until ACK or NACK is received.

SDMA   Spatial Division Multiple Access

SFN   System Frame Number. Frame number used by the UE to synchronize with Node B.

SG	Serving Grant. An grant access value calculated using AG and RG, which determines the maximum transmit power of the E-DCH scheduled data.
SI	Scheduling Information. An UE's report of its total buffer occupancy, highest priority channel ID and its buffer occupancy and power headroom.
SINR	Signal to Interference plus Noise Ratio, same as signal to noise ratio (SNR) when there is no interference.
SIR	Signal to Interference Ratio
SNR	Signal to Noise Ratio, noise energy compared to the signal energy
TDD	Time Division Duplex
TDMA	Time Division Multiple Access
TFRC	Transport Format and Resource Combination. A functionality in MAC-hs to select transport block size and allocate resources.
TTA	Telecommunications Technology Association. A standardization organization in South Korea.
TTC	Telecommunication Technology Committee. A standardization organization in Japan.
TTI	Transmission Time Interval. The length of transmission period for a packet of data from higher layer.
UE	User Equipment. A term used in 3G for mobile phone.
UL	Uplink. Transmission from UE to Node B.
UMTS	Universal Mobile Telecommunication System
VPMD	Viterbi's Path Metric Difference. A HS-SCCH detection algorithm by comparing the path metric difference of the last trellis stage to a threshold.
WCDMA	Wideband Code Division Multiple Access. It is a wideband spread-spectrum mobile air interface that utilizes the direct sequence Code Division Multiple Access (CDMA).
YI	Yamamoto–Itoh. A HS-SCCH detection algorithm by comparing the path metric difference of the every trellis stage to a threshold.

Bibliography

[1] M. Barrett and R. Arnott, "Adaptive antennas for mobile communications," *IEE Electronics & Communications Engineering Journal*, pp. 203–214, August 1994.

[2] S. C. Swales, M. A. Beach, D. J. Edwards and J. P. McGeehan, "The Performance Enhancement of Multibeam Adaptive Base-Station Antennas for Cellular Land Mobile Radio Systems," *IEEE Transactions on Vehicular Technology*, vol. 39, pp. 56–67, February 1990.

[3] J. Litva and T. Lo, *Digital Beamforming in Wireless Communications*. Artech House, London, 1996.

[4] A. B. Carlson, *Communication Systems*. McGraw-Hill, 1986.

[5] J. G. Proakis, *Digital Communications*. McGraw-Hill International Editions, 3rd ed., 1995.

[6] L. C. Godara, "Applications of Antenna Arrays to Mobile Communications, Part I: Performance Improvement, Feasibility, and System Considerations," *Proceedings of the IEEE*, vol. 85, pp. 1029–1060, July 1997.

[7] G. V. Tsoulos and M. A. Beach, "Calibration and Linearity issues for an Adaptive Antenna System," in *IEEE Proceedings of Vehicular Technology Conference*, pp. 1597–1600, 1997.

[8] B. D. V. Veen and K. M. Buckley, "Beamforming: A Versatile Approach to Spatial Filtering," *IEEE ASSP Magazine*, pp. 4–24, April 1988.

[9] A. J. Paulraj and B. C. Ng, "Space-Time Modems for Wireless Personal Communications," *IEEE Personal Communications*, pp. 36–48, February 1998.

[10] A. J. Paulraj and E. Lindskog, "Taxonomy of space-time processing for wireless networks," *IEEE Proceedings on Radar, Sonar and Navigation*, vol. 145, pp. 25–31, February 1998.

[11] R. Steele and L. Hanzo, *Mobile Radio Communications*. IEEE Press-John Wiley, 2nd ed., 1999.

[12] W. T. Webb and L. Hanzo, *Modern Quadrature Amplitude Modulation: Principles and Applications for Wireless Communications*. IEEE Press-Pentech Press, 1994. ISBN 0-7273-1701-6.

[13] L. Hanzo, W. T. Webb and T. Keller, *Single- and Multi-Carrier Quadrature Amplitude Modulation*. John Wiley, IEEE Press, 2000.

[14] N. Anderson and P. Howard, "Technology and Transceiver Architecture Considerations for Adaptive Antenna Systems," in *Proceedings of ACTS Summit*, pp. 965–970, 1997.

[15] J. Strandell, M. Wennstrom, A. Rydberg, T. Oberg, O. Gladh, L. Rexberg, E. Sandberg, B. V. Andersson and M. Appelgren, "Experimental Evaluation of an Adaptive Antenna for a TDMA Mobile Telephony System," in *Proceedings of PIMRC*, pp. 79–84, 1997.

[16] J. J. Monot, J. Thibault, P. Chevalier, F. Pipon, S. Mayrargue and A. Levy, "A fully programmable prototype for the experimentation of the SDMA concept and use of smart antennas for UMTS and GSM/DCS1800 networks," in *Proceedings of PIMRC*, (Helsinki, Finland), pp. 534–538, September 1997.

[17] M. Mizuno and T. Ohgane, "Application of Adaptive Array Antennas to Radio Communications," *Electronics and Communications in Japan, Part 1*, vol. 77, no. 2, pp. 48–56, 1994.

[18] Y. Ogawa and T. Ohgane, "Adaptive Antennas for Future Mobile Radio," *IEICE Trans. Fundamentals,* vol. E79-A, pp. 961–967, July 1996.

[19] G. V. Tsoulos, M. A. Beach and S. C. Swales, "On the Sensitivity of the Capacity Enhancement of a TDMA system with Adaptive Multibeam Antennas," in *IEEE VTC Proceedings,* pp. 165–169, 1997.

[20] P. Leth-Espensen, P. E. Mogensen, F. Frederiksen, K. Olesen and S. L. Larsen, "Performance of Different Combining Algorithms for a GSM System applying Antenna Arrays," in *Proceedings of ACTS Summit,* 1997.

[21] W. Jakes, ed., *Microwave Mobile Communications.* Wiley-Interscience, 1974.

[22] T. Ojanpera and R. Prasad, "An overview of third-generation wireless personal communications: a European perspective," *IEEE Personal Communications,* vol. 5, pp. 59–65, December 1998.

[23] E. Dahlman, B. Gudmundson, M. Nilsson and A. Skold, "UMTS/IMT-2000 based on wideband CDMA," *IEEE Communications Magazine,* vol. 39, pp. 70–80, September 1998.

[24] B. Brand and A. Aghvami, "Multidimensional PRMA with prioritized Bayesian broadcast-a MAC strategy for multiservice traffic over UMTS," *IEEE Transactions on Vehicular Technology,* vol. 47, pp. 1148–1161, November 1998.

[25] J. Markoulidakis, R. Menolascino, F. Galliano and M. Pizarroso, "Network planning methodology applied to the UTRA specifications," in *Proceedings of the IEEE International Symposium on Personal, Indoor and Mobile Radio Communications,* vol. 2, (Boston, USA), pp. 1009–1013, September 1998.

[26] X. Mestre, M. Najar, C. Anton and J. Fonollosa, "Adaptive beamforming for high bit rate services in the FDD mode of UTRA," in *Proceedings of the IEEE Vehicular Technology Conference 1999 Fall,* vol. 4, (Amsterdam, Netherlands), pp. 1951–1955, 19–22 September 1999.

[27] S. Akhtar and D. Zeghlache, "Capacity evaluation of the UTRA WCDMA interface," in *Proceedings of the IEEE Vehicular Technology Conference 1999 Fall,* vol. 2, (Amsterdam, Netherlands), pp. 914–918, 19–22 September 1999.

[28] F. Berens, T. Bing, H. Michel, A. Worm and P. Baier, "Performance of low complexity turbo-codes in the UTRA-TDD-mode," in *Proceedings of the IEEE Vehicular Technology Conference 1999 Fall,* vol. 5, (Amsterdam, Netherlands), pp. 2621–2625, 19–22 September 1999.

[29] N. Haardt and W. Mohr, "The complete solution for third-generation wireless communications: two modes on air, one winning strategy," *IEEE Personal Communications,* vol. 7, pp. 18–24, December 2000.

[30] H. Holma, S. Heikkinen, O.-A. Lehtinen and A. Toskala, "Interference considerations for the time division duplex mode of the UMTS Terrestrial Radio Access," *IEEE Journal on Selected Areas in Communications,* vol. 18, pp. 1386–1393, August 2000.

[31] L. Aguado, T. O'Farrell and J. Harris, "Evaluation of impact of mixed traffic on UTRA performance," *IEE Electronics Letters,* vol. 36, pp. 1876–1877, October 2000.

[32] H. Haas and S. McLaughlin, "A Dynamic Channel Assignment Algorithm for a Hybrid TDMA/CDMA-TDD Interface Using the TS-Opposing Technique," *IEEE Journal on Selected Areas in Communications,* vol. 19, pp. 1831–1846, October 2001.

[33] M. Guenach and L. Vandendorpe, "Downlink performance analysis of a BPSK-based WCDMA using conventional RAKE receivers with channel estimation," *IEEE Journal on Selected Areas in Communications,* vol. 19, pp. 2165–2176, November 2001.

[34] M. Poza, A. Heras, J. Lablanca and E. Lopez, "Downlink interference estimation in UMTS," *IEE Electronics Letters,* vol. 37, pp. 590–591, April 2001.

[35] J. Perez-Romero, O. Sallent, R. Agusti and J. Sanchez, "Managing radio network congestion in UTRA-FDD," *IEE Electronics Letters,* vol. 38, pp. 1384–1386, October 2002.

[36] B. Allen, M. Beach and P. Karlsson, "Analysis of smart antenna outage in UTRA FDD networks," *IEE Electronics Letters,* vol. 38, pp. 2–3, January 2002.

[37] M. Ruiz-Garcia, J. Romero-Jerez and A. Diaz-Estrella, "Quality of service support of MAC protocols for multimedia traffic in UTRA," *IEE Electronics Letters,* vol. 38, pp. 139–141, January 2002.

[38] A. Ebner, H. Rohling, R. Halfmann and M. Lott, "Synchronization in ad hoc networks based on UTRA TDD," in *Proceedings of the IEEE International Symposium on Personal, Indoor and Mobile Radio Communications,* vol. 4, (Lisbon, Portugal), pp. 1650–1654, 15–18 September 2002.

[39] A. Agnetis, G. Brogi, G. Ciaschetti, P. Detti and G. Giambene, "Optimal packet scheduling in UTRA-TDD," *IEEE Communications Letters*, vol. 7, pp. 112–114, March 2003.

[40] C. Kao and J. Mar, "Intelligent MBWIMA/UMTS protocol using cascade fuzzy logic control for UTRA TDD mode," *IEEE Transactions on Vehicular Technology*, vol. 52, pp. 1663–1674, November 2003.

[41] J. Blogh and L. Hanzo, "Adaptive modulation and adaptive antenna array assisted network performance of multi-user detection aided UTRA-like FDD/CDMA," in *Proceedings of the IEEE Vehicular Technology Conference 2002 Fall*, vol. 3, (Vancouver, Canada), pp. 1806–1810, 24–28 September 2002.

[42] R. Rummler, Y. W. Chung and H. Aghvami, "A new multicast protocol for UMTS," in *Proceedings of the IEEE Global Telecommunications Conference*, vol. 2, (San Francisco, USA), pp. 687–691, 1–5 December 2003.

[43] Y. Yang and T.-S. Yum, "Maximally flexible assignment of orthogonal variable spreading factor codes for multirate traffic," *IEEE Transactions on Wireless Communications*, vol. 3, pp. 781–792, May 2004.

[44] K. Sivarajah and H. Al-Raweshidy, "Dynamic channel allocation for ongoing calls in UTRA TDD system," *IEE Electronics Letters*, vol. 40, pp. 1197–1198, September 2004.

[45] Y. Yang and T.-S. P. Yum, "UTRA TDD handover performance," in *Proceedings of the IEEE Global Telecommunications Conference*, vol. 5, (Dallas, TX, USA), pp. 3305–3309, 29 November–3 December 2004.

[46] S. Ni and L. Hanzo, "Genetic algorithm aided timeslot scheduling for UTRA TDD CDMA networks," *IEE Electronics Letters*, vol. 41, pp. 422–424, March 2005.

[47] T. Rouse, S. McLaughlin and I. Band, "Congestion-based routing strategies in multihop TDD-CDMA networks," *IEEE Journal on Selected Areas in Communications*, vol. 23, pp. 668–681, March 2005.

[48] P. Zhang, X. Tao, J. Zhang, Y. Wang, L. Li and Y. Wang, "A vision from the future: beyond 3g tdd," *IEEE Communications Magazine*, vol. 43, pp. 38–44, January 2005.

[49] J. S. Blogh, P. J. Cherriman and L. Hanzo, "Adaptive Beamforming Assisted Dynamic Channel Allocation," in *Proceedings of VTC*, (Houston, USA), pp. 199–203, May 1999.

[50] J. S. Blogh, P. J. Cherriman and L. Hanzo, "Comparative Study of Dynamic Channel Allocation Algorithms," *IEEE Transactions on Vehicular Technology*, 2001.

[51] J. S. Blogh, P. J. Cherriman and L. Hanzo, "Dynamic Channel Allocation Using Adaptive Antennas and Power Control," in *Proceedings of ACTS Mobile Communications Summit*, (Sorrento), pp. 943–948, June 1999.

[52] J. S. Blogh, P. J. Cherriman and L. Hanzo, "Dynamic Channel Allocation Techniques using Adaptive Modulation and Adaptive Antennas," *Accepted for publication in IEEE Journal on Selected Areas in Communications*, 2001.

[53] J. S. Blogh, P. J. Cherriman and L. Hanzo, "Dynamic Channel Allocation Techniques using Adaptive Modulation and Adaptive Antennas," in *Proceedings of VTC Fall*, (Amsterdam, The Netherlands), pp. 2348–2352, September 1999.

[54] J. Rapeli, "UMTS: Targets, system concept, and standardization in a global framework," *IEEE Personal Communications*, vol. 2, pp. 20–28, February 1995.

[55] L. Hanzo and J. Stefanov, "The Pan-European Digital Cellular Mobile Radio System — known as GSM," in Steele [155], ch. 8, pp. 677–765.

[56] P.-G. Andermo and L.-M. Ewerbring, "A CDMA-based radio access design for UMTS," *IEEE Personal Communications*, vol. 2, pp. 48–53, February 1995.

[57] E. Nikula, A. Toskala, E. Dahlman, L. Girard and A. Klein, "FRAMES multiple access for UMTS and IMT-2000," *IEEE Personal Communications*, vol. 5, pp. 16–24, April 1998.

[58] T. Ojanperä and R. Prasad, ed., *Wideband CDMA for 3rd Generation Mobile Communications*. Artech House Publishers, 1998.

[59] H. Holma and A. Toskala, eds., *WCDMA for UMTS: Radio Access for Third Generation Mobile Communications*. John Wiley & Sons, Ltd., 2000.

[60] E. Berruto, M. Gudmundson, R. Menolascino, W. Mohr and M. Pizarroso, "Research activities on UMTS radio interface, network architectures, and planning," *IEEE Communications Magazine*, vol. 36, pp. 82–95, February 1998.

[61] M. Callendar, "Future public land mobile telecommunication systems," *IEEE Personal Communications*, vol. 12, no. 4, pp. 18–22, 1994.

[62] W. Lee, "Overview of cellular CDMA," *IEEE Transactions on Vehicular Technology*, vol. 40, pp. 291–302, May 1991.

[63] K. Gilhousen, I. Jacobs, R. Padovani, A. Viterbi, L. Weaver Jr. and C. Wheatley III, "On the capacity of a cellular CDMA system," *IEEE Transactions on Vehicular Technology*, vol. 40, pp. 303–312, May 1991.

[64] R. Pickholtz, L. Milstein and D. Schilling, "Spread spectrum for mobile communications," *IEEE Transactions on Vehicular Technology*, vol. 40, pp. 312–322, May 1991.

[65] R. Kohno, *Wireless Communications: TDMA versus CDMA*, ch. 1. Spatial and Temporal Communication Theory using Software Antennas for Wireless Communications, pp. 293–321. Kluwer Academic Publishers, 1997.

[66] A. Viterbi, *CDMA: Principles of Spread Spectrum Communication*. Addison-Wesley, June 1995. ISBN 0201633744.

[67] S. Glisic and B. Vucetic, *Spread Spectrum CDMA Systems for Wireless Communications*. Artech House, April 1997. ISBN 0890068585.

[68] R. Prasad, *CDMA for Wireless Personal Communications*. Artech House, May 1996. ISBN 0890065713.

[69] V. Garg, K. Smolik, J. Wilkes and K. Smolik, *Applications of CDMA in Wireless/Personal Communications*. Englewood Cliffs NJ: Prentice-Hall, 1996.

[70] R. Price and E. Green Jr., "A communication technique for multipath channels," *Proceedings of the IRE*, vol. 46, pp. 555–570, March 1958.

[71] B. Sklar, "Rayleigh fading channels in mobile digital communication systems part I : Characterization," *IEEE Communications Magazine*, vol. 35, pp. 90–100, July 1997.

[72] B. Sklar, "Rayleigh fading channels in mobile digital communication systems part II: Mitigation," *IEEE Communications Magazine*, vol. 35, pp. 148–155, July 1997.

[73] F. Amoroso, "Use of DS/SS signaling to mitigate Rayleigh fading in a dense scatterer environment," *IEEE Personal Communications*, vol. 3, pp. 52–61, April 1996.

[74] W. C. Jakes, ed., *Microwave Mobile Communications*. John Wiley and Sons, 1974. ISBN 0-471-43720-4.

[75] M. Nakagami, "The m-distribution-a general formula of intensity distribution of fading," *Statistical Methods in Radio Wave Propagation*, 1960. W. C. Hoffman, ed., New York: Pergamon.

[76] H. Suzuki, "A statistical model for urban multipath propagation," *IEEE Transactions on Communications*, vol. COM-25, pp. 673–680, July 1977.

[77] "COST 207: Digital land mobile radio communications, final report." Office for Official Publications of the European Communities, 1989. Luxembourg.

[78] M. Whitmann, J. Marti and T. Kürner, "Impact of the power delay profile shape on the bit error rate in mobile radio systems," *IEEE Transactions on Vehicular Technology*, vol. 46, pp. 329–339, May 1997.

[79] D. Greenwood and L. Hanzo, "Characterisation of mobile radio channels," in Steele [155], ch. 2, pp. 92–185.

[80] T. Eng, N. Kong and L. Milstein, "Comparison of diversity combining techniques for Rayleigh-fading channels," *IEEE Transactions on Communications*, vol. 44, pp. 1117–1129, September 1996.

[81] M. Kavehrad and P. McLane, "Performance of low-complexity channel coding and diversity for spread spectrum in indoor, wireless communications," *AT&T Technical Journal*, vol. 64, pp. 1927–1965, October 1985.

[82] K.-T. Wu and S.-A. Tsaur, "Selection diversity for DS-SSMA communications on Nakagami fading channels," *IEEE Transactions on Vehicular Technology*, vol. 43, pp. 428–438, August 1994.

[83] L.-L. Yang and L. Hanzo, "Serial acquisition techniques for DS-CDMA signals in frequency-selective multi-user mobile channels," in *Proceedings of VTC'98 (Spring)* [557].

[84] L.-L. Yang and L. Hanzo, "Serial acquisition of DS-CDMA signals in multipath fading mobile channels." submitted to IEEE Transactions on Vehicular Technology, 1998.

[85] R. Ziemer and R. Peterson, *Digital Communications and Spread Spectrum System*. New York: Macmillan Publishing Company, 1985.

[86] R. Pickholtz, D. Schilling and L. Milstein, "Theory of spread-spectrum communications — a tutorial," *IEEE Transactions on Communications*, vol. COM-30, pp. 855–884, May 1982.

[87] S. Rappaport and D. Grieco, "Spread-spectrum signal acquisition: Methods and technology," *IEEE Communications Magazine*, vol. 22, pp. 6–21, June 1984.

[88] E. Ström, S. Parkvall, S. Miller and B. Ottersten, "Propagation delay estimation in asynchronous direct-sequence code division multiple access systems," *IEEE Transactions on Communications*, vol. 44, pp. 84–93, January 1996.

[89] R. Rick and L. Milstein, "Optimal decision strategies for acquisition of spread-spectrum signals in frequency-selective fading channels," *IEEE Transactions on Communications*, vol. 46, pp. 686–694, May 1998.

[90] J. Lee, *CDMA Systems Engineering Handbook*. London: Artech House Publishers, 1998.

[91] M. Varanasi and B. Aazhang, "Multistage detection in asynchronous code-division multiple-access communications," *IEEE Transactions on Communications*, vol. 38, pp. 509–519, April 1990.

[92] S. Moshavi, "Multi-user detection for DS-CDMA communications," *IEEE Communications Magazine*, vol. 34, pp. 124–136, October 1996.

[93] S. Verdú, *Multiuser Detection*. Cambridge University Press, 1998.

[94] L. Hanzo, C. H. Wong and M. S. Yee, *Adaptive Wireless Transceivers*. John Wiley, IEEE Press, 2002. (For detailed contents please refer to http://www-mobile.ecs.soton.ac.uk.)

[95] E. Kuan and L. Hanzo, "Joint detection CDMA techniques for third-generation transceivers," in *Proceedings of ACTS Mobile Communication Summit '98*, (Rhodes, Greece), pp. 727–732, ACTS, 8–11 June 1998.

[96] E. Kuan, C. Wong and L. Hanzo, "Burst-by-burst adaptive joint detection CDMA," in *Proceedings of VTC'98 (Spring)* [557].

[97] S. Verdú, *Multiuser Detection*. Cambridge: Cambridge University Press, 1998.

[98] F. Simpson and J. Holtzman, "Direct sequence CDMA power control, interleaving, and coding," *IEEE Journal on Selected Areas in Communications*, vol. 11, pp. 1085–1095, September 1993.

[99] M. Pursley, "Performance evaluation for phase-coded spread-spectrum multiple-access communication-part I: System analysis," *IEEE Transactions on Communications*, vol. COM-25, pp. 795–799, August 1977.

[100] R. Morrow Jr., "Bit-to-bit error dependence in slotted DS/SSMA packet systems with random signature sequences," *IEEE Transactions on Communications*, vol. 37, pp. 1052–1061, October 1989.

[101] J. Holtzman, "A simple, accurate method to calculate spread-spectrum multiple-access error probabilities," *IEEE Transactions on Communications*, vol. 40, pp. 461–464, March 1992.

[102] U.-C. Fiebig and M. Schnell, "Correlation properties of extended m-sequences," *Electronic Letters*, vol. 29, pp. 1753–1755, September 1993.

[103] J. McGeehan and A. Bateman, "Phase-locked transparent tone in band (TTIB): A new spectrum configuration particularly suited to the transmission of data over SSB mobile radio networks," *IEEE Transactions on Communications*, vol. COM-32, no. 1, pp. 81–87, 1984.

[104] A. Bateman, G. Lightfoot, A. Lymer and J. McGeehan, "Speech and data transmissions over a 942MHz TAB and TTIB single sideband mobile radio system," *IEEE Transactions on Vehicular Technology*, vol. VT-34, pp. 13–21, February 1985.

[105] F. Davarian, "Mobile digital communications via tone calibration," *IEEE Transactions on Vehicular Technology*, vol. VT-36, pp. 55–62, May 1987.

[106] M. Moher and J. Lodge, "TCMP — a modulation and coding strategy for Rician fading channels," *IEEE Journal on Selected Areas in Communications*, vol. 7, pp. 1347–1355, December 1989.

[107] G. Irvine and P. McLane, "Symbol-aided plus decision-directed reception for PSK/TCM modulation on shadowed mobile satellite fading channels," *IEEE Journal on Selected Areas in Communications*, vol. 10, pp. 1289–1299, October 1992.

[108] A. Baier, U.-C. Fiebig, W. Granzow, W. Koch, P. Teder and J. Thielecke, "Design study for a CDMA-based third-generation mobile system," *IEEE Journal on Selected Areas in Communications*, vol. 12, pp. 733–743, May 1994.

[109] P. Rapajic and B. Vucetic, "Adaptive receiver structures for asynchronous CDMA systems," *IEEE Journal on Selected Areas in Communications*, vol. 12, pp. 685–697, May 1994.

[110] M. Benthin and K.-D. Kammeyer, "Influence of channel estimation on the performance of a coherent DS-CDMA system," *IEEE Transactions on Vehicular Technology*, vol. 46, pp. 262–268, May 1997.

[111] M. Sawahashi, Y. Miki, H. Andoh and K. Higuchi, "Pilot symbol-assisted coherent multistage interference canceller using recursive channel estimation for DS-CDMA mobile radio," *IEICE Transactions on Communications*, vol. E79-B, pp. 1262–1269, September 1996.

[112] J. Torrance and L. Hanzo, "Comparative study of pilot symbol assisted modem schemes," in *Proceedings of IEE Conference on Radio Receivers and Associated Systems (RRAS'95)*, (Bath), pp. 36–41, IEE, 26–28 September 1995.

[113] J. Cavers, "An analysis of pilot symbol assisted modulation for Rayleigh fading channels," *IEEE Transactions on Vehicular Technology*, vol. 40, pp. 686–693, November 1991.

[114] S. Sampei and T. Sunaga, "Rayleigh fading compensation for QAM in land mobile radio communications," *IEEE Transactions on Vehicular Technology*, vol. 42, pp. 137–147, May 1993.

[115] *The 3GPP1 website.* http://www.3gpp.org.

[116] *The 3GPP2 website.* http://www.3gpp2.org.

[117] T. Ojanperä and R. Prasad, *Wideband CDMA for Third Generation Mobile Communications.* London: Artech House, 1998.

[118] E. Dahlman, B. Gudmundson, M. Nilsson and J. Sköld, "UMTS/IMT-2000 based on wideband CDMA," *IEEE Communications Magazine*, vol. 36, pp. 70–80, September 1998.

[119] T. Ojanperä, "Overview of research activities for third generation mobile communications," in Glisic and Leppanen [348], ch. 2 (Part 4), pp. 415–446. ISBN 0792380053.

[120] European Telecommunications Standards Institute, *The ETSI UMTS Terrestrial Radio Access (UTRA) ITU-R RTT Candidate Submission*, June 1998. ETSI/SMG/SMG2.

[121] Association of Radio Industries and Businesses, *Japan's Proposal for Candidate Radio Transmission Technology on IMT-2000: W-CDMA*, June 1998.

[122] F. Adachi, M. Sawahashi and H. Suda, "Wideband DS-CDMA for next-generation mobile communications systems," *IEEE Communications Magazine*, vol. 36, pp. 56–69, September 1998.

[123] F. Adachi and M. Sawahashi, "Wideband wireless access based on DS-CDMA," *IEICE Transactions on Communications*, vol. E81-B, pp. 1305–1316, July 1998.

[124] A. Sasaki, "Current situation of IMT-2000 radio transmission technology study in Japan," *IEICE Transactions on Communications*, vol. E81-B, pp. 1299–1304, July 1998.

[125] P. Baier, P. Jung and A. Klein, "Taking the challenge of multiple access for third-generation cellular mobile radio systems — a European view," *IEEE Communications Magazine*, vol. 34, pp. 82–89, February 1996.

[126] J. Schwarz da Silva, B. Barani and B. Arroyo-Fernández, "European mobile communications on the move," *IEEE Communications Magazine*, vol. 34, pp. 60–69, February 1996.

[127] F. Ovesjö, E. Dahlman, T. Ojanperä, A. Toskala and A. Klein, "FRAMES multiple access mode 2 — wideband CDMA," in *Proceedings of IEEE International Symposium on Personal, Indoor and Mobile Radio Communications, PIMRC'97* [558].

[128] *The UMTS Forum website.* http://www.umts-forum.org/.

[129] M. Sunay, Z.-C. Honkasalo, A. Hottinen, H. Honkasalo and L. Ma, "A dynamic channel allocation based TDD DS CDMA residential indoor system," in *IEEE 6th International Conference on Universal Personal Communications, ICUPC'97*, (San Diego, CA), pp. 228–234, October 1997.

[130] F. Adachi, M. Sawahashi and K. Okawa, "Tree-structured Generation of Orthogonal Spreading Codes with Different Lengths for Forward Link of DS-CDMA Mobile," *Electronics Letters*, vol. 33, no. 1, pp. 27–28, 1997.

[131] F. Adachi, K. Ohno, A. Higashi, T. Dohi and Y. Okumura, "Coherent multicode DS-CDMA mobile Radio Access," *IEICE Transactions on Communications*, vol. E79-B, pp. 1316–1324, September 1996.

[132] L. Hanzo, C. Wong and P. Cherriman, "Channel-adaptive wideband video telephony," *IEEE Signal Processing Magazine*, vol. 17, pp. 10–30, July 2000.

[133] P. Cherriman and L. Hanzo, "Programmable H.263-based wireless video transceivers for interference-limited environments," *IEEE Trans. on Circuits and Systems for Video Technology*, vol. 8, pp. 275–286, June 1998.

[134] C. Berrou and A. Glavieux, "Near optimum error correcting coding and decoding: turbo codes," *IEEE Transactions on Communications*, vol. 44, pp. 1261–1271, October 1996.

[135] A. Fujiwara, H. Suda and F. Adachi, "Turbo codes application to DS-CDMA mobile radio," *IEICE Transactions on Communications*, vol. E81A, pp. 2269–2273, November 1998.

[136] M. Juntti, "System concept comparison for multirate CDMA with multiuser detection," in *Proceedings of IEEE Vehicular Technology Conference (VTC'98)* [559], pp. 18–21.

[137] S. Ramakrishna and J. Holtzman, "A comparison between single code and multiple code transmission schemes in a CDMA system," in *Proceedings of IEEE Vehicular Technology Conference (VTC'98)* [559], pp. 791–795.

[138] M. K. Simon, J. K. Omura, R. A. Scholtz and B. K. Levitt, *Spread Spectrum Communications Handbook*. McGraw-Hill, 1994.

[139] T. Kasami, *Combinational Mathematics and its Applications*. University of North Carolina Press, 1969.

[140] A. Brand and A. Aghvami, "Multidimensional PRMA with prioritized Bayesian broadcast — a MAC strategy for multiservice traffic over UMTS," *IEEE Transactions on Vehicular Technology*, vol. 47, pp. 1148–1161, November 1998.

[141] R. Ormondroyd and J. Maxey, "Performance of low rate orthogonal convolutional codes in DS-CDMA," *IEEE Transactions on Vehicular Technology*, vol. 46, pp. 320–328, May 1997.

[142] A. Chockalingam, P. Dietrich, L. Milstein and R. Rao, "Performance of closed-loop power control in DS-CDMA cellular systems," *IEEE Transactions on Vehicular Technology*, vol. 47, pp. 774–789, August 1998.

[143] R. Gejji, "Forward-link-power control in CDMA cellular-systems," *IEEE Transactions on Vehicular Technology*, vol. 41, pp. 532–536, November 1992.

[144] K. Higuchi, M. Sawahashi and F. Adachi, "Fast cell search algorithm in DS-CDMA mobile radio using long spreading codes," in *Proceedings of IEEE VTC '97*, vol. 3, (Phoenix, Arizona, USA), pp. 1430–1434, IEEE, 4–7 May 1997.

[145] M. Golay, "Complementary series," *IRE Transactions on Information Theory*, vol. IT-7, pp. 82–87, 1961.

[146] V. Tarokh, H. Jafarkhani and A. Calderbank, "Space-time block codes from orthogonal designs," *IEEE Transactions on Information Theory*, vol. 45, pp. 1456–1467, May 1999.

[147] W. Lee, *Mobile Communications Engineering*. New York: McGraw-Hill, 2nd ed., 1997.

[148] H. Wong and J. Chambers, "Two-stage interference immune blind equaliser which exploits cyclostationary statistics," *Electronics Letters*, vol. 32, pp. 1763–1764, September 1996.

[149] C. Lee and R. Steele, "Effects of Soft and Softer Handoffs on CDMA System Capacity," *IEEE Transactions on Vehicular Technology*, vol. 47, pp. 830–841, August 1998.

[150] M. Gustafsson, K. Jamal and E. Dahlman, "Compressed mode techniques for inter-frequency measurements in a wide-band DS-CDMA system," in *Proceedings of IEEE International Symposium on Personal, Indoor and Mobile Radio Communications, PIMRC'97* [558], pp. 231–235.

[151] D. Knisely, S. Kumar, S. Laha and S. Nanda, "Evolution of wireless data services : IS-95 to cdma2000," *IEEE Communications Magazine*, vol. 36, pp. 140–149, October 1998.

[152] Telecommunications Industry Association (TIA), *The cdma2000 ITU-R RTT Candidate Submission*, 1998.

[153] D. Knisely, Q. Li and N. Rames, "cdma2000: A third generation radio transmission technology," *Bell Labs Technical Journal*, vol. 3, pp. 63–78, July–September 1998.

[154] Y. Okumura and F. Adachi, "Variable-rate data transmission with blind rate detection for coherent DS-CDMA mobile radio," *IEICE Transactions on Communications*, vol. E81B, pp. 1365–1373, July 1998.

[155] R. Steele, ed., *Mobile Radio Communications*. IEEE Press-Pentech Press, 1992.

[156] M. Raitola, A. Hottinen and R. Wichman, "Transmission diversity in wideband CDMA," in *Proceedings of VTC'98 (Spring)* [557], pp. 1545–1549.

[157] J. Liberti Jr. and T. Rappaport, "Analytical results for capacity improvements in CDMA," *IEEE Transactions on Vehicular Technology*, vol. 43, pp. 680–690, August 1994.

[158] J. Winters, "Smart antennas for wireless systems," *IEEE Personal Communications*, vol. 5, pp. 23–27, February 1998.

[159] T. Lim and L. Rasmussen, "Adaptive symbol and parameter estimation in asynchronous multiuser CDMA detectors," *IEEE Transactions on Communications*, vol. 45, pp. 213–220, February 1997.

[160] T. Lim and S. Roy, "Adaptive filters in multiuser (MU) CDMA detection," *Wireless Networks*, vol. 4, pp. 307–318, June 1998.

[161] L. Wei, "Rotationally-invariant convolutional channel coding with expanded signal space, part I and II," *IEEE Transactions on Selected Areas in Comms*, vol. SAC-2, pp. 659–686, September 1984.

[162] T. Lim and M. Ho, "LMS-based simplifications to the kalman filter multiuser CDMA detector," in *Proceedings of IEEE Asia-Pacific Conference on Communications/International Conference on Communication Systems*, (Singapore), November 1998.

[163] D. You and T. Lim, "A modified blind adaptive multiuser CDMA detector," in *Proceedings of IEEE International Symposium on Spread Spectrum Techniques and Application (ISSSTA'98)* [560], pp. 878–882.

[164] S. Sun, L. Rasmussen, T. Lim and H. Sugimoto, "Impact of estimation errors on multiuser detection in CDMA," in *Proceedings of IEEE Vehicular Technology Conference (VTC'98)* [559], pp. 1844–1848.

[165] Y. Sanada and Q. Wang, "A co-channel interference cancellation technique using orthogonal convolutional codes on multipath rayleigh fading channel," *IEEE Transactions on Vehicular Technology*, vol. 46, pp. 114–128, February 1997.

[166] P. Patel and J. Holtzman, "Analysis of a simple successive interference cancellation scheme in a DS/CDMA system," *IEEE Journal on Selected Areas in Communications*, vol. 12, pp. 796–807, June 1994.

[167] P. Tan and L. Rasmussen, "Subtractive interference cancellation for DS-CDMA systems," in *Proceedings of IEEE Asia-Pacific Conference on Communications/International Conference on Communication Systems*, (Singapore), November 1998.

[168] K. Cheah, H. Sugimoto, T. Lim, L. Rasmussen and S. Sun, "Performance of hybrid interference canceller with zero-delay channel estimation for CDMA," in *Proceedings of Globecom'98*, (Sydney, Australia), pp. 265–270, IEEE, 8–12 November 1998.

[169] S. Sun, L. Rasmussen and T. Lim, "A matrix-algebraic approach to hybrid interference cancellation in CDMA," in *Proceedings of IEEE International Conference on Universal Personal Communications '98*, (Florence, Italy), pp. 1319–1323, October 1998.

[170] A. Johansson and L. Rasmussen, "Linear group-wise successive interference cancellation in CDMA," in *Proceedings of IEEE International Symposium on Spread Spectrum Techniques and Application (ISSSTA'98)* [560], pp. 121–126.

[171] S. Sun, L. Rasmussen, H. Sugimoto and T. Lim, "A hybrid interference canceller in CDMA," in *Proceedings of IEEE International Symposium on Spread Spectrum Techniques and Application (ISSSTA'98)* [560], pp. 150–154.

[172] D. Guo, L. Rasmussen, S. Sun, T. Lim and C. Cheah, "MMSE-based linear parallel interference cancellation in CDMA," in *Proceedings of IEEE International Symposium on Spread Spectrum Techniques and Application (ISSSTA'98)* [560], pp. 917–921.

[173] L. Rasmussen, D. Guo, Y. Ma and T. Lim, "Aspects on linear parallel interference cancellation in CDMA," in *Proceedings of IEEE International Symposium on Information Theory*, (Cambridge, MA), p. 37, August 1998.

[174] L. Rasmussen, T. Lim, H. Sugimoto and T. Oyama, "Mapping functions for successive interference cancellation in CDMA," in *Proceedings of IEEE Vehicular Technology Conference (VTC'98)* [559], pp. 2301–2305.

[175] S. Sun, T. Lim, L. Rasmussen, T. Oyama, H. Sugimoto and Y. Matsumoto, "Performance comparison of multi-stage SIC and limited tree-search detection in CDMA," in *Proceedings of IEEE Vehicular Technology Conference (VTC'98)* [559], pp. 1854–1858.

[176] H. Sim and D. Cruickshank, "Chip based multiuser detector for the downlink of a DS-CDMA system using a folded state-transition trellis," in *Proceedings of VTC'98 (Spring)* [557], pp. 846–850.

[177] H. Holma and A. Toskala, *WCDMA for UMTS*. John Wiley and Sons, 2001.

[178] 3GPP, *Physical Layer Procedures (FDD)*. TS 25.214.

[179] 3GPP, *Medium Access Control (MAC) protocol specification*. TS 25.321.

[180] 3GPP, *UE Radio Access Capability*. TS 25.306.

[181] D. Chase, "Code combining: a maximum-likelihood decoding approach for combining an arbitrary number of noisy packets," *IEEE Transactions on Communications*, vol. 33, pp. 593–607, 1985.

[182] L. Hanzo, S.-X. Ng, T. Keller and W. Webb, *Quadrature Amplitude Modulation: From Basics to Adaptive Trellis-Coded, Turbo-Equalised and Space-Time Coded OFDM, CDMA and MC-CDMA Systems*. John Wiley and IEEE Press, 2004.

[183] 3GPP, *Multiplexing and channel coding (FDD)*. TS 25.212.

[184] L. Hanzo, M. Münster, B.-J. Choi and T. Keller, *OFDM versus MC-CDMA for Broadband Multi-User Communications, WLANs and Broadcasting*. John Wiley and IEEE press, 2003.

[185] 3GPP, *Physical channels and mapping of transport channels onto physical channels (FDD)*. TS 25.211.

[186] 3GPP TSG-RAN WG1 No. 25, *Performance of HS-SCCH*. R1-02-0649.

[187] H. Yamamoto and K. Itoh, "Viterbi decoding algorithm for convolutional codes with repeat request," *IEEE Transaction on Information Theory*, vol. IT-26, pp. 540–547, September 1980.

[188] 3GPP, *User Equipment (UE) radio transmission and reception (FDD)*. TS 25.101.

[189] M. Harteneck, M. Boloorian, S. Georgoulis and R. Tanner, "Practical aspects of an HSDPA 14 Mbps terminal," *Signals, Systems and Computers*, vol. 1, pp. 799–803, 7–10 November 2004.

[190] M. Harteneck and C. Luschi, "Practical implementation aspects of MMSE equalisation in a 3GPP HSDPA terminal," *VTC Spring*, vol. 1, pp. 445–449, 17–19 May 2004.

[191] J. Blogh and L. Hanzo, *3G Systems and Intelligent Networking*. John Wiley and IEEE Press, 2002. (For detailed contents please refer to http://www-mobile.ecs.soton.ac.uk.)

[192] L. Hanzo, P. Cherriman and J. Streit, *Video Compression and Communications over Wireless Channels: From Second to Third Generation Systems and Beyond*. IEEE Press and John Wiley, 2001. (For detailed contents please refer to http://www-mobile.ecs.soton.ac.uk.)

[193] L. Hanzo, "Bandwidth-efficient wireless multimedia communications," *Proceedings of the IEEE*, vol. 86, pp. 1342–1382, July 1998.

[194] S. Nanda, K. Balachandran and S. Kumar, "Adaptation techniques in wireless packet data services," *IEEE Communications Magazine*, vol. 38, pp. 54–64, January 2000.

[195] Research and Development Centre for Radio Systems, Japan, *Public Digital Cellular (PDC) Standard, RCR STD-27*.

[196] Telcomm. Industry Association (TIA), Washington, DC, USA, *Dual-mode subscriber equipment — Network equipment compatibility specification, Interim Standard IS-54*, 1989.

[197] Telcomm. Industry Association (TIA), Washington, DC, USA, *Mobile station — Base station compatibility standard for dual-mode wideband spread spectrum cellular system, EIA/TIA Interim Standard IS-95*, 1993.

[198] T. Ojanperä and R. Prasad, *Wideband CDMA for Third Generation Mobile Communications*. Artech House, Inc., 1998.

[199] W. Webb and R. Steele, "Variable rate QAM for mobile radio," *IEEE Transactions on Communications*, vol. 43, no. 7, pp. 2223–2230, 1995.

[200] S. Sampei, S. Komaki and N. Morinaga, "Adaptive Modulation/TDMA scheme for large capacity personal multimedia communications systems," *IEICE Transactions on Communications*, vol. E77-B, pp. 1096–1103, September 1994.

[201] J. M. Torrance and L. Hanzo, "Upper bound performance of adaptive modulation in a slow Rayleigh fading channel," *Electronics Letters*, vol. 32, pp. 718–719, 11 April 1996.

[202] C. Wong and L. Hanzo, "Upper-bound performance of a wideband burst-by-burst adaptive modem," *IEEE Transactions on Communications*, vol. 48, pp. 367–369, March 2000.

[203] J. M. Torrance and L. Hanzo, "Optimisation of switching levels for adaptive modulation in a slow Rayleigh fading channel," *Electronics Letters*, vol. 32, pp. 1167–1169, 20 June 1996.

[204] H. Matsuoka, S. Sampei, N. Morinaga and Y. Kamio, "Adaptive modulation system with variable coding rate concatenated code for high quality multi-media communications systems," in *Proceedings of IEEE VTC'96* [561], pp. 487–491.

[205] A. J. Goldsmith and S. G. Chua, "Variable Rate Variable Power MQAM for Fading Channels," *IEEE Transactions on Communications*, vol. 45, pp. 1218–1230, October 1997.

[206] S. Otsuki, S. Sampei and N. Morinaga, "Square-qam adaptive modulation/TDMA/TDD systems using modulation estimation level with walsh function," *IEE Electronics Letters*, vol. 31, pp. 169–171, February 1995.

[207] J. Torrance and L. Hanzo, "Demodulation level selection in adaptive modulation," *Electronics Letters*, vol. 32, pp. 1751–1752, 12 September 1996.

[208] Y. Kamio, S. Sampei, H. Sasaoka and N. Morinaga, "Performance of modulation-level-control adaptive-modulation under limited transmission delay time for land mobile communications," in *Proceedings of IEEE Vehicular Technology Conference (VTC'95)*, (Chicago, USA), pp. 221–225, IEEE, 15–28 July 1995.

[209] J. M. Torrance and L. Hanzo, "Latency and Networking Aspects of Adaptive Modems over Slow Indoors Rayleigh Fading Channels," *IEEE Transactions on Vehicular Technology*, vol. 48, pp. 1237–1251, July 1999.

[210] T. Ue, S. Sampei and N. Morinaga, "Symbol rate controlled adaptive modulation/TDMA/TDD for wireless personal communication systems," *IEICE Transactions on Communications*, vol. E78-B, pp. 1117–1124, August 1995.

[211] T. Suzuki, S. Sampei and N. Morinaga, "Space and path diversity combining technique for 10 Mbits/s adaptive modulation/TDMA in wireless communications systems," in *Proceedings of IEEE VTC '96* [561], pp. 1003–1007.

[212] K. Arimochi, S. Sampei and N. Morinaga, "Adaptive modulation system with discrete power control and predistortion-type non-linear compensation for high spectral efficient and high power efficient wireless communication systems," in *Proceedings of IEEE International Symposium on Personal, Indoor and Mobile Radio Communications, PIMRC'97* [558], pp. 472–477.

[213] T. Ikeda, S. Sampei and N. Morinaga, "TDMA-based adaptive modulation with dynamic channel assignment (AMDCA) for high capacity multi-media microcellular systems," in *Proceedings of IEEE Vehicular Technology Conference*, (Phoenix, USA), pp. 1479–1483, May 1997.

[214] T. Ue, S. Sampei and N. Morinaga, "Adaptive modulation packet radio communication system using NP-CSMA/TDD scheme," in *Proceedings of IEEE VTC '96* [561], pp. 416–421.

[215] M. Naijoh, S. Sampei, N. Morinaga and Y. Kamio, "ARQ schemes with adaptive modulation/TDMA/TDD systems for wireless multimedia communication systems," in *Proceedings of IEEE International Symposium on Personal, Indoor and Mobile Radio Communications, PIMRC'97* [558], pp. 709–713.

[216] S. Sampei, T. Ue, N. Morinaga and K. Hamguchi, "Laboratory experimental results of an adaptive modulation TDMA/TDD for wireless multimedia communication systems," in *Proceedings of IEEE International Symposium on Personal, Indoor and Mobile Radio Communications, PIMRC'97* [558], pp. 467–471.

[217] J. Torrance and L. Hanzo, "Interference aspects of adaptive modems over slow Rayleigh fading channels," *IEEE Transactions on Vehicular Technology*, vol. 48, pp. 1527–1545, September 1999.

[218] L. Hanzo, T. H. Liew and B. L. Yeap, *Turbo Coding, Turbo Equalisation and Space-Time Coding*. John Wiley, IEEE Press, 2002. (For detailed contents please refer to http://www-mobile.ecs.soton.ac.uk.)

[219] J. Cheung and R. Steele, "Soft-decision feedback equalizer for continuous-phase modulated signals in wide-band mobile radio channels," *IEEE Transactions on Communications*, vol. 42, pp. 1628–1638, February/March/April 1994.

[220] M. Yee, T. Liew and L. Hanzo, "Radial basis function decision feedback equalisation assisted block turbo burst-by-burst adaptive modems," in *Proceedings of VTC '99 Fall*, (Amsterdam, Holland), pp. 1600–1604, 19–22 September 1999.

[221] M. S. Yee, B. L. Yeap and L. Hanzo, "Radial basis function assisted turbo equalisation," in *Proceedings of IEEE Vehicular Technology Conference*, (Japan, Tokyo), pp. 640–644, IEEE, 15–18 May 2000.

[222] L. Hanzo, F. Somerville and J. Woodard, *Voice Compression and Communications: Principles and Applications for Fixed and Wireless Channels.* IEEE Press and John Wiley, 2002. (For detailed contents, please refer to http://www-mobile.ecs.soton.ac.uk.)

[223] ITU-T, *Recommendation H.263: Video coding for low bitrate communication*, March 1996.

[224] A. Klein, R. Pirhonen, J. Skoeld and R. Suoranta, "FRAMES multiple access mode 1 — wideband TDMA with and without spreading," in *Proceedings of IEEE International Symposium on Personal, Indoor and Mobile Radio Communications, PIMRC'97* [558], pp. 37–41.

[225] P. Cherriman, C. Wong and L. Hanzo, "Turbo- and BCH-coded wide-band burst-by-burst adaptive H.263-assisted wireless video telephony," *IEEE Transactions on Circuits and Systems for Video Technology*, vol. 10, pp. 1355–1363, December 2000.

[226] T. Keller and L. Hanzo, "Adaptive multicarrier modulation: A convenient framework for time-frequency processing in wireless communications," *Proceedings of the IEEE*, vol. 88, pp. 611–642, May 2000.

[227] A. Klein and P. Baier, "Linear unbiased data estimation in mobile radio systems applying CDMA," *IEEE Journal on Selected Areas in Communications*, vol. 11, pp. 1058–1066, September 1993.

[228] K. Gilhousen, I. Jacobs, R. Padovani, A. Viterbi, L. Weaver Jr. and C. Wheatley III, "On the capacity of a cellular CDMA system," *IEEE Transactions on Vehicular Technology*, vol. 40, pp. 303–312, May 1991.

[229] S. Kim, "Adaptive rate and power DS/CDMA communications in fading channels," *IEEE Communications Letters*, vol. 3, pp. 85–87, April 1999.

[230] T. Ottosson and A. Svensson, "On schemes for multirate support in DS-CDMA systems," *Wireless Personal Communications (Kluwer)*, vol. 6, pp. 265–287, March 1998.

[231] S. Ramakrishna and J. Holtzman, "A comparison between single code and multiple code transmission schemes in a CDMA system," in *Proceedings of IEEE Vehicular Technology Conference (VTC'98)* [559], pp. 791–795.

[232] M. Saquib and R. Yates, "Decorrelating detectors for a dual rate synchronous DS/CDMAchannel," *Wireless Personal Communications (Kluwer)*, vol. 9, pp. 197–216, May 1999.

[233] A.-L. Johansson and A. Svensson, "Successive interference cancellation schemes in multi-rate DS/CDMA systems," in *Wireless Information Networks (Baltzer)*, pp. 265–279, 1996.

[234] S. Abeta, S. Sampei and N. Morinaga, "Channel activation with adaptive coding rate and processing gain control for cellular DS/CDMA systems," in *Proceedings of IEEE VTC '96* [561], pp. 1115–1119.

[235] M. Hashimoto, S. Sampei and N. Morinaga, "Forward and reverse link capacity enhancement of DS/CDMA cellular system using channel activation and soft power control techniques," in *Proceedings of IEEE International Symposium on Personal, Indoor and Mobile Radio Communications, PIMRC'97* [558], pp. 246–250.

[236] T. Liew, C. Wong and L. Hanzo, "Block turbo coded burst-by-burst adaptive modems," in *Proceedings of Microcoll'99, Budapest, Hungary*, pp. 59–62, 21–24 March 1999.

[237] L. Hanzo, P. Cherriman and J. Streit, *Wireless Video Communications: From Second to Third Generation Systems, WLANs and Beyond.* IEEE Press-John Wiley, 2001. IEEE Press, 2001. (For detailed contents please refer to http://www-mobile.ecs.soton.ac.uk.)

[238] V. Tarokh, N. Seshadri and A. Calderbank, "Space-time codes for high data rate wireless communication: Performance criterion and code construction," *IEEE Transactions on Information Theory*, vol. 44, pp. 744–765, March 1998.

[239] V. Tarokh, A. Naguib, N. Seshadri and A. Calderbank, "Space-time codes for high data rate wireless communication: Performance criteria in the presence of channel estimation errors, mobility, and multiple paths," *IEEE Transactions on Communications*, vol. 47, pp. 199–207, February 1999.

[240] V. Tarokh, N. Seshadri and A. Calderbank, "Space-time codes for high data rate wireless communications: Performance criterion and code construction," in *Proc IEEE International Conference on Communications'97*, (Montreal, Canada), pp. 299–303, 1997.

[241] V. Tarokh, H. Jafarkhani and A. Calderbank, "Space-time block codes from orthogonal designs," *IEEE Transactions on Information Theory*, vol. 45, pp. 1456–1467, July 1999.

[242] T. Rappaport, ed., *Smart Antennas: Adaptive Arrays, Algorithms and Wireless Position Location.* IEEE, 1998.

[243] B. Widrow, P. E. Mantey, L. J. Griffiths and B. B. Goode, "Adaptive Antenna Systems," in *Proceedings of the IEEE*, vol. 55, pp. 2143–2159, December 1967.

[244] S. P. Applebaum, "Adaptive Arrays," *IEEE Transactions on Antennas and Propagation*, vol. AP-24, pp. 585–598, September 1976.

[245] O. L. Frost III, "An Algorithm for Linearly Constrained Adaptive Array Processing," in *Proceedings of the IEEE*, vol. 60, pp. 926–935, August 1972.

[246] I. S. Reed, J. D. Mallett and L. E. Brennan, "Rapid Convergence Rate in Adaptive Arrays," *IEEE Transactions on Aerospace Electronic System*, vol. AES-10, pp. 853–863, November 1974.

[247] J. Fernandez, I. R. Corden and M. Barrett, "Adaptive Array Algorithms for Optimal Combining in Digital Mobile Communications Systems," *IEE 8th International Conference on Antennas and Propagation*, pp. 983–986, 1993.

[248] L. C. Godara, "Applications of Antenna Arrays to Mobile Communications, Part I: Performance Improvement, Feasibility, and System Considerations," in *Proceedings of the IEEE*, vol. 85, pp. 1031–1060, July 1997.

[249] L. C. Godara, "Application of Antenna Arrays to Mobile Communications, Part II: Beam-Forming and Direction-of-Arrival Considerations," in *Proceedings of the IEEE*, vol. 85, pp. 1195–1245, August 1997.

[250] W. F. Gabriel, "Adaptive Processing Array Systems," in *Proceedings of the IEEE*, vol. 80, pp. 152–162, January 1992.

[251] A. J. Paulraj and C. B. Papadias, "Space Time Processing for Wireless Communications," *IEEE Personal Communications*, vol. 14, pp. 49–83, November 1997.

[252] J. H. Winters, "Smart Antennas for Wireless Systems," *IEEE Personal Communications*, vol. 1, pp. 23–27, February 1998.

[253] R. Kohno, "Spatial and Temporal Communication Theory Using Adaptive Antenna Array," *IEEE Personal Communications*, vol. 1, pp. 28–35, February 1998.

[254] H. Krim and M. Viberg, "Two Decades of Array Signal Processing Research," *IEEE Signal Processing Magazine*, pp. 67–94, July 1996.

[255] G. V. Tsoulos, "Smart Antennas for Mobile Communication Systems: Benefits and Challenges," *IEE Electronics and Communication Engineering Journal*, vol. 11, pp. 84–94, April 1999.

[256] Special Issue on Active and Adaptive Antennas, *IEEE Transactions on Antennas and Propagation*, vol. AP-12, March 1964.

[257] Special Issue on Adaptive Antennas, *IEEE Transaction on Antennas and Propagation*, vol. AP-24, September 1976.

[258] Special Issue on Adaptive Antennas, *IEEE Transaction on Antennas and Propagation*, vol. AP-34, March 1986.

[259] A. Paulraj, R. Roy and T. Kailath, "A Subspace Rotation Approach to Signal Parameter Estimation," in *Proceedings of the IEEE*, vol. 74, pp. 1044–1046, July 1986.

[260] J. H. Winters, "Signal Acquisition and Tracking with Adaptive Arrays in the Digital Mobile Radio System IS-54 with Flat Fading," *IEEE Transactions on Vehicular Technology*, vol. 42, pp. 373–384, November 1993.

[261] L. C. Godara and D. B. Ward, "A General Framework for Blind Beamforming," *IEEE TENCON*, pp. 1240–1243, June 1999.

[262] W. Pora, J. A. Chambers and A. G. Constantinides, "A Combined Kalman Filter and Constant Modulus Algorithm Beamformer for Fast Fading Channels," *IEEE International Conference on Acoustics, Speech and Signal Processing*, vol. 5, pp. 2925–2928, March 1999.

[263] J. H. Winters, J. Salz and R. D. Gitlin, "The Impact of Antenna Diversity on the Capacity of Wireless Communications System," *IEEE Transactions on Communications*, vol. 42, February/March/April 1994.

[264] G. V. Tsoulos, M. A. Beach and S. C. Swales, "Adaptive Antennas for Third Generation DS-CDMA Cellular Systems," *IEEE Vehicular Technology Conference*, vol. 45, pp. 45–49, 1995.

[265] L. E. Brennan and I. S. Reed, "Theory of Adaptive Radar," *IEEE Transactions on Aerospace and Electronic Systems*, vol. AES-9, pp. 237–252, March 1973.

[266] L. E. Brennan, E. L. Pugh and I. S. Reed, "Control Loop Noise in Adaptive Array Antennas," *IEEE Transactions on Aerospace and Electronic Systems*, vol. AES-7, pp. 254–262, March 1971.

[267] L. E. Brennan and I. S. Reed, "Effect of Envelope Limiting in Adaptive Array Control Loops," *IEEE Transactions on Aerospace and Electronic Systems*, vol. AES-7, pp. 698–700, July 1971.

[268] L. E. Brennan, J. Mallet and I. S. Reed, "Adaptive Arrays in Airborne MTI Radar," *IEEE Transactions on Antennas and Propagation*, vol. AP-24, pp. 607–615, September 1976.

[269] B. Widrow and J. M. McCool, "A Comparison of Adaptive Algorithms Based on the Method of Steepest Descent and Random Search," *IEEE Transactions on Antennas and Propagation*, vol. AP-24, pp. 615–637, September 1976.

[270] R. O. Schmidt, "Multiple Emitter Location and Signal Parameter Estimation," *IEEE Transactions on Antennas Propagation*, vol. AP-34, pp. 276–280, July 1986.

[271] R. Roy and T. Kailath, "ESPRIT-Estimation of Signal Parameters via Rotational Invariance Techniques," *IEEE Transactions on Acoustic, Speech and Signal Processing*, vol. ASSP-37, pp. 984–995, July 1989.

[272] L. J. Griffiths, "A Simple Adaptive Algorithm for Real-time Processing in Antenna Arrays," in *Proceedings of the IEEE*, vol. 57, pp. 1696–1704, October 1969.

[273] B. D. Van Veen and K. M. Buckley, "Beamforming: A Versatile Approach to Spatial Filtering," *IEEE Acoustic, Speech and Signal Processing Magazine*, pp. 4–24, April 1988.

[274] T. Chen, "Highlights of Statistical Signal and Array Processing," *IEEE Signal Processing Magazine*, vol. 15, pp. 21–64, September 1998.

[275] C. Q. Xu, C. L. Law and S. Yoshida, "On Nonlinear Beamforming for Interference Cancellation," in *IEEE Vehicular Technology Conference*, May 2001.

[276] A. Margarita, S. J. Flores, L. Rubio, V. Almenar and J. L. Corral, "Application of MUSIC for Spatial Reference Beamforming for SDMA in a Smart Antenna for GSM and DECT," *IEEE Vehicular Technology Conference*, May 2001.

[277] J. E. Hudson, *Adaptive Array Principles*. New York: Peter Peregrinus, Ltd., 1981.

[278] S. Haykin, *Array Signal Processing*. NJ: Prentice-Hall, Inc., 1985.

[279] S. P. Applebaum, "Adaptive Arrays," tech. rep., Syracuse University Research Corporation, 1965. Reprinted in IEEE Transactions on Antennas and Propagation, September 1976.

[280] B. Widrow, P. E. Mantey, L. J. Griffiths and B. B. Goode, "Adaptive Antenna Systems," *Proceedings of the IEEE*, vol. 55, pp. 2143–2159, December 1967.

[281] O. L. Frost III, "An Algorithm for Linearly Constrained Adaptive Array Processing," *Proceedings of the IEEE*, vol. 60, pp. 926–935, August 1972.

[282] L. J. Griffiths, "A Simple Adaptive Algorithm for Real-Time Processing in Antenna Arrays," *Proceedings of the IEEE*, vol. 57, pp. 1696–1704, October 1969.

[283] L. C. Godara, "Applications of Antenna Arrays to Mobile Communications, Part II: Beam-Forming and Direction-of-Arrival Considerations," *Proceedings of the IEEE*, vol. 85, pp. 1193–1245, August 1997.

[284] J. Capon, "High-resolution frequency-wavenumber spectrum analysis," *Proceedings of the IEEE*, vol. 57, pp. 1408–1418, August 1969.

[285] I. S. Reed, J. D. Mallett and L. E. Brennan, "Rapid Convergence Rate in Adaptive Arrays," *IEEE Transactions on Aerospace and Electronic Systems*, vol. AES-10, pp. 853–863, November 1974.

[286] A. Paulraj and C. B. Papadias, "Space-Time Processing for Wireless Communications," *IEEE Signal Processing Magazine*, pp. 49–83, November 1997.

[287] J. E. Hudson, *Adaptive Array Principles*. Peregrinus, London, 1981.

[288] S. Haykin, *Adaptive Filter Theory*. Prentice-Hall International, 1991.

[289] B. Widrow and S. Steams, *Adaptive Signal Processing*. Prentice-Hall, 1985.

[290] R. A. Monzingo and T. W. Miller, *Introduction to Adaptive Arrays*. John Wiley & Sons, Inc., 1980.

[291] J. H. Winters, "Smart Antennas for Wireless Systems," *IEEE Personal Communications*, vol. 5, pp. 23–27, February 1998.

[292] U. Martin and I. Gaspard, "Capacity Enhancement of Narrowband CDMA by Intelligent Antennas," in *Proceedings of PIMRC*, pp. 90–94, 1997.

[293] A. R. Lopez, "Performance Predictions for Cellular Switched-Beam Intelligent Antenna Systems," *IEEE Communications Magazine*, pp. 152–154, October 1996.

[294] C. M. Simmonds and M. A. Beach, "Active Calibration of Adaptive Antenna Arrays for Third Generation Systems," in *Proceedings of ACTS Summit*, 1997.

[295] H. Steyskal, "Digital Beamforming Antennas: An Introduction," *Microwave Journal*, pp. 107–124, January 1987.

[296] A. Mammela, *Diversity receivers in a fast fading multipath channel*. VTT Publications, 1995.

[297] W. Hollemans, "Performance Analysis of Cellular Digital Mobile Radio Systems including Diversity Techniques," in *Proceedings of PIMRC*, pp. 266–270, 1997.

[298] W. Tuttlebee, ed., *Cordless telecommunications in Europe: The Evolution of Personal Communications*. London: Springer-Verlag, 1990. ISBN 3540196331.

[299] H. Ochsner, "The digital European cordless telecommunications specification, DECT," in Tuttlebee [298], pp. 273–285. ISBN 3540196331.

[300] P. Petrus, J. H. Reed and T. S. Rappaport, "Effects of Directional Antennas at the Base Station on the Doppler Spectrum," *IEEE Communications Letters*, vol. 1, pp. 40–42, March 1997.

[301] H. Krim and M. Viberg, "Two Decades of Array Signal Processing Research," *IEEE Signal Processing Magazine*, pp. 67–94, July 1996.

[302] S. Ponnekanti, A. Pollard, C. Taylor and M. G. Kyeong, "Flexibility for the deployment of adaptive antennas in the IMT-2000 framework and enhanced interference cancellation," in *Proceedings of ACTS Summit*, 1997.

[303] J. H. Winters, J. Salz and R. D. Gitlin, "The Impact of Antenna Diversity on the Capacity of Wireless Communication Systems," *IEEE Transactions on Communications*, vol. 42, pp. 1740–1751, February/March/April 1994.

[304] M. Barnard and S. McLaughlin, "Reconfigurable terminals for mobile communication systems," *IEE Electronics and Communication Engineering Journal*, vol. 12, pp. 281–292, December 2000.

[305] S. M. Leach, A. A. Agius and S. R. Saunders, "The intelligent quadrifilar helix antenna," *IEE Proceedings of Microwave Antennas Propagation*, pp. 219–223, June 2000.

[306] P. Petrus, R. B. Ertel and J. H. Reed, "Capacity Enhancement Using Adaptive Arrays in an AMPS System," *IEEE Transactions on Vehicular Technology*, vol. 47, pp. 717–727, August 1998.

[307] J. Laurila and E. Bonek, "SDMA Using Blind Adaption," in *Proceedings of ACTS Summit*, 1997.

[308] M. C. Wells, "Increasing the capacity of GSM cellular radio using adaptive antennas," *IEE Proceedings on Communications*, vol. 143, pp. 304–310, October 1996.

[309] J. H. Winters, "Signal Acquisition and Tracking with Adaptive Arrays in the Digital Mobile Radio System IS-54 with Flat Fading," *IEEE Transactions on Vehicular Technology*, vol. 42, November 1993.

[310] B. Widrow and E. Walach, "On the statistical efficiency of the LMS algorithm with nonstationary inputs," *IEEE Trans. Information Theory - Special Issue on Adaptive Filtering*, vol. 30, pp. 211–221, March 1984.

[311] Z. Raida, "Steering an Adaptive Antenna Array by the Simplified Kalman Filter," *IEEE Trans. on Antennas and Propagation*, vol. 43, pp. 627–629, June 1995.

[312] M. W. Ganz, R. L. Moses and S. L. Wilson, "Convergence of the SMI Algorithms with Weak Interference," *IEE Trans. Antenna Propagation*, vol. 38, pp. 394–399, March 1990.

[313] H. Steyskal, "Array Error Effects in Adaptive Beamforming," *Microwave Journal*, September 1991.

[314] M. C. Vanderveen, C. B. Papadias and A. Paulraj, "Joint Angle and Delay Estimation (JADE) for Multipath Signals Arriving at an Antenna Array," *IEEE Communications Letters*, vol. 1, pp. 12–14, January 1997.

[315] C. Passman and T. Wixforth, "A Calibrated Phased Array Antenna with Polarization Flexibility for the Tsunami (II) SDMA Field Trial," in *Proceedings of ACTS Summit*, 1997.

[316] D. N. Godard, "Self-Recovering Equalization and Carrier Tracking in Two-Dimensional Data Communication Systems," *IEEE Transactions on Communications*, vol. COM-28, pp. 1876–1875, November 1980.

[317] Z. Ding, R. A. Kennedy, B. D. O. Anderson and C. R. Johnson Jr, "Ill-Convergence of Godard Blind Equalizers in Data Communication Systems," *IEEE Transactions on Communications*, vol. 39, pp. 1313–1327, September 1991.

[318] J. E. Mazo, "Analysis of decision-directed equalizer convergence," *Bell Systems Technical Journal*, 1980.

[319] D. Gerlach and A. Paulraj, "Adaptive Transmitting Antenna Arrays with Feedback," *IEEE Signal Processing Letters*, vol. 1, pp. 150–152, October 1994.

[320] D. Gerlach and A. Paulraj, "Base station transmitting antenna arrays for multipath environments," *Signal Processing*, pp. 59–73, 1996.

[321] T. Kanai, "Autonomous Reuse Partitioning in Cellular Systems," in *IEEE Proceedings of Vehicular Technology Conference*, vol. 2, pp. 782–785, 1992.

[322] I. Katzela and M. Naghshineh, "Channel Assignment Schemes for Cellular Mobile Telecommunication Systems: A Comprehensive Survey," *IEEE Personal Communications Magazine*, vol. 3, pp. 10–31, June 1996.

[323] M. Dell'Anna and A. H. Aghvami, "Performance of optimum and sub-optimum combining at the antenna array of a W-CDMA system," *IEEE Journal on Selected Areas in Communications*, pp. 2123–2137, December 1999.

[324] I. Howitt and Y. M. Hawwar, "Evaluation of Outage Probability Due to Cochannel Interference in Fading for a TDMA System with a Beamformer," in *Proceedings of VTC*, pp. 520–524, 1998.

[325] L. Ortigoza-Guerrero and A. H. Aghvami, "A self-adaptive prioritised hand-off DCA strategy for a microcellular environment," in *Proceedings of PIMRC*, (Helsinki, Finland), pp. 401–405, September 1997.

[326] L. Ortigoza-Guerrero and A. H. Aghvami, "A prioritised hand-off dynamic channel allocation strategy for PCS," *IEEE Transactions on Vehicular Technology*, pp. 1203–1215, July 1999.

[327] T. H. Le and H. Aghvami, "Fast channel access and DCA scheme for connection and connectionless-oriented services in UMTS," *Electronics Letters*, pp. 1048–1049, June 1999.

[328] L. Anderson, "A simulation study of some dynamic channel assignment algorithms in a high capacity mobile telecommunications system," *IEEE Trans. on Communication*, vol. 21, pp. 1294–1301, November 1973.

[329] J. I. Chuang, "Performance issues and algorithms for dynamic channel assignment," *IEEE JSAC*, vol. 11, pp. 955–963, August 1993.

[330] J. I. Chuang and N. Sollenberger, "Performance of autonomous dynamic channel assignment and power control for TDMA/FDMA wireless access," *IEEE JSAC*, vol. 12, pp. 1314–1323, October 1994.

[331] M. L. Cheng and J. I. Chuang, "Performance evaluation of distributed measurement-based dynamic channel assignment in local wireless communications," *IEEE JSAC*, vol. 14, pp. 698–710, May 1996.

[332] Chihlin I and C. PiHui, "Local packing - distributed dynamic channel allocation at cellular base station," in *Proceedings of IEEE Globecom '93*, vol. 1, (Houston, TX, USA), pp. 293–301, November 29–December 2 1993.

[333] G. L. Stüber, *Principles of Mobile Communication*. Kluwer Academic Publishers, 1996.

[334] A. Baiocchi, F. Delli-Priscoli, F. Grilli and F. Sestini, "The geometric dynamic channel allocation as a practical strategy in mobile networks with bursty user mobility," *IEEE Trans. on Vech. Tech.*, vol. 44, pp. 14–23, February 1995.

[335] F. D. Priscoli, N. P. Magnani, V. Palestini and F. Sestini, "Application of dynamic channel allocation strategies to the GSM cellular network," *IEEE Journal on Selected Areas in Comms.*, vol. 15, pp. 1558–1567, October 1997.

[336] P. J. Cherriman, F. Romiti and L. Hanzo, "Channel Allocation for Third-generation Mobile Radio Systems," in *ACTS '98, Rhodes, Greece*, pp. 255–260, June 1998.

[337] R. B. Ertel, P. Cardieri, K. W. Sowerby, T. S. Rappaport and J. H. Reed, "Overview of Spatial Channel Models for Antenna Array Communications Systems," *IEEE Personal Communications*, pp. 10–22, February 1998.

[338] J. C. Liberti and T. S. Rappaport, "A Geometrically Based Model for Line-Of-Sight Multipath Radio Channels," in *VTC Proceedings*, pp. 844–848, 1996.

[339] S. W. Wales, "Technique for cochannel interference suppression in TDMA mobile radio systems," *IEE Proc. Communication*, vol. 142, no. 2, pp. 106–114, 1995.

[340] J. Litva and T. Lo, *Digital Beamforming in Wireless Communications*. Artech House, London, 1996.

[341] L. Godara, "Applications of antenna arrays to mobile communications, part I: Performance improvement, feasibility, and system considerations," *Proceedings of the IEEE*, vol. 85, pp. 1029–1060, July 1997.

[342] L. Godara, "Applications of antenna arrays to mobile communications, part II: Beam-forming and direction-of-arrival considerations," *Proceedings of the IEEE*, vol. 85, pp. 1193–1245, August 1997.

[343] E. Sourour, "Time slot assignment techniques for TDMA digital cellular systems," *IEEE Trans. Vech. Tech.*, vol. 43, pp. 121–127, February 1994.

[344] D. Wong and T. Lim, "Soft handoffs in CDMA mobile systems," *IEEE Personal Comms.*, pp. 6–17, December 1997.

[345] S. Tekinay and B. Jabbari, "A measurement-based prioritisation scheme for handovers in mobile cellular networks," *IEEE JSAC*, vol. 10, no. 8, pp. 1343–1350, 1992.

[346] G. P. Pollini, "Trends in handover design," *IEEE Comms. Mag.*, vol. 34, pp. 82–90, March 1996.

[347] R. C. Bernhardt, "Timeslot re-assignment in a frequency reuse TDMA portable radio system," *IEEE Tr. on Vech. Tech.*, vol. 41, pp. 296–304, August 1992.

[348] S. Glisic and P. Leppanen, eds., *Wireless Communications : TDMA versus CDMA*. Kluwer Academic Publishers, June 1997. ISBN 0792380053.

[349] A. H. M. Ross and K. S. Gilhousen, "CDMA technology and the IS-95 north American standard," in Gibson [562], ch. 27, pp. 430–448.

[350] ETSI, *Universal Mobile Telecommunications Systems (UMTS); UMTS Terrestrial Radio Access (UTRA); Concept evaluation*, December 1997. TR 101 146 V3.0.0.

[351] I. Katzela and M. Naghshineh, "Channel assignment schemes for cellular mobile telecommunication systems: A comprehensive survey," *IEEE Personal Comms.*, pp. 10–31, June 1996.

[352] S. Tekinay and B. Jabbari, "Handover and channel assignment in mobile cellular networks," *IEEE Comms. Mag.*, pp. 42–46, November 1991.

[353] B. Jabbari, "Fixed and dynamic channel assignment," in Gibson [562], ch. 83, pp. 1175–1181.

[354] J. Zander, "Radio resource management in future wireless networks: Requirements and limitations," *IEEE Comms. Magazine*, pp. 30–36, August 1997.

[355] D. Everitt, "Traffic engineering of the radio interface for cellular mobile networks," *Proceedings of the IEEE*, vol. 82, pp. 1371–1382, September 1994.

[356] J. Dahlin, "Ericsson's multiple reuse pattern for DCS1800," *Mobile Communications International*, November 1996.

[357] M. Madfors, K. Wallstedt, S. Magnusson, H. Olofsson, P. Backman and S. Engström, "High capacity with limited spectrum in cellular systems," *IEEE Comms. Mag.*, vol. 35, pp. 38–45, August 1997.

[358] A. Safak, "Optimal channel reuse in cellular radio systems with multiple correlated log-normal interferers," *IEEE Tr. on Vech. Tech*, vol. 43, pp. 304–312, May 1994.

[359] H. Jiang and S. S. Rappaport, "Prioritized channel borrowing without locking: a channel sharing strategy for cellular communications," *IEEE/ACM Transactions on Networking*, vol. 43, pp. 163–171, April 1996.

[360] J. Engel and M. Peritsky, "Statistically optimum dynamic server assignment in systems with interfering servers," *IEEE Trans. on Vehicular Tech.*, vol. 22, pp. 203–209, November 1973.

[361] M. Zhang and T. Yum, "Comparisons of channel assignment strategies in cellular mobile telephone systems," *IEEE Trans. on Vehicular Tech.*, vol. 38, pp. 211–215, November 1989.

[362] S. M. Elnoubi, R. Singh and S. Gupta, "A new frequency channel assignment algorithm in high capacity mobile communications systems," *IEEE Trans. on Vehicular Tech.*, vol. 31, pp. 125–131, August 1982.

[363] M. Zhang and T. Yum, "The non-uniform compact pattern allocation algorithm for cellular mobile systems," *IEEE Trans. on Vehicular Tech.*, vol. 40, pp. 387–391, May 1991.

[364] S. S. Kuek and W. C. Wong, "Ordered dynamic channel assignment scheme with reassignment in highway microcell," *IEEE Trans. on Vehicular Tech.*, vol. 41, pp. 271–277, August 1992.

[365] T. Yum and W. Wong, "Hot spot traffic relief in cellular systems," *IEEE Journal on selected areas in Comms.*, vol. 11, pp. 934–940, August 1993.

[366] J. Tajima and K. Imamura, "A strategy for flexible channel assignment in mobile communication systems," *IEEE Trans. on Vehicular Tech.*, vol. 37, pp. 92–103, May 1988.

[367] ETSI, *Digital European Cordless Telecommunications (DECT)*, 1st ed., October 1992. ETS 300 175-1 – ETS 300 175-9.

[368] R. Steele, "Digital European Cordless Telecommunications (DECT) systems," in Steele [155], ch. 1.7.2, pp. 79–83.

[369] S. Asghar, "Digital European Cordless Telephone," in Gibson [562], ch. 30, pp. 478–499.

[370] A. Law and L. B. Lopes, "Performance comparison of DCA call assignment algorithms within DECT," in *Proceedings of IEEE VTC '96* [561], pp. 726–730.

[371] H. Salgado-Galicia, M. Sirbu and J. M. Peha, "A narrowband approach to efficient PCS spectrum sharing through decentralized DCA access policies," *IEEE Personal Communications*, pp. 24–34, February 1997.

[372] R. Steele, J. Whitehead and W. C. Wong, "System aspects of cellular radio," *IEEE Communications Magazine*, vol. 33, pp. 80–86, January 1995.

[373] D. Cox and D. Reudink, "The behavior of dynamic channel-assignment mobile communications systems as a function of number of radio channels," *IEEE Trans. on Communications*, vol. 20, pp. 471–479, June 1972.

[374] D. D. Dimitrijević and J. Vučerić, "Design and performance analysis of the algorithms for channel allocation in cellular networks," *IEEE Trans. on Vehicular Tech.*, vol. 42, pp. 526–534, November 1993.

[375] D. C. Cox and D. O. Reudink, "Increasing channel occupancy in large scale mobile radio systems: Dynamic channel reassignment," *IEEE Trans. on Vehicular Tech.*, vol. 22, pp. 218–222, November 1973.

[376] D. C. Cox and D. O. Reudink, "A comparison of some channel assignment strategies in large-scale mobile communications systems," *IEEE Trans. on Communications*, vol. 20, pp. 190–195, April 1972.

[377] S. A. Grandhi, R. D. Yates and D. Goodman, "Resource allocation for cellular radio systems," *IEEE Trans. Vech. Tech.*, vol. 46, pp. 581–587, August 1997.

[378] M. Serizawa and D. Goodman, "Instability and deadlock of distributed dynamic channel allocation," in *Proceedings of IEEE VTC '93*, (Secaucus, NJ, USA), pp. 528–531, May 18–20 1993.

[379] Y. Akaiwa and H. Andoh, "Channel segregation - a self-organized dynamic channel allocation method: Application to tdma/fdma microcellular system," *IEEE Journal on Selected Areas in Comms.*, vol. 11, pp. 949–954, August 1993.

[380] E. D. Re, R. Fantacci and G. Giambene, "Handover and dynamic channel allocation techniques in mobile cellular networks," *IEEE Trans. on Vech. Tech.*, vol. 44, pp. 229–237, May 1995.

[381] T. Kahwa and N. Georganas, "A hybrid channel assignment scheme in large-scale, cellular-structured mobile communication systems," *IEEE Trans. on Communications*, vol. 26, pp. 432–438, April 1978.

[382] J. S. Sin and N. Georganas, "A simulation study of a hybrid channel assignment scheme for cellular land-mobile radio systems with erlang-c service," *IEEE Trans. on Communications*, vol. 29, pp. 143–147, February 1981.

[383] S.-H. Oh and D.-W. Tcha, "Prioritized channel assignment in a cellular radio network," *IEEE Trans. on Communications*, vol. 40, pp. 1259–1269, July 1992.

[384] D. Hong and S. Rappaport, "Traffic model and performance analysis for cellular mobile radio telephone systems with prioritizes and nonprioritized handoff procedures," *IEEE Trans. on Vehicular Technology*, vol. 35, pp. 77–92, August 1986.

[385] R. Guérin, "Queueing-blocking system with two arrival streams and guard channels," *IEEE Trans. on Communications*, vol. 36, pp. 153–163, February 1988.

[386] S. Grandhi, R. Vijayan, D. Goodman and J. Zander, "Centralized power control in cellular radio systems," *IEEE Trans. Vech. Tech.*, vol. 42, pp. 466–468, November 1993.

[387] J. Zander, "Performance of optimum transmitter power control in cellular radio systems," *IEEE Tr. on Vehicular Technology*, vol. 41, pp. 57–62, February 1992.

[388] J. Zander, "Distributed cochannel interference control in cellular radio systems," *IEEE Tr. on Vehicular Technology*, vol. 41, pp. 305–311, August 1992.

[389] A. Tanenbaum, "Introduction to queueing theory," in *Computer Networks*, pp. 631–641, Prentice-Hall, 2nd ed., 1989. ISBN 0131668366.

[390] D. C. Cox and D. O. Reudink, "Effects of some nonuniform spatial demand profiles on mobile radio system performance," *IEEE Trans. on Vehicular Tech.*, vol. 21, pp. 62–67, May 1972.

[391] P. J. Cherriman, *Mobile Video Communications*. PhD thesis, University of Southampton, 1998.

[392] R. C. French, "The Effect of Fading and Shadowing on Channel Reuse in Mobile Radio," *IEEE Transactions on Vehicular Technology*, vol. 28, pp. 171–181, August 1979.

[393] W. Gosling, "A simple mathematical model of co-channel and adjacent channel interference in land mobile radio," *The Radio and Electronic Engineer*, vol. 48, pp. 619–622, December 1978.

[394] R. Muammar and S. C. Gupta, "Cochannel Interference in High-Capacity Mobile Radio Systems," *IEEE Transactions on Communications*, vol. 30, pp. 1973–1978, August 1982.

[395] P. J. Cherriman and L. Hanzo, "Error-rate-based power-controlled multimode H.263-assisted video telephony," *IEEE Transactions on Vehicular Technology*, vol. 48, pp. 1726–1738, September 1999.

[396] G. Foschini and Z. Miljanic, "Distributed Autonomous Wireless Channel Assignment Algorithm with Power Control," *IEEE Transactions on Vehicular Technology*, vol. 44, pp. 420–429, August 1995.

[397] J.-I. Chuang and N. Sollenberger, "Spectrum Resource Allocation for Wireless Packet Access with Application to Advanced Cellular Internet Service," *IEEE Journal On Selected Areas in Communications*, vol. 16, pp. 820–829, August 1998.

[398] J. M. Torrance, L. Hanzo and T. Keller, "Interference Aspects of Adaptive Modems over Slow Rayleigh Fading Channels," *IEEE Transactions on Vehicular Technology*, vol. 48, pp. 1527–1545, September 1999.

[399] C. H. Wong and L. Hanzo, "Upper-bound performance of a wideband burst-by-burst adaptive mode," *IEEE Transactions on Communications*, March 2000.

[400] "3rd Generation Partnership Project; Technical Specification Group Radio Access Network; Spreading and modulation (FDD)." 3G TS 25.213 V3.2.0 (2000-03).

[401] "3rd Generation Partnership Project; Technical Specification Group Radio Access Network; Physical layer procedures (FDD)." 3G TS 25.214 V3.2.0 (2000-03).

[402] A. D. Whalen, *Detection of Signals in Noise*. Academic Press, 1971.

[403] W. T. Webb and L. Hanzo, *Modern Quadrature Amplitude Modulation: Principles and Applications for Fixed and Wireless Channels*. John Wiley and IEEE Press, 1994.

[404] R. L. Pickholtz, L. B. Milstein and D. L. Schilling, "Spread spectrum for mobile communications," *IEEE Transactions on Vehicular Technology*, vol. 40, pp. 313–322, May 1991.

[405] K. S. Gilhousen, I. M. Jacobs, R. Padovani, A. J. Viterbi, L. A. Weaver and C. E. Wheatley, "On the capacity of a cellular CDMA system design," *IEEE Transactions on Vehicular Technology*, vol. 40, pp. 303–312, May 1991.

[406] L. Wang and A. H. Aghvami, "Optimal power allocation based on QoS balance for a multi-rate packet CDMA system with multi-media traffic," in *Proceedings of Globecom*, (Rio de Janeiro, Brazil), pp. 2778–2782, December 1999.

[407] D. Koulakiotis and A. H. Aghvami, "Data detection techniques for DS/CDMA mobile systems: A review," *IEEE Personal Communications*, pp. 24–34, June 2000.

[408] P. J. Cherriman, E. L. Kuan and L. Hanzo, "Burst-by-burst adaptive joint-detection CDMA/H.263 based video telephony," in *Proceedings of the ACTS Mobile Communications Summit, Sorrento, Italy*, pp. 715–720, June 1999.

[409] J. Laiho-Steffens, A. Wacker and P. Aikio, "The Impact of the Radio Network Planning and Site Configuration on the WCDMA Network Capacity and Quality of Service," in *IEEE Proceedings of Vehicular Technology Conference*, (Tokyo, Japan), pp. 1006–1010, 2000.

[410] R. D. Kimmo Hiltunen, "WCDMA Downlink Capacity Estimation," in *IEEE Proceedings of Vehicular Technology Conference*, (Tokyo, Japan), pp. 992–996, 2000.

[411] K. Sipilä, Z.-C. Honkasalo, J. Laiho-Steffens and A. Wacker, "Estimation of Capacity and Required Transmission Power of WCDMA Downlink Based on a Downlink Pole Equation," in *IEEE Proceedings of Vehicular Technology Conference*, (Tokyo, Japan), pp. 1002–1005, 2000.

[412] "GSM 06.90: Digital cellular telecommunications system (Phase 2+)." Adaptive Multi-Rate (AMR) speech transcoding, version 7.0.0, Release 1998.

[413] S. Bruhn, E. Ekudden and K. Hellwig, "Adaptive Multi-Rate: A new speech service for GSM and beyond," in *Proceedings of 3rd ITG Conference on Source and Channel Coding*, (Technical Univ. Munich, Germany), pp. 319–324, 17–19 January 2000.

[414] S. Bruhn, P. Blocher, K. Hellwig and J. Sjoberg, "Concepts and Solutions for Link Adaptation and Inband Signalling for the GSM AMR Speech Coding Standard," in *Proceedings of VTC*, (Houston, Texas, USA), 16–20 May 1999.

[415] R. Owen, P. Jones, S. Dehgan and D. Lister, "Uplink WCDMA capacity and range as a function of inter-to-intra cell interference: theory and practice," in *IEEE Proceedings of Vehicular Technology Conference*, vol. 1, (Tokyo, Japan), pp. 298–303, 2000.

[416] J. S. Blogh and L. Hanzo, *Third-Generation Systems and Intelligent Wireless Networking - Smart Antennas and Adaptive Modulation*. John Wiley and IEEE Press, 2002.

[417] J. S. Blogh and L. Hanzo, "Adaptive Antenna Assisted Network Performance of FDD-Mode UMTS," in *Proceedings of the IEEE Vehicular Technology Conference 2001 Spring*, (Rhodes, Greece), pp. 2455–2459, May, 2001.

[418] J. S. Blogh and L. Hanzo, "The Network Performance of Multi-Rate FDD-Mode UMTS," in *Proceedings of the IEEE Vehicular Technology Conference 2001 Fall*, (Atlantic City, USA), pp. 1294–1298, October, 2001.

[419] L. Hanzo, C. Wong and M. Yee, *Adaptive wireless transceivers: Turbo-Coded, Turbo-Equalised and Space-Time Coded TDMA, CDMA and OFDM systems*. John Wiley and IEEE Press, 2002.

[420] F. Adachi, M. Sawahashi and K. Okawa, "Tree-structured Generation of Orthogonal Spreading Codes with Different Lengths for Forward Link of DS-CDMA Mobile," *IEE Electronics Letters*, vol. 33, no. 1, pp. 27–28, 1997.

[421] S. Stańczak, H. Boche and M. Haardt, "Are LAS-codes a miracle?," in *Proceedings of the IEEE Global Telecommunications Conference, GLOBECOM '01*, (San Antonio, Texas, USA), pp. 589–593, November, 2001.

[422] D. Li, "A high spectrum efficient multiple access code," *Chinese Journal of Electronics*, vol. 8, pp. 221–226, July 1999.

[423] N. B. Mehta, L. J. Greenstein, T. M. Willis and Z. Kostic, "Analysis and Results for the Orthogonality Factor in WCDMA Downlinks," in *Proceedings of the IEEE Vehicular Technology Conference 2002 Spring*, vol. 1, (Birmingham, AL, USA), pp. 100–104, 6–9 May 2002.

[424] M. Hunukumbure, M. Beach and B. Allen, "Downlink Orthogonality Factor in UTRA FDD Systems," *IEE Electronics Letters*, vol. 38, pp. 196–197, February 2002.

[425] R. Assarut, K. Kawanishi, R. Deshpande, U. Yamamoto and Y. Onozato, "Performance evaluation of Orthogonal Variable-Spreading-Factor Code Assignment Schemes in W-CDMA," in *Proceeding of the IEEE International Conference on Communications ICC'2002*, vol. 5, (New York, USA), pp. 3050–3054, 28 April–2 May 2002.

[426] K. I. Pedersen and P. E. Mogensen, "The Downlink Orthogonality Factors Influence on WCDMA System Performance," in *Proceedings of the IEEE Vehicular Technology Conference 2002 Fall*, vol. 4, (Vancouver, Canada), pp. 2061–2065, 24–28 September 2002.

[427] S. Ni, H. Wei, J. S. Blogh and L. Hanzo, "Network Performance of Asynchronous UTRA-like FDD/CDMA Systems using Loosely Synchronised Spreading Codes," in *Proceedings of the IEEE Vehicular Technology Conference 2003 Fall*, vol. 2, (Orlando, USA), pp. 1359–1363, October, 2003.

[428] B. J. Choi and L. Hanzo, "On the Design of LAS Spreading Codes," in *Proceedings of the IEEE Vehicular Technology Conference 2002 Fall*, vol. 4, (Vancouver, Canada), pp. 2172–2176, 24–28 September 2002.

[429] H. Wei, L. Yang and L. Hanzo, "Interference-free broadband single- and multicarrier ds-cdma," *IEEE Communications Magazine*, vol. 43, pp. 68–73, February 2005.

[430] H. Wei and L. Hanzo, "On the Uplink Performance of LAS-CDMA," in *To appear in the IEEE Vehicular Technology Conference 2005 Spring*, (Stockholm, Sweden), 30 May–1 June 2005.

[431] C.-C. Tseng and C. L. Liu, "Complementary Sets of Sequences," *IEEE Transactions on Information Theory*, vol. 18, pp. 644–652, September 1972.

[432] R. L. Frank, "Polyphase Complementary Codes," *IEEE Transactions on Information Theory*, vol. 26, pp. 641–647, November 1980.

[433] R. Sivaswamy, "Multiphase Complementary Codes," *IEEE Transactions on Information Theory*, vol. 24, pp. 546–552, September 1987.

[434] L. Hanzo, L. L. Yang, E. L. Kuan and K. Yen, *Single- and Multi-carrier DS-CDMA*. John Wiley and IEEE Press, 2003.

[435] H. Wei, L. Yang and L. Hanzo, "Interference-free broadband single- and multicarrier ds-cdma," *IEEE Communications Magazine*, vol. 43, pp. 68–73, February 2005.

[436] L. Hanzo, P. J. Cherriman and J. Streit, *Wireless Video Communications*. John Wiley and IEEE Press, New York, 2001.

[437] B. Walke, *Mobile Radio Networks - Networking, Protocols and Traffic Performance, 2nd Edition*. John Wiley and Sons, IEEE Press, New York, 2002.

[438] M. Soleimanipour, W. Zhuang and G. H. Freeman, "Modelling and resource allocation in wireless multimedia CDMA system," in *Proceedings of the IEEE Vehicular Technology Conference 1998 Spring*, vol. 2, (Ottawa, Canada), pp. 1279–1283, 18–21 May 1998.

[439] A. S. Acampora and M. Naghshinen, "Control and Quality-of-Service Provisioning in High-Speed Microcellular Networks," *IEEE Personal Communications*, vol. 1, pp. 36–43, 2nd Quarter 1994.

[440] L. Hanzo, M. Muenster, B. Choi and T. Keller, *OFDM and MC-CDMA for Broadband Multi-user Communications, WLANs and Broadcasting*. John Wiley and IEEE Press, 2003.

[441] N. Yee, J.-P. Linnartz and G. Fettweis, "Multicarrier CDMA in indoor wireless radio networks," in *Proceedings of the IEEE Personal, Indoor and Mobile Radio Communications, PIMRC'93*, vol. 1, pp. 109–113, 1993.

[442] K. Fazel and L. Papke, "On the performance of convolutionally-coded CDMA/OFDM for mobile communication system," in *Proceedings of the IEEE Personal, Indoor and Mobile Radio Communications, PIMRC'93*, vol. 1, pp. 468–472, 1993.

[443] A. Chouly, A. Brajal and S. Jourdan, "Orthogonal multicarrier techniques applied to direct sequence spread spectrum CDMA systems," in *Proceedings of the IEEE Global Telecommunications Conference 1993*, (Houston, TX, USA), pp. 1723–1728, 29 November–2 December 1993.

[444] V. M. DaSilva and E. S. Sousa, "Performance of orthogonal CDMA codes for quasi-synchronous communication systems," in *Proceedings of the IEEE International Conference on Universal Personal Communications, ICUPC '93*, (Ottawa, Canada), pp. 995–999, October 1993.

[445] L. Vandendorpe, "Multitone direct sequence CDMA system in an indoor wireless environment," in *Proceedings of the IEEE Symposium on Communications and Vehicular Technology, SCVT1993*, (Delft, The Netherlands), pp. 4.1.1–4.1.8, October 1993.

[446] R. Prasad and S. Hara, "Overview of multicarrier CDMA," *IEEE Communications Magazine*, vol. 35, pp. 126–133, December 1997.

[447] R. Prasad and S. Hara, "Overview of multi-carrier CDMA," in *Proceedings of the IEEE International Symposium on Spread Spectrum Techniques and Applications ISSSTA'1996*, (Mainz, Germany), pp. 107–114, 22–25 September 1996.

[448] D. I. Scott, P. M. Grant, S. McLaughlin, G. Povey and D. Cruickshank, "Research in reconfigurable terminal design for mobile and personal communications," tech. rep., Department of Electrical Engineering, The University of Edinburgh, March 1997.

[449] N. Yee and J. P. Linnartz, "MICRO 93-101: Multi-carrier CDMA in an indoor wireless radio channel," tech. rep., University of California at Berkeley, 1994.

[450] "COST 207 : Digital land mobile radio communications, final report," tech. rep., Office for Official Publications of the European Communities, Luxembourg, 1989.

[451] M. Haardt, A. Klein, R. Koehn, S. Oestreich, M. Purat, V. Sommer and T. Ulrich, "The TD-CDMA Based UTRA TDD Mode," *IEEE Journal on Selected Areas in Communications*, vol. 18, pp. 1375–1385, August 2000.

[452] J. Gibson, *The Mobile Communications Handbook*. CRC PRESS, 1996.

[453] W. Lu, B. Walke and X. Shen, "4G mobile communications: toward open wireless architecture," *IEEE Personal Communications*, vol. 11, pp. 4–6, April 2004.

[454] *The UMTS Forum website.* http://www.umts-forum.org/.

[455] T. Le and H. Aghvami, "Fast channel access and DCA scheme for connection and connectionless-oriented services in UMTS," *IEE Electronics Letters*, vol. 35, pp. 1048–1049, June 1999.

[456] B. Walke, M. Althoff and P. Seidenberg, *UMTS - A Comprehensive Introduction.* John Wiley and Sons, IEEE Press, New York, 2002.

[457] R. Steele and L. Hanzo, *Mobile Radio Communications.* Piscataway, NJ: IEEE Press, 1999.

[458] T. Kriengchaiyapruk and I. Forkel, "Adaptive switching point allocation in TD/CDMA systems," in *Proceedings of the IEEE Vehicular Technology Conference 2002 Fall*, vol. 3, (Vancouver, Canada), pp. 1456–1460, 24–28 September 2002.

[459] J. Laiho, A. Wacker and T. Novosad, *Radio Network Planning and Optimisation for UMTS.* John Wiley & Sons, Ltd. and IEEE Press, 2002.

[460] H. Holma, G. J. R. Povey and A. Toskala, "Evaluation Interference Between Uplink and Downlink in UTRA/TDD," in *Proceedings of the IEEE Vehicular Technology Conference 1999 Fall*, vol. 5, (Amsterdam, the Netherlands), pp. 2616–2620, 19–22 September 1999.

[461] H. Haas, S. McLaughlin and G. J. R. Povey, "A Novel Interference Resolving Algorithm for the TDD TD-CDMA Mode in UMTS," in *Proceedings of the IEEE International Symposium on Personal, Indoor and Mobile Radio Communications*, vol. 2, (London, UK), pp. 1231–1235, September 2000.

[462] T. Kriengchaiyapruk, *Dynamic Channel Allocation in UTRA-TDD.* PhD thesis, Lehrstuhl für Kommunikationsnetze, July 2004.

[463] Q. Y. Miao, W. B. Wang and D. C. Yang, "The Coexistence of UTRA-TDD and FDD System In the Adjacent Channel," in *Proceedings of the IEEE Global Telecommunications Conference, Globecom 2001*, vol. 6, (San Antonio, USA), pp. 3714–3718, November 2001.

[464] T. L. Lee, C. Faure and D. Grandblaise, "Impact of FDD/TDD Co-Existence on Overall UMTS System Performance," in *Proceedings of the IEEE Vehicular Technology Conference 2001 Spring*, vol. 4, (Rhodes, Greece), pp. 2655–2659, May 2001.

[465] H. Haas, S. McLaughlin and G. Povey, "Capacity-coverage analysis of TDD and FDD Mode in UMTS at 1920MHz," *IEE Electronics and Communication Engineering Journal*, vol. 149, pp. 51–57, February 2002.

[466] I. Forkel, P. Seidenberg, R. Pabst and G. Heidelberger, "Performance Evaluation of Power Control Algorithms in Cellular UTRA Systems," *3G Mobile Communication Technologies (3G 2001)*, pp. 11–15, March 2001.

[467] J. Kurjenniemi, S. Hämäläinen, T. Ristaniemi, O. Lehtinen and P. Patronen, "Convergence of UTRA TDD Uplink Power Control," in *Proceedings of the IEEE Vehicular Technology Conference 2001 Spring*, vol. 4, (Rhodes, Greece), pp. 2908–2912, May 2001.

[468] J. Wen, J. Sheu and J. Chen, "An Optimum Downlink Power Control Method for CDMA Cellular Mobile Systems," in *Proceedings of the IEEE International Conference on Communication, ICC2001*, vol. 6, (Helsinki, Finland), pp. 1738–1742, June 2001.

[469] J. Kurjenniemi, S. Hämäläinen and T. Ristaniemi, "Uplink Power Control in UTRA TDD," in *Proceedings of the IEEE International Conference on Communication, ICC2001*, vol. 5, (Helsinki, Finland), June 2001.

[470] Q. Wu, "Performance of optimum transmitter power control in CDMA cellular mobile systems," *IEEE Transactions on Vehicular Technology*, vol. 48, pp. 571–575, March 1999.

[471] R. Choi and R. Murch, "Evaluation of a Pre-Rake Smart Antenna System for TDD CDMA Systems," in *Proceedings of the IEEE Vehicular Technology Conference 2002 Fall*, vol. 1, (Vancouver, Canada), pp. 346–350, September 2002.

[472] R. Machauer, M. Iurascu and F. Jondral, "FFT Speed Multiuser Detection for High Rate Data Mode in UTRA-FDD," in *Proceedings of the IEEE Vehicular Technology Conference 2001 Fall*, vol. 1, (Atlantic City, NJ, USA), pp. 502–505, October 2001.

[473] S. Ni, J. S. Blogh and L. Hanzo, "On the network performance of UTRA-like TDD and FDD CDMA systems using adaptive modulation and adaptive beamforming," in *Proceedings of the IEEE Vehicular Technology Conference 2003 Spring*, vol. 1, (Jeju, Korea), pp. 606–610, April, 2003.

[474] X. Wu, L.-L. Yang and L. Hanzo, "Uplink capacity investigations of TDD/CDMA," in *Proceedings of the IEEE Vehicular Technology Conference, VTC Spring 2002*, vol. 2, (Birmingham, Alabama, USA), pp. 997–1001, 6–9 May 2002.

[475] S. Tekinay and B. Jabbari, "Handover and channel assignment in mobile cellular networks," *IEEE Communications Magazine*, vol. 29, pp. 42–46, November 1991.

[476] E. Berruto, M. Gudmundson, R. Menolascino, W. Mohr and M. Pizarroso, "Research activities on UMTS radio interface, network architectures, and planning," *IEEE Communications Magazine*, vol. 36, pp. 82–95, February 1998.

[477] I. Katzela and M. Naghshineh, "Channel assignment schemes for cellular mobile telecommunication systems: a comprehensive survey," *IEEE Personal Communications*, vol. 3, pp. 10–31, June 1996.

[478] B. Jabbari, "Teletraffic aspects of evolving and next-generation wireless communication networks," *IEEE Personal Communications*, vol. 3, pp. 4–9, December 1996.

[479] N. D. Tripathi, J. H. Reed and H. F. VanLandinoham, "Handoff in cellular systems," *IEEE Personal Communications*, vol. 5, pp. 26–37, December 1998.

[480] R. Vijayan and J. Holtzman, "A model for analyzing handoff algorithms," *IEEE Transactions on Vehicular Technology*, vol. 42, pp. 351–356, August 1993.

[481] J. Wang, J. Liu and Y. Cen, "Handoff algorithms in dynamic spreading WCDMA system supporting multimedia traffic," *IEEE Journal on Selected Areas in Communications*, vol. 21, pp. 1652–1662, December 2003.

[482] D. Lugara, J. Tartiere and L. Girard, "Performance of UMTS to GSM handover algorithms," in *Proceedings of the IEEE International Symposium on Personal, Indoor and Mobile Radio Communications*, vol. 1, (Barcelona, Spain), pp. 444–448, 5–8 September 2004.

[483] O. Grimlund and B. Gudmundson, "Handoff strategies in microcellular," in *Proceedings of the IEEE Vehicular Technology Conference*, pp. 505–510, 19–22 May 1991.

[484] A. Viterbi, K. Gilhousen and E. Zehavi, "Soft handoff extends CDMA cell coverage and increases reverse link capacity," *IEEE Journal on Selected Areas in Communications*, vol. 12, pp. 1281–1288, October 1994.

[485] E. Nakano, N. Umeda and K. Ohno, "Performance of diversity handover in DS-CDMA cellular systems," in *Proceedings of the IEEE 4th International Conference on Universal Personal Communications*, (Tokyo, Japan), pp. 421–425, November 1995.

[486] Y.-I. Kim, K.-J. Lee and Y.-O. Chin, "Analysis of multi-level threshold handoff algorithm," in *Proceedings of the IEEE Global Telecommunications Conference, GLOBECOM '96*, vol. 2, (London, UK), pp. 1141–1145, November 1996.

[487] D. Wong and D. Cox, "A handoff algorithm using pattern recognition," in *Proceedings of the IEEE International Conference on Universal Personal Communications*, vol. 1, (Florence, Italy), pp. 759–763, October 1998.

[488] S. Tekinay and B. Jabbari, "A measurement-based prioritization scheme for handovers in mobile cellular networks," *IEEE Journal on Selected Areas in Communications*, vol. 10, pp. 1343–1350, October 1992.

[489] D. Calin and D. Zeghlache, "Performance and handoff analysis of an integrated voice-data cellular system," in *Proceedings of the IEEE 8th International Symposium on Personal, Indoor and Mobile Radio Communications*, vol. 2, (Helsinki, Finland), pp. 386–390, 1997.

[490] N. Benvenuto and F. Santucci, "A least squares path-loss estimation approach to handover algorithms," *IEEE Transactions on Vehicular Technology*, vol. 48, pp. 437–447, March 1999.

[491] F. Santucci, M. Pratesi, M. Ruggieri and F. Graziosi, "A general analysis of signal strength handover algorithms with cochannel interference," *IEEE Transactions on Communications*, vol. 48, pp. 231–241, February 2000.

[492] X. Yang, S. Ghaheri-Niri and R. Tafazolli, "Performance of power-triggered and E_c/N_0-triggered soft handover algorithms for UTRA," in *Second International Conference on 3G Mobile Communication Technologies*, no. 477, (London, UK), pp. 7–10, 2001.

[493] S. Wang, S. Sridhar and M. Green, "Adaptive soft handoff method using mobile location information," in *Proceedings of the IEEE Vehicular Technology Conference 2002 Spring*, vol. 4, (Birmingham, Alabama, USA), pp. 1936–1940, 6–9 May 2002.

[494] H. Persson and J. Karlsson, "Maintaining QoS by utilizing hierarchical wireless systems," in *Proceedings of The 9th Asia-Pacific Conference on Communications*, vol. 1, (Penang, Malaysia), pp. 292–296, 21–24 September 2003.

[495] A. Tolli, I. Barbancho, J. Gomez and P. Hakalin, "Intra-system load balancing between adjacent GSM cells," in *Proceedings of the IEEE Vehicular Technology Conference*, vol. 1, (Jeju Island, Korea), pp. 393–397, 22–25 April 2003.

[496] J. Kim, D. H. Kim, P. jung Song and S. Kim, "Design of optimum parameters for handover initiation in WCDMA," in *Proceedings of the IEEE Vehicular Technology Conference*, vol. 4, (Atlantic City, NJ, USA), pp. 2768–2772, 7–11 October 2001.

[497] B. Singh, K. Aggarwal and S. Kumar, "An analytical model for intersystem handover," in *Proceedings of the IEEE Conference on Convergent Technologies for Asia-Pacific Region*, vol. 4, (Bangalore, India), pp. 1311–1315, 15–17 October 2003.

[498] G. Durastante and A. Zanella, "An efficient monitoring strategy for intersystem handover from TD-SCDMA to GSM networks," in *Proceedings of the IEEE International Symposium on Personal, Indoor and Mobile Radio Communications (PIMRC) 2002*, vol. 4, (Lisbon, Portugal), pp. 1555–1560, 15–18 September 2002.

[499] M. Iwamura and A. Aghvami, "Impact of handover blocking on control load, capacity and coverage of W-CDMA downlink," in *Proceedings of the IEEE International Symposium on Personal, Indoor and Mobile Radio Communications (PIMRC) 2004*, vol. 1, (Barcelona, Spain), pp. 350–354, 5–8 September 2004.

[500] K. Hiramatsu, K. Miya, O.Kato and K. Homma, "Transmit diversity applied on the CDMA/TDD cellular systems," in *Proceedings of the IEEE Vehicular Technology Conference*, vol. 2, (Tokyo, Japan), pp. 1170–1174, 15–18 May 2000.

[501] P. Chaudhury, W. Mohr and S. Onoe, "The 3gpp proposal for imt-2000," *IEEE Communications Magazine*, vol. 37, pp. 72–81, December 1999.

[502] J. Rodriguez, X. Yang, R. Tafazolli, D. Huy, V. Monteiro and A. Gameiro, "Dynamic System Level Performance for MC-CDMA Scheme," in *Proceedings of the European Wireless Conference 2004*, (Barcelona, Spain), 24–27 February 2004.

[503] S. H. Hwang and L. Hanzo, "Effects of multipath propagation delay on uplink performance of synchronous DS-CDMA systems communicating in dispersive Rayleigh fading channels," *IEE Electronics Letters*, vol. 40, pp. 1589–1591, December 2004.

[504] TS25.201, "Technical Specification Group Radio Access Network; Physical layer - General description," *3GPP TSG R6, v6.0.0*, 2003-12.

[505] X. Yang, S. Ghaheri-Niri and R. Tafazolli, "Evaluation of soft handover algorithms for UMTS," in *Proceedings of the IEEE International Symposium on Personal, Indoor and Mobile Radio Communications (PIMRC) 2000*, vol. 2, (London, UK), pp. 772–776, 18–21 September 2000.

[506] N. Binucci, K. Hiltunen and M. Caselli, "Soft handover gain in WCDMA," in *Proceedings of the IEEE Vehicular Technology Conference*, vol. 3, (Boston, MA, USA), pp. 1467–1472, 24–28 September 2000.

[507] K. Sipila, M. Jasberg, J. Laiho-Steffens and A. Wacker, "Soft handover gains in a fast power controlled WCDMA uplink," in *Proceedings of the IEEE Vehicular Technology Conference 1999 Spring*, vol. 2, (Houston, TX, USA), pp. 1594–1598, 16–20 May 1999.

[508] W. L. Lim, Y. C. Foo and R. Tafazolli, "Adaptive softer handover algorithm for high altitude platform station UMTS with onboard power resource sharing," in *Proceedings of the IEEE International Symposium on Wireless Personal Multimedia Communications*, vol. 1, (Hawaii, USA), pp. 52–56, 27–30 October 2002.

[509] W.-U. Pistelli and R. Verdone, "Downlink capacity for WCDMA with soft and softer handover; advantages of unbalanced received powers," in *Proceedings of the IEEE International Symposium on Wireless Personal Multimedia Communications*, vol. 1, (Hawaii, USA), pp. 77–81, 27–30 October 2002.

[510] R. Verdone and A. Zanella, "Performance of received power and traffic driven handover algorithms in urban cellular networks," *IEEE Transactions on Wireless Communications*, vol. 9, pp. 60–70, February 2002.

[511] TS25.331, "Technical Specification Group Radio Access Network; Radio Resource Control (RRC)," *3GPP TSG R6, v6.0.1*, 2004-01.

[512] R. Owen, P. Jones, S. Dehghan and D. Lister, "Uplink WCDMA capacity and range as a function of inter-to-intra cell interference: theory and practice," in *Proceedings of the IEEE Vehicular Technology Conference 2000 Spring*, vol. 1, (Tokyo, Japan), pp. 298–303, 2000.

[513] D. Calin and M. Areny, "Impact of Radio Resource Allocation Policies on the TD-CDMA System Performance: Evaluation of Major Critical Parameters," *IEEE Journal on Selected Areas in Communications*, vol. 19, pp. 1847–1859, October 2001.

[514] TS25.105, "Technical Specification Group Radio Access Network; Base Station (BS) radio transmission and reception (TDD)," *3GPP TSG R6, v6.0.0*, 2003-12.

[515] J. Kurjenniemi, S. Hämäläinen and T. Ristanlemi, "UTRA TDD handover performance," in *Proceedings of the IEEE Global Telecommunications Conference, GLOBECOM '01*, vol. 1, (San Antonio, Texas, USA), pp. 533–537, 2001.

[516] TS25.123, "Technical Specification Group Radio Access Network; Requirements for support of radio resource management," *3GPP TSG R6, v6.0.0*, 2003-12.

[517] A. Chockalingam, P. Dietrich, L. Milstein and R. Rao, "Performance of closed-loop power control in DS-CDMA cellular systems," *IEEE Transactions on Vehicular Technology*, vol. 47, pp. 774–789, August 1998.

[518] R. Gejji, "Forward-link-power control in CDMA cellular-systems," *IEEE Transactions on Vehicular Technology*, vol. 41, pp. 532–536, November 1992.

[519] J. Dunlop, J. Irvine and B. Manzanedo, "Power control in a cellular adaptive air interface," *IEE Electronics Letters*, vol. 32, pp. 1968–1969, October 1996.

[520] B. Manzanedo, P. Cosimini, J. Irvine and J. Dunlop, "Implementation and assessment of adaptive power control in a 3rd generation cellular system," in *Proceedings of the IEEE Vehicular Technology Conference 1996 Spring*, vol. 2, (Atlanta, USA), pp. 1135–1139, 28 April–1 May 1996.

[521] TS25.104, "Technical Specification Group Radio Access Network; Base Station (BS) radio transmission and reception (FDD)," *3GPP TSG R6, v6.7.0*, 2004-09.

[522] TS25.222, "Technical Specification Group Radio Access Network; Multiplexing and channel coding (TDD)," *3GPP TSG R6, v6.0.0*, 2003-12.

[523] TS25.224, "Technical Specification Group Radio Access Network; Physical layer procedures (TDD)," *3GPP TSG R6, v6.2.0*, 2004-09.

[524] J. Kurjenniemi, O. Lehtinen and T. Ristaniemi, "Signaled step size for downlink power control of dedicated channels in UTRA TDD," in *4th International Workshop On Mobile and Wireless Communications Network, 2002*, vol. 1, (Stockholm, Sweden), pp. 675–679, 2002.

[525] J. D. Gibson, *The Communications Handbook*. CRC Press and IEEE Press, Boca Raton, Florida, 1997.

[526] "3rd Generation Partnership Project; Technical Specification Group Radio Access Network; Physical layer procedures (TDD)." 3G TS 25.224 V4.3.0 (2001-12).

[527] D. Goldberg, *Genetic Algorithms in Search, Optimization, and Machine Learning*. ISBN 0201157675, MA USA: Addison-Wesley, August 2001.

[528] J. Holland, *Adaptation in Natural and Artificial Systems*. Ann Arbor, Michigan: University of Michigan Press, 1975.

[529] M. Mitchell, *An Introduction to Genetic Algorithms*. Cambridge, Massachusetts: MIT Press, 1996.

[530] D. Whitley, "A Genetic Algorithm Tutorial," *Statistics and Computing*, vol. 4, pp. 65–85, June 1994.

[531] S. Forrest, "Genetic algorithms: Principles of natural selection applied to computation," *Science*, vol. 261, pp. 872–878, August 1993.

[532] H. Mühlenbein, *Foundations of Genetic Algorithms*, ch. Evolution in time and space – The Parallel Genetic Algorithm, pp. 316–337. California, USA: G. Rawlins, ed., Morgan Kaufmann, 1991.

[533] J. J. Grefenstette and J. E. Baker, "How genetic algorithms work: A critical look at implicit parallelism," in *Proceedings of the IEEE Third International Conference on Genetic Algorithms* (J. D. Schaffer, ed.), (California, USA), pp. 20–27, Morgan Kaufmann, 1989.

[534] A. Wolfgang, N. N. Ahmad, S. Chen and L. Hanzo, "Genetic algorithm assisted minimum bit error rate beamforming," in *CDROM of the IEEE 59th Vehicular Technology Conference*, vol. 1, (Milan Italy), pp. 142–146, 17–19 May, 2004.

[535] M. Alias, A. Samingan, S. Chen and L. Hanzo, "Multiple antenna aided OFDM employing minimum bit error rate multiuser detection," *IEE Electronics Letters*, vol. 39, pp. 1769–1770, November 2003.

[536] E. Hou, N. Ansari and H. Ren, "A Genetic Algorithm for Multiprocessor Scheduling," *IEEE Transaction on Parallel and Distributed Systems*, vol. 5, pp. 113–120, February 1994.

[537] K.-S. Tang, K.-T. Ko, K. F. Man and S. Kwong, "Topology Design and Bandwidth Allocation of Embedded ATM Networks Using Genetic Algorithms," *IEEE Communications Letters*, vol. 2, pp. 171–173, June 1998.

[538] C. Y. Ngo and V. O. K. Li, "Fixed Channel Assignment in Cellular Radio Network Using a Modified Genetic Algorithm," *IEEE Transactions on Vehicular Technology*, vol. 47, pp. 163–172, February 1998.

[539] D. Sadot, U. Mahlab and V. Natan-Bar, "New method for developing optical CDMA address code sequences using the genetic algorithm," in *Proceedings of the IEEE International Conference on Communication, ICC1998*, vol. 3, (Atlanta, GA, USA), pp. 7–11, June 1998.

[540] Y. H. Lee, S. T. Sheu and M. H. Chen, "Application of Genetic Algorithm for Optimal Simultaneous Code Set Selection in Optical CDMA," in *Modelling and Simulation, MS99*, vol. 3, (Philadelphia, USA), pp. 5–8, May 1999.

[541] S. H. Hwang and L. Hanzo, "Reverse-link performance of synchronous DS-CDMA systems in dispersive Rician multipath fading channels," *IEE Electronics Letters*, vol. 39, pp. 1682–1684, November 2003.

[542] B. G. Evans and K. Baughan, "Visions of 4G," *IEE Electronics and Communication Engineering Journal*, vol. 12, pp. 293–303, December 2000.

[543] S. Hara and R. Prasad, "Overview of Multicarrier CDMA," *IEEE Communications Magazine*, vol. 35, pp. 126–133, December 1997.

[544] B. Abdool-Rassool, F. Heliot, L. Revelly, R. Nakhai and H. Aghvami, "4-PSK space-time trellis codes with five and six transmit antennas for slow Rayleigh fading channels," *IEE Electronics Letters*, vol. 39, pp. 297–299, February 2003.

[545] B. Rassool, B. Allen, R. Nakhai, R. Roberts and P. Sweeney, "Error statistics of optimal and sub-optimal space-time trellis codes: concatenation requirements," in *Proceedings of the IEEE International Symposium on Personal, Indoor and Mobile Radio Communications*, vol. 2, (Beijing, China), pp. 1007–1011, 7–10 September 2003.

[546] B. Rassool, F. Heliot, L. Revelly, M. Dohler, R. Nakhai and H. Aghvami, "Fast search techniques for obtaining space-time trellis codes for Rayleigh fading channels and its performance in CDMA systems," in *Proceedings of the IEEE Vehicular Technology Conference 2003 Spring*, vol. 1, (Jeju, Korea), pp. 66–69, 22–25 April 2003.

[547] F. Heliot, M. Ghavami, R. Nakhai and A. Aghvami, "Performance of space-time block coding and space-time trellis coding for impulse radio," in *Proceedings of the IEEE Global Telecommunications Conference*, vol. 5, (Dallas, Texas USA), pp. 3225–3229, 29 November–3 December 2004.

[548] C. K. Toh, *Wireless ATM and AD-HOC Networks Protocols and Architectures*. Kluwer Academic Publishers, 1997.

[549] A. Jardosh, E. Belding-Royer, K. Almeroth and S. Suri, "Real-world environment models for mobile network evaluation," *IEEE Journal on Selected Areas in Communications*, vol. 23, pp. 622–632, March 2005.

[550] X. Liu and L. Hanzo, "Effects of rate adaptation on the throughput of random ad hoc networks," in *To appear in the IEEE Vehicular Technology Conference 2005 Fall*, (Dallas, Texas, USA), 25–28 September 2005.

[551] H. Luo and S. Lu, "A topology-independent wireless fair queueing model in ad hoc networks," *IEEE Journal on Selected Areas in Communications*, vol. 23, pp. 585–597, March 2005.

[552] V. Srinivasan, P. Nuggehalli, C.-F. Chiasserini and R. Rao, "An Analytical Approach to the Study of Cooperation in Wireless Ad Hoc Networks," *IEEE Transaction on Wireless Communications*, vol. 4, pp. 722–733, March 2005.

[553] M. Ghassemian, P. Hofmann, C. Prehofer, V. Friderikos and H. Aghvami, "Performance analysis of Internet gateway discovery protocols in ad hoc networks," in *Proceedings of the IEEE Wireless Communications and Networking Conference*, vol. 1, pp. 120–125, 21–25 March 2004.

[554] L. Yang, H. Wei and L. Hanzo, "A Multicarrier DS-CDMA System Using Both Time-Domain and Frequency-Domain Spreading," in *Proceedings of the IEEE Vehicular Technology Conference 2003 Fall*, vol. 4, (Orlando, Florida, USA), pp. 2426–2430, 6–9 October 2003.

[555] M. Wall, *GAlib Genetic Algorithm Package version 2.45*. Massachusetts Institute of Technology: http://lancet.mit.edu/ga/, 1996.

[556] W. Spears and K. DeJong, *Foundations of Genetic Algorithms*. California, USA: G. Rawlins ed., Morgan Kaufmann, 1991.

[557] IEEE, *Proceedings of VTC'98 (Spring)*, (Houston, Texas, USA), 16–20 May 1999.

[558] IEEE, *Proceedings of IEEE International Symposium on Personal, Indoor and Mobile Radio Communications, PIMRC'97*, (Marina Congress Centre, Helsinki, Finland), 1–4 September 1997.

[559] IEEE, *Proceedings of IEEE Vehicular Technology Conference (VTC'98)*, (Ottawa, Canada), 18–21 May 1998.

[560] IEEE, *Proceedings of IEEE International Symposium on Spread Spectrum Techniques and Application (ISSSTA'98)*, (Sun City, South Africa), September 1998.

[561] IEEE, *Proceedings of IEEE VTC '96*, (Atlanta, GA, USA), 1996.

[562] J. D. Gibson, ed., *The Mobile Communications Handbook*. CRC Press and IEEE Press, 1996.

Subject Index

Author Index

L

X

Y

Z